Applied Probability and Statistics (Cc...

continued on back

D1190473

Methods for Statistical
Analysis of
Reliability and Life Data

Methods for Statistical Analysis of Reliability and Life Data

NANCY R. MANN

Rocketdyne, Rockwell International Corporation

RAY. E. SCHAFER

Hughes Aircraft Company

NOZER D. SINGPURWALLA

The George Washington University

JOHN WILEY & SONS   New York · London · Sydney · Toronto

Library of Congress Cataloging in Publication Data:

Mann, Nancy R.
  Methods for statistical analysis of reliability and life data.

  (Wiley series in probability and mathematical statistics) (A Wiley publication in applied statistics)
    Includes bibliographies.
    1.   Reliability (Engineering)—Statistical methods.
I.   Schafer, Ray E., joint author.   II.   Singpurwalla, Nozer D., joint author.   III.   Title.

TS173.M36   620′.004′4   73-20461
ISBN 0-471-56737-X

Printed in the United States of America
10 9 8 7 6 5 4 3 2

Preface

In recent years reliability has been formulated as the science of predicting, estimating, or optimizing the probability of survival, the mean life, or, more generally, the life distribution of components or systems. During the past decade, the development of reliability as a discipline has been rapid because of the ever increasing emphasis on reliability by the government and the consumer. Recently, interest in reliability has been manifested by mathematicians, economists, and those concerned with the environmental and life sciences. An integral part of this development and this interest has been the interplay between reliability and statistics.

Reliability, though considered by many to be merely an application of probability and statistics to a special class of problems, has stimulated the development of new areas in statistics. One such development is the nonparametric characterization of distribution functions and the study of renewal theory, estimation theory, ranking and selection theory, and so forth under such characterizations. Evidence of the intimate relationship between reliability and statistics is available in the significant number of papers written on statistical methods in reliability, and in the fact that more and more graduate-level courses are offered in the statistics of reliability to students in statistics, operations research, industrial engineering, systems engineering, and civil engineering. Our objective in writing this book is not only to provide these students with a textbook commensurate with the standards of graduate-level education, but also to provide the professional reliability engineer with a comprehensive source of available statistical methods commonly required in reliability. We believe that this book could serve as a one-semester graduate-level text in reliability, requiring of the reader some mathematical maturity or exposure to the calculus and, perhaps, to probability and statistics.

Chapters 2 and 3 present the required probabilistic and statistical methods as a base for developing the rest of the text. The heart of the book begins with Chapter 4 with the presentation and development of failure models. The rest of the book deals essentially with the various methods and techniques of estimation. We include much material that either has not previously appeared in textbook form or, if it has appeared, is not generally available. Chapters 8 and 9 discuss Bayesian methods in reliability, and accelerated life testing, respectively. Both subjects are currently being emphasized in the literature and, we believe, are therefore important enough to be a part of this book.

At the end of each chapter (or, in some cases, after each section) are problems for solution. These problems have been constructed to elicit a good blend of both the theory and the practice of reliability. At the end of each chapter we provide a list of references, which, because of space restrictions, does not include all that has been written on the topics considered but hopefully encompasses the basic literature. The interested reader may use these references to develop his own bibliography to suit his specific needs.

As one might expect, a large number of authors have contributed to the development of the statistical methods useful in reliability. We owe a debt to all these authors. We have tried to discharge our obligation by citing specific contributions, by author name, in the text material. Yet there may be instances in which this was not done; if so, the omission was inadvertent.

Parts of the research of Dr. Mann appearing herein were orginally supported by the Aerospace Research Laboratories, Wright-Patterson Air Force Base, the Air Force Office of Scientific Research, and the Interdivisional Technology Program of the Aerospace and Systems Group of Rockwell International.

NANCY R. MANN
RAY E. SCHAFER
NOZER D. SINGPURWALLA

Canoga Park, California
Fullerton, California
Washington, D.C.
August 1973

Contents

Methods for Statistical
Analysis of
Reliability and Life Data

CHAPTER 1

Introduction

1.1 STATISTICAL METHODS IN RELIABILITY AND LIFE TESTING

This book deals primarily with statistical methods for solving problems associated with the concept of reliability. Throughout the book the word *reliability* should be understood to mean "the probability of a device (or item or organism) performing its (or his or her) defined purpose adequately for a specified period of time, under the operating conditions encountered." The statistical theory underlying a method presented for the analysis and solution of reliability problems is included when it provides further insight to develop, understand, and apply the method.

The data to which statistical methods are applied, in order that parameters of interest can be estimated in this reliability context, usually result from *life tests*. A life test is one in which prototypes of the item or organism of interest are subjected to stresses and environmental conditions that typify the intended operating conditions. During the test, successive times to failure are noted. Since the failures occur in order, the theory of *order statistics* plays an important role in the analysis of the life-test data. Also, because testing of this sort is often very costly in one sense or another, large-sample theory and associated limit theorems, which can be conveniently invoked in several fields of application, must be used very judiciously here.

Literature related to statistical methods used in the analysis of life-test data lies scattered in a number of professional journals: *Annals of Mathematical Statistics, Journal of the American Statistical Association, Technometrics, Naval Research Logistics Quarterly,* and *I.E.E.E. Transactions on Reliability,* among others. Barlow and Proschan (1965) trace the historical development of reliability theory and allude to these references. The journals cited above cater to audiences requiring different levels of mathematical sophistication, with the consequence that much of the body

1

of knowledge concerning reliability is inaccessible to the practicing engineer, the statistician, the student, or even the research worker. This book has been written to serve as a reasonable compromise in the presentation of statistical methods that make efficient use of failure-time or life-time data.

1.2   PURPOSE AND SCOPE

This book is meant to be (a) a textbook for students taking a first-level graduate course in statistical methods in reliability, and (b) a reference book for practitioners and research workers in reliability or for individuals concerned with the analysis of life data. Within the scope of our objectives, we have tried to make the book as self contained and up to date as possible. Some recent developments in the statistical theory of reliability have not been included or even alluded to here, however, because of their advanced and rather specialized nature.

The text of the book consists of nine chapters. Chapters 2 and 3 attempt to provide rather compact treatments of probability and statistics, respectively. These two chapters are included mainly to keep the book somewhat self-contained, even though many readers will have had some previous preparation in probability and statistics. Chapter 3 includes special topics from statistical theory that are relevant to the development of the subsequent text. Users who have had a limited exposure to statistics may benefit by selectively reading topics from Chapter 3.

The heart of the book begins with Chapter 4, entitled Statistical Failure Models. A special feature of this chapter is that many of the frequently used failure (probability) distributions are developed from consideration of some underlying physical process. A study of this chapter should equip the reader with the capability of finding or developing failure distributions to suit many specific situations.

In Chapter 5, methods of point and interval estimation for the parameters of many of the models of Chapter 4 are given. Chapter 6 deals with the testing of reliability hypotheses. It is closely related to Chapter 5, but treats in particular the problem of determining sample sizes in the presence of prespecified risks.

Chapter 7 considers some special topics and nonparametric techniques in reliability. Nonparametric techniques play an important role in reliability, and a very comprehensive mathematical theory of reliability has been developed on the basis of the nonparametric approach. A summary of this theory is given in Chapter 7. Also presented are methods of testing the fit of data to various failure distributions with unknown parameters and techniques to be used in computer-generating statistical distributions by Monte Carlo methods.

Chapters 8 and 9 include material that has not as yet appeared in a textbook or other comprehensive form. Chapter 8 deals with Bayesian methods in reliability, and Chapter 9 discusses inference from accelerated life tests. Bayesian methods are beginning to be used rather widely in reliability analysis today, especially in agencies of the Department of Defense. Hence it was felt necessary to devote a chapter to this topic. Accelerated life tests are quite important from an economical point of view, and it is anticipated that, with the development of improved techniques to analyze the results of such tests, life tests of the future will be, in large measure, accelerated. Not a great deal has been written on the topic of accelerated life tests, and it is hoped that Chapter 9 represents a fairly complete coverage of the significant results up to the present time.

Chapter 10 on system reliability contains some material that can be considered by many as purely probabilistic. This material has been included for completeness and to provide the reader with some insight into the analysis of system-reliability problems. The rest of Chapter 10 deals with the statistical techniques that are useful in system-reliability analysis.

At the end of each chapter (and sometimes after a section) problems for solution have been provided. Some of the problems are meant to assist the reader in becoming more facile in applying the methods in the text; others, to provide proofs of some of the material; and a third group, to expand the text of the material. Because of the wealth of literature that is available, it has been necessary to relegate some of the material to the problems.

REFERENCES

Barlow, R. E. and F. Proschan (1965), *Mathematical Theory of Reliability*, John Wiley and Sons, New York.

CHAPTER 2

Elements of Probability Theory

A definition of reliability was given in Section 1.1 of this book. From this definition, it is clear that probability theory will play an important role in the study of reliability. The purpose of this chapter is to familiarize the reader with some commonly used techniques in probability theory. What is presented here is in no way complete or detailed; moreover, it is presented with reliability applications in mind. It is hoped that a review of this chapter will give the reader a working knowledge of probability theory so that he can proceed rapidly to the heart of the book.

Probability theory is concerned with the methods of analysis that can be used in the study of random phenomena. "A *random phenomenon* is an empirical phenomenon characterized by the property that its observation under a given set of circumstances does not always lead to the same observed outcome but rather to different outcomes in such a way that there is *statistical regularity*" (Parzen, 1960). In effect, probability theory is concerned with the statements one can make about a random phenomenon that has certain known or assumed properties. In order to formulate postulates concerning a random phenomenon, the concept of a *sample description space* of a random phenomenon is introduced. A sample description space is a *set* of descriptions of all possible outcomes of the phenomenon.

It is clear, from the above, that one needs to know more about sets and their properties, and Section 2.1 is provided with this in mind. Before we proceed to Section 2.1, an example of a random phenomenon that is of interest in reliability will be pointed out.

The lifetime of a particular unit of a product is a random phenomenon. The key idea here is that before a life test is conducted on a particular device the exact outcome of the test is unpredictable.

Another related example of a random phenomenon is whether or not the above random lifetime exceeds some fixed number T. Of course this means

4

that the unit has survived for time T if the event is observed.

Both of the above types of random phenomena will be studied in detail in this book.

2.1 FUNDAMENTALS OF SET THEORY

A *set* is a collection of objects called *elements*. The word *collection* is defined implicitly in terms of the defining property for membership in the set, or collection. Throughout, a set is designated by a capital letter, such as A, whereas its elements will be denoted by small letters.

Example. Let A be the set of even positive integers, that is

$$A = \{2, 4, 6, 8, \ldots\}.$$

A can also be written as

$$A = \{y : y = 2I, I \text{ a positive integer}\}.$$

In any case the defining property of A, that is, the property which defines the collection is that y be an even positive integer.

Definitions

1. y is an element of a set A and is written $y \in A$.
2. If every element of a set A_1 is also an element of another set A_2, then A_1 is called a *subset* of A_2 and is written as $A_1 \subset A_2$.
3. *Equality* of sets. If $A_1 \subset A_2$ and $A_2 \subset A_1$, it is said that $A_1 = A_2$.
4. The totality of all elements under discussion is sometimes called the *universal set* or *space* and is denoted by S.

Example. Let S be the set of positive and negative integers, that is,

$$S = \{y : y = \pm I, I \text{ a positive integer}\}.$$

Then the set of positive integers, say A_1, is such that $A_1 \subset S$. The set A of the previous example is such that $A \subset A_1 \subset S$.

5. The set $\overline{A}$ represents the elements of S that are not in A and is called the *complement* of A with respect to S.

Example. In the preceding example the complement of $A_1, \overline{A}_1$, is the set of negative integers. In discussing a given set A and its complement $\overline{A}$ it is important to specify S. For example, if $A = \{y : 0 \leqslant y \leqslant \frac{1}{2}\}$ and $S = \{y : 0 \leqslant y \leqslant 2\}$, then $\overline{A} = \{y : \frac{1}{2} < y \leqslant 2\}$. But if $S = \{y : 0 \leqslant y \leqslant 1\}$, then $\overline{A} = \{y : \frac{1}{2} < y \leqslant 1\}$.

6. The set with no elements is called the empty or *null set* and is denoted by $\varnothing$.

Usually the "number" of elements in a set is of little importance; however, three classes should be mentioned.

a. *Finite sets.* The elements of a finite set can be put into a one-to-one correspondence with a set of positive integers, $1, 2, \ldots, n$, where n is finite.

Example. $A = \{y : y = 2, 4, 6, 8, 10\}$; then the number of elements is 5.

b. *Denumerably infinite sets.* The elements of this set can be put into a one-to-one correspondence with the positive integers.

Example. $A = \{y : y = 2I, I \text{ a positive integer}\}$, that is,

$$A = \{2, 4, 6, 8, 10, \ldots\}.$$

Letting I correspond to $2I$, we see that the elements of A can be matched one to one with the set of positive integers I; hence A is denumerably infinite or simply *denumerable* or *countable*.

c. *Nondenumerable sets.* These are sets that do not fall into either of the two classes given above.

Example. $A = \{y : 0 \leqslant y \leqslant 1\}$.

Two basic operations are defined for sets.
1. *Set union.* $A_1 \cup A_2$ is the set of elements that belong to A_1 or to A_2 or to both.

Example. If $A_1 = \{0 \leqslant y \leqslant 1\}$ and $A_2 = \{\frac{1}{2} \leqslant y \leqslant \frac{3}{2}\}$, then

$$A_1 \cup A_2 = \{0 \leqslant y \leqslant \frac{3}{2}\}.$$

Note that the union operation is a way of forming new sets. When the union operator is applied to arbitrary sets $A_1, A_2, \ldots, A_K$, the set

$$A = \bigcup_{i=1}^{K} A_i = A_1 \cup A_2 \cup \ldots \cup A_K$$

is well defined; and, if there are a denumerably infinite number of sets A_i, then

$$A = \bigcup_{i=1}^{\infty} A_i$$

is still well defined.

2. *Set intersection.* $A_1 \cap A_2$ is the set of elements that belong to both A_1 and A_2.

Example. If $A_1 = \{0 \leqslant y \leqslant 1\}$ and $A_2 = \{\frac{1}{2} \leqslant y \leqslant \frac{3}{2}\}$, then

$$A_1 \cap A_2 = \{\frac{1}{2} \leqslant y \leqslant 1\}.$$

Again the intersections

$$A = \bigcap_{i=1}^{K} A_i \quad \text{and} \quad A = \bigcap_{i=1}^{\infty} A_i$$

are well-defined sets consisting of the elements that belong to every set A_i.

Using the definitions given above and the two operations just defined, we can easily establish the following results:

a. $A \cup A = A$.

b. $A \cap A = A$.

c. $A \cup S = S$.

d. $A \cap S = A$.

e. If $A_1 \subset A_2$, then $A_1 \cap A_2 = A_1$.

f. $A \cup \overline{A} = S$.

g. If $A_1 \subset A_2$, then $\overline{A}_2 \subset \overline{A}_1$.

h. $A \cap \overline{A} = \varnothing$.

i. $\varnothing \subset A$, where A is any set.

j. $A \cup \varnothing = A$.

k. $A \cap \varnothing = \varnothing$.

The proofs of the above are left as an exercise for the reader.

Successful manipulation with sets requires the use of orderly notation. For example, if $A = \{1, 2, 3\}$, then A has three and only three elements. The set $A_1 = \{1, 2\}$ is a subset of A but is not an element of A. In fact, there are 2^n (see the problems at the end of the chapter) subsets of a set with n

elements. For $A = \{1,2,3\}$ there are $2^3 = 8$ subsets:

$$A_1 = \{1,2,3\}, \quad A_5 = \{1\},$$

$$A_2 = \{1,3\}, \quad A_6 = \{2\},$$

$$A_3 = \{1,2\}, \quad A_7 = \{3\},$$

$$A_4 = \{2,3\}, \quad A_8 = \varnothing.$$

Notice that there is a basic difference between the element 3, which is a member of A, and the set $\{3\}$, which is a subset of A. The element 3 has no subsets, whereas the set $\{3\}$ has one element, namely 3, and $2^1 = 2$ subsets, namely, $\{3\}$ and $\varnothing$. An important operation with sets is that any set can be written as the union of disjoint sets. For example, if A_1 and A_2 are two sets, it is easy to show that

$$A_2 = (A_2 \cap A_1) \cup \left(A_2 \cap \overline{A}_1\right).$$

2.2   SAMPLE DESCRIPTION SPACES, RANDOM VARIABLES, AND THE DEFINITION OF PROBABILITY

2.2.1 Sample Description Spaces and Random Variables

We will assume that, for any given experiment or trial, it is possible to identify but not necessarily enumerate the set of all possible outcomes. This set of outcomes will be called the *sample description space S*. It should be noted that S is exhaustive for the experiment, that is, S contains all the possible outcomes. The outcomes in S are also to be thought of as "elementary outcomes" since no single outcome involves more than one observable outcome. The outcomes of S are also mutually exclusive, since for any one given experiment one and only one of the outcomes of S is possible. When the outcomes of S are not numerical-valued, it is often useful to define a numerical-valued function on the sample description space S.

Example 1.   Suppose that a unit of a product is turned on at time zero and is observed for T units of time to see whether it operated throughout T or did not operate throughout T. Then S consists of two outcomes:

$A_1 =$ product operated for T units of time;

$A_2 =$ product did not operate for T units of time.

A numerical-valued function can be defined on S as follows:

$$f(A_1) = 1 \quad \text{(product operated throughout } T),$$

$$f(A_2) = 0 \quad \text{(product did not operate throughout } T);$$

and a new sample description space is obtained, consisting of the elements $(1, 0)$. Whether or not it is necessary to construct such a function depends entirely on the situation at hand. This point will be discussed in more detail later.

Example 2. Suppose that a unit of product is turned on and monitored continuously in time and that its lifetime x is recorded. Then

$$S = \{ x : 0 \leqslant x < \infty \}.$$

This merely expresses the fact that the lifetime can be any real number between zero and infinity. It is to be noted that (a) the outcomes of S are numerical-valued and (b) the set S has a nondenumerable infinity of points.

These two examples illustrate two types of recurring reliability problems. The first is the type of problem in which the only concern about lifetime is whether it exceeds some fixed number T, and in the second the exact value of lifetime is of concern.

Definitions

A *random variable* is a numerical-valued function defined on the sample description space.

Often the function is the identity function. In this case the values which the random variable can assume are the elements of the sample description space.

An *event* is any subset of the sample description space.

The events A_1 and A_2 are said to be *mutually exclusive* if the occurrence of one precludes the occurrence of the other, that is, if $A_1 \cap A_2 = \varnothing$

If A_1 and A_2 are any two arbitrary events, $A_1 \cup A_2$ means either event A_1 or event A_2 or both A_1 and A_2, and $A_1 \cap A_2$ means the event A_1 and A_2.

Example. Suppose that two six-faced dice are tossed. Let A_1 be the event that the sum of the numbers on the faces is 7, and let A_2 be the event that exactly one of the faces shows a 2. Also, $A_1 \cup A_2$ is the event that either the sum is 7 or exactly one of the faces shows a 2, or both, and

$A_1 \cap A_2$ is the event that the sum of the faces is 7 and exactly one face is a 2. In this example it should be noted that the elements of S_2 are pairs of random variables.

The event A_1 is a subset of the sample description space

$$S_1 = \{ x : x = 2, 3, 4, \ldots, 12 \},$$

that is, x is the sum of the two faces. The event A_2 is a subset of the sample description space

$$S_2 = \{ (x_1, x_2) : (1,1), (1,2), \ldots, (6,6) \},$$

where x_1 represents the number showing on the first die, and x_2 represents the number showing on the second die. It is easy to see that S_2 consists of 36 elements, whereas S_1 consists of 11 elements. A closer look at S_1 reveals that it is a *derived* sample description space, that is, the dice do not come up showing sums and a mechanical operation (addition) is involved in obtaining S_1. In fact, the points in S_1 can be represented by points in S_2. Consider the accompanying table.

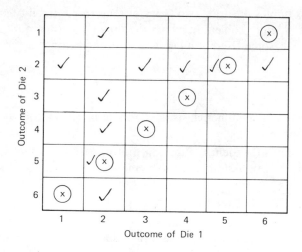

This table is a tabular display of S_2, and the $\otimes$'s indicate those of the 36 possible elements that result in a sum of 7 (i.e., in event A_1). Event A_2 is represented by the check marks. Thus there are 6 elements (the 6 $\otimes$'s) of the 36 that represent A_1, and 10 elements of the 36 that represent A_2. Also it can be seen that the sample description space S_2 serves to define the possible outcomes quite nicely. Turning to the two events $A_1 \cap A_2$

and $A_1 \cup A_2$, we can see from the table that $A_1 \cap A_2$ is represented by the squares containing both a check and an $\otimes$ and that there are 2 such elements, namely, $(2,5)$ and $(5,2)$. The event $A_1 \cup A_2$ is represented by the squares containing either a check or an $\otimes$ or both. There are 14 such elements, which can be counted directly from the table or may be calculated as

$$(10 \text{ checks}) + (6 \otimes\text{'s}) - (2 \text{ checks and } \otimes\text{'s}),$$

the last 2 being subtracted because the squares containing both a check and an $\otimes$ were counted twice—once in counting the checks and once in counting the $\otimes$'s. In Problem 2.4 it is noted that (De Morgan's laws) $(\overline{A_1 \cup A_2}) = \overline{A}_1 \cap \overline{A}_2$. This rule is easily verified here. $(\overline{A_1 \cup A_2})$ is the set of elements $(36 - 14 = 22$ in number) represented by the squares containing no marks in the table.

2.2.2 Definition of Probability

From a practical point of view, the following interpretation of probability seems to be the most useful.

Let S be a sample description space and let A be an event of S. Consider now a number n of repeated experiments whose outcomes are described by S. Let X be the number of times that A occurs in the n repeated experiments. Then the probability of event A, written as $P(A)$, is defined as

$$P(A) \equiv \lim_{n \to \infty} \left(\frac{X}{n} \right). \qquad (2.1)$$

It is to be noted that for fixed n the quantity (X/n) is the relative frequency of the occurrence of A. Since it is impossible to physically let $n \to \infty$ (i.e., to conduct an infinite number of trials), many of the methods of probability and statistics deal with the question of how to estimate $P(A)$. In many cases it is obvious from inspecting the sample description space S what the limit (2.1) will be.

Example. Suppose that the experiment is the random tossing of a well-balanced six-sided die. Then the random variable is the number showing and

$$S = \{ x : x = 1, 2, \ldots, 6 \}.$$

If $A \equiv$ the event that a 2 turns up on a given toss, then

$$P(A) = \lim_{n \to \infty} \left(\frac{X}{n} \right) = \frac{1}{6}.$$

In other words, if each outcome of the sample description space is equally likely, then

$$P(A) = \frac{\text{number of elements of } S \text{ that corresponds to } A}{\text{total number of elements of } S}. \quad (2.2)$$

In the die example,

$$P(A) = \tfrac{1}{6}.$$

Mathematically, the probability of an event $A, P(A)$, is a function defined on any subset A of S, such that the following are satisfied:

(i) $P(A) \geqslant 0$;

(ii) $P(\bigcup_{i=1}^{\infty} A_i) = \sum_{i=1}^{\infty} P(A_i)$, provided that for every $i, j, i \neq j, A_i \cap A_j = \varnothing$;

(iii) $P(S) = 1$.

It is clear that definition (2.1) satisfies these three properties.

Stated in words, the above conditions imply that the probability function always lies between 0 and 1, that the probability of the union of mutually exclusive events is the sum of the probabilities of the individual events, and that an outcome in S always occurs.

Conditions (i), (ii), and (iii) above are taken as the axioms of probability theory. These axioms lead to several important rules.

RULE 1.   ADDITION. $P(A_1 \cup A_2) = P(A_1) + P(A_2) - P(A_1 \cap A_2)$ for any two events A_1 and A_2.

Proof. Events A_1, A_2, and their complements can be observed simultaneously and exhaustively as $A_1 \cap A_2$, $A_1 \cap \overline{A}_2$, $\overline{A}_1 \cap A_2$, or $\overline{A}_1 \cap \overline{A}_2$. These four events are also mutually exclusive and hence

$$A_1 \cup A_2 = (A_1 \cap A_2) \cup (A_1 \cap \overline{A}_2) \cup (\overline{A}_1 \cap A_2).$$

Applying axiom (ii), we obtain

$$P(A_1 \cup A_2) = P(A_1 \cap A_2) + P(A_1 \cap \overline{A}_2) + P(\overline{A}_1 \cap A_2). \quad (2.3)$$

Now note that (this will be the same operation as decomposing a set A_1 with respect to a set A_2)

$$A_1 = (A_1 \cap \overline{A}_2) \cup (A_1 \cap A_2)$$

and

$$A_2 = (A_2 \cap A_1) \cup (A_2 \cap \overline{A}_1).$$

Thus

$$P(A_1) = P(A_1 \cap \overline{A}_2) + P(A_1 \cap A_2), \qquad (2.4)$$

$$P(A_2) = P(A_2 \cap A_1) + P(A_2 \cap \overline{A}_1), \qquad (2.5)$$

and hence from (2.3) and (2.4)

$$P(A_1 \cup A_2) = P(A_1) + P(\overline{A}_1 \cap A_2).$$

But from (2.5) $P(\overline{A}_1 \cap A_2) = P(A_2) - P(A_1 \cap A_2)$, so that

$$P(A_1 \cup A_2) = P(A_1) + P(A_2) - P(A_1 \cap A_2).$$

This simple proof has been given in detail to illustrate an important method of finding the probability of an event: decomposing it into its mutually exclusive forms.

Example.   Consider the dice example of Section 2.2.1. As before, let A_1 denote the event that the sum of the faces is 7, and A_2 the event that exactly one of the faces shows a 2. Then applying the addition rule directly yields

$$P(A_1 \cup A_2) = P(A_1) + P(A_2) - P(A_1 \cap A_2)$$

and using equation (2.2) gives

$$P(A_1 \cup A_2) = \tfrac{6}{36} + \tfrac{10}{36} - \tfrac{2}{36}$$

$$= \tfrac{7}{18}.$$

The answer could also have been developed by using the decomposition equation (2.3):

$$P(A_1 \cup A_2) = \tfrac{2}{36} + \tfrac{4}{36} + \tfrac{8}{36}$$

$$= \tfrac{7}{18}.$$

RULE 2.   INEQUALITY. If $A_1 \subset A_2$, that is, if event A_1 implies event A_2, then $P(A_1) \leqslant P(A_2)$. This result may give a useful bound on the probability of an event when the exact probability is unknown or cannot be calculated.

Proof. The proof again indicates the importance of decomposition of an event into its mutually exclusive forms:

$$A_2 = (A_2 \cap A_1) \cup (A_2 \cap \overline{A}_1),$$

but clearly, since by hypothesis $A_1 \subset A_2$, then $A_1 \subset A_2 \cap A_1$. Also, whenever $A_1 \cap A_2$ occurs, so does A_1 so that

$$A_1 = A_2 \cap A_1 \quad \text{and} \quad A_2 = A_1 \cup (A_2 \cap \overline{A}_1).$$

Hence, by applying axiom (ii), $P(A_2) = P(A_1) + P(A_2 \cap \overline{A}_1)$. This means $P(A_2) - P(A_1) = P(A_2 \cap \overline{A}_1)$. But by axiom (i) every P is such that $0 \leqslant P$. Thus $P(A_2) - P(A_1) = P(A_2 \cap \overline{A}_1) \geqslant 0$ and hence $P(A_2) \geqslant P(A_1)$.

RULE 3.   For any event A; $P(A) + P(\overline{A}) = 1$.

Proof. The sample description space S can be decomposed as $S = A \cup \overline{A}$, and A and $\overline{A}$ are mutually exclusive. Thus, applying axiom (ii) gives

$$P(S) = P(A) + P(\overline{A}),$$

but from axiom (iii) $P(S) = 1$, so that

$$1 = P(A) + P(\overline{A}).$$

RULE 4. $P(A \cap \overline{A}) = 0$. This event, $A \cap \overline{A}$, is sometimes called the impossible event.

Proof. Applying Rule 1, we obtain

$$P(A \cup \overline{A}) = P(S) = P(A) + P(\overline{A}) - P(A \cap \overline{A}).$$

Now using axiom (iii) and Rule 3 gives

$$P(S) = 1 = 1 - P(A \cap \overline{A})$$

and hence

$$P(A \cap \overline{A}) = 0.$$

RULE 5. If two events, A_1 and A_2, are mutually exclusive, $P(A_1 \cap A_2) = 0$.

Proof. Applying Rule 1 gives

$$P(A_1 \cup A_2) = P(A_1) + P(A_2) - P(A_1 \cap A_2),$$

but using axiom (ii) yields

$$P(A_1 \cup A_2) = P(A_1) + P(A_2).$$

Hence

$$P(A_1 \cap A_2) = 0.$$

Definitions

The *conditional probability* of event A_1, given that A_2 has occurred, $P(A_1|A_2)$, is defined as

$$P(A_1|A_2) = \frac{P(A_1 \cap A_2)}{P(A_2)}, \quad \text{if } P(A_2) \neq 0.$$

If $P(A_1 \cap A_2) = P(A_1)P(A_2)$, the events A_1 and A_2 are said to be *independent*.

Example. Let an ordinary six-sided die be tossed. Let A_1 be the event that the upturned face shows a number ≥ 2. Let A_2 be the event that the upturned face shows exactly 5. Compute $P(A_2|A_1)$. Applying the definition, we obtain

$$P(A_2|A_1) = \frac{P(A_1 \cap A_2)}{P(A_1)}, \quad P(A_1) \neq 0.$$

Now clearly

$$P(A_1) = P(\text{face shows } 2) + P(\text{face shows } 3) + \cdots + P(\text{face shows } 6)$$

$$= \tfrac{5}{6}.$$

But also

$$P(A_1 \cap A_2) = P(A_1|A_2)P(A_2)$$

and

$$P(A_2) = \tfrac{1}{6}; \ P(A_1|A_2) = 1.$$

Thus

$$P(A_1 \cap A_2) = \tfrac{1}{6} \quad \text{and} \quad P(A_2|A_1) = \frac{\tfrac{1}{6}}{\tfrac{5}{6}} = \frac{1}{5}.$$

This example illustrates a frequently occurring situation: in calculating $P(A_2|A_1)$ it is often difficult to see what $P(A_1 \cap A_2)$ is, but by using the relation

$$P(A_1 \cap A_2) = P(A_1|A_2)P(A_2)$$

$P(A_1 \cap A_2)$ can be calculated.

2.3   ELEMENTARY COMBINATORIAL ANALYSIS

Because of Equation (2.2) it is often useful to be able to count the elements in a sample description space S and the elements corresponding to an arbitrary event A. For this purpose some counting rules are needed. They all follow from one rule.

RULE.   If event A_1 can occur in a_1 ways and event A_2 can occur in a_2 ways, event $A_1 \cap A_2$ can occur in $a_1 a_2$ ways.

Example.   Consider a system comprised of N units, each with two states: operative and nonoperative. The total number of different states in which the system can be with respect to its N units is

$$2 \times 2 \times 2 \cdots N \text{ times} = 2^N.$$

2.3.1   Permutations and Combinations

Consider N distinguishable objects and suppose that $x \, (0 \leqslant x \leqslant N)$ of these objects are selected so as to form a set. How many different sets can be selected? If the above rule is applied repeatedly, the first object can be selected in N ways, the second object in $N-1$ ways (since there are only $N-1$ left from which to choose the second object), and so on. Finally, the xth object must be chosen from the remaining $N-(x-1)=(N-x+1)$ objects. Thus the number of different ways of selecting x objects from N distinguishable objects is

$$N(N-1)(N-2) \cdots (N-x+1). \tag{2.6}$$

It is important to define what is meant by "different." Under the above construction two sets may be different even if they contain the same objects. If the order of selection was different, the sets are counted as different sets.

Example 1. Suppose that the distinguishable objects are the letters a, b, c, d. Suppose further that $x = 2$, that is, two objects are to be taken from the $N = 4$ objects and the question is how many different sets can be formed. Equation (2.6) gives the answer as

$$N(N-1) \cdots (N-x+1) = 4 \times 3 = 12.$$

The twelve sets can be identified as

$$ab, \quad ba,$$
$$ac, \quad ca,$$
$$ad, \quad da,$$
$$bc, \quad cb,$$
$$bd, \quad db,$$
$$cd, \quad dc.$$

Two aspects of Example 1 are important.

1. Sets are counted as different if they contain different objects or if they contain the same objects but selected in different order. Such sets are called *permutations*.
2. Often it is virtually impossible to enumerate permutations and (2.6) must be used instead.

Equation (2.6) is used so often that it has been denoted as

$$(N)_x = N(N-1) \ldots (N-x+1)$$

= the number of permutations of N distinguishable objects taken x at a time.

Example 2. A board of directors must select from $N = 8$ candidates a president and a vice president. What is the probability that a particular candidate will be given an office? The number of elements in S is

$$(N)_2 = 8 \cdot 7 = 56.$$

Of these 56 different selections a particular candidate can be president with 7 different candidates and vice president with 7 different candidates so that, when Equation (2.2) is applied, the required probability is

$$P = \frac{14}{56} = \frac{1}{4}.$$

When, for reasons of the particular problem at hand, it is desired to count as different only the sets that contain different objects, the different sets are called *combinations*. The number of combinations of N objects taken x at a time is denoted as $\begin{pmatrix} N \\ x \end{pmatrix}$. Certainly the number of permutations is $x!$ times the number of combinations, so that

$$\begin{pmatrix} N \\ x \end{pmatrix} = \frac{(N)_x}{x!} = \frac{N!}{x!(N-x)!}.$$

We will take $0! = 1$ and $\begin{pmatrix} N \\ x \end{pmatrix} = 0$ when $x > N$.

2.3.2 Simple Random Sampling: with and without Replacement

Suppose that a sample of size x is obtained from a population of N distinguishable objects by drawing one object at a time and replacing it into the population before the next draw. This is called *sampling with replacement*. The first member of the sample can be drawn in N ways, the second in N ways, and so on, so that N^x different samples of size x can be obtained.

If the same situation as above occurs, but if each object is withheld after being drawn, the first object can be drawn in N ways, the second in $(N-1)$ ways, and so on, so that

$$N(N-1)(N-2)\dots[N-(x-1)] = (N)_x$$

different samples of size x are obtainable.

Definition

A *simple random sample* is one that is obtained in such a way that each of the possible different samples [a total of N^x when sampling with replacement and a total of $(N)_x$ when sampling without replacement] is equally likely.

The following situation illustrates some ideas that are useful in sampling: for a population of N distinguishable objects, what is the probability that a sample of size x will include a particular object, say the ith?

(a) Sampling without Replacement

Method 1.   Let A be the event that a particular object, say the ith, is included in the sample. As before, the sample description space S consists of a total of $(N)_x = N(N-1)\ldots(N-x+1)$ elements. If the elements of S corresponding to A can be counted, Equation (2.2) can be applied. As is often the case, event A can be decomposed in x mutually exclusive events:

$$A = A_1 \cup A_2 \cup \ldots \cup A_x,$$

where A_j = the event that the ith object is drawn on the jth draw, with $j = 1,2,\ldots,x$. Now for each A_j the other $(x-1)$ objects may be selected from the remaining $(N-1)$ objects in $(N-1)_{x-1}$ ways. Thus the total elements of S corresponding to A are $x(N-1)_{x-1}$, and applying Equation (2.2) gives

$$P(\text{a particular object selected}) = \frac{x(N-1)_{x-1}}{(N)_x}$$

$$= x\frac{(N-1)!}{(N-x)!}\frac{(N-x)!}{N!}$$

$$= \frac{x}{N}, \quad 0 \leqslant x \leqslant N.$$

Method 2.   Again, event A can occur in x mutually exclusive ways:

$$A = A_1 \cup A_2 \cup \ldots \cup A_x,$$

where the A_j, with $j = 1,2,\ldots,x$, are as defined in Method 1. Now, if each $P(A_j)$ could be calculated, axiom (ii) could be applied, that is,

$$P(A) = P(A_1) + P(A_2) + \ldots + P(A_x).$$

Now $P(A_1)$ = the probability that a particular object is obtained on the first draw. The first draw can occur in N ways, one of which results in the particular object being selected, so that applying Equation (2.2) yields

$$P(A_1) = \frac{1}{N}.$$

For the object to be selected on the second draw means that it was not selected on the first but was selected on the second. Thus

$$A_2 = \overline{A}_1 \cap A_2,$$

and applying Equation (2.2) twice and the definition of conditional probability yields

$$P(A_2) = P(\overline{A}_1 \cap A_2) = P(A_2|\overline{A}_1)P(\overline{A}_1)$$

$$= \frac{1}{N-1} \frac{N-1}{N}$$

$$= \frac{1}{N}.$$

In general

$$A_j = A_j \cap \overline{A}_{j-1} \cap \overline{A}_{j-2} \cap \ldots \cap \overline{A}_1,$$

so that

$$P(A_j) = P(A_j \cap \overline{A}_{j-1} \cap \overline{A}_{j-2} \cap \ldots \cap \overline{A}_1)$$

$$= P(A_j|\overline{A}_{j-1} \cap \overline{A}_{j-2} \cap \ldots \cap \overline{A}_1)P(\overline{A}_{j-1} \cap \overline{A}_{j-2} \cap \ldots \cap \overline{A}_1).$$

By a repeated application of the definition of conditional probability

$$P(\overline{A}_2 \cap \overline{A}_1) = P(\overline{A}_2|\overline{A}_1)P(\overline{A}_1) = \left(\frac{N-2}{N-1}\right)\left(\frac{N-1}{N}\right) = \left(\frac{N-2}{N}\right)$$

and

$$P(\overline{A}_3 \cap \overline{A}_2 \cap \overline{A}_1) = P(\overline{A}_3|\overline{A}_2 \cap \overline{A}_1)P(\overline{A}_2 \cap \overline{A}_1)$$

$$= \left(\frac{N-3}{N-2}\right)\left(\frac{N-2}{N}\right) = \left(\frac{N-3}{N}\right).$$

It can be seen that

$$P(\overline{A}_{j-1} \cap \overline{A}_{j-2} \cap \ldots \cap \overline{A}_1) = \frac{N-(j-1)}{N} = \frac{N-j+1}{N},$$

so that

$$P(A_j) = \left(\frac{1}{N-j+1}\right)\left(\frac{N-j+1}{N}\right)$$

$$= \frac{1}{N}.$$

Hence

$$P(A) = \underbrace{\frac{1}{N} + \frac{1}{N} + \ldots + \frac{1}{N}}_{x}$$

$$= \frac{x}{N}.$$

Three observations can be made on the basis of the above problem.

1. Simple random sampling implies that each object in the population has an equal chance, (x/N), of becoming a member of the sample.

2. Conditional probability, although defined in terms of only two events, can easily be extended to more than two events by repeated application of the definition.

3. Since $P(A_1)P(A_2) = (1/N)^2$, events A_1, A_2 are not independent; in fact, $P(A_1 \cap A_2) = 0$.

(b) Sampling with Replacement. Again the probability of event A (i.e., of the inclusion of a particular object in the sample) is sought. As before, N^x different samples are possible. It is noted that in sampling with replacement a particular object may be included more than once.

Method 1. The total number of sample space points is N^x. For the event $\bar{A}$ (a given object not included) there are

$$\underbrace{(N-1)(N-1)\ldots(N-1)}_{x} = (N-1)^x$$

points. Thus applying Equation (2.2) gives

$$P(\bar{A}) = \left(\frac{N-1}{N}\right)^x.$$

Using Rule 3, we obtain

$$P(A) = 1 - \left(\frac{N-1}{N} \right)^x.$$

Method 2. This method is left as an exercise for the reader.

2.4 PROBABILITY DISTRIBUTIONS

In the balance of this chapter the sample description space S should be thought of as consisting of the values which the random variable X assumes. If X is not the identity function then S is really a derived sample description space.

A random variable X is said to be *discrete* if its sample description space consists of a finite or a denumerably infinite number of points. In this case, each element of S can be identified with a positive integer and S can be represented by

$$S = \{ x_1, x_2, \dots, x_N \} \qquad \text{or} \qquad S = \{ x_1, x_2, \dots \}.$$

Here X is used to denote the generic random variable, whereas x_i is a particular value that X takes on. If a function $f_X(x_i)$ is defined on S such that

(i) $P(X = x_i) = f_X(x_i)$;

(ii) $f_X(x_i) \geqslant 0$;

(iii) $\displaystyle\sum_{i=1}^{N} f_X(x_i) = 1$ if S contains a finite number of elements, N;

or $\displaystyle\sum_{i=1}^{\infty} f_X(x_i) = 1$ if S contains a denumerably infinite number of ele-

ments; then $f_X(x_i)$ is called a *discrete probability distribution* (or probability mass function). The X in $f_X(x_i)$ is often omitted when it is clear which X is under discussion.

If a function $f_X(x)$ is defined on S such that

(i) $P(a < X < b) = \displaystyle\int_a^b f_X(x)\, dx, \ a < b$.

(ii) $f_X(x) \geqslant 0$;

(iii) $\displaystyle\int_{-\infty}^{\infty} f_X(x)\, dx = 1$; then $f_X(x)$ is called a *continuous probability distri-bution* (or probability density function) and X is termed a *continuous* random variable.

Even though X may be restricted to a finite interval, $f(x)$ can be defined to be zero outside the interval, so that the infinite limits in (i) and (iii) need cause no confusion. Notice that the subscript i on x has to be dropped for a continuous random variable because the values that X takes on can no longer be counted. In fact the sample description space S for a continuous random variable is a continuum of points. The requirement that $f_X(x)$ be a continuous function could be weakened but the return in utility would be small since most of the probability distributions useful in Reliability are continuous (or discrete).

2.5 SAMPLING FROM A PROBABILITY DISTRIBUTION

In Section 2.4 two types of probability distributions were defined which describe the probabilistic behavior of a random variable. The elements in the sample description space serve to identify all the possible outcomes of an experiment, and the probability distribution is used to assign probabilities to the individual outcomes or to some collection of them. Of course, in the continuous random variable case, the probability of a particular outcome, say $X = a$, is zero since

$$P(X=a) = \int_a^a f(x)\,dx = 0, \quad a \in S,$$

but nonetheless the continuous probability distribution describes the probability of events like $(a < X < b)$ so that

$$P(a<X<b) = P(a \leqslant X \leqslant b) = \int_a^b f(x)\,dx \quad , \quad a<b, \quad (a,b) \subset S.$$

Given a probability distribution for a random variable X and a sample description space S, we can consider a sequence of observations on X.

Definition

For a given sample description space S, random variable X, and probability distribution $f(x)$ defined on (X,S), a particular sequence of n observations on X (n performances of the experiment) with outcomes

$$(X=x_1, X=x_2, \ldots, X=x_n) = (x_1, x_2, \ldots, x_n)$$

is called a *sample* from the probability distribution $f(x)$.

The idea of taking a sample of size n from a probability distribution (discrete or continuous) is of fundamental importance to virtually all statistical methods. For this reason it is explored in more detail at this point.

The sample description space S describes the possible outcomes for one conduct of a particular experiment. Thus each of the n outcomes, $x_1, x_2, \ldots,$ x_n, belongs to S, that is,

$$x_i \in S, \quad i = 1, 2, \ldots, n.$$

However, the *set* of outcomes $(x_1, x_2, \ldots, x_n)$ does not belong to S in the sense of being an element of S as defined in Section 2.1. In fact, the set of outcomes $(x_1, x_2, \ldots, x_n)$ belongs to the n-dimensional sample description space

$$\underbrace{S \times S \times \ldots \times S}_{n.} = S^{(n)}$$

Example 1. Suppose that S consists of the two points $S = \{0, 1\}$ and the probability distribution is such that

$$P(X=1) = f(1) = p, \qquad P(X=0) = f(0) = 1 - p.$$

This discrete probability distribution can be written more compactly as

$$f_X(x) = p^x (1-p)^{1-x}, \quad x = 0, 1; \ 0 < p < 1.$$

Now suppose that this experiment is conducted $n = 3$ times. There will then be 2^3 possible outcomes for the three x's (x_1, x_2, x_3):

$$(1,1,1), (1,1,0), (1,0,1), (1,0,0), (0,1,1), (0,1,0), (0,0,1), (0,0,0).$$

These points do not belong to $S = \{0, 1\}$, which contains only two points, but they belong to $S \times S \times S$, which contains the eight points shown.

Example 2. Suppose that $S = \{x : 0 \leqslant x \leqslant 1\}$ and

$$f_X(x) = 1, \quad 0 \leqslant x \leqslant 1,$$

$$= 0, \quad \text{elsewhere.}$$

Now suppose that $n = 2$ observations are made on X, that is, suppose that two samples are drawn from the $f(x)$ above. The possible outcomes are described by $S \times S$, that is, by pairs of points, (x_1, x_2) say, such that $x_1 \in [0, 1], x_2 \in [0, 1]$. Figure 2.1 is a graphical display of $S \times S$.

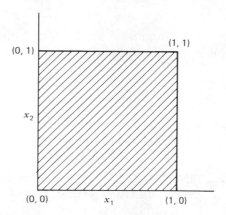

Figure 2.1.

In the general case of n samples the space $S^{(n)} = S \times S \times \ldots \times S$ is an n-dimensional cube. The important point to note is that the result of n conducts of an experiment (whose possible outcomes are represented by S), represented by $(x_1, x_2, \ldots, x_n)$, is a point or outcome in $S^{(n)}$.

Now $S^{(n)}$ is again a sample description space, and a probability distribution can be defined on it. In fact, the n observations could also have arisen from a single conduct of n different experiments, each having a sample space S_i and a probability distribution $f_i, i = 1, 2, \ldots, n$. In such a case the sample description space $S^{(n)}$ is sometimes called the *Cartesian product* of the involved sample description spaces.

Definition

A probability distribution defined on the Cartesian product of two or more sample description spaces (and hence defined for two or more random variables) is called a *joint probability distribution*.

To avoid notational difficulties, the random variables corresponding to each sample space concerned are denoted as $X_1, X_2, \ldots, X_n$, and the notation for the joint probability distribution is

$$f_{X_1, X_2, \ldots, X_n}(x_1, x_2, \ldots, x_n).$$

Joint probability distributions are discussed in more detail in the next section. This section will close with an important definition.

Definition

A sample of size n taken from a probability distribution $f_X(x)$ is called a *random sample* if the joint probability distribution of the n observations $(x_1, x_2, \ldots, x_n)$ is

$$f_{X_1, X_2, \ldots, X_n}(x_1, x_2, \ldots, x_n) = \prod_{i=1}^{n} f_X(x_i). \qquad (2.7)$$

That is, the joint probability distribution is the product of the (identical) individual probability distributions.

2.6   JOINT, MARGINAL, AND CONDITIONAL PROBABILITY DISTRIBUTIONS

2.6.1   Joint Probability Distributions

In this section the case considered is limited to continuous random variables, the development for discrete random variables being much the same. The joint probability distribution was introduced in Section 2.5 as the probability distribution of the n random variables $X_1, X_2, \ldots, X_n$, and all these random variables may have (*a*) the same probability distribution, in which case they are said to be *identically distributed*, or (*b*) different probability distributions, or (*c*) any combination of (*a*) and (*b*). In any event the joint probability distribution must satisfy three properties:

   (i) $f(x_1, x_2, \ldots, x_n) \geqslant 0$;
   (ii) $P(a_1 < X_1 < b_1, a_2 < X_2 < b_2, \ldots, a_n < X_n < b_n)$
      $= \int_{a_n}^{b_n} \cdots \int_{a_1}^{b_1} f(x_1, x_2, \ldots, x_n) \, dx_1 \, dx_2 \ldots dx_n$;
   (iii) $\int_{-\infty}^{\infty} \cdots \int_{-\infty}^{\infty} f(x_1, x_2, \ldots, x_n) \, dx_1 \, dx_2 \ldots dx_n = 1$.

Property (ii) is read as the probability that $a_1 < X_1 < b$, $a_2 < X_2 < b_2$, and so on. Independence of two events has already been defined as

$$P(A \cap B) = P(A)P(B).$$

An analog to this definition exists for random variables.

Definition

The random variables $X_1, X_2, \ldots, X_n$ are said to be *independent* (or independently distributed or statistically independent or stochastically independent) if $f(x_1, x_2, \ldots, x_n) = f_1(x_1)f_2(x_2)\ldots f_n(x_n)$, where the subscript i on f is a reminder that the X_i may have different probability distributions.

2.6.2   Marginal Probability Distributions

Consider, as before, a set of random variables, $X_1, X_2, \ldots, X_n$. The probability distribution of X_i will be denoted by $f_i(x_i)$ and, of course,

$$P(a_i < X_i < b_i) = \int_{a_i}^{b_i} f_i(x_i)\, dx_i. \tag{2.8}$$

Now the event $(a_i < X_i < b_i)$ implies the event $(a_i < X_i < b_i, -\infty < X_1 < \infty, \ldots, -\infty < X_{i-1} < \infty, -\infty < X_{i+1} < \infty, \ldots, -\infty < X_n < \infty)$ and conversely, so that

$$P(a_i < X_i < b_i) = \int_{-\infty}^{\infty} \cdots \int_{-\infty}^{\infty} \int_{a_i}^{b_i} \int_{-\infty}^{\infty} \cdots \int_{-\infty}^{\infty} f(x_1, x_2, \ldots, x_n)\, dx_1\, dx_2 \ldots dx_n.$$

$$\tag{2.9}$$

The multiple integral

$$\int_{-\infty}^{\infty} \cdots \int_{-\infty}^{\infty} f(x_1, x_2, \ldots, x_n)\, dx_1\, dx_2 \ldots dx_{i-1}\, dx_{i+1}\, dx_n$$

is a function of x_i only, and Equations (2.8) and (2.9) imply that

$$f_i(x_i) = \int_{-\infty}^{\infty} \cdots \int_{-\infty}^{\infty} f(x_1, x_2, \ldots, x_n)\, dx_1\, dx_2 \ldots dx_{i-1}\, dx_{i+1} \ldots dx_n.$$

When $f_i(x_i)$ is obtained in this way, it is often called a *marginal* probability distribution. The adjective marginal is necessary only when discussing jointly distributed random variables.

A word of caution is in order at this point. In writing down a probability distribution, it is always necessary to specify the region of definition of the random variable X. This is particularly true when working with joint probability distributions because the region of definition of X_i may depend on one or more of the other random variables.

Example.   Suppose that two random variables X_1, X_2 have a joint probability distribution:

$$f(x_1, x_2) = 15(x_1 x_2^2), \quad 0 \leqslant x_1 \leqslant 1, 0 \leqslant x_2 \leqslant x_1,$$

$$= 0, \quad \text{elsewhere.}$$

The sample description space is shown in Figure 2.2.

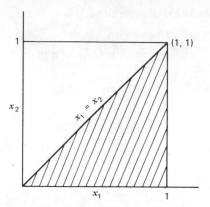

Figure 2.2.

The marginal distributions are as follows:

$$f_1(x_1) = \int_{-\infty}^{\infty} 15(x_1 x_2^2)\, dx_2 = \int_0^{x_1} 15(x_1 x_2^2)\, dx_2$$

$$= 5x_1^4, \quad 0 \leqslant x_1 \leqslant 1.$$

When finding $f_2(x_2)$, reference to Figure 2.2 is a reminder that x_1 ranges between x_2 and 1. Thus

$$f_2(x_2) = \int_{-\infty}^{\infty} 15(x_1 x_2^2)\, dx_1 = \int_{x_2}^1 15(x_1 x_2^2)\, dx_1$$

$$= \frac{15}{2}(x_2^2 - x_2^4), \quad 0 \leqslant x_2 \leqslant 1.$$

It is easily verified that

$$\int_0^1 f_1(x_1)\, dx_1 = 1 \quad \text{and} \quad \int_0^1 f_2(x_2)\, dx_2 = 1,$$

so that $f(x_1, x_2) = 15(x_1 x_2^2)$ is a probability distribution. Clearly X_1 and X_2 are not independent, since the region of definition of X_2 depends on the value observed for X_1. It should be noted that, with reference to Figure 2.2, the regions of definition could have been stated as

$$f(x_1, x_2) = 15(x_1 x_2^2), \quad 0 \leqslant x_2 \leqslant 1, \, x_2 \leqslant x_1 \leqslant 1,$$

$$= 0, \quad \text{elsewhere.}$$

2.6.3   Conditional Probability Distributions

Given a joint probability distribution $f(x_1, x_2)$ for two random variables X_1, X_2, we find that the ratio $f(x_1, x_2)/f_2(x_2)$ satisfies three conditions:

(i) $\dfrac{f(x_1, x_2)}{f_2(x_2)} \geqslant 0$ at points of S_2 for which $f_2(x_2) \neq 0$;

(ii) $P(a_1 < X_1 < b_1 | X_2 = x_2) = \displaystyle\int_{a_1}^{b_1} \dfrac{f(x_1, x_2)}{f_2(x_2)} \, dx_1$;

(iii) $\displaystyle\int_{-\infty}^{\infty} \dfrac{f(x_1, x_2)}{f_2(x_2)} \, dx_1 = \dfrac{1}{f_2(x_2)} \int_{-\infty}^{\infty} f(x_1, x_2) \, dx_1 = 1.$

Thus $f(x_1, x_2)/f_2(x_2)$ is a probability distribution and is called the *conditional distribution* of the random variable X_1, given X_2. It is usually denoted by

$$f_{X_1 | X_2}(x_1 | x_2) = f(x_1 | x_2) = \frac{f(x_1, x_2)}{f_2(x_2)}, \quad f_2(x_2) \neq 0.$$

A similar argument leads to

$$f(x_2 | x_1) = \frac{f(x_1, x_2)}{f_1(x_1)}.$$

The sampling interpretation is clear enough. When sampling from a conditional probability distribution in two random variables, the outcomes for, say, X_1 in the sample space are only those that can occur with a particular fixed value of X_2, say x_2. As will be seen in Chapter 8, in developing Bayesian methods for reliability it is sometimes very difficult to obtain random samples from conditional probability distributions. Finally, it should be noted that by repeated application of the definitions conditional distributions in more than two random variables can be created, for example,

$$f(x_1 | x_2, x_3, \ldots, x_n).$$

Example.   This example illustrates, among other things, that both x_1 and x_2 need not appear explicitly in $f(x_1, x_2)$ for X_1 and X_2 to be nonindependent. Suppose that

$$f(x_1, x_2) = 4x_1^2, \quad 0 < x_1 < 1, \, 0 < x_2 < x_1,$$

$$= 0, \quad \text{elsewhere.}$$

The sample description space for $f(x_1, x_2)$ is the triangle in the (x_1, x_2) plane bounded by the lines $x_2 = 0$, $x_1 = 1$, $x_1 = x_2$:

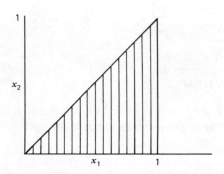

The sample description $S_{1|2}$ for $f(x_1|x_2)$ is the interval $(x_1 = x_2, 1)$:

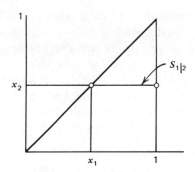

The sample description space $S_{2|1}$ for $f(x_2|x_1)$ is the interval $(0, x_2 = x_1)$:

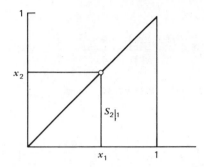

and the sample description space S_1 is the interval $(0, 1)$, and likewise for S_2. Clearly,

$$f_1(x_1) = \int_0^{x_1} 4x_1^2 \, dx_2$$

$$= 4x_1^2 x_2 \big]_0^{x_1} = 4x_1^3, \quad 0 < x_1 < 1,$$

$$= 0, \quad \text{elsewhere};$$

$$f_2(x_2) = \int_{x_2}^1 4x_1^2 \, dx_1$$

$$= \tfrac{4}{3}x_1^3 \big]_{x_2}^1 = \tfrac{4}{3}(1 - x_2^3), \quad 0 < x_2 < 1,$$

$$= 0, \text{ elsewhere};$$

$$f(x_1|x_2) = \frac{f(x_1, x_2)}{f_2(x_2)} = \frac{4x_1^2}{\tfrac{4}{3}(1 - x_2^3)}, \quad 0 < x_2 < x_1 < 1,$$

$$= 0, \text{ elsewhere}.$$

Note that

$$\int_{x_2}^1 f(x_1|x_2) \, dx_1 = \left[\tfrac{4}{3}(1 - x_2^3)\right]^{-1} \int_{x_2}^1 4x_1^2 \, dx_1$$

$$= \left[\tfrac{4}{3}(1 - x_2^3)\right]^{-1} \left[\tfrac{4}{3}(1 - x_2^3)\right] = 1.$$

Also

$$f(x_2|x_1) = \frac{f(x_1,x_2)}{f_1(x_1)} = \frac{4x_1^2}{4x_1^3} = \frac{1}{x_1}, \quad 0<x_2<x_1<1,$$

$$=0, \quad \text{elsewhere},$$

and

$$\int_0^{x_1} f(x_2|x_1)\, dx_2 = \int_0^{x_1} \frac{dx_2}{x_1} = \frac{x_2}{x_1}\Big]_0^{x_1} = 1.$$

Thus the random variables X_1, X_2 are not independent.

2.7   CUMULATIVE DISTRIBUTION FUNCTIONS

An event constantly recurring in statistical methods is that a random variable X is less than or equal to some number x, that is, the event $(X \leqslant x)$.

A discrete random variable has at most a denumerably infinite number of points in its sample description space; and since the points are numerical-valued,

$$S = \{ x : x_1, x_2, \dots \}.$$

Then

$$P(X \leqslant x) \equiv F_X(x) = \sum_{x_i \leqslant x} f_X(x_i).$$

The function of x, $F_X(x)$, is called the *cumulative distribution function* or, more simply, the *distribution function*. The subscript X will often be suppressed.

For continuous random variables the distribution function is given by

$$F(x) \equiv P(X \leqslant x) = \int_{-\infty}^x f(s)\, ds.$$

Distribution functions satisfy four properties worth noting.

1. $0 \leqslant F(x) \leqslant 1$ because of probability axioms (i) and (iii).
2. $F(x)$ is a never-decreasing function of x, that is, $F(x'') \geqslant F(x')$ whenever $x'' \geqslant x'$. This result follows by noting that the event $(X \leqslant x')$ implies the event $(X \leqslant x'')$.
3. $F(\infty) = 1; F(-\infty) = 0$. These results follow from the definitions.

4. $F(x)$ is a continuous function of x and always exists in the continuous case. For a discrete random variable $F(x)$ is continuous from the right.

In the continuous random variable case

$$\frac{dF(x)}{dx} = \frac{d\int_{-\infty}^{x} f(s)\,ds}{dx} = f(x).$$

In the case of a joint probability distribution

$$F(x_1, x_2, \ldots, x_n) = \int_{-\infty}^{x_n} \cdots \int_{-\infty}^{x_1} f(s_1, s_2, \ldots, s_n)\,ds_1\,ds_2 \ldots ds_n$$

with properties similar to those above.

Example 1.   Discrete Case. Suppose that

$$f(x) = \frac{1}{N+1}, \quad x = 0, 1, \ldots, N,$$

$$= 0, \quad \text{elsewhere;}$$

$$F(4) = P(X \leqslant 4) = \sum_{x=0}^{4} f(x) = \sum_{x=0}^{4} \frac{1}{N+1}$$

$$= \frac{5}{N+1};$$

$$F(N) = P(X \leqslant N) = \sum_{x=0}^{N} \frac{1}{N+1} = \frac{N+1}{N+1} = 1.$$

In summary, (where $[x]$ is the integer part of x)

$$F(x) = 0, \quad x < 0,$$

$$F(x) = \sum_{0}^{[x]} \frac{1}{N+1} = \frac{[x]+1}{N+1}, \quad [x] = 0, 1, 2, \ldots, N,$$

$$F(x) = 1, \quad x > N.$$

Example 2.   Discrete Case. Suppose that

$$f(x) = (\tfrac{1}{2})^x, \quad x = 1, 2, \ldots,$$

$$= 0, \quad \text{elsewhere.}$$

$$F(x) = 0, \quad x < 1,$$

$$F(x) = \sum_{i=1}^{[x]} (\tfrac{1}{2})^i$$

$$= \sum_{i=1}^{\infty} (\tfrac{1}{2})^i - \sum_{i=[x]+1}^{\infty} (\tfrac{1}{2})^i,$$

but

$$\sum_{i=1}^{\infty} (\tfrac{1}{2})^i = 1,$$

because $f(x)$ is a probability distribution, and by summing a geometric series it is seen that

$$\sum_{i=[x]+1}^{\infty} (\tfrac{1}{2})^i = (\tfrac{1}{2})^{[x]},$$

so that

$$F(x) = 1 - (\tfrac{1}{2})^{[x]}, \quad [x] = 1, 2, \ldots,$$

and thus

$$F(\infty) = 1; \quad F(-\infty) = F(0) = 0.$$

Example 3.   Continuous Case. Suppose that

$$f(x) = \tfrac{3}{4} x^2 e^{-x^3/4}, \quad x > 0$$

$$= 0, \quad \text{elsewhere.}$$

Then

$$F(x) = \int_{-\infty}^{x} \tfrac{3}{4} s^2 e^{-s^3/4} \, ds = \int_{0}^{x} \tfrac{3}{4} s^2 e^{-s^3/4} \, ds$$

$$= 1 - e^{-x^3/4};$$

and hence

$$F(x) = 0, \quad x \leqslant 0,$$

$$F(x) = 1 - e^{-x^3/4}, \quad x > 0,$$

and, of course,

$$F(\infty) = 1, \qquad F(-\infty) = 0.$$

It should be noted that the distribution function $F(x)$ does not always exist in closed form.

2.8 PARAMETERS IN PROBABILITY DISTRIBUTIONS

Consider the following probability distribution:

$$f_X(x) = \alpha e^{-\alpha x}, \quad \alpha, x > 0. \tag{2.10}$$

This is the exponential distribution, which will be studied in some detail in Chapter 4 and other parts of the book. The value that $f(x)$ takes is determined by (*a*) the value x taken by X, (*b*) the known constant e, and (*c*) α.

Here α is called a *parameter*, and probability distributions often contain more than one parameter, although a number larger than three is rare. The important idea is that parameters are usually unknown, and many of the methods presented here are aimed at estimating the values of these parameters, given some observed data. In fact, it is probably fair to say that a central contribution of statistics is the development of methods for estimating unknown parameters. Two broad classes of estimating situations are considered and will be illustrated in terms of the probability distribution (2.10).

Sampling Situation 1. The parameter α is regarded as fixed, but unknown and (random) samples of size n are obtained from $f(x) = \alpha e^{-\alpha x}$ to assist the estimation of α. This will be called the classical situation. The only probability distribution involved is $f(x)$.

Sampling Situation 2. The parameter α is regarded as fixed and unknown, but the fixed unknown value of α was the result of a random observation from a probability distribution defined on a sample description space S_u, which describes the outcomes for α. Thus there is a *prior probability distribution* on A, say $g(\alpha)$, and then it is assumed that the sample size of n is obtained from the *conditional* probability distribution $f(x|\alpha)$. This situation is called the Bayes situation, and Bayesian methods have been developed to handle it. The Bayes methods are discussed briefly in the next section and in some detail in Chapter 8.

In any event, because of the two situations described above and because parameters are usually unknown, it seems clear that great care must be exercised with notation. This will be done as follows. Let $\alpha_1, \ldots, \alpha_n$ be a set of parameters in a probability distribution $f(x)$. The notation $f(x; \alpha_1, \ldots, \alpha_n)$ means that the set of parameters is fixed and is never regarded as a random variable (the classical situation). On the other hand, the notation $f(x|\alpha_1, \ldots, \alpha_n)$ means that the set of parameters, although fixed, has entered the probability distribution as a result of a random experiment conducted on some joint *prior* probability distribution $g(\alpha_1, \ldots, \alpha_n)$. There may be, of course, situations in which some parameters are not random variables, whereas others are. If $A_1, \ldots, A_i$ are random variables and $\alpha_{i+1}, \ldots, \alpha_n$ are fixed, the notation will be

$$f(x|\alpha_1, \ldots, \alpha_i; \alpha_{i+1}, \ldots, \alpha_n).$$

The important idea is that sometimes random samples are obtained directly from marginal distributions (the classical idea), and sometimes the sampling process is two stage, that is, first a sample is taken from a prior distribution on the parameters(s), and then samples are taken from a conditional distribution with the parameter(s) fixed (the Bayes idea).

The reason that it is of critical importance to distinguish between these two situations is simple: they often require different methods of estimation. In the $f(x; \alpha)$ situation no other probability distributions are involved and inferences about α must be made by samples from $f(x; \alpha)$. In the $f(x|\alpha)$ situation a prior distribution $g(\alpha)$ is involved. Hence there is a joint probability distribution

$$h(x, \alpha) = f(x|\alpha)g(\alpha) \tag{2.11}$$

and a marginal probability distribution $k(x)$. We then have

$$g(\alpha|x) = \frac{f(x|\alpha)g(\alpha)}{k(x)}, \quad k(x) \neq 0. \tag{2.12}$$

Here $g(\alpha|x)$ is called the *posterior* distribution of A and is very useful because it is about A that knowledge is desired, and x, the conditioning variable, is observable.

It is often useful to be able to interpret parameters in probability distributions in terms of the physics underlying the process of the experiment that generates the random variable. This idea is explored in detail with some success in Chapter 4. It is always comforting to be able to justify the view that a given probability distribution describes a random variable on more than just empirical evidence. To assist in this task, *reparameterization* is sometimes desirable. As an example, consider the probability distribution in (2.10):

$$f(x;\alpha) = \alpha e^{-\alpha x}, \quad \alpha, x > 0,$$

$$= 0, \quad \text{elsewhere.}$$

1. It is easy to show that the first moment, called the *mean* (see Chapter 3), is

$$E(X) = \frac{1}{\alpha}.$$

There is no reason why the parameter cannot be replaced by

$$\theta = \frac{1}{\alpha},$$

and thus

$$f(x;\theta) = \left(\frac{1}{\theta}\right) e^{-x/\theta}, \quad \theta, x > 0,$$

$$= 0, \quad \text{elsewhere.}$$

2. The value x_p, such that

$$P(X < x_p) = p, \quad 0 < p < 1,$$

is called the pth quantile of a continuous distribution. For

$$f(x;\alpha) = \alpha e^{-\alpha x},$$

x_p is the solution to the equation

$$\int_0^{x_p} f(x;\alpha)\, dx = p = 1 - e^{-\alpha x} \Big]_0^{x_p},$$

which means that

$$p = 1 - e^{-\alpha x_p}.$$

Solving for α yields

$$\alpha = -\frac{\log(1-p)}{x_p},$$

and $f(x;\alpha)$ can be written as

$$f(x;x_p) = -\frac{\log(1-p)}{x_p}e^{[\log(1-p)/x_p]x},$$

where, of course, x_p is a parameter now (and unknown if α is). The idea of reparameterization is sometimes a matter of convenience. For example, if the random variable is lifetime, it may be important to estimate the mean or some other quantile.

2.9 BAYES' THEOREM

The conditional probability of event B, given that event A has occurred, has already been defined as

$$P(B|A) = \frac{P(A\cap B)}{P(A)}, \quad P(A) \neq 0. \tag{2.13}$$

This probability is correctly interpreted as the limiting value of the ratio of the number of times that B occurs (with A) to the total number of times that A is observed (with or without B). Of course,

$$P(\bar{B}|A) + P(B|A) = 1 \tag{2.14}$$

because

$$P(\bar{B}|A) = \frac{P(\bar{B}\cap A)}{P(A)}$$

and

$$P(\bar{B}|A) + P(B|A) = \frac{P(\bar{B}\cap A)}{P(A)} + \frac{P(A\cap B)}{P(A)} = \frac{P(\bar{B}\cap A)+P(A\cap B)}{P(A)}$$

but $A = (A\cap B)\cup(A\cap\bar{B})$, so that

$$P(\bar{B}|A) + P(B|A) = \frac{P(A)}{P(A)} = 1.$$

Suppose now that an event A can occur with any one of n mutually exclusive events, $B_1, B_2, \ldots, B_n$ and that the B_i are exhaustive for A, that is, whenever A occurs it must occur with some one B_i. Then decomposing A gives

$$A = (A \cap B_1) \cup (A \cap B_2) \cup \ldots \cup (A \cap B_n), \qquad (2.15)$$

and by axiom (ii) of Section 2.2.2

$$P(A) = P(A \cap B_1) + P(A \cap B_2) + \cdots + P(A \cap B_n). \qquad (2.16)$$

Applying Equation (2.13), we obtain

$$P(A) = P(A|B_1)P(B_1) + P(A|B_2)P(B_2) + \cdots + P(A|B_n)P(B_n). \qquad (2.17)$$

Applying Equation (2.13) again gives

$$P(B_i|A) = \frac{P(A|B_i)P(B_i)}{\sum_{i=1}^{n} P(A|B_i)P(B_i)}, \quad P(A) \neq 0, \qquad (2.18)$$

and Equation (2.18) is the celebrated Bayes theorem. This theorem was achieved by purely formal operations. Thus any controversy about its utility centers on how it is applied rather than whether it is true.

Usually, in situations where the theorem is applicable, event A is observed and it is desired to determine which B_i caused A. The unfortunate point is that usually the $P(B_i)$ are difficult, if not impossible, to estimate.

Example. Suppose that a system consists of three units, each failing independently of the others. Suppose further that the system fails when and only when the first unit failure has occurred. Let

$$B_i = \text{the } i\text{th unit failed};$$

$$A = \text{system has failed};$$

$$P(B_1) = .01, \quad P(B_2) = .03, \quad P(B_3) = .04;$$

and, of course, $P(A|B_i) = 1$. Thus

$$P(B_1|A) = \frac{.01}{.08} = .125, \qquad P(B_2|A) = \frac{.03}{.08} = .375,$$

$$P(B_3|A) = \frac{.04}{.08} = .50 \quad \text{and} \quad \sum_{i=1}^{3} P(B_i|A) = 1.$$

Bayes' theorem is particularly useful in some reliability applications when a parameter in a failure distribution (i.e., the random variable is lifetime) can be regarded as a random variable with prior distribution $g(\alpha)$. When n random samples have been obtained from $f(x|\alpha)$, then

$$g(\alpha|x_1,\ldots,x_n) = \frac{h(x_1,\ldots,x_n,\alpha)}{f(x_1,\ldots,x_n)} \qquad (2.19)$$

Equation (2.19) can be made to look more like Equation (2.18) by noting that

$$f(x_1,\ldots,x_n) = \int_{-\infty}^{\infty} f(x_1,\ldots,x_n|\alpha) g(\alpha)\, d\alpha$$

and

$$h(x_1,\ldots,x_n,\alpha) = f(x_1,\ldots,x_n|\alpha) g(\alpha).$$

Thus

$$g(\alpha|x_1,\ldots,x_n) = \frac{f(x_1,\ldots,x_n|\alpha) g(\alpha)}{\int_{-\infty}^{\infty} f(x_1,\ldots,x_n|\alpha) g(\alpha)\, d\alpha}. \qquad (2.20)$$

Here $g(\alpha|x_1,\ldots,x_n)$ is called the posterior distribution, and $g(\alpha)$ the prior distribution; both are defined on the random variable A.

Equations (2.19) and (2.20) are particularly useful since, if the prior distribution $g(\alpha)$ and the conditional distribution $f(x|\alpha)$ are known, a great deal is known about A. For example,

$$P(\alpha_* < A < \alpha^*) = \int_{\alpha_*}^{\alpha^*} g(\alpha|x_1,\ldots,x_n)\, d\alpha. \qquad (2.21)$$

Bayesian methods useful in reliability will be discussed in considerable detail in Chapter 8.

PROBLEMS

2.1. Two sets, A_1 and A_2, are said to be *disjoint* if $A_1 \cap A_2 = \varnothing$. If $A_2 = \{0 \leqslant y \leqslant 3\}$ and $A_1 = \{\frac{1}{2} \leqslant y \leqslant 3\}$, write A_2 as the union of two disjoint sets, one of which is A_1.

2.2. Show that a set A with a finite number of elements n has 2^n subsets. *Hint:* Use the binomial expansion.

2.3. Show that $\overline{(\overline{A})} = A$.

2.4. Show that

$$\overline{\left(\bigcup_{i=1}^{K} A_i\right)} = \bigcap_{i=1}^{K} \overline{A}_i \quad \text{and} \quad \overline{\left(\bigcap_{i=1}^{K} A_i\right)} = \bigcup_{i=1}^{K} \overline{A}_i.$$

These are sometimes called *De Morgan's laws*.

2.5. The elements of three sets, A_1, A_2, A_3, all belong to S. Write A_3 as the union of disjoint sets.

2.6. For any two sets, A_1 and A_2, whose elements belong to S show that

$$S = (A_1 \cap A_2) \cup (A_1 \cap \overline{A}_2) \cup (\overline{A}_1 \cap A_2) \cup (\overline{A}_1 \cap \overline{A}_2).$$

2.7. A two-unit system operates if and only if one or the other or both units are operating. At any instant of time the state of each unit is observed, and let
$S = A_1$: unit 1 operating, unit 2 down;
  A_2: unit 1 operating, unit 2 operating;
  A_3: unit 1 down, unit 2 operating;
  A_4: unit 1 down, unit 2 down.
Construct a function f on S that defines a numerical-valued random variable and represents the two system states (up and down).

2.8. In the two-dice example in Section 2.2.1 with a sample description space of 36 elements,

$$S = \{(x_1, x_2):(1,1),(1,2),\dots,(6,6)\},$$

let $A_1 = $ the event that the sum of the faces is odd;
  $A_2 = $ the event that at least one of the faces is a 4.
  a. How many elements represent A_1 and A_2, respectively?
  b. How many elements represent $A_1 \cup A_2$ and $A_1 \cap A_2$, respectively?
  c. Illustrate the rule that $\overline{(A_1 \cup A_2)} = \overline{A}_1 \cap \overline{A}_2$.
  d. Illustrate the rule that $\overline{(A_1 \cap A_2)} = \overline{A}_1 \cup \overline{A}_2$.
  e. How many elements represent $\overline{A}_1$?

2.9. Show by decomposition that

$$P(A_1 \cup A_2 \cup A_3) = P(A_1) + P(A_2) + P(A_3) - P(A_1 \cap A_2)$$

$$- P(A_1 \cap A_3) - P(A_2 \cap A_3) + P(A_1 \cap A_2 \cap A_3).$$

2.10. If two events, A_1 and A_2, are independent and if $P(A_1)$ and $P(A_2) \neq 0$, show that

$$P(A_1|A_2) = P(A_1) \quad \text{and} \quad P(A_2|A_1) = P(A_2).$$

2.11. If $P(A_1|A_2) = 1$, show that

$$P(A_2|A_1) = \frac{P(A_2)}{P(A_1)}.$$

2.12. Suppose that an N-unit system is composed of two types of units: N_1 units, which have two states: operate/failed, and N_2 units, which have three states: operate / failed /standby. How many total system states exist?

2.13. Show that

$$\binom{N}{x} = \binom{N}{N-x}.$$

2.14. Let it be desired to decompose a population of N distinguishable objects into K subsets: x_1 objects in the first subset, x_2 in the second, and so on, so that

$$x_1 + x_2 + \ldots + x_K = N.$$

Show that the number of different decompositions is

$$\frac{N!}{x_1! x_2! \ldots x_K!}$$

Hint: Show that

$$\frac{N!}{x_1! x_2! \ldots x_K!} = \binom{N}{x_1}\binom{N-x_1}{x_2}\binom{N-x_1-x_2}{x_3}\ldots\binom{N-\sum_{i=1}^{K-2} x_i}{x_{K-1}}.$$

What does the right-hand side represent?

2.15. Show that

$$\binom{N}{x-1} + \binom{N}{x} = \binom{N+1}{x}.$$

2.16. The digits $1,2,3,\ldots,N$ are arranged in random order. What is the probability that the numbers 1 and N are next to each other?

2.17. Show that a fancy way of writing 32 is

$$32 = \binom{5}{0} + \binom{5}{1} + \binom{5}{2} + \binom{5}{3} + \binom{5}{4} + \binom{5}{5}.$$

2.18. Generalize Problem 2.17 to

$$2^N = \sum_{i=0}^{N} \binom{N}{i}.$$

2.19.a. Suppose S is such that $x_1 = 1, x_2 = 2, x_3 = 3, \ldots$ and

$$P(X = x_i = i) = f(x_i) = (\tfrac{1}{2})^i, \quad i = 1,2,3,\ldots.$$

Verify that $f(x_i)$ is a discrete probability distribution.

b. Suppose S is such that $x_1 = 1, x_2 = 2, \ldots, x_N = N$ and

$$P(X = x_i = i) = f(x_i) = \frac{1}{N}, \quad i = 1,2,\ldots,N.$$

Verify that $f(x_i)$ is a discrete probability distribution.

c. Suppose S is such that $0 \leqslant x \leqslant 1$ and

$$f_X(x) = 1, \quad 0 \leqslant x \leqslant 1,$$

$$f_X(x) = 0, \quad \text{elsewhere.}$$

Verify that $f(x)$ is a continuous probability distribution.

2.20. Show that, if $X_1, X_2, \ldots, X_n$ are independent, then

$$P(a_1 < X_1 < b_1, a_2 < X_2 < b_2, \ldots, a_n < X_n < b_n) = P(a_1 < X_1 < b_1)P(a_2 < X_2 < b_2)\ldots P(a_n < X_n < b_n).$$

2.21. Let

$$f(x_1, x_2) = e^{-(x_1 + x_2)}, \quad x_1, x_2 > 0,$$

$$= 0, \quad \text{elsewhere.}$$

a. Describe the sample description space for (X_1, X_2).
b. Verify that $f(x_1, x_2)$ is a probability distribution.
c. Write $f(x_1, x_2)$ as the product of two identical probability distributions.

2.22. Let

$$f(x_1, x_2) = p^{x_1 + x_2}(1 - p)^{2 - x_1 - x_2}, \quad 0 < p < 1; \; x_1 = 0, 1; \; x_2 = 0, 1.$$

a. Describe the sample space for (X_1, X_2).
b. Verify that $f(x_1, x_2)$ is a probability distribution.
c. Write $f(x_1, x_2)$ as the product of two identical probability distributions.

2.23. Suppose that

$$f(x_1, x_2) = e^{-(x_1 + x_2)}, \quad x_1, x_2 > 0,$$

$$= 0, \quad \text{elsewhere.}$$

Find the two marginal and two conditional probability distributions. Are X_1, X_2 independent?

2.24. Suppose that

$$f(x) = Ke^{-.05x}, \quad x > 0,$$

$$= 0, \quad \text{elsewhere.}$$

Find K so that $f(x)$ is a probability distribution.

2.25. Suppose that

$$f(x) = Kx^2, \quad 1 < x < 2, \quad 4 < x < 6,$$

$$= 0, \quad \text{elsewhere.}$$

Find K so that $f(x)$ is a probability distribution.

2.26. Suppose that

$$f(x_1, x_2) = 4x_1(1 - x_2), \quad 0 < x_1 < 1, 0 < x_2 < 1,$$

$$= 0, \quad \text{elsewhere.}$$

Find the two conditional and two marginal distributions. Are X_1, X_2 independent?

2.27. Suppose that

$$f(x_1, x_2, x_3) = 8x_1 x_2 x_3, \quad 0 < x_1 < 1, \quad 0 < x_2 < 1, \quad 0 < x_3 < 1,$$

$$= 0, \quad \text{elsewhere.}$$

Find $f(x_3|x_1,x_2)$ and the corresponding sample description space, $S_{3|12}$.

2.28. Let

$$f(x) = \frac{3x^2}{8}, \quad 0 < x < 2,$$

$$= 0, \quad \text{elsewhere.}$$

Find the following:

(a) $P(X < \frac{1}{2})$;

(b) $P(X \geqslant \frac{1}{2})$;

(c) $P(X > \frac{3}{2})$;

(d) the value of x such that $P(X < x) = \frac{1}{2}$;

(e) the value of x such that $P(X < x) = \frac{3}{4}$;

(f) $P(X < 1$ and $X < 2)$;

(g) $P(0 < X < 1$ or $X > \frac{1}{4})$.

2.29. Let

$$f(x) = \binom{3}{x} p^x (1-p)^{3-x}, \quad x = 0, 1, 2, 3,$$

$$= 0, \quad \text{elsewhere.}$$

Find the following:

(a) $P(X = 1)$;

(b) $P(X \leqslant 2)$;

(c) $P(X < 2)$;

(d) $P(X = 2)$.

2.30. Let

$$f(x_1, x_2) = 4x_1(1 - x_2), \quad 0 < x_1 < 1, \quad 0 < x_2 < 1,$$

$$= 0, \quad \text{elsewhere.}$$

Find the following:

(a) $P(X_1 > \frac{1}{2})$;

(b) $P(X_2 < \frac{1}{4})$;

(c) $P(X_1 < \frac{1}{2}, X_2 < \frac{1}{4})$;

(d) $P(X_1 > \frac{1}{2}|X_2 = \frac{1}{2})$;

(e) $P(X_1 < \frac{1}{4}|X_2 < \frac{1}{2})$.

2.31. Let

$$f(x_1, x_2) = 3x_1, \quad 0 < x_1 < 1, \quad 0 < x_2 < x_1,$$

$$= 0, \quad \text{elsewhere.}$$

Find the following:

(a) $P(X_1 > \frac{3}{4})$;

(b) $P(X_2 < \frac{1}{2})$;

(c) $P(X_1 < \frac{1}{4}, X_2 < \frac{1}{2})$;

(d) $P(X_2 < \frac{1}{2}|X_1 = \frac{1}{4})$.

2.32. Given the probability distribution

$$f(x; \alpha, \beta) = \frac{\beta}{\alpha} x^{\beta - 1} e^{-x^\beta / \alpha}, \quad x, \ \beta, \alpha > 0,$$

reparameterize it to read $f(x; x_p, \beta)$.

2.33. (In applying Bayes theorem in particular and conditional probabilities in general, it is of vital importance to use the conditioning information exactly as given.) A box contains four transistors, exactly two of which are good.

a. Two transistors are selected (at random and without replacement) and tested. You are told that at least one is good. What is the probability that the other (in the sample) is good?

b. Two transistors are selected (at random and without replacement). One of these two is selected (at random and without replacement), tested, and found good. What is the probability that the other is good?

c. Two transistors are selected (at random and without replacement). The first one selected is tested and found good. What is the probability that the second is good?

d. Change (b) so that, when the transistor is selected for testing, the probability that it is the first transistor is $1/3$ (and hence $2/3$ for the second transistor).

2.34. Four machines, M_1, M_2, M_3, M_4, produce $10, 20, 25, 45$ % of the output of a certain product. The probability of a defective product from each machine is $p_1 = .030, p_2 = .005, p_3 = .020, p_4 = .010$. What is the probability that a given defective product came from M_4?

2.35. Two fair, ordinary six-sided dice are tossed randomly. What is the probability of obtaining a 6 before a 7?

REFERENCE

Parzen, Emanuel (1960), *Modern Probability Theory and Its Applications*, John Wiley and Sons, New York.

CHAPTER 3

Elements of Statistical Theory

In Chapter 2 the elements of probability theory necessary for an understanding of some techniques in reliability theory were presented. In this chapter the elements of statistical theory are similarly discussed. The material presented in this chapter is in no way intended to be a complete or a detailed treatment of statistical theory. For a more complete presentation the reader is referred to some of the references cited in the text.

3.1 MOMENTS, GENERATING FUNCTIONS AND TRANSFORMS

Although a probability distribution provides a complete representation of a random variable, often, for practical purposes, certain compact representations are useful. The purpose of this section is to present some of these compact representations and to discuss several methods by which they can be obtained.

3.1.1 The Expected Value

Consider a discrete random variable X which can take values x_i with corresponding probabilities p_i, $i = 1, 2, \ldots,$. If

$$\sum_i |x_i| p_i < +\infty,$$

then the *expected value* of X, henceforth denoted by $E(X)$, is defined as

$$E(X) = \sum_i x_i p_i. \tag{3.1}$$

Similarly, if X is continuous with a distribution function $F_X(x)$, then $E(X)$ is defined as the Lebesgue-Stieltjes integral

46

$$E(X) = \int_{-\infty}^{\infty} x \, dF_X(x).$$

If the derivative of $F_X(x)$, $f_X(x)$, exists for all values of x, then $E(X)$ is the Riemann integral

$$E(X) = \int_{-\infty}^{\infty} x f_X(x) \, dx. \tag{3.2}$$

The expected value of X, $E(X)$, is often referred to as the *mean* of X, and is a measure of central tendency. This definition of the expected value has been extended to include all functions of X, say $g(X)$. The *expected value* of $g(X)$, denoted by $E[g(X)]$, is defined as

$$E[g(X)] = \sum_i g(x_i) p_i, \quad \text{if } X \text{ is discrete,}$$

$$= \int_{-\infty}^{\infty} g(x) f_X(x) \, dx, \quad \text{if } X \text{ is continuous.} \tag{3.3}$$

From this result it follows that the expected value of a constant is the constant itself. The concept of jointly distributed random variables was introduced in Section 2.6. By use of the notation established therein, it is possible to define the expected values of functions of jointly distributed random variables.

Let the n continuous random variables $X_1, X_2, \ldots, X_n$ be jointly distributed with a probability density function given by $f(x_1, x_2, \ldots, x_n)$. Let $g(X_1, X_2, \ldots, X_n)$ be any function of these n random variables. Then the expected value of $g(X_1, X_2, \ldots, X_n)$ is defined as

$$E[g(X_1, X_2, \ldots X_n)]$$

$$= \int_{-\infty}^{\infty} \cdots \int_{-\infty}^{\infty} g(x_1, x_2, \ldots, x_n) f(x_1, x_2, \ldots, x_n) \, dx_1, dx_2, \ldots, dx_n. \tag{3.4}$$

If n random variables, $X_1 \ldots X_n$, are discrete with a probability mass function given by $p(x_1, x_2, \ldots, x_n)$, the expected value of any function of the random variables $g(X_1, X_2, \ldots, X_n)$ is defined as

$$E[g(X_1, X_2, \ldots, X_n)] = \sum_{x_1} \cdots \sum_{x_n} g(x_1, x_2, \ldots, x_n) p(x_1, x_2, \ldots, x_n). \tag{3.5}$$

Several important properties of the expected value can be summarized by the following two theorems, the proofs of which are left as an exercise for the reader.

1. Let the n random variables $X_1, X_2, \ldots, X_n$ be jointly distributed with a probability density given by $f(x_1, x_2, \ldots, x_n)$. The random variables can be either discrete or continuous. Let there be m constants, $C_1, C_2, \ldots, C_m$, and let $g_i(X_1, X_2, \ldots, X_n)$, $i = 1, 2, \ldots, m$, be m functions of these random variables. Then

$$E[C_1 g_1(X_1, X_2, \ldots, X_n) + C_2 g_2(X_1, X_2, \ldots, X_n) + \cdots + C_m g_m(X_1, X_2, \ldots, X_n)]$$

$$= C_1 E[g_1(X_1, X_2, \ldots, X_n)] + C_2 E[g_2(X_1, X_2, \ldots, X_n)]$$

$$+ \cdots + C_m E[g_m(X_1, X_2, \ldots, X_n)]. \tag{3.6}$$

2. If the n random variables $X_1, X_2, \ldots, X_n$ are mutually independent, and if g_k is any one of the $g_i(X_1, X_2, \ldots, X_n) = C_1 X_1 \cdot C_2 X_2 \cdot \cdots \cdot C_n X_n$, where any or all of the C_i's can be equal to 1, then

$$E[g_k(X_1, X_2, \ldots, X_n)] = C_1 E(X_1) \cdot C_2 E(X_2) \cdot \cdots \cdot C_m E(X_n). \tag{3.7}$$

3.1.2   The Moments

The *Kth moment* about a constant C of the distribution of a random variable X (discrete or continuous), denoted by $Z_K(C)$, is defined as

$$Z_K(C) = E(X - C)^K.$$

If $C = 0$, the corresponding Kth moment is also called the Kth moment about zero and will be denoted by Z_K. If, on the other hand, the constant $C = E(X)$, the corresponding Kth moment is referred to as the *Kth central moment* and will be denoted by μ_K. The second central moment, μ_2, is used to evaluate the spread of the values of a random variable about its mean; it is referred to as the *variance* of X and will be denoted by Var(X). Hence

$$\mu_2 = E[X - E(X)]^2 = E(X^2) - [E(X)]^2.$$

It is easy to verify that the variance of a constant is zero. Again, some properties of the variance can be summarized by the following theorem, the proof of which is left as an exercise to the reader.

Let $X_1, X_2, \ldots, X_n$ be a set of n mutually independent random variables (discrete or continuous), each having a variance denoted by Var(X_i), $i = 1, 2, \ldots, n$. Let there be n constants, $C_1, C_2, \ldots, C_n$, where any or all of the C_i's can be equal to 1. Then

$$\text{Var}(C_1 X_1 + C_2 X_2 + \cdots + C_n X_n)$$

$$= C_1^2 \text{Var}(X_1) + C_2^2 \text{Var}(X_2) + \cdots + C_n^2 \text{Var}(X_n). \qquad (3.8)$$

The preceding notions have been extended to distributions of several variates as follows.

Let the n continuous random variables $X_1, X_2, \ldots, X_n$ have a joint probability density function given by $f(x_1, x_2, \ldots, x_n)$. Then, for any set of constants $C_i, i = 1, 2, \ldots, n$, and n constants, K_i, that are any positive integers or zero, the *joint* (or product) *moments* of $X_1, X_2, \ldots, X_n$, about $C_1, C_2, \ldots, C_n$, respectively, are defined by

$$E\Big[(X_1 - C_1)^{K_1}(X_2 - C_2)^{K_2}\ldots(X_n - C_n)^{K_n}\Big] = \int_{-\infty}^{\infty}\cdots\int_{-\infty}^{\infty}(X_1 - C_1)^{K_1}$$

$$\times (X_2 - C_2)^{K_2}\ldots(X_n - C_n)^{K_n}f(x_1, x_2, \ldots, x_n)\, dx_1, dx_2, \ldots, dx_n.$$

An important joint (product) moment defined for two random variables X_i and X_j, with expected values $E(X_i)$, $E(X_j)$ and variances $\text{Var}(X_i)$, $\text{Var}(X_j)$, respectively, is the *covariance*.

The covariance between X_i and X_j, denoted by $\text{Cov}(X_i, X_j)$, is defined as

$$\text{Cov}(X_i, X_j) = E\big\{[X_i - E(X_i)][X_j - E(X_j)]\big\}.$$

The *correlation* between the variables X_i and X_j, to be denoted by $\rho_{X_i X_j}$, is defined as

$$\rho_{X_i X_j} = \frac{\text{Cov}(X_i, X_j)}{[\text{Var}(X_i)\text{Var}(X_j)]^{1/2}}$$

Tables 3.1 and 3.2 at the end of Section 3.3 give the means and the variances of some well-known distributions.

3.1.3 Generating Functions

Since generating functions play an important role in reliability problems, some basic concepts of generating functions and their important properties are discussed here.

Let X be a discrete random variable with a probability mass function given by $P_X(x), x = 0, 1, 2, \ldots$. The function

$$G_X(t) = \sum_{x=0}^{\infty} P_X(x)t^x$$

is called the *generating function* of the sequence $P_X(x)$, if $G_X(t)$ converges in some interval $-t^* < t < t^*$. Since $P_X(x)$ is bounded for all $x \geqslant 0$, a comparison with the geometric series shows that $G_X(t)$ converges, at least for $|t| < 1$.

The following results, which can be verified easily, are left as an exercise for the reader:

$$E(X) = \frac{d}{dt} G_X(t)\big|_{t=1}$$

$$E(X^2) = \left[\frac{d^2}{dt^2} G_X(t) + \frac{d}{dt} G_X(t) \right]_{t=1}$$

and

$$\mathrm{Var}(X) = \left\{ \frac{d^2}{dt^2} G_X(t) + \frac{d}{dt} G_X(t) - \left[\frac{d}{dt} G_X(t) \right]^2 \right\}_{t=1}.$$

The generating function of the sum of independent random variables can be obtained by multiplying the generating functions of the random variables. More specifically, let $X_1, X_2, \ldots, X_n$ be n independent random variables with generating functions $G_{X_1}(t), G_{X_2}(t), \ldots, G_{X_n}(t)$, respectively. Define a new random variable Y_n, where

$$Y_n = X_1 + X_2 + \cdots X_n.$$

Then $G_{Y_n}(t)$, the generating function of $P_{Y_n}(x)$, is given by

$$G_{Y_n}(t) = \prod_{i=1}^{n} G_{X_i}(t).$$

This result can be proved by noting that it is true for the case of two random variables, X_1 and X_2, and then extending it to the case of n random variables. For the case of two independent random variables, X_1 and X_2, the distribution of Y_2 is the *convolution* of $P_{X_1}(x)$ and $P_{X_2}(x)$, denoted by $P_{X_1}(x) * P_{X_2}(x)$, that is,

$$P_{Y_2}(x) = \sum_{j=0}^{x} P_{X_1}(j) P_{X_2}(x-j) = P_{X_1}(x) * P_{X_2}(x).$$

Termwise multiplication of the power series for $G_{X_1}(t)$ and $G_{X_2}(t)$ gives the product $G_{Y_2}(t)$.

3.1.4 Moment Generating Functions

Often the generating function does not have a simple expression, and in such cases the *moment generating function* is more convenient.

Let X be a random variable with a probability density (mass) function given by $f(x)$ (p_i). Then the expected value of e^{tx} is called the moment generating function of X, if the expected value converges in some interval $-t^* < t < t^*$. The moment generating function is denoted by $M_X(t)$, and

$$M_X(t) = \int_{-\infty}^{\infty} e^{tx} f(x)\, dx, \quad \text{if } X \text{ is continuous,}$$

$$= \sum_i e^{tx} p_i, \quad \text{if } X \text{ is discrete.}$$

Some properties of the moment generating function are now discussed. By using a series expansion of e^{tx}, one can express the moment generating function as

$$M_X(t) = E\left(1 + tX + \frac{t^2 X^2}{2!} + \cdots + \frac{t^n X^n}{n!} \cdots \right), \quad \text{or}$$

$$M_X(t) = 1 + tE(X) + \frac{t^2}{2!} E(X^2) + \frac{t^3}{3!} E(X^3) + \cdots + \frac{t^n}{n!} E(X^n) + \cdots,$$

$$= \sum_{j=0}^{\infty} \frac{t^j}{j!} E(X^j).$$

From this result it is evident that $E(X^j)$ can be obtained by using a series expansion of $M_X(t)$ and noting the coefficient of the term $t^j/j!$. Alternatively, $E(X^j)$ can also be obtained by differentiating $M_X(t)$ j times and then setting $t=0$. This can be easily verified, because

$$M_X(t) = \int_{-\infty}^{\infty} e^{tx} f(x)\, dx$$

or

$$\frac{d^j M_X(t)}{dt^j} = \int_{-\infty}^{\infty} x^j e^{tx} f(x)\, dx$$

and

$$\left. \frac{d^j M_X(t)}{dt^j} \right|_{t=0} = \int_{-\infty}^{\infty} x^j f(x)\, dx = E(X^j).$$

A useful property of the moment generating function is the following.

Let $M_Z(t)$ be the moment generating function of Z. Define a new random variable X such that $X = aZ + b$, where a and b are constants. Then the moment generating function of X is

$$M_X(t) = M_{aZ+b}(t) = e^{tb}M_Z(at).$$

The proof of this is left as an exercise to the reader.

The concept of the moment generating function can be easily extended to the case of multivariate distributions. Let n continuously distributed random variables, $X_1, X_2, \ldots, X_n$, have a probability density function given by $f(x_1, x_2, \ldots, x_n)$. Let there exist n constants t_i, $i = 1, 2, \ldots, n$, such that

$$M(t_1, t_2, \ldots, t_n) = E\left[\exp\left(\sum_{i=1}^{n} t_i x_i \right) \right]$$

exists for all t_i's in the interval $-t^* < t_i < t^*$, for some t^*. Then $M(t_1, t_2, \ldots, t_n)$ is called the *joint moment generating* function of the n variables $X_1, X_2, \ldots, X_n$. It follows that the ith moment of X_j can be obtained by differentiating $M(t_1, t_2, \ldots, t_n)$ i times with respect to t_j andthen setting all the t's equal to 0.

The usefulness of the moment generating function is indicated by the following two theorems, which are stated here without proof (Cramér, 1946, p. 89). These theorems, given here for the univariate case, apply also to multivariate cases.

(a) **The Uniqueness Theorem.** Let X_1 and X_2 be two random variables, discrete or continuous, with probability density (mass) functions $f_{X_1}(x)$ and $f_{X_2}(x)$, respectively. Suppose that $M_{X_1}(t)$ and $M_{X_2}(t)$ are the moment generating functions of X_1 and X_2 and that $M_{X_1}(t) = M_{X_2}(t)$ for all t's in the interval $-t^* < t < t^*$. Then $f_{X_1}(x) = f_{X_2}(x)$ for all x, except possibly at points of discontinuity, if X_1 and X_2 are continuous random variables. This theorem is used in Sections 3.3.4 and 3.3.5.

(b) **The Continuity Theorem.** Let $X_1, X_2, \ldots, X_n$ be a sequence of random variables with corresponding distribution functions given by the sequence $\{F_n(x)\}$ and moment generating functions given by the sequence $\{M_{X_n}(t)\}$. A necessary and sufficient condition for the convergence of the sequence $\{F_n(x)\}$ to a distribution function $F_X(x)$ is that, for every t, the sequence $\{M_{X_n}(t)\}$ converges to a limit $M_X(t)$.

When this condition is satisfied, the limit $M_X(t)$ is identical with the moment generating function of the limiting distribution function, $F_X(x)$.

For an example illustrating the use of this theorem see Problem 3.6.

The moment generating functions for some of the well-known distributions are given in Tables 3.1 and 3.2 at the end of Section 3.3.

3.1.5 The Laplace and Mellin Transforms

Let $F(x)$ be a known function of x, defined for all $x > 0$, and let $K(t,x)$ be a known function of the two variables t and x. If the integral

$$\varphi(t) = \int_0^\infty K(t,x) F(x) \, dx$$

is convergent, $\varphi(t)$ defines a function of the variable t; $\varphi(t)$ is called the *integral transform* of the function $F(x)$ with the *kernel* $K(t,x)$.

If

$$K(t,x) = e^{-tx}$$

$\varphi(t)$ is called the *Laplace transform* and will be denoted by $L[F(x)]$:

$$L[F(x)] = \int_0^\infty e^{-tx} F(x) \, dx.$$

On the other hand, if the kernel $K(t,x)$ is

$$K(t,x) = x^{t-1}$$

$\varphi(t)$ is called the *Mellin transform* and will be denoted by $M[F(x)]$:

$$M[F(x)] = \int_0^\infty x^{t-1} F(x) \, dx.$$

The Laplace transform and sometimes the Mellin transform serve as convenient tools in the study of reliability. A few elementary properties of the Laplace transform are presented here; for a more detailed discussion of these properties the reader is referred to Widder (1941).

1. The Laplace transform is a linear operator.

For any two functions, $f(x)$ and $g(x)$, defined for all $x > 0$, and any two constants, C_1 and C_2,

$$L[C_1 f(x) + C_2 g(x)] = C_1 L[f(x)] + C_2 L[g(x)].$$

2. The Laplace transform of the derivative and the integral of a function $f(x)$ can be obtained as follows. Integrating by parts, one can easily verify

that

$$L[f'(x)] = tL[f(x)] - f(0)$$

and

$$L\left[\int_0^x f(k)\,dk\right] = \frac{1}{t}L[f(x)].$$

3. The Laplace transform of the convolution of two functions is the product of their Laplace transforms.

The convolution for discrete variables was defined in Section 3.1.3. For any two continuous functions, $f(x)$ and $g(x)$, defined for $x > 0$, the convolution of $f(x)$ and $g(x)$ is defined as the integral

$$\int_0^x f(x-\lambda)g(\lambda)\,d\lambda = f(x) * g(x).$$

Now

$$L[f(x) * g(x)] = \int_0^\infty e^{-tx}\left[\int_0^x f(x-\lambda)g(\lambda)\,d\lambda\right]dx$$

$$= \int_0^\infty g(\lambda)e^{-t\lambda}d\lambda\int_\lambda^\infty f(x-\lambda)e^{-t(x-\lambda)}dx$$

$$= \int_0^\infty g(\lambda)e^{-t\lambda}d\lambda\,\{L[f(x)]\}$$

$$= L[f(x)]L[g(x)].$$

This result can be similarly extended to the case of three or more functions.

It is to be remarked that, if the function $f(x)$ [or, alternatively, the function $F'(x)$] is a probability density function of a continuous random variable X, for $X > 0$, then $L[f(x)]$ is merely the expected value of e^{-xt}.

The similarity between the Laplace transform and the moment generating function is apparent from the fact that the former gives the expected value of e^{-xt}, where $x > 0$, whereas the latter gives the expected value of e^{xt}, where x can take any value on the real line.

With the aid of the above, it should be easy for the reader to verify the convolution property of the moment generating function.

The use of the Laplace transform in reliability problems is illustrated in Chapter 10, Sections 10.1, 10.2, and 10.3.

3.1.6 Other Measures of Central Tendency

In addition to the mean, two other measures of central tendency are often used in applied statistics; these are the median and the mode.

The *median* of a distribution is the value of the random variable that bisects the area under the probability density function, if X is continuous. For discrete values of X, the median need not necessarily exist because it may not be possible to find an admissible value of X that bisects the probability mass function.

Specifically, let X be a random variable with a distribution function given by $F_X(x)$. The root of the equation

$$F_X(x) = p, \quad 0 < p < 1,$$

is called the pth quantile of the distribution $F_X(x)$. Clearly, if $F_X(x)$ is discontinuous, the pth quantile need not exist for all values of p. The root of the above equation, for $p = \frac{1}{2}$, is called the median of the distribution of X. The pth quantile is also called the $100p$th percentile.

The *mode* of a distribution is the value of a random variable at which the probability density function or the probability mass function attains its maximum value. Clearly, the mode need not be unique.

3.2 SOME BASIC DISTRIBUTIONS

In this section certain probability distributions fundamental to reliability methodology are briefly discussed. These distributions are fundamental in the sense that they have not been derived by considerations of some physical failure process (as will be done in Chapter 4), or by transformations on random variables (as will be done in Section 3.3).

3.2.1 Discrete Distributions

(a) The Geometric Distribution. Suppose that an experiment consists of independent trials, called Bernoulli trials, such that there are only two outcomes, a_1 and a_2; a_1 can be identified with the occurrence of a particular event, and a_2 with its nonoccurrence. Thus the sample description space of this experiment is

$$S = \{a_1, a_2\}.$$

Define a random variable X_i, such that

$$X_i = 0 \quad \text{if } a_1 \text{ occurs on the } i\text{th trial,}$$

$$= 1 \quad \text{if } a_2 \text{ occurs on the } i\text{th trial.}$$

Also, let

$$P(X_i=0)=p \quad \text{for all } i$$

and

$$P(X_i=1)=1-p \quad \text{for all } i.$$

Define another random variable X to denote the number of independent trials to the first occurrence of, say, a_1. The sample description space of X is

$$S=\{x:x=1,2,\dots\}$$

and

$$P(X=x)=(1-p)^{x-1}p, \quad x=1,2,\dots.$$

The random variable X is said to have a geometric distribution with a parameter p, and in reliability theory it can arise in the following situation. Suppose that every period of operation of a device is identified as a trial. If a_1 (a_2) is identified with failure-free operation (operation with a failure), $X-1$ is the number of successive periods of failure-free operation, whereas X is the period at which failure occurs for the first time.

(b) The Binomial Distribution. Consider a series of n independent Bernoulli trials, where n is fixed. Define a random variable X, where X denotes the total number of occurrences of the event a_1. The sample description space of X is

$$S: \{x:x=0,1,2,\dots,n\},$$

and

$$P(X=x)=\binom{n}{x}p^x(1-p)^{n-x}, \quad x=0,1,2,\dots,n.$$

The random variable X is said to have a binomial distribution, with a parameter p and an index n, and in reliability theory it can be visualized in the situation described above, if n can be considered as the total number of periods for which the device is observed. If the failed device can be instantly repaired at the end of each period, X is the number of periods of failure-free operation. Some methods for dealing with binomial data are discussed in Sections 7.4 and 10.4.

(c) The Negative Binomial Distribution. The negative binomial distribution is a generalization of the geometric distribution, in the sense that the random variable X stands for the number of trials up to and including the rth occurrence of a_1, where r is fixed:

$$P(X=x) = \binom{x-1}{r-1} p^{r-1}(1-p)^{x-r} p, \quad x \geqslant r, r = 1, 2, \ldots .$$

Here the parameters are p and r.

An application of the negative binomial distribution to reliability problems is discussed in Chapter 8.

(d) The Hypergeometric Distribution. The random variable X is said to have a hypergeometric distribution if

$$P(X=x) = \frac{\binom{Np}{x} \binom{N-Np}{n-x}}{\binom{N}{n}}, \quad x = 0, 1, 2, \ldots, \min(n, Np),$$

where N, n, and p are the parameters.

Physical situations in which this distribution can be easily visualized exist in quality control and acceptance sampling. The hypergeometric distribution occurs in practice whenever there is sampling from finite lots.

(e) The Poisson Distribution. The Poisson distribution can be derived by considering the random occurrences of an event according to a Poisson process. This process is described in some detail in Chapter 4, but for reasons of completeness the Poisson distribution is presented briefly here.

A random variable X is said to have a Poisson distribution with a parameter μ if

$$P(X=x) = \frac{e^{-\mu}\mu^{x}}{x!}, \quad x = 0, 1, 2, \ldots .$$

The Poisson distribution also arises often as the limiting form of a binomial distribution when n becomes large and p is small, so that $\mu = np$ is a constant. The derivation of the Poisson distribution as the limiting form of a binomial is left as an exercise to the reader (see Problem 3.6).

(f) The Multinomial Distribution. This distribution is the multivariate analog of a binomial distribution in the sense that repeated trials of an experiment lead to more than two outcomes, and one is interested in the number of occurrences of each outcome in n trials of the experiment.

Specifically, suppose that each experiment leads to k outcomes, $a_1, a_2, \ldots, a_k$, such that outcome a_i occurs with probability $p_i, i = 1, 2, \ldots, k$, on any particular trial. The random variable $X_i, i = 1, 2, \ldots, k$, denotes the number of occurrences of event a_i in n independent trials of the experiment. Then the random variables $X_i, i = 1, \ldots, k$, are said to have a multinomial distribution if

$$P(X_1 = x_1, X_2 = x_2, \ldots, X_k = x_k) = \frac{n!}{x_1! x_2! \ldots x_k!} p_1{}^{x_1} p_2{}^{x_2} \ldots p_k{}^{x_k}$$

for

$$\sum p_i = 1, \quad x_i = 0, 1, \ldots, n, \quad \text{and} \quad \sum x_i = n.$$

3.2.2 Continuous Distributions

(a) The Uniform Distribution. The uniform distribution is mathematically very simple and is particularly useful in applied statistics. A random variable X is said to have a uniform distribution with parameters a and b, if its probability density function is given by

$$f_X(x) = \frac{1}{b-a}, \quad a \leqslant x \leqslant b,$$

$$= 0, \quad \text{otherwise.}$$

The usefulness of the uniform distribution is enhanced by the fact that any continuous variate can be transformed to the uniform density. This result provides the basis for a Monte Carlo generation of failure distributions (see Section 7.2).

(b) The Beta Distribution. The beta distribution is also known as the Pearson Type I distribution, and its usefulness in reliability and quality control work has been well discussed by Kao (1961). Applications of this distribution are given in Chapters 8 and 10.

The generalized beta distribution is a four-parameter family of distributions for a continuous random variable X, defined in a finite interval on the real line with its probability density function given by

$$f_X(x) = \frac{\Gamma(\alpha + \beta)}{\Gamma(\alpha)\Gamma(\beta)\eta^{\alpha + \beta - 1}} (x - \gamma)^{\alpha - 1}(\gamma + \eta - x)^{\beta - 1},$$

$$\gamma \leqslant x \leqslant (\gamma + \eta), \gamma \text{ real, and } \alpha, \beta, \eta > 0,$$

$$= 0, \quad \text{otherwise.}$$

If $Y = (X - \gamma)/\eta$, it can be shown that the probability density function of Y is

$$f_Y(y) = \frac{\Gamma(\alpha+\beta)}{\Gamma(\alpha)\Gamma(\beta)} y^{\alpha-1}(1-y)^{\beta-1}, \quad 0 \leqslant y \leqslant 1, \ \alpha, \beta > 0,$$

$$= 0, \quad \text{otherwise.}$$

Here

$$\Gamma(x) = \int_0^\infty z^{x-1} e^{-z} \, dz.$$

It is left as an exercise for the reader to verify that the uniform distribution is a special case of the beta distribution. The relationship between the beta distribution and the binomial distribution is given in Section 7.4.3.

(c) The Normal Distribution. A random variable X is said to have a normal or Gaussian distribution with parameters μ and σ if the probability density function is given by

$$f_X(x) = \frac{1}{\sqrt{2\pi}\,\sigma} \exp\left[-\frac{1}{2}\left(\frac{x-\mu}{\sigma}\right)^2\right], \quad -\infty < x < \infty.$$

This distribution is symmetric about its mean, denoted by the parameter μ, which for the normal distribution is also the value of the mode and the median. Its variance is given by σ^2.

If the random variable X is identified as time to failure, the validity of the normal distribution as a failure model is questionable because the range of X includes negative values and this does not seem to be reasonable for time as a random variable. It is for this reason that the truncated normal distribution is more palatable. Consequently, X is said to have a *truncated normal distribution* if

$$f_X(x) = \frac{1}{\kappa\sqrt{2\pi}\,\sigma} \exp\left[-\frac{1}{2}\left(\frac{x-\mu}{\sigma}\right)^2\right], \quad 0 \leqslant x < \infty,$$

where κ is referred to as the *normalizing constant* and is such that

$$\int_0^\infty f_X(x)\, dx = 1.$$

Some of the continuous distributions discussed here are illustrated in Figure 3.1; their moment generating functions, their means, and their variances are given in Table 3.2 at the end of Section 3.3.

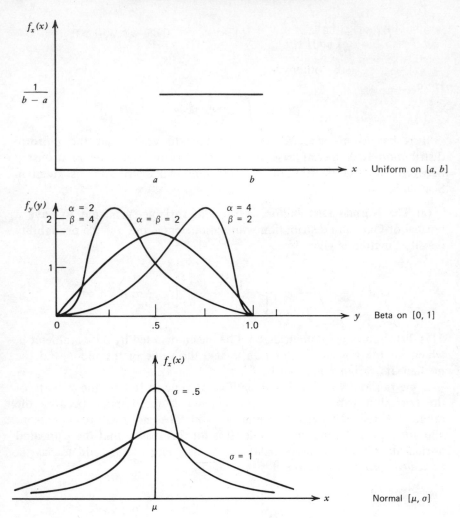

Figure 3.1

The other well-known continuous distributions, such as the exponential, the gamma, and the lognormal, will be derived in Chapter 4 from physical considerations commonly occurring in reliability.

3.3 DERIVED DISTRIBUTIONS

In Section 3.2 some fundamental probability distributions were briefly discussed. Using these fundamental distributions, one can derive other probability distributions that are useful in reliability methodology. Here such distributions will be termed *derived distributions*, and it is the purpose of this section to present a few of them.

In order to be able to develop derived distributions, it is first necessary to become familiar with a few techniques to be described in the following.

3.3.1 Transformation of Variables

Suppose that a random variable X has a density $f_X(x)$ for $a < x < b$, and 0 elsewhere. Suppose also that Y is function of X, say $Y = u(X)$, where $u'(X)$ is continuous and positive for $a < x < b$. Let $X = v(Y)$ be the solution of $Y = u(X)$ for X. It is desired to find the density of Y, $f_Y(y)$. Now

$$F_Y(y) = P(Y \leqslant y) = P[u(X) \leqslant y] = P[X \leqslant v(y)]$$

$$= \int_a^{v(y)} f_X(x)\, dx;$$

$$F_Y(y) = F_X[v(y)] - F_X(a), \quad u(a) < y < u(b),$$

$$= 0, \quad y \leqslant u(a),$$

$$= 1, \quad y \geqslant u(b).$$

The probability density of Y, $f_Y(y)$, is obtained by taking the derivative of $F_Y(y)$. Thus

$$f_Y(y) = F_Y'(y) = \frac{d}{dy}\{F_X[v(y)] - F_X(a)\}, \quad u(a) < y < u(b),$$

$$= 0, \quad \text{otherwise.}$$

If $u'(X)$ is continuous and negative, a similar proof holds.

Example. Suppose that $f_X(x) = e^{-x}$, for $0 \leqslant x < \infty$, and that $Y = \sqrt[3]{X}$. It is desired to obtain the probability density function of Y, say $f_Y(y)$. Since $d(x^{1/3})/dx = (1/3)x^{-2/3}$ is continuous and positive for $0 < x < \infty$, the required density can be found as follows:

$$F_Y(y) = P(Y \leqslant y) = P(X \leqslant y^3) = \int_0^{y^3} e^{-x}\, dx$$

$$= 1 - e^{-y^3}, \quad 0 < y < \infty,$$

$$= 0, \quad y \leqslant 0.$$

The probability density function of Y, $f_Y(y)$, is

$$f_Y(y) = F_Y'(y) = 3y^2 e^{-y^3}, \quad 0 < y < \infty,$$

$$= 0, \quad \text{otherwise.}$$

3.3.2   Functions of Independent Variates

Often it is necessary to determine the distribution function of functions of independent random variables. Although the principles presented here are general, the exposition is given in terms of two independent random variables, X and Y, with marginal density functions, $f_X(x)$ and $f_Y(y)$. Let $T = U(X, Y)$ be monotonic in the sense that $U(X, Y)$ is a monotonic function of X when Y is held constant, and vice versa. Examples of this are $T = X + Y$, $T = X - Y$, $T = XY$, and $T = X/Y$. It is necessary to find the distribution of T, and to do this one of the variates, say X, is assumed to be fixed and Y is allowed to vary over its range according to its marginal distribution. The procedure is as follows:

Since X and Y are independent,

$$f_{X,Y}(x,y) = f_X(x) f_Y(y).$$

Since $T = U(X, Y)$, let $Y = U^{-1}(T, X)$; also let $F_{Y|X}(y|x) = P(Y \leqslant y | X = x)$. Then

$$F_{T|X}(t|x) = P(T \leqslant t | X = x) = P(U(X, Y) \leqslant t | X = x).$$

Since $Y = U^{-1}(T, X)$, it follows that

$$F_{T|X}(t|x) = P[Y \leqslant U^{-1}(t,x) | X = x].$$

Differentiating $F_{T|X}(t|x)$ with respect to t gives the conditional probability density function (p.d.f.) of $(T|X)$:

$$f_{T|X}(t|x) = F_{T|X}'(t|x).$$

From the theorem on conditional probabilities, it is clear that

$$f_{T,X}(t,x) = f_{T|X}(t|x) f_X(x),$$

where $f_{T,X}(t,x)$ is the joint p.d.f. of T and X.
The marginal p.d.f. of T is then

$$f_T(t) = \int_{\text{all } x} f_{T|X}(t|x) f_X(x) \, dx.$$

The preceding discussion can be clarified by the following example.

Example. Let

$$f_X(x) = 1, \quad 0 \leqslant x \leqslant 1.$$

Suppose that two observations, x_1 and x_2, are independently drawn from $f_X(x)$, and it is desired to find the probability density function of $T = X_1 + X_2$. Since X_1 and X_2 are independent, it follows that

$$f_{X_1, X_2}(x, y) = 1, \quad 0 \leqslant x \leqslant 1, \quad 0 \leqslant y \leqslant 1.$$

Since $X_2 = T - X_1$,

$$F_{T|X_1}(t|x) = P(T \leqslant t | X_1 = x) = P(X_2 \leqslant t - x | X_1 = x)$$

$$= \int_0^{t-x} f_{X_2}(x) \, dx = t - x$$

since X_1 and X_2 are independent. Therefore

$$f_{T|X_1}(t|x_1) = 1, \quad x_1 \leqslant t \leqslant x_1 + 1.$$

Consequently

$$f_{T, X_1}(t, x_1) = 1, \quad 0 \leqslant x_1 \leqslant 1, \quad x_1 \leqslant t \leqslant x_1 + 1.$$

The ranges of the two random variables are shown in the following figure. The marginal density of T is obtained by integration as follows:

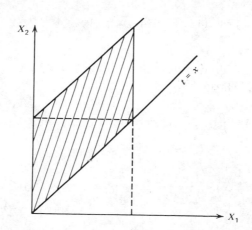

$$f_T(t) = \int_0^t f_{T,X_1}(t,x)\,dx = t, \quad 0 \leqslant t \leqslant 1,$$

$$= \int_{t-1}^1 f_{T,X_1}(t,x)\,dx = 2 - t, \quad 1 \leqslant t \leqslant 2.$$

The approach presented here can be generalized to include functions of more than two independent variables, and the interested reader is urged to work the applicable problems given at the end of this chapter.

3.3.3 Functions of Nonindependent Variates

The principles discussed in Section 3.3.2 can be extended to nonindependent variates by substituting conditional distributions for marginal distributions.

Reverting to the notations used in the preceding section, suppose now that X and Y are nonindependent. Then

$$f_{X,Y}(x,y) = f_X(x) f_{Y|X}(y|x).$$

With X held fixed, the conditional density of T, given X, is derived from $F_{T|X}(t|x)$ by the usual method. Next,

$$f_{T,X}(t,x) = f_{T|X}(t|x) f_X(x),$$

and the remaining steps are the same as before.

The following example illustrates the foregoing procedure.

Example. Given that X and Y are jointly distributed so that

$$f_{X,Y}(x,y) = 120y(x-y)(1-x), \quad 0 \leqslant x \leqslant 1, 0 \leqslant y \leqslant x,$$

find the probability density function of $T = Y/\sqrt{X}$. The marginal density of X, $f_X(x)$, is

$$f_X(x) = \int_0^x f_{X,Y}(x,y)\,dy = 20x^3(1-x), \quad 0 \leqslant x \leqslant 1,$$

and

$$f_{Y|X}(y|x) = \frac{f_{X,Y}(x,y)}{f_X(x)} = \frac{6y}{x^3}(x-y), \quad 0 \leqslant y \leqslant x.$$

Then,

$$F_{T|X}(t|x) = P(T \leqslant t | X = x)$$

$$= P(Y \leqslant t\sqrt{x} \,|X = x) = \int_0^{t\sqrt{x}} f_{Y|X}(y|x)\, dy$$

$$= \frac{3t^2}{x} - \frac{2t^3}{x^{3/2}}, \quad 0 \leqslant t \leqslant \sqrt{x},$$

and

$$f_{T|X}(t|x) = F'_{T|X}(t|x) = \frac{6t}{x} - \frac{6t^2}{x^{3/2}}, \quad 0 \leqslant t \leqslant \sqrt{x}.$$

Consequently,

$$f_{T,X}(t,x) = 20x^3(1-x) \cdot \frac{6t}{x^{3/2}}(\sqrt{x} - t)$$

$$= 120x^{3/2}(1-x)t(\sqrt{x} - t), \quad 0 \leqslant x \leqslant 1, 0 \leqslant t \leqslant \sqrt{x}.$$

The range of the two variables is shown in the following figure, and integration of $f_{T,X}(t,x)$ from t^2 to 1 yields the desired result.

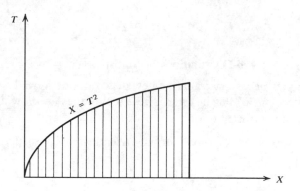

3.3.4 Use of Moment Generating Functions

As the reader may have noticed, the above methods of deriving distributions can often become cumbersome. The use of moment generating functions, however, may simplify the effort.

Let $X_1, X_2, \ldots, X_n$ be n jointly distributed continuous random variables with a probability density function given by $f(x_1, x_2, \ldots, x_n)$. Let $T = U(X_1, X_2, \ldots, X_n)$ be some function of these random variables, whose p.d.f. is desired.

If the moment generating function of T exists, then by definition

$$M_T(t) = E(e^{tT}) = \int_{-\infty}^{\infty} \cdots \int_{-\infty}^{\infty} e^{tU(x_1, x_2, \ldots, x_n)} f(x_1, x_2, \ldots, x_n) \prod_{i=1}^{n} dx_i.$$

From the uniqueness theorem of the moment generating function it is possible to determine the probability density function of T.

Example. Suppose that $T = U(X_1, X_2, \ldots, X_n) = X_1 + X_2 + \cdots X_n$. Then, if the X_i's are independent,

$$M_T(t) = \int_{-\infty}^{\infty} \cdots \int_{-\infty}^{\infty} e^{tx_1} e^{tx_2} \ldots e^{tx_n} f(x_1) f(x_2) \ldots f(x_n) \prod_{i=1}^{n} dx_i$$

$$= \prod_{i=1}^{n} M_{X_i}(t).$$

If all the X_i's have the same distribution,

$$M_T(t) = [M_X(t)]^n.$$

3.3.5 Some Derived Distributions in Statistics

The techniques necessary to derive new distributions were developed in the preceding sections of this chapter. These techniques are now used to derive some well-known distributions.

(a) The Sample Mean from Normal Densities. Suppose that the p.d.f. of X is

$$f_X(x) = \frac{1}{\sqrt{2\pi}\,\sigma} \exp\left[-\frac{1}{2}\left(\frac{x-\mu}{\sigma} \right)^2 \right], \quad -\infty < x < \infty.$$

Here $X_1, X_2, \ldots, X_n$ is a random sample of size n which can be drawn from $f_X(x)$, and the sample mean (or the arithmetic average)

$$\overline{X} = \frac{\sum\limits_{i=1}^{n} X_i}{n}$$

is computed. It is desired to find the p.d.f. of $\overline{X}$.

The fact that the X_i's, $i = 1, 2, \ldots, n$, constitute a random sample from $f_X(x)$ implies that the X_i's are independent of each other and that the p.d.f. of any X_i is

$$f_{X_i}(x) = \frac{1}{\sqrt{2\pi}\,\sigma} \exp\left[-\frac{1}{2}\left(\frac{x-\mu}{\sigma} \right)^2 \right], \quad -\infty < x < \infty.$$

Since

$$\overline{X} = \frac{X_1}{n} + \frac{X_2}{n} + \cdots + \frac{X_n}{n},$$

the distribution of $\overline{X}$ will be given by the n-fold convolution of the distributions of the X_i/n. Since the X_i's are independent, the moment generating function of the convolution will be given by the product of the moment generating functions of all X_i/n (see Section 3.1.5). To be able to find the moment generating function of X_i/n, it is first necessary to find the moment generating function of X_i.

Let

$$Z = \frac{X_i - \mu}{\sigma}$$

so that

$$X_i = \mu + Z\sigma.$$

It is easy to see, using the transformation of variables technique discussed in Section 3.3.1, that

$$f_Z(z) = \frac{1}{\sqrt{2\pi}} e^{-z^2/2}, \quad -\infty < z < \infty.$$

The moment generating function of $f_Z(z)$, $M_Z(t)$, is

$$M_Z(t) = \int_{-\infty}^{\infty} e^{tz} \frac{1}{\sqrt{2\pi}} e^{-z^2/2}\, dz = e^{t^2/2} \int_{-\infty}^{\infty} e^{-(1/2)(z^2 - 2tz + t^2)} \frac{1}{\sqrt{2\pi}}\, dz$$

$$= e^{t^2/2} \int_{-\infty}^{\infty} \frac{1}{\sqrt{2\pi}} e^{-(1/2)(z-t)^2}\, dz = e^{t^2/2}.$$

Since $X_i = \mu + Z\sigma$, it is easy to see, using a property of the moment generating function presented in Section 3.1.4, that

$$M_{X_i}(t) = e^{t\mu} \cdot e^{t^2\sigma^2/2} = e^{\mu t + (\sigma^2 t^2)/2}$$

and

$$M_{X_i/n}(t) = \exp\left(\frac{\mu t}{n} + \frac{\sigma^2 t^2}{2n^2} \right).$$

By applying the convolution property, it can be seen that the moment generating function of $\bar{X}$ is

$$M_{\bar{X}}(t) = \left[\exp\left(\frac{\mu t}{n} + \frac{\sigma^2 t^2}{2n^2} \right) \right]^n = \exp\left(\mu t + \frac{\sigma^2 t^2}{2n} \right).$$

The uniqueness property of the moment generating function (Section 3.1.4) asserts that the above expression can be the moment generating function of only the normal distribution. Hence the sample mean $\bar{X}$ is normally distributed with a mean μ and a variance given by σ^2/n, that is,

$$f_{\bar{X}}(x) = \frac{\sqrt{n}}{\sqrt{2\pi\sigma^2}} \exp\left[-\frac{n}{2}\left(\frac{x-\mu}{\sigma} \right)^2 \right].$$

(b) The Chi-Square Distribution. Let X_i, $i = 1, 2, \ldots, \nu$, be normally and independently distributed with means μ_i and variances σ_i^2. Define a variable

$$\chi^2(\nu) = \sum_{i=1}^{\nu} \left(\frac{X_i - \mu_i}{\sigma_i} \right)^2.$$

It is desired to obtain the probability density function of $\chi^2(\nu)$. The method of moment generating functions will be used to determine this probability density function.

Suppose that a random variable Y, with a parameter $n = 1, 2, 3, \ldots$, has a probability density function given by

$$f_Y(y) = \frac{e^{-y/2} y^{(n/2)-1}}{2^{n/2}\Gamma(n/2)}, \quad y > 0.$$

The above form of a probability density is often referred to as a chi-square density with *n degrees of freedom* (d.f.). Percentage points, originally tabulated in *Biometrika*, Vol. 32 (1941), are given in the Appendix.

The moment generating function of Y, $M_Y(t)$, is

$$M_Y(t) = \int_0^\infty e^{ty} f_Y(y)\, dy = \frac{1}{2^{n/2}\Gamma(n/2)} \int_0^\infty y^{(n/2)-1} e^{-(1/2)(1-2t)y}\, dy.$$

To evaluate the above integral, the substitution

$$k = \left(\frac{1-2t}{2}\right) y$$

has to be used. Thus

$$M_Y(t) = \frac{1}{2^{n/2}\Gamma(n/2)} \int_0^\infty e^{-k} k^{(n/2)-1} \left(\frac{2}{1-2t}\right)^{n/2} dk$$

$$= \left(\frac{1}{1-2t}\right)^{n/2} = (1-2t)^{-n/2}$$

since

$$\int_0^\infty e^{-y} y^{x-1}\, dy = \Gamma(x).$$

Since X_i is distributed normally with a mean μ_i and a variance σ_i^2, it is clear that

$$Z_i = \frac{X_i - \mu_i}{\sigma_i}$$

is distributed normally with a mean 0 and a variance 1. Hence the moment generating function of Z_i^2, $M_{Z_i^2}(t)$, is

$$M_{Z_i^2}(t) = E(e^{tZ_i^2}) = \int_0^\infty e^{tz_i^2} \frac{1}{\sqrt{2\pi}} e^{-z_i^2/2}\, dz_i$$

$$= \frac{1}{\sqrt{2\pi}} \int_0^\infty e^{-(1/2)(1-2t)z_i^2}\, dz_i$$

$$= (1-2t)^{-1/2}.$$

Since $\chi^2(\nu) = \sum Z_i^2$, and since the Z_i's are independent, the moment generating function of $\chi^2(\nu)$ is

$$M_{\chi^2(\nu)}(t) = (1 - 2t)^{-\nu/2}.$$

By the uniqueness theorem of the moment generating function, it can then be inferred that the probability density function of $\chi^2(\nu)$ is a chi-square distribution with ν degrees of freedom. Several applications of the chi-square distribution to reliability problems appear in Chapters 5 and 9.

(c) **The Fisher-Snedecor Distribution.** This distribution, known also as the F distribution, is of considerable practical interest.

Let the random variables X and Y be independent, and let each have a probability density function given by the chi-square law, with γ_1 and γ_2 degrees of freedom, respectively. Define a random variable F, where

$$F = \frac{X/\gamma_1}{Y/\gamma_2}.$$

The probability density function of F can be found using the methods discussed previously. It has been shown (Mood and Graybill, 1963, p. 232), that $f_F(f)$ is

$$f_F(f) = \frac{\Gamma[\frac{1}{2}(\gamma_1 + \gamma_2)]}{\Gamma(\gamma_1/2)\Gamma(\gamma_2/2)} \left(\frac{\gamma_1}{\gamma_2}\right)^{\gamma_1/2} f^{(1/2)\gamma_1 - 1} \left(1 + \frac{\gamma_1}{\gamma_2}f\right)^{-(1/2)(\gamma_1 + \gamma_2)} \qquad f > 0,$$

$$= 0, \quad \text{otherwise.}$$

Here γ_1 and γ_2 are the parameters of this distribution, and the percentage points of the F distribution are given in *Biometrika*, Vol. 33 (1943). Applications of this distribution to reliability problems appear in Chapter 5, and portions of the *Biometrika* tables of percentage points are given in the Appendix.

(d) **Student's t Distribution.** Let X and Y be two random variables that are independently distributed. Also let the p.d.f. of X be given by a normal law with a mean of μ and a variance of σ^2, and let the p.d.f. of Y be given by a chi-square law with ν degrees of freedom.

Define a random variable T, where

$$T = \frac{(X - \mu)/\sigma}{\sqrt{Y/\nu}}.$$

The probability density function of T is given by Student's law, and it has been shown (Mood and Graybill, 1963, p. 233) that

$$f_T(t) = \frac{[(\nu-1)/2]!}{\sqrt{n\pi}\,[(\nu-2)/2]!}\left(1+\frac{t^2}{\nu}\right)^{-(\nu+1)/2}, \quad -\infty < t < \infty.$$

Here ν is the number of degrees of freedom, and the percentage points of this distribution are given in Fisher and Yates (1938). It is easy to verify that T^2 has an F distribution with 1 and ν degrees of freedom.

Tables 3.1 and 3.2 summarize some of the discrete and some of the continuous distributions, respectively, discussed so far. The reader is advised to verify the moment generating functions of these distributions.

Table 3.1   Some discrete distributions

Discrete Probability Distribution	Probability Mass Function	Moment Generating Function	Mean	Variance
Geometric with parameter p $0 \leqslant p \leqslant 1 : \nu \equiv 1-p$	$p(x)=(1-p)^{x-1}p$ $x=1,2,\ldots$	$\dfrac{p}{e^{-t}-\nu}$	$\dfrac{1}{p}$	$\dfrac{\nu}{p^2}$
Binomial with parameters n and p $0 \leqslant p \leqslant 1 : \nu \equiv 1-p$	$p(x)=\dbinom{n}{x}p^x(1-p)^{n-x}$ $x=0,1,2,\ldots,n$	$(\nu+pe^t)^n$	np	$np\nu$
Negative binomial with parameters r and p	$p(x)=\dbinom{x-1}{r-1}(1-p)^{x-r}p^r$ $x=r,r+1,\ldots;\, r=1,2,\ldots$	$\left(\dfrac{p}{e^{-t}-\nu}\right)^r$	$\dfrac{r}{p}$	$\dfrac{r\nu}{p^2}$
Poisson with parameter $\mu>0$	$p(x)=\dfrac{e^{-\mu}\mu^x}{x!}$ $x=0,1,2,\ldots$	$e^{\mu}(e^t-1)$	μ	μ

Table 3.2   Some continuous distributions[a]

Continuous Probability[a] Distribution	Probability Density Function	Moment Generating Function	Mean	Variance
Uniform over $[a,b]$	$f(x)=(b-a)^{-1}$ $a \leqslant x \leqslant b$	$\dfrac{e^{bt}-e^{at}}{t(b-a)}$	$\dfrac{(a+b)}{2}$	$\dfrac{(b-a)^2}{12}$
Normal with parameters μ and σ^2	$f(x)=\dfrac{1}{\sqrt{2\pi\sigma^2}}\exp\left[-\dfrac{1}{2}\left(\dfrac{x-\mu}{\sigma}\right)^2\right]$ $-\infty < x < \infty$	$e^{\mu t + \sigma^2 t^2/2}$	μ	σ^2
Chi square with γ degrees of freedom	$f(x)=\dfrac{e^{-x/2}(x)^{\gamma/2-1}}{2^{\gamma/2}\Gamma(\gamma/2)}$ $x \geqslant 0$	$(1-2t)^{-\gamma/2}$	γ	2γ

[a]A similar table is given in Chapter 4 for other continuous distributions, which can be derived by considering a physical failure process.

3.4 ESTIMATION

The basic and derived distributions presented in Sections 3.2 and 3.3 have parameters in their probability density functions. This is true also of the failure models that will be presented in Chapter 4. Often these parameters are unknown to the user, and the objective of this section is to present methods by which they can be estimated from sample data. The topic of estimation covers both point and interval procedures, and for a more detailed discussion of estimation the reader is referred to Cramér (1946) or to any of several other books on the subject.

3.4.1 Point Estimates

Suppose that one is given a sample from a population having a probability density function of known form, but involving a certain number of unknown parameters. The objective here is to obtain estimates of these unknown parameters by using some function of the sample values. Since there could be many such functions that could serve as candidates for the estimation problem, guidelines should be developed for the selection procedure so that it yields estimates that are *best* in some sense. To formalize the discussion, attention will be focused on the following simple situation.

Let $f_X(x;\theta)$ be the density function of a random variable X and θ be an unknown parameter of $f_X(x;\theta)$ that can take on all possible values in Ω. An estimate of θ is desired. Suppose a random sample of size n will be drawn from $f_X(x;\theta)$, and let this sample be denoted by $X_1, X_2, \ldots, X_n$. Let $\hat{\theta}_1, \hat{\theta}_2, \ldots$ be several functions of the X_i's, and find the $\hat{\theta}_i$ that is a best estimator of θ in some sense.

(a) Criteria of Good Estimators. The following criteria can be individually or collectively used to characterize good estimators.

1. Invariance and Unbiasedness.
2. Consistency.
3. Efficiency.
4. Minimum risk invariant and minimum variance unbiased.

It is often desirable for an estimator to have some additional asymptotic properties, and these will be discussed in detail later. These criteria are utilized in Chapter 5, which deals with the estimation problems in reliability.

To formalize the above criteria, it is first necessary to introduce the concepts of a loss function and a risk function. Let $\hat{\theta}$ be an estimator of θ; then the *loss function* is defined as a real-valued nonnegative function that reflects the loss in choosing $\hat{\theta}$ as an estimator of θ. Since $\hat{\theta}$ is random, it follows that the loss function is a random function. The expected value of the loss function is known as the *risk function*. A best estimator of θ is one that will in some fashion minimize the risk for every value of θ in Ω.

The specification of a realistic loss function is often difficult. A reasonable choice of the loss function is to assume that the loss is proportional to the squared error. Then the loss function, $L(\hat{\theta}, \theta)$, is given by

$$L(\hat{\theta}, \theta) = k(\hat{\theta} - \theta)^2, \qquad k \geqslant 0,$$

where k is the constant of proportionality. Consequently, the risk function, $R(\hat{\theta}, \theta)$, is

$$R(\hat{\theta}, \theta) = E\left[k(\hat{\theta} - \theta)^2 \right] = kE\left[(\hat{\theta} - \theta)^2 \right];$$

$E[(\hat{\theta}-\theta)^2]$ is also known as the mean-squared error.

INVARIANCE AND UNBIASEDNESS. When the loss function is taken to be $k(\hat{\theta}-\theta)^2$, an estimator $\hat{\theta}$ is said to have bias independent of any parameter ξ (which may or may not be θ) if $E[\sqrt{k}\,(\hat{\theta}-\theta)]$ is independent of ξ. For a distribution with only a location parameter θ_1 and a scale parameter θ_2 and $\theta \equiv \theta_1$ or $\theta \equiv \theta_2$, the bias of $\hat{\theta}$, $E[\sqrt{k}\,(\hat{\theta}-\theta)]$, is said to be invariant under transformations of location and scale if it is independent of both θ_1 and θ_2. The location and scale parameters, θ_1 and θ_2, respectively, for a distribution $F(x)$ are such that $F(x)=G[(x-\theta_1)/\theta_2]$ for some G.

If $\sqrt{k}\,E(\hat{\theta}-\theta)$ is invariant under transformations of location and/or scale, then so is the risk function, $E[k(\hat{\theta}-\theta)^2]$, since

$$E\left[k(\hat{\theta}-\theta)^2\right]=k\,\mathrm{Var}(\hat{\theta})+\left\{E\left[\sqrt{k}\,(\hat{\theta}-\theta)\right]\right\}^2$$

and since $\mathrm{Var}(\hat{\theta})$ is proportional to the square of the scale parameter. If $E[k(\hat{\theta}-\theta)^2]$ is independent of all parameters, $\hat{\theta}$ is said to have *constant risk*. If $E[k(\hat{\theta}-\theta)]=0$, then $\hat{\theta}$ is said to be an *unbiased estimator* of θ.

CONSISTENCY. *a. Squared-error consistency.* A sequence of estimators of θ, $\{\hat{\theta}_i\}$, $i=1,2,\ldots,n$, is said to be a *squared-error consistent* estimator of θ if

$$\lim_{n\to\infty} E\left[\left(\hat{\theta}_n-\theta\right)^2\right]=0, \quad \text{for all } \theta \text{ in } \Omega.$$

b. Simple consistency. The sequence $\{\hat{\theta}_i\}$ is a *simple consistent* estimator of θ if, for every $\epsilon>0$,

$$\lim_{n\to\infty} P\left\{\theta-\epsilon<\hat{\theta}_n<\theta+\epsilon\right\}=1, \quad \text{for all } \theta \text{ in } \Omega.$$

EFFICIENCY. When the loss function is taken to be squared error, an estimator of θ, $\hat{\theta}_1$, based on a sample of size n, is more efficient than another estimator, $\hat{\theta}_2$, based on a sample of the same size if

$$E\left[\left(\hat{\theta}_1-\theta\right)^2\right]\leqslant E\left[\left(\hat{\theta}_2-\theta\right)^2\right]$$

with the strict inequality for some θ. If the estimators are unbiased, the efficient estimator is the one that has a smaller variance.

The absolute measure of efficiency of an estimate would require that its mean-squared error be compared with a lower bound on the absolute minimum of such mean-squared errors, if one that is nonzero exists. For unbiased estimators, when the bound exists (i.e. the estimator is *regular*), it is the *information inequality* or what is also known as the *Cramér-Rao lower bound*.

Let the joint density function of the sample observations be denoted by

$$f(x_1, x_2, \ldots, x_n; \theta) = \phi(\mathbf{x}; \theta).$$

Then the *information* in the sample is given by

$$I(\theta) = E\left\{\left[\frac{d}{d\theta}\log\phi(\mathbf{x};\theta)\right]^2\right\}.$$

The efficiency of any unbiased estimator $\hat{\theta}$ (a function of the sample observations) is defined as

$$e(\hat{\theta}) = \frac{1/I(\theta)}{\mathrm{Var}(\hat{\theta})}.$$

If the efficiency is 1, that is, if the variance of the estimator $\hat{\theta}$ attains the information in the sample, the estimator $\hat{\theta}$ is said to be efficient.

Cramér-Rao lower bounds for the mean-squared errors of invariant estimators of location and scale parameters can be obtained from results discussed below under "Completeness," applied to the lower bounds for the variances of regular unbiased estimators.

MINIMUM RISK INVARIANT AND MINIMUM VARIANCE UNBIASED. If a squared error loss function is assumed, a minimum-risk invariant estimator is also known as the minimum-mean-squared error (MSE) invariant estimator.

A statistic $\hat{\theta}$ is the minimum-risk invariant estimator of θ when the loss is $k(\hat{\theta} - \theta)^2$ if

(a) $E[k(\hat{\theta} - \theta)^2]$ is independent of all parameters, that is, $\hat{\theta}$ has constant risk (this will often be true when distribution parameters are of location and/or scale, so that the risk function is invariant under transformations of location and/or scale); and

(b) $E[(\hat{\theta} - \theta)^2]$ is less than the expected squared error of any other constant-risk estimator of θ.

A statistic $\hat{\theta}$ is the minimum-variance unbiased estimator (MVUE) of θ when loss is $k(\hat{\theta} - \theta)^2$ if

(a) $E[\sqrt{k}\,(\hat{\theta}-\theta)]=0$; and

(b) $E[(\hat{\theta}-\theta)^2]=\text{Var}(\hat{\theta})$ is less than the variance of any other unbiased estimator of θ.

Two other important concepts in the theory of estimation also need to be emphasized: sufficiency and completeness.

SUFFICIENCY. Since $\hat{\theta}$, an estimator of θ, is a function of the n random variables X_i, a question arises as to whether, in the process of condensation of the n random variables to a single random variable $\hat{\theta}$, any information about θ is lost. Sufficiency attempts to address itself to this question, and $\hat{\theta}$ is said to be a *sufficient* statistic of θ if it contains all the information about θ that is in the sample.

A simple criterion proposed by Neyman for examining the sufficiency of a set of statistics is stated below for the multiparameter case.

Let $X_1, X_2, \ldots, X_n$ be a random sample of size n from the density $f(x; \theta_1, \theta_2, \ldots, \theta_k)$, for $a < x < b$, where a and b do not involve the θ's. Let $\hat{\theta}_i$, $i = 1, 2, \ldots, m$, be m functions (statistics) of the X_i's. If the joint density of a random sample from the above density can be factored as

$$g(x_1, x_2, \ldots, x_n; \theta_1, \theta_2, \ldots, \theta_k) = h\big(\hat{\theta}_1, \hat{\theta}_2, \ldots, \hat{\theta}_m; \theta_1, \theta_2, \ldots, \theta_k\big) \cdot q(x_1, x_2, \ldots x_n),$$

where $q(x_1, x_2, \ldots, x_n)$ does not contain the θ_i, then $\hat{\theta}_1, \ldots, \hat{\theta}_m$ is a set of m sufficient statistics.

Example.   Let $X_1, X_2, \ldots, X_n$ be a random sample from

$$f_X(x) = \frac{1}{\sqrt{2\pi}\,\sigma} \exp\left[-\frac{1}{2}\left(\frac{x-\mu}{\sigma}\right)^2\right], \quad -\infty < x < \infty, \ -\infty < \mu < \infty, \ \sigma^2 > 0.$$

Consider the two statistics $\hat{\mu}$ and $\hat{\sigma}^2$, where

$$\hat{\mu} = \overline{X} = \frac{\displaystyle\sum_{i=1}^{n} X_i}{n} \quad \text{and} \quad \hat{\sigma}^2 = \frac{1}{n-1} \sum_{i=1}^{n} \left(X_i - \overline{X}\right)^2.$$

Now

$$g(x_1, x_2, \ldots, x_n; \mu, \sigma^2) = \frac{1}{\left(\sigma\sqrt{2\pi}\,\right)^n} \exp\left[-\frac{1}{2\sigma^2} \sum_{i}^{n} (x_i - \mu)^2\right]$$

can be written as

$$\frac{1}{\left(\sigma\sqrt{2\pi}\,\right)^n}\exp\left\{-\frac{1}{2\sigma^2}\left[(n-1)\hat{\sigma}^2+n(\hat{\mu}-\mu)^2\right]\right\}.$$

Thus

$$g(x_1,x_2,\ldots,x_n;\mu,\sigma^2)=h(\hat{\sigma}^2,\hat{\mu};\sigma^2,\mu)q(x_1,x_2,\ldots,x_n),$$

where $q(x_1,x_2,\ldots,x_n)=1$, implies that $\hat{\sigma}^2$ and $\hat{\mu}$ are sufficient for the estimation of σ^2 and μ in the complete sample case.

COMPLETENESS. Let $f_X(x;\theta)$ be a family of density functions defined for all θ in the interval $\alpha<\theta<\beta$, such that

$$(i)\quad f_X(x;\theta)>0,\quad a<x<b,$$

$$=0,\quad\text{otherwise,}$$

and

$$(ii)\quad a\text{ and }b\text{ are independent of }\theta.$$

If there exists no function of X, say $u(X)$, continuous in the interval $a<x<b$, for which $E[u(X)]=0$ for all θ in the interval $\alpha<\theta<\beta$ except the zero function, $u_0(X)=0$, then $f_X(x;\theta)$ is defined to be a *complete* family of densities.

To prove that a density function is complete requires mathematics that becomes rather involved. However, it is often possible to detect density functions that are not complete by merely finding a nonzero function of the x, say $u(X)$, such that $E[u(X)]=0$, for all values of the parameter in a specified interval, $\alpha<\theta<\beta$.

The concepts of sufficiency and completeness immediately allow one to state the Léhman-Schéffé-Blackwell theorem for minimum variance unbiased estimators.

Let $\hat{\theta}$ be a sufficient estimator of θ whose density is complete. Also, let $v(\hat{\theta})$ be some function of $\hat{\theta}$ such that $E[v(\hat{\theta})]=u(\theta)$. Then $v(\hat{\theta})$ is the unique MVUE of $v(\theta)$.

This theorem can be illustrated by means of the following example. Let $X_1,X_2,\ldots,X_n$ be a random sample from

$$f_X(x)=\frac{1}{\sqrt{2\pi}}\exp\left[-\tfrac{1}{2}(x-\mu)^2\right],\quad \alpha_0<\mu<\beta_0,\ -\infty<x<\infty,$$

$$=0,\quad\text{otherwise.}$$

The statistic

$$\overline{X} = \frac{\sum\limits_{i=1}^{n} X_i}{n}$$

was proved to be sufficient by means of a previous example.
The probability density function of $\overline{X}$ is

$$f_{\overline{X}}(x) = \frac{\sqrt{n}}{\sqrt{2\pi}} \exp\left[-\frac{n}{2}(x-\mu)^2 \right], \quad \alpha_0 < \mu < \beta_0, \; -\infty < x < \infty,$$

$$= 0, \quad \text{otherwise.}$$

It can be proved that the family of densities $f_{\overline{X}}(x)$ is complete for all values of μ in the specified interval. Thus $\overline{X}$ is unbiased and sufficient, and its density is complete.

It is easy to verify by the Léhman-Schéffé-Blackwell theorem that the MVUE of $\mu^2 + 6\mu$ is

$$\overline{X}^2 + 6\overline{X} - \frac{1}{n}$$

since

$$E\left[\overline{X}^2 + 6\overline{X} - \frac{1}{n} \right] = \frac{1}{n} + \mu^2 + 6\mu - \frac{1}{n}$$

$$= \mu^2 + 6\mu.$$

Consider a distribution $F(x)$ with unknown parameter vector $\boldsymbol{\theta} = (\theta_1, \theta_2)$, where θ_1 is a location parameter and θ_2 is a scale parameter, that is, $F(x) = G[(x - \theta_1)/\theta_2]$ for some G. Let $\hat{\theta}_1$ and $\hat{\theta}_2$ be the unique MVUE's specified by the Léhmann-Schéffé-Blackwell theorem for θ_1 and θ_2, respectively. Also, let $A\theta_2^2$ and $C\theta_2^2$ be the variances of θ_1 and θ_2, respectively, and let $B\theta_2^2$ be their covariance. Then $\tilde{\theta}_2 = \hat{\theta}_2/(1 + C)$ and $\tilde{\theta}_1 = \hat{\theta}_1 - B\tilde{\theta}_2$ are the unique minimum mean-squared error estimators of θ_2 and θ_1, respectively, with MSE $(\tilde{\theta}_2) = C\theta_2^2/(1 + C)$ and MSE $(\tilde{\theta}_1) = [A - B^2/(1 + C)]\theta_2^2$. If loss is defined as squared error divided by θ_2^2, then $\tilde{\theta}_1$ and $\tilde{\theta}_2$ are the minimum-risk invariant estimators of θ_1 and θ_2. A proof of this result is given by Mann (1969), and examples of its use are presented in Section 5.1.

If the parameters of interest are those of location and/or scale, the results given above can be used to obtain a lower bound on the mean-squared error of an invariant estimator directly from the Cramér-Rao lower bound on the variance of a regular unbiased estimator. This can be done when maximum-likelihood estimators are unbiased or asymptotically unbiased. Specific examples of Cramér-Rao bounds for regular invariant estimators are given in Sections 5.1 and 5.2.

(b) Asymptotic Properties of Estimators. The following large-sample properties of estimators are often useful.

SQUARED-ERROR ASYMPTOTICALLY EFFICIENT ESTIMATORS. A sequence of squared-error consistent estimators of θ, $\{\hat{\theta}_i\}$, $i=1,2,\ldots,n$, is called a *squared-error asymptotically efficient estimator* of θ if there is no other sequence of squared-error consistent estimators, $\{\hat{\theta}_i^*\}$, $i=1,2,\ldots,n$, for which

$$\lim_{n\to\infty} \frac{E\left[\left(\hat{\theta}_n - \theta\right)^2\right]}{E\left[\left(\hat{\theta}_n^* - \theta\right)^2\right]} > 1 \quad \text{for all } \theta \text{ in } \Omega.$$

BEST ASYMPTOTICALLY NORMAL (BAN) ESTIMATORS. The sequence of estimators $\{\hat{\theta}_i\}$, $i=1,\ldots,n$, is a *best asymptotically normal* (BAN) *estimator* of θ if three conditions are satisfied.

1. As n approaches infinity, the distribution function of $\sqrt{n}\,(\hat{\theta}_n - \theta)$ approaches a normal distribution with mean 0 and variance $\sigma^2(\theta)$. The symbol $\sigma^2(\theta)$ indicates that the variance may depend on θ.
2. For every $\epsilon > 0$,

$$\lim_{n\to\infty} P\left\{|\hat{\theta}_n - \theta| > \epsilon\right\} = 0 \quad \text{for all } \theta \text{ in } \Omega.$$

3. There does not exist another sequence of consistent estimators, $\{\theta_i^*\}$, $i=1,2,\ldots,n$, for which the distribution of $\sqrt{n}\,(\theta_n^* - \theta)$ approaches a normal distribution with mean 0 and variance $\sigma^{*2}(\theta)$ and such that

$$\frac{\sigma^2(\theta)}{\sigma^{*2}(\theta)} > 1 \quad \text{for all } \theta \text{ in } \Omega.$$

3.4.2 Methods of Obtaining Point Estimators

Several methods that can produce estimators with some of the properties discussed in the preceding section, are available.

(a) Method of Moments. Let a random variable X have a probability density function given by $f_X(x; \theta_1, \theta_2, \ldots, \theta_k)$, where the θ_i's represent the k parameters.

Let μ_t' be the tth moment of $f_X(x; \theta_1, \ldots, \theta_k)$ about zero, that is,

$$\mu_t' = \int_{-\infty}^{\infty} x^t f_X(x; \theta_1, \ldots, \theta_k)\, dx.$$

Clearly μ_t' is a function of the k parameters, and hence μ_t' can be written as

$$\mu_t' = \mu_t'(\theta_1, \theta_2, \ldots, \theta_k).$$

Let $X_1, X_2, \ldots, X_n$ be a random sample of size n from $f_X(x; \theta_1, \ldots, \theta_k)$. Form the first k sample moments, m_t',

$$m_t' = \frac{1}{n} \sum_{i=1}^{n} X_i^t.$$

The moment estimators, $\hat{\theta}_i, i = 1, 2, \ldots, k$, of the k θ_i's are obtained by solving the following k equations for the θ_i:

$$\mu_t' = m_t', \quad t = 1, 2, \ldots, k.$$

Example. If the random variable X has a p.d.f. given by

$$f_X(x) = \frac{1}{\sqrt{2\pi}\,\sigma} \exp\left\{ -\frac{1}{2}\left[\left(\frac{x - \mu}{\sigma}\right)^2 \right] \right\}, \quad -\infty < x < \infty$$

then, by use of the moment generating function, it is easy to verify that

$$\mu_1' = \mu \qquad \text{and} \qquad \mu_2' = \mu^2 + \sigma^2.$$

The sample moments are

$$m_1' = \frac{1}{n} \sum_{i=1}^{n} X_i \qquad \text{and} \qquad m_2' = \frac{1}{n} \sum_{i=1}^{n} X_i^2.$$

Setting $m_1' = \mu_1' = \mu$ and $m_2' = \mu^2 + \sigma^2$, yields

$$\hat{\mu} = \overline{X} \qquad \text{and} \qquad \hat{\sigma}^2 = \frac{1}{n}\left(\Sigma X_i^2 - n\overline{X}^2 \right)$$

as the moment estimators of μ and σ^2, respectively.

Under some very general conditions (Cramér, 1946, p.354), it has been shown that the moment estimators are (*a*) simple and squared-error consistent; (*b*) asymptotically normal, but not, in general, asymptotically efficient or best asymptotically normal.

(b) Method of Maximum Likelihood. The method of maximum likelihood is a commonly used procedure because maximum-likelihood estimates given below often have very desirable properties.

Let $X_1, X_2, \ldots, X_n$ be a random sample of size n drawn from a probability density function $f_X(x; \theta)$, where θ is an unknown parameter. The *likelihood function* of this random sample is the joint density of the n random variables and is a function of the unknown parameter. Thus

$$L = \prod_{i=1}^{n} f_{X_i}(x_i; \theta)$$

is the likelihood function. If the X_i's are discrete random variables with a probability mass function $p(x_i; \theta)$,

$$L = \prod_{i=1}^{n} p(x_i; \theta).$$

The *maximum-likelihood estimator* (MLE) of θ, say $\hat{\theta}$, is the value of θ that maximizes L or, equivalently, the logarithm of L. Often, but not always, the MLE of θ is a solution of

$$\frac{d \log L}{d\theta} = 0,$$

where solutions that are not functions of the sample values $X_1, X_2, \ldots, X_n$ are not admissible, nor are solutions which are not in the parameter space Ω. An example requiring one to obtain an MLE by an alternative procedure is given by Problem 3.18a.

In effect

$$\hat{\theta} = h(X_1, X_2, \ldots, X_n).$$

If the X_i's are chosen from a multiparameter density, say $f_X(x; \theta_1, \ldots, \theta_k)$, the likelihood function is

$$L = \prod_{i=1}^{n} f_{X_i}(x_i; \theta_1, \theta_2, \ldots, \theta_k).$$

The MLE's of the parameters are the random variables $\hat{\theta}_i = h_i(X_1, X_2, \ldots, X_n)$, $i = 1, 2, \ldots, k$, where the $\hat{\theta}_i$'s are the values of the θ_i that maximize log L.

Again, the $\hat{\theta}_i$'s are often the solutions of the following k equations:

$$\frac{\partial \log L}{\partial \theta_i} = 0, \quad i = 1, 2, \dots, k.$$

The important properties of MLE's are the following.

1. If an efficient estimator of $\theta, \hat{\theta}$, exists, the likelihood equation applying to θ will have a unique solution that is $\hat{\theta}$.

2. If the number of sufficient statistics is equal to the number of unknown parameters, the MLE's are the minimum-variance estimators of their respective expected values, as shown by Rao (1949).

3. Under certain general conditions, the likelihood equation applying to θ has a solution that converges in probability† to the true value of θ, as $n \to \infty$. This solution is an asymptotically normal and an asymptotically efficient estimate of θ.

In addition, maximum-likelihood estimates possess the property of *invariance*. In other words, if $\hat{\theta}$ is the MLE of θ, and if $u(\theta)$ is a function of θ with a single-valued inverse, $u(\hat{\theta})$ is the MLE of $u(\theta)$. (This type of invariance, of course, is not related to the concept of invariant estimators discussed in Section 3.4.1.)

LARGE-SAMPLE PROPERTIES OF MAXIMUM-LIKELIHOOD ESTIMATORS In Section 3.3.5, some derived distributions were discussed. In particular, the distribution of $\overline{X}$, the sample mean from a normal density, was derived. It can be shown that the sample mean $\overline{X}$ is also the maximum-likelihood estimator of the parameter μ, and hence it can be concluded that the MLE of μ is distributed normally with a mean μ and a variance σ^2/n. This result is true for any value of n.

In many other cases the distribution of the MLE cannot be easily derived. However, under some general conditions, its distribution for large values of n can be found. This is referred to as the *asymptotic distribution* of the MLE, which was briefly described in item 3 above. It is now discussed in more detail.

1. *Single-parameter case.* Let θ be an unknown parameter in a probability density function given by $f_X(x; \theta)$. Let $\hat{\theta}$ be the maximum-likelihood estimator of θ, based on a random sample of size n.

Then, if $f_X(x; \theta)$ is such that

$$E\left\{ \frac{d^3}{d\theta^3} \left[\log f_X(x; \theta) \right] \right\} < \infty,$$

---

†For convergence in probability see Section 3.7.

$\hat{\theta}$ is asymptotically $(n\rightarrow\infty)$ normally distributed with a mean θ and a variance given by $1/n\tau^2$, where

$$\tau^2 = -E\left\{\frac{d^2}{d\theta^2}[\log f_X(x;\theta)]\right\}.$$

2. *Multiparameter case.* Let $\theta_1,\theta_2,\dots,\theta_k$ be k unknown parameters in a probability density function given by $f_X(x;\theta_1,\theta_2,\dots,\theta_k)$. Let $\hat{\theta}_1,\hat{\theta}_2,\dots,\hat{\theta}_k$ be the maximum-likelihood estimators of the θ_i, $i=1,2,\dots,k$, based on a sample of size n.

Then, as $n\rightarrow\infty$, the joint distribution of $\hat{\theta}_1,\hat{\theta}_2,\dots,\hat{\theta}_k$ is multivariate normal with means θ_i, $i=1,2,\dots,k$, respectively, and variances and covariances given by $\mathbf{V}/n$, where $\mathbf{V}=\mathbf{R}^{-1}$, and $\mathbf{R}$ is a matrix whose elements, r_{ij}, are given by

$$r_{ij} = -E\left[\frac{\partial^2}{\partial\theta_i\,\partial\theta_j}\log f_X(x;\theta_1,\theta_2,\dots,\theta_k)\right], \quad i,j=1,2,\dots,k.$$

The p.d.f. of the k-variate normal distribution can be written as follows. Let

$$\boldsymbol{\theta}:[\,\theta_1,\theta_2,\dots,\theta_k\,], \text{ that is, a } k\times1 \text{ vector}$$

and

$$\hat{\boldsymbol{\theta}}:\left[\hat{\theta}_1,\hat{\theta}_2,\dots,\hat{\theta}_k\right], \text{ that is, a } k\times1 \text{ vector};$$

then

$$f\left(\hat{\theta}_1,\hat{\theta}_2,\dots,\hat{\theta}_k;\theta_1,\theta_2,\dots,\theta_k\right) = \frac{|n\mathbf{R}|^{1/2}}{(2\pi)^{k/2}}\exp\left\{-\tfrac{1}{2}\left[(\hat{\boldsymbol{\theta}}-\boldsymbol{\theta})'n\mathbf{R}(\hat{\boldsymbol{\theta}}-\boldsymbol{\theta})\right]\right\},$$

$$-\infty<\hat{\theta}_i<\infty, \, i=1,2,\dots,k.$$

For a proof of these results, the reader is referred to Cramér (1946, p. 500).

Example. The following example is meant to clarify the foregoing ideas. Let $X_1,X_2,\dots,X_n$ be a random sample from

$$f_X(x;\mu,\sigma) = \frac{1}{\sqrt{2\pi}\,\sigma}\exp\left[-\frac{1}{2}\left(\frac{x-\mu}{\sigma}\right)^2\right], \quad -\infty<x<\infty.$$

Then

$$\log L = -\frac{n}{2}\log 2\pi - \frac{n}{2}\log\sigma^2 - \frac{1}{2\sigma^2}\sum_{i=1}^{n}(x_i-\mu)^2$$

and

$$\frac{\partial\log L}{\partial\mu}=0 \quad\text{and}\quad \frac{\partial\log L}{\partial\sigma}=0,$$

giving

$$\hat{\mu}=\overline{X}=\frac{1}{n}\sum_{i=1}^{n}X_i \quad\text{and}\quad \hat{\sigma}^2=\frac{1}{n}\sum_{i=1}^{n}\left(X_i-\overline{X}\right)^2$$

as the maximum-likelihood estimators of μ and σ^2. Also, since $E(\hat{\mu})=\mu$, $\hat{\mu}$ is an unbiased estimator of μ. However, $E(\hat{\sigma}^2)=[(n-1)/n]\sigma^2$, and hence $\hat{\sigma}^2$ is a biased estimator of σ^2, the bias of which can be easily removed since it is in terms of a known quantity. It is to be noted here that the MLE's are generally biased; and it can be easily verified that $\text{Var}(\hat{\sigma}^2)=2(n-1)\sigma^4/n^2$.

The large-sample properties of MLE's can be used to obtain the following results:

$$r_{ij}=-E\left[\frac{\partial^2\log f}{\partial\mu\,\partial(\sigma^2)}\right], \quad i,j=1,2,$$

where

$$\log f = -\tfrac{1}{2}\log 2\pi - \tfrac{1}{2}\log\sigma^2 - \frac{1}{2\sigma^2}(x-\mu)^2.$$

The required derivatives are

$$\frac{\partial^2\log f}{\partial\mu^2}=-\frac{1}{\sigma^2}, \qquad \frac{\partial^2\log f}{\partial\mu\,\partial(\sigma^2)}=-\frac{x-\mu}{\sigma^4}$$

and

$$\frac{\partial^2\log f}{\partial(\sigma^2)\cdot\partial(\sigma^2)}=\frac{1}{2\sigma^4}-\frac{(x-\mu)^2}{\sigma^6}.$$

Since $E(X) = \mu$, and $E(X - \mu)^2 = \sigma^2$, the matrix R is

$$R = \begin{bmatrix} \dfrac{1}{\sigma^2} & 0 \\[2ex] 0 & \dfrac{1}{2\sigma^4} \end{bmatrix},$$

and R^{-1}, the inverse of R, is given as

$$R^{-1} = \begin{bmatrix} \sigma^2 & 0 \\ 0 & 2\sigma^4 \end{bmatrix}.$$

The asymptotic variances and the asymptotic covariance of the estimators $\hat{\mu}$ and $\hat{\sigma}^2$ are obtained from the matrix V/n, where $V = R^{-1}$. Thus for large n

$$\mathrm{Var}(\hat{\mu}) = \frac{\sigma^2}{n}$$

a result which according to Section 3.3.5, is also true for small n. Also, for large n

$$\mathrm{Var}(\hat{\sigma}^2) = \frac{2\sigma^4}{n} \quad \text{and} \quad \mathrm{Cov}(\hat{\mu}, \hat{\sigma}^2) = 0.$$

It can also be shown that $\hat{\mu}$ and $\hat{\sigma}^2$ have zero covariance for any sample size.

The joint asymptotic distribution of these estimators is a bivariate normal and can be written as

$$f(\hat{\mu}, \hat{\sigma}^2; \mu, \sigma^2) = \frac{1}{2\pi\sqrt{\mathrm{Var}(\hat{\mu})\,\mathrm{Var}(\hat{\sigma}^2)}} \exp\left\{ -\frac{1}{2}\left[\frac{(\hat{\mu}-\mu)^2}{\mathrm{Var}(\hat{\mu})} + \frac{(\hat{\sigma}^2-\sigma^2)^2}{\mathrm{Var}(\hat{\sigma}^2)} \right] \right\}.$$

(c) **Method of Least Squares.** This method of estimation requires the use of order statistics, which are discussed in Section 3.5. For this reason a discussion of least-squares estimation has been deferred till Section 3.6.

(d) **Bayes Estimators.** Bayes methods are becoming quite popular in reliability theory. Chapter 8 of this book is devoted exclusively to this topic.

3.4.3 Interval Estimators and Tests of Hypotheses

Since point estimators are random variables, it is important to incorporate into them some measure of the error of estimation. In practice, this is accomplished by constructing an interval, sometimes of the form $\hat{\theta} - a$ to $\hat{\theta} + b$ about $\hat{\theta}$, an estimator of θ. More generally the interval will be of the form $[L, U]$, depending upon the sample data but not upon θ. If the method of determining L and U is such that at least $100(1 - \alpha)\%$ of the random intervals so constructed will contain the true parameter θ, the interval is called the $100(1 - \alpha)\%$ *confidence interval*, and $(1 - \alpha)$ is termed the *confidence coefficient*. If *exactly* $100(1 - \alpha)\%$ of the random intervals contain the true value θ, the confidence interval is *exact*. If $\theta > L$ $(\theta < U)$ with probability 1, then U (L) is a one-sided *upper* (*lower*) *confidence bound* or *confidence limit* for θ. A one-sided confidence bound for θ is based on the acceptance region of a test of a hypothesis $H_0 : \theta = \theta_0$ against an alternative such as $\theta \geqslant \theta_0$. Similarly, a two-sided confidence interval (involving both an upper and a lower confidence bound) is based on the acceptance region of a test of a hypothesis $H_0 : \theta = \theta_0$ against an alternative $\theta \neq \theta_0$. See Section 6.5.1.

In certain cases in which confidence intervals can be found only by iterative solution of an equation, the only point estimator, in any way associated with and contained in the interval, might be a corresponding 50% lower (or upper) confidence bound for θ, a *median unbiased* point estimator of θ. An example of this type of confidence interval is discussed in Section 5.2.4.

A general procedure for obtaining confidence intervals is to find a function of the sample observations and the parameter to be estimated that has a distribution independent of the parameter in question and any other unknown parameters. Then any probability statement about the function will give a probability statement about the interval. This technique can be successfully applied in many situations, though there are a number of cases in which it cannot be used because it is impossible to find the desired function.

If more than one such function can be found, it is preferable to base a lower (upper) bound for θ on the function giving the most *accurate* bounds [those having the lowest probability of including a value less than (greater than) any $\theta' < (>)\theta$]. From such functions one can form tests of hypotheses with highest *power*, or probability of rejecting a false hypothesis. For two-sided confidence intervals (or tests of hypotheses) or cases in which nuisance parameters exist, intervals that are most accurate (or tests that are most powerful) can often be found in classes of intervals (or tests) such as the class containing those that are *invariant* under transformations of location and/or scale in some space, or those that are *unbiased*. A confidence interval for θ is said to be unbiased if the probability of the interval's covering the true value of θ is greater than or equal to the probability of its covering any false value. A test of a hypothesis is unbiased if its probability of rejecting a false hypothesis is greater than or equal to its probability of rejecting one that is true. Ferguson (1967) gives conditions under which hypothesis tests are *uniformly most powerful* (*UMP*) in a given class of tests. A UMP test in any specified class is the one that rejects every false hypothesis with highest probability among all tests in the class. A *uniformly most accurate* (UMA) confidence interval in the specified class of intervals is based on the acceptance region of a hypothesis test that is UMP in the corresponding class of tests.

For the most part, the confidence intervals and tests discussed in Chapters 5 and 6, respectively, are associated with point estimators which have, or tend to have, smallest mean squared error among all estimators of the parameter of interest that have invariant risk, in some space, under transformations of location and/or scale. In other words they tend to be the minimum-variance estimators of their expected values. Usually these are functions of maximum-likelihood estimators. Optimum (non-randomized) lower confidence bounds, which are based only on discrete data and which are discussed in Sections 7.4.3 and 10.4.2, are not exact and are *shortest* on the average among bounds associated with the specified confidence coefficient, or *confidence level*. For most of the goodness-of-fit tests recommended in Section 7.1, the power has been compared with that of other candidate tests by Monte Carlo simulation techniques.

In classical statistical theory, it is assumed that θ is fixed and the interval $[L, U]$ is random. However, it is useful to note that any UMA confidence interval for θ in any specified class of intervals has associated with it a particular *prior density function* and a particular *posterior density function* for θ. The concepts of prior density function and posterior density function of a parameter are explained in Section 2.8 which describes Bayesian and decision-theoretic approaches to estimation; the relationship between the classical and the decision-theoretic approaches is used implicitly in Sections 5.1.2 and 10.4.1.

The concept of confidence intervals can be expanded to that of *confidence regions* for two parameters. A $100(1 - \alpha)\%$ confidence region is a region constructed from the sample in such a way that, if samples were repeatedly drawn and a region constructed for each sample, $100(1 - \alpha)\%$ of these regions, on the average, would include the true parameters (Mood and Graybill, 1963, p. 251).

Example 1.   Confidence intervals for the Mean of a Normal Distribution.
Let $X_1, X_2, \ldots, X_n$ be a random sample of size n from

$$f_X(x;\mu,\sigma^2) = \frac{1}{\sqrt{2\pi}\,\sigma}\exp\left\{-\frac{1}{2}\left[\frac{x-\mu}{\sigma}\right]^2\right\}, \quad -\infty < x < \infty,$$

where μ is an unknown parameter and σ^2 is assumed to be known. It was shown before that $\overline{X}$ is the maximum-likelihood estimator of μ, and that the p.d.f. of $\overline{X}$ is

$$f_{\overline{X}}(x;\mu,\sigma^2,n) = \frac{\sqrt{n}}{\sqrt{2\pi}\,\sigma}\exp\left\{-\frac{1}{2}\left[\frac{(x-\mu)\sqrt{n}}{\sigma}\right]^2\right\}, \quad -\infty < x < \infty;$$

that is, $(\overline{X}-\mu)\sqrt{n}/\sigma$ is distributed normally with a mean 0 and a variance 1. Consequently,

$$P\left\{U_{\alpha/2} \leqslant \frac{(\overline{X}-\mu)\sqrt{n}}{\sigma} \leqslant U_{1-\alpha/2}\right\} = 1-\alpha,$$

where $U_{\alpha/2}$ and $U_{1-\alpha/2}$ are two points on the abscissa of a normal distribution (mean 0 and variance 1) that cut off an area $\alpha/2$ from either tail of the distribution.

$$\therefore \qquad P\left\{\overline{X} - U_{\alpha/2}\frac{\sigma}{\sqrt{n}} \leqslant \mu \leqslant \overline{X} + U_{1-\alpha/2}\frac{\sigma}{\sqrt{n}}\right\} = 1-\alpha.$$

Thus a $100(1-\alpha)\%$ confidence interval for μ has been obtained. From results given in Ferguson (1967) it can be shown that two-sided confidence intervals based on $\overline{X}$ are uniformly most accurate among unbiased intervals.

Example 2. Confidence Intervals for the Variance of a Normal Distribution. Suppose now that σ^2 is unknown; then consider the quantity

$$W\sigma^2 = \sum_{i=1}^{n}\left(X_i - \overline{X}\right)^2;$$

that is,

$$W\sigma^2 = \sum_{i}^{n}\left(X_i - \mu - \overline{X} + \mu\right)^2$$

$$= \sum_{i}^{n}\left[(X_i-\mu)^2 - (\overline{X}-\mu)^2\right]$$

$$= \sum_{i}^{n} (X_i - \mu)^2 - n(\overline{X} - \mu)^2$$

or

$$W = \sum_{i}^{n} \left(\frac{X_i - \mu}{\sigma} \right)^2 - \left(\frac{\overline{X} - \mu}{\sigma/\sqrt{n}} \right)^2.$$

In view of the fact that the X_i's and the $\overline{X}$ are normally distributed with mean μ and variances σ^2 and σ^2/n, respectively, one can use the methodology of Section 3.3.5(b) to show that W has a chi-square distribution with $n-1$ degrees of freedom. Thus

$$P\left\{ \chi^2_{\alpha/2} \leqslant W \leqslant \chi^2_{1-\alpha/2} \right\} = 1 - \alpha$$

or

$$P\left\{ \chi^2_{\alpha/2} \leqslant \frac{\Sigma(X_i - \overline{X})^2}{\sigma^2} \leqslant \chi^2_{1-\alpha/2} \right\} = 1 - \alpha$$

or

$$P\left\{ \frac{\Sigma(X_i - \overline{X})^2}{\chi^2_{1-\alpha/2}} \leqslant \sigma^2 \leqslant \frac{\Sigma(X_i - \overline{X})^2}{\chi^2_{\alpha/2}} \right\} = 1 - \alpha,$$

where $\chi^2_{\alpha/2}$ and $\chi^2_{1-\alpha/2}$ are points on the abscissa of a chi-square distribution with $(n-1)$ degrees of freedom which cut off an area $\alpha/2$ on either tail of the distribution.

INTERVAL ESTIMATES FOR LARGE SAMPLES In Section 3.4.2 the large-sample properties of maximum-likelihood estimators were discussed. It was stated therein that $\hat{\theta}$, the MLE of θ, for some very general conditions on $f_X(x; \theta)$, was for large values of n approximately normally distributed about θ with a variance $V(\hat{\theta})$, where

$$V(\hat{\theta}) = -\frac{1}{n} \frac{1}{E\left\{ (d^2/d\theta^2)[\log f_X(x; \theta)] \right\}}.$$

Thus, for large samples, a $100(1-\alpha)\%$ confidence interval for θ can be constructed using the procedure discussed before.

In Section 3.4.2 it was also stated that the joint distribution of $\hat{\theta}_1, \hat{\theta}_2, \ldots,$ $\hat{\theta}_k$, the MLE's of θ_i, $i = 1, 2, \ldots, k$, for some very general conditions on $f_X(x; \theta_1, \theta_2, \ldots, \theta_k)$ and large values of n, was multivariate normal with means θ_i, $i = 1, 2, \ldots, k$ and a variance-covariance matrix given by $\mathbf{V}/n$. Also, $\mathbf{V}$

$= \mathbf{R}^{-1}$, and the elements of $\mathbf{R}$, r_{ij}, were given as

$$r_{ij} = - E \left[\frac{\partial^2}{\partial \theta_i \, \partial \theta_j} \log f_X(x; \theta_1, \theta_2, \ldots, \theta_k) \right], \qquad i, j = 1, 2, \ldots, k.$$

To obtain a confidence region for the parameters $\theta_1, \theta_2, \ldots, \theta_k$, one can make use of the fact that the following quantity has a chi-square distribution with k degrees of freedom:

$$Z = \sum_{i=1}^{k} \sum_{j=1}^{k} r_{ij} \left(\hat{\theta}_i - \theta_i \right) \left(\hat{\theta}_j - \theta_j \right).$$

The variate Z allows one to set up a confidence region for the θ_i, using the procedure discussed before.

This section can be concluded by quoting the following statement (Mood and Graybill, 1963, p. 267): "Large-sample confidence intervals and regions based on the maximum likelihood estimators will be smaller, on the average, than intervals and regions determined by any other estimators of the parameters." These results are used in the methods of Chapters 5 and 9.

3.5   ORDER STATISTICS

Order statistics play an important role in reliability theory, especially in life testing.

Let $X_1, X_2, \ldots, X_n$ be a random sample from a probability density function $f_X(x; \theta)$. Suppose now that the n observations are arranged in ascending order so that $X_{(1)} \leqslant X_{(2)} \leqslant \cdots \leqslant X_{(n)}$, where $X_{(1)}$ is the smallest observation and $X_{(n)}$ the largest. Clearly $X_{(1)}$ can be any one of the n X_i's. Then $X_{(1)}$ is called the *first order statistic*, whereas $X_{(n)}$ is called the nth *order statistic*. In general $X_{(i)}$ is called the ith *order statistic*, and it has $i - 1$ observations preceding it. The objective of this section is to study the properties of the ith order statistic, namely, its distribution and its moments. In life testing, where the random sample constitutes the various times to failure, $X_{(1)} \leqslant X_{(2)} \leqslant \cdots \leqslant X_{(n)}$, the observations do appear ordered, that is, the first observation is the smallest time to failure, and hence each observation is an order statistic.

3.5.1 Distribution of the ith Order Statistic

Consider the ith order statistic, $X_{(i)}$, which has arisen from a probability density $f_X(x; \theta)$ and a distribution function $F_X(x; \theta)$. Also, it is assumed that n observations have been recorded, and that one needs to find the probability density function of $X_{(i)}$, say $f_{X_{(i)}}(x;\,\boldsymbol{\theta})$.

Let E denote the event that the ith ordered observation $X_{(i)}$ lies between x and $x+dx$. This implies that $i-1$ observations occur before x, and $n-i$ observations after $x+dx$. Using the multinomial probabiliy mass function, discussed in Section 3.2.1, one has

$$P\{E\}=\mathrm{P}\{x\leqslant X_{(i)}\leqslant x+dx\}$$

$$=\frac{n!}{(i-1)!1!(n-i)!}[F_X(x)]^{i-1}f_X(x)dx[1-F_X(x)]^{n-i}.$$

In the limit, when $dx\rightarrow0$, one has by definition

$$f_{X_{(i)}}(x)=i\binom{n}{i}[F_X(x)]^{i-1}[1-F_X(x)]^{n-i}f_X(x).$$

Specifically, if $i=1$, $f_{X_{(1)}}(x)$ gives the probability density function of the first (smallest) order statistic as

$$f_{X_{(1)}}(x)=n[1-F_X(x)]^{n-1}f_X(x),$$

and, if $i=n$, $f_{X_{(n)}}(x)$ gives the probability density function of the last (largest) order statistic as

$$f_{X_{(n)}}(x)=n[F_X(x)]^{n-1}f_X(x).$$

The distribution functions of $X_{(1)}$ and $X_{(n)}$ can be easily obtained since

$$F_{X_{(1)}}(x)=P(X_{(1)}\leqslant x)=1-P(X_{(1)}\geqslant x).$$

But

$$P(X_{(1)}\geqslant x)=P(X_{(1)}\geqslant x,X_{(2)}\geqslant x,\ldots,X_{(n)}\geqslant x).$$

$$\therefore\qquad F_{X_{(1)}}(x)=1-[1-F_X(x)]^{n}.$$

Now consider $F_{X_{(n)}}(x)$.

$$F_{X_{(n)}}(x)-\mathrm{Pr}(X_{(n)}\leqslant x)-\mathrm{Pr}(X_{(1)}\leqslant x,X_{(2)}\leqslant x,\ldots,X_{(n)}\leqslant x).$$

$$\therefore\quad F_{X_{(n)}}(x)=[F_X(x)]^{n}.$$

To illustrate the application of the results developed above to reliability problems, the following model is discussed.

A CHAIN MODEL. For a chain that consists of n small links, it is obvious that the strength of the chain cannot be greater than the strength of the weakest link. In effect, the chain breaks when its weakest link breaks. Suppose that the life-length distribution of each link, $f_X(x;\lambda)$, is an exponential distribution (discussed in Chapters 4 and 5). Then the life distribution of the chain $F_{X_1}(x)$ will be given by the distribution of the smallest order statistic. Thus

$$F_{X_1}(x) = 1 - [1 - F_X(x)]^n.$$

Since $1 - F_X(x) = e^{-\lambda x}$, it follows that

$$F_{X_1}(x) = 1 - e^{-\lambda n x}.$$

By taking a first derivative, one can obtain the probability density function of the life-length distribution as

$$f_X(x) = \lambda n e^{-\lambda n x}.$$

This is again an exponential distribution with a parameter λn. This property—that the smallest order statistic from an exponential distribution also has an exponential distribution—is often referred to as the *self-reproducing* property of the exponential distribution. Other distributions (e.g., the Weibull; see Chapter 4) also have this property.

It is of interest to compute the moments of $f_{X_{(1)}}(x)$ and to compare them with the moments of $f_X(x)$.

Recall that when X has an exponential density the mean and the variance of X (see Problem 3.3) are

$$E(X) = \frac{1}{\lambda} = \theta \quad \text{and} \quad \text{Var}(X) = \frac{1}{\lambda^2} = \theta^2.$$

It can easily be shown that the mean and the variance of $X_{(1)}$ are

$$E(X_{(1)}) = \frac{1}{n\lambda} = \frac{\theta}{n} \quad \text{and} \quad \text{Var}\left(X_{(1)} = \frac{1}{n^2\lambda^2}\right) = \left(\frac{\theta}{n}\right)^2.$$

Hence

$$E(X) = nE(X_{(1)}) \quad \text{and} \quad \text{Var}(X) = n^2\text{Var}(X_{(1)}).$$

It can also be shown that the life-length distribution of the strongest link, $X_{(n)}$, is given by

$$F_{X_{(n)}}(x) = [F_X(x)]^n = (1 - e^{-\lambda x})^n$$

and

$$f_{X_{(n)}}(x) = n\lambda(1 - e^{-\lambda x})^{n-1}e^{-\lambda x}.$$

3.5.2 Joint Distribution of the First r Order Statistics

Using the notation established in Section 3.5.1, one obtains

$$P\{E\} = P\{x_1 \leqslant X_{(1)} \leqslant x_1 + dx_1,$$

$$x_2 \leqslant X_{(2)} \leqslant x_2 + dx_2, \ldots, x_n \leqslant X_{(r)} \leqslant x_r + dx_r\},$$

where the event E here denotes the occurrence of no observations prior to x_1, the occurrence of the first observation between x_1 and $x_1 + dx_1$, no observations between $x_1 + dx_1$ and x_2, and so on, and finally the occurrence of $n - r$ observations after $x_r + dx_r$. Again, using the multinomial probability mass function, one has

$$P(E) = \frac{n!}{0!1!0!\ldots(n-r)!} f_X(x_1)dx_1 f_X(x_2)dx_2, \ldots f_X(x_r)dx_r[1 - F(x_r)]^{n-r}$$

or

$$f_{X_{(1)}, X_{(2)}, \ldots, X_{(r)}}(x_1, x_2, \ldots, x_r) = \frac{n!}{(n-r)!}[1 - F(x_r)]^{n-r}\prod_{i=1}^{r} f_X(x_i).$$

If one chooses to order all the n observations, that is, if $r = n$, then

$$f_{X_{(1)}, X_{(2)}, \ldots, X_{(n)}}(x_1, x_2, \ldots, x_n) = n!\prod_{i=1}^{n} f_X(x_i).$$

This result implies that ordering destroys the independence property of a random sample.

3.6 THE GENERALIZED GAUSS-MARKOV THEOREM AND THE LEAST-SQUARES PRINCIPLE

As mentioned earlier, the method of least squares is another means of parameter estimation. This method was not presented in Section 3.4 because it was necessary to discuss order statistics before considering certain aspects of least-squares estimation. The generalized Gauss-Markov

Theorem, discussed here, allows for estimation of parameters under conditions for which the methods of Section 3.4 are often not applicable. In this section it is assumed that the reader has some knowledge of matrix algebra.

Suppose that there exists a vector of n observable variates, $Y' = (Y_1, Y_2, \ldots, Y_n)$, with covariance matrix Σ. (In other words, the ith diagonal element of Σ is the variance of Y_i, $i = 1, 2, \ldots, n$, while the element in the ith row and jth column, $i \neq j$, is the covariance between Y_i and $Y_j, i, j = 1, 2, \ldots, n$.) Also suppose that Σ is of the form $\theta^2 \mathbf{B}$ with $\mathbf{B}$ known and θ an unknown scalar factor. The matrix $\mathbf{B}$ is symmetric, positive definite, and nonsingular.

Now define

$$e_i = \frac{Y_i - \sum_{j=1}^{p} x_{ji} \beta_j}{\theta}, \quad i = 1, 2, \ldots, n,$$

where the $\{x_{ji}\}$ (with x_{ji} the element in the ith row and jth column of a matrix $\mathbf{x}$) are known constants, and $\beta_1, \beta_2, \ldots, \beta_p$ are p unknown quantities. The following additional assumptions are made:

$$Y_1 = \beta_1 x_{11} + \beta_2 x_{21} + \cdots + \beta_p x_{p1} + e_1 \theta,$$

$$Y_2 = \beta_1 x_{12} + \beta_2 x_{22} + \cdots + \beta_p x_{p2} + e_2 \theta,$$

$$\cdot$$
$$\cdot$$
$$\cdot$$

$$Y_n = \beta_1 x_{1n} + \beta_2 x_{2n} + \cdots + \beta_p x_{pn} + e_n \theta,$$

with

$$E(e_i) \equiv x_{p+1, i}, \quad i = 1, 2, \ldots, n.$$

(In other words, it is assumed that

$$E(Y_i) = \sum_{j=1}^{p+1} x_{ji} \beta_j, \quad i = 1, 2, \ldots, n,$$

with $\beta_{p+1} = \theta$.) One can now determine *linear estimators* of the elements of β, that is, estimators that are linear in the observable Y_i's.

Under the preceding assumptions, the unique estimators of the elements of $\beta' = (\beta_1, \beta_2, \ldots, \beta_p, \beta_{p+1})$ which minimize the expression $(\mathbf{Y} - \mathbf{x}\beta)'$ $\mathbf{B}^{-1}(\mathbf{Y} - \mathbf{x}\beta)$ are the Gauss-Markov *least-squares* estimators given by $\beta^* = (\mathbf{x}'\mathbf{B}^{-1}\mathbf{x})^{-1}\mathbf{x}'\mathbf{B}^{-1}\mathbf{Y}$, where $\mathbf{x}'$ denotes the transpose of $\mathbf{x}$. The covariance matrix of $\boldsymbol{\beta}^*$ is equal to $(\mathbf{x}'\mathbf{B}^{-1}\mathbf{x})^{-1}\theta^2$, where $\boldsymbol{\beta}^*$ consists of minimum-variance unbiased linear estimators of β_j, $j = 1, \ldots, p+1$. A proof of this result, the generalized Gauss-Markov theorem, is given by Aitken (1935).

Special cases of this theorem are now considered, and simplified expressions given for $\boldsymbol{\beta}^*$ in each case.

3.6.1 A Linear Regression Model

Often the $\{e_i\}$ are assumed to be errors in observing Y_i, with $E(e_i) = 0$, $i = 1, 2, \ldots, n$, whereas the $\{x_{ji}\}$ are assumed to be observed without error. Usually for this model, one supposes that the covariance matrix of the vector of observable variates, $\mathbf{Y}' = (Y_1, Y_2, \ldots, Y_n)$, is of the form $\sigma^2 \mathbf{I}$, with $\mathbf{I}$, the identity matrix, consisting of 1's in the diagonal and 0's elsewhere. In other words, assume that each Y_i, $i = 1, 2, \ldots, n$, is an independent observable variate with variance σ^2. In such cases the expression for $\boldsymbol{\beta}^*(p \times 1)$ reduces to $(\mathbf{x}'\mathbf{x})^{-1}\mathbf{x}'\mathbf{Y}$ (with $\mathbf{x}$ $n \times p$), and the covariance matrix of $\boldsymbol{\beta}^*$ is $(\mathbf{x}'\mathbf{x})^{-1}\sigma^2$. If the model is simply $Y_i = \beta_1 + \beta_2 x_i + e_i \sigma$, with $E(e_i) = 0$ or, equivalently, $E(Y_i) = \beta_1 + \beta_2 x_i$, $i = 1, 2, \ldots, n$, then

$$\beta_1^* = \frac{\sum_{i=1}^{n} x_i^2 \sum_{i=1}^{n} Y_i + \sum_{i=1}^{n} x_i \sum_{i=1}^{n} x_i Y_i}{n \sum_{i=1}^{n} x_i^2 - \left(\sum_{i=1}^{n} x_i\right)^2} \quad \text{and} \quad \beta_2^* = \frac{\sum_{i=1}^{n} x_i Y_i - \sum_{i=1}^{n} x_i \sum_{i=1}^{n} Y_i / n}{\sum_{i=1}^{n} x_i^2 - \left(\sum_{i=1}^{n} x_i\right)^2 / n}.$$

If one lets $u_i = x_i - \sum_{k=1}^{n} x_k / n$, so that $\sum_{i=1}^{n} u_i = 0$ and the model becomes $E(Y_i) = \beta_1 + \beta_2 u_i$, $i = 1, \ldots, n$, then

$$\beta_1^* = \frac{\sum_{i=1}^{n} Y_i}{n} \quad \text{and} \quad \beta_2^* = \frac{\sum_{i=1}^{n} u_i Y_i}{\sum_{i=1}^{n} u_i^2}.$$

When more generally $\mathbf{x}$ is an $n \times p$ matrix, the optimum unbiased estimator of σ^2 (see Hsu, 1938) is

$$s_{y \cdot x}^2 = \frac{\sum_{i=1}^{n} \left(Y_i - \sum_{j=1}^{p} \beta_j^* x_{ji}\right)^2}{n - p}.$$

If it is known that the covariance between each pair of observations is zero but an assumption of equal variances cannot be made for $Y_1, Y_2, \ldots, Y_n$, the generalized Gauss-Markov theorem applies with $\mathbf{B}$ a diagonal matrix. If $\theta = 1$ and the ith diagonal element of $\mathbf{B}$ is denoted by σ_i^2, and if $E(e_i) = 0$, $i = 1, 2, \ldots, n$, the expression to be minimized to obtain the minimum-variance unbiased linear estimator of β_j, $j = 1, 2, \ldots, p$, is

$$\sum_{i=1}^{n} \sigma_i^{-2} \left(Y_i - \sum_{j=1}^{p} \beta_j x_{ji} \right)^2 .$$

In practice the $\{\sigma_i^2\}$ will normally not be known, but each σ_i^2 can be estimated by a sample variance, $\hat{\sigma}_i^2$, if two or more observations are made for each distinct value of σ_i^2. (See Section 5.4.1 for a discussion of the optimum estimation of σ_i^2.) Then one can obtain an unbiased weighted least-squares estimate of β_j, $j = 1, 2, \ldots, p$, by solving the system of equations given by

$$\sum_i (\hat{\sigma}_i^2)^{-1} Y_i x_{ki} - \sum_i \sum_j \beta_j^* x_{ji} x_{ki} (\hat{\sigma}_i^2)^{-1} = 0, \quad k = 1, 2, \ldots, p,$$

for β_j^*, $j = 1, 2, \ldots, p$. The variance of this estimator, however, is equal to $(\mathbf{x}'\mathbf{B}^{-1}\mathbf{x})^{-1}\theta^2$ only when $\hat{\sigma}_i^2 = \sigma_i^2$, $i = 1, 2, \ldots, n$, that is, when the $\{\sigma_i\}$ are known. An example of the use of weighted least-squares estimates is given in Chapter 9.

3.6.2 Estimates Based on Order Statistics from Distributions of Known Form

If a sample of size n is selected from a distribution of known form with a location and a scale parameter only, and if the first two moments of the order statistics for a size-n sample from the reduced (parameter-free) distribution have been calculated and are available, the generalized Gauss-Markov theorem is applicable. Here the assumption of a location parameter μ and a scale parameter σ implies that the distribution of the observable variate Y is such that

$$F_Y(y) = G\left(\frac{y - \mu}{\sigma} \right)$$

for some G, and that the ordered variate $Z_i = (Y_{(i)} - \mu)/\sigma$, with $Z_{i-1} \leqslant Z_i$, $i = 2, 3, \ldots, n$, has a parameter-free distribution. Thus

$$E(Y_{(i)}) = \mu + E(Z_i)\sigma$$

and

$$\mathrm{Cov}(Y_{(i)}, Y_{(j)}) = \mathrm{Cov}(Z_i, Z_j)\sigma^2, \quad i = 1, 2, \ldots, n; \ j = 1, 2, \ldots, n.$$

Therefore, if $E(Z_i)$ and $\text{Cov}(Z_i, Z_j)$, $i,j = 1, 2, \ldots, n$, are known, all conditions are satisfied for application of the generalized Gauss-Markov theorem. In applying the theorem, one can obtain sets of weights $\{a_{i,n}\}$ and $\{c_{i,n}\}$ such that

$$\mu^* = \sum_i a_{i,n} Y_{(i)}, \qquad \sigma^* = \sum_i c_{i,n} Y_{(i)}, \qquad \text{and} \qquad l\mu^* + m\sigma^*$$

are the unique minimum-variance linear unbiased estimators of μ, σ, and $\Phi = l\mu + m\sigma$, based on any specified selection of r of the n order statistics. An alternative approach is simply to minimize the expression

$$\sum_i \sum_j w_{i,n} w_{j,n} \text{Cov}(Y_{(i)}, Y_{(j)}),$$

the variance of

$$\sum_i w_{i,n} Y_{(i)} = l \sum_i a_{i,n} Y_{(i)} + m \sum_i c_{i,n} Y_{(i)},$$

under the constraint that

$$E\left[\sum_i w_{i,n} Y_{(i)}\right] = l\mu + m\sigma.$$

This approach (used first by Lieblein, 1954) leads to a system of $r + 2$ equations in $r + 2$ unknowns, the r values of the $\{w_{i,n}\}$ plus two Lagrange multipliers specifying the constraints

$$\sum_i w_{i,n} = l \qquad \text{and} \qquad E\left[\sum_i w_{i,n} Z_i\right] = m.$$

These two constraints represent four constraints on the coefficients of μ^* and σ^*, namely,

$$\sum_i a_{i,n} = 1, \quad E\left[\sum_i a_{i,n} Z_i\right] = 0, \quad \sum_i c_{i,n} = 0, \quad \text{and} \quad E\left[\sum_i c_{i,n} Z_i\right] = 1.$$

Coefficients $\{a_{i,n}\}$ and $\{c_{i,n}\}$ for obtaining Gauss-Markov best linear unbiased estimates of linear functions of location and scale parameters of several distributions useful in reliability analysis have been calculated for various sample sizes. These are discussed in Sections 5.2.2 and 5.4.2.

3.6.3. Improvement of the Estimators

The generalized Gauss-Markov theorem specifies unbiased linear estimators of linear functions of the unknown parameters that have the smallest expected squared deviation from the true parameter values. Suppose that we consider a class of linear estimators which contains as a subclass all unbiased linear estimators. We define this class as that containing all linear estimators with mean-squared error invariant under translations, that is, independent of all location parameters [where a location parameter μ of the distribution of X is such that $F_X(x) = G(x - \mu)$ for some G]. In the regression model given above, $\beta_1, \beta_2, \ldots, \beta_p$ are all location parameters and θ is a scale parameter.

Let us define as a general linear function $\Phi = l'\mu + m\sigma$, $l \neq 0$, which applies to the general model. The parameter Φ can represent $E(Y_i)$, $i = 1, 2, \ldots, n$, for the regression model; or, for the order-statistic model of Section 3.6.2, $\Phi = l\mu + m\sigma$, for l equal to 1 and m the $100P$ % point of the reduced variate Z, represents the $100P$ % point of the distribution of the random variable Y. For the order-statistic model, if m is zero, Φ is simply μ.

Let $A\sigma^2$ be the variance of $\Phi^* = l'\mu^* + m\sigma^*$; $C\sigma^2$ the variance of σ^*, the best linear unbiased estimator of σ; and $B\sigma^2$ the covariance between Φ^* and σ^*. Under these conditions, the following result holds. The estimators

$$\tilde{\sigma} = \frac{\sigma^*}{1 + C} \quad \text{and} \quad \tilde{\Phi} = \Phi^* - \left(\frac{B}{1 + C}\right)\sigma^*$$

are the unique minimum-mean-squared-error linear estimators of Φ and σ, respectively, with mean-squared error independent of μ. The mean-squared errors for $\tilde{\sigma}$ and $\tilde{\Phi}$ are $[C/(1 + C)]\sigma^2$ and $[A - B^2/(1 + C)]\sigma^2$. A proof of this is given by Mann (1969), and the result is applied in Sections 5.1.1, 5.1.2, and 5.2.2. It is particularly useful in certain cases when the sample size is very small or there is a great deal of censoring of the sample.

3.7 SOME LIMIT THEOREMS USEFUL IN RELIABILITY THEORY

Limit theorems play an important role in statistical theory and its applications; hence a brief review of some of these theorems is presented here. For a detailed discussion of limit theorems and their proofs, the reader is referred to Feller (1965).

Before these limit theorems are discussed, a description of the different types of convergence is useful.

CONVERGENCE ALMOST SURELY. Consider a sample description space S on which a probability function $P[\cdot]$ is defined. Let $X_1, X_2, \ldots, X_n$, and X be random variables defined on S.

Then the sequence of jointly distributed random variables $X_1, X_2, \ldots, X_n$ is said to converge *almost surely* (or to converge with probability 1) if

$$P\left[\lim_{n \to \infty} X_n = X\right] = 1.$$

In other words, for almost all members x of S on which the random variables are defined,

$$\lim_{n \to \infty} X_n(s) = X(s).$$

This type of convergence is often denoted as

$$\{X_n\} \overset{\text{a.s.}}{\to} X.$$

CONVERGENCE IN PROBABILITY. The sequence $X_1, X_2, \ldots, X_n, \ldots$ is said to converge in probability to X if, for all $\epsilon > 0$,

$$\lim_{n \to \infty} P\{|X_n - X| > \epsilon\} = 0.$$

This type of convergence is often denoted as

$$\{X_n\} \overset{P}{\to} X.$$

It can be proved (Parzen, 1960, p. 416) that, if

$$\{X_n\} \overset{\text{a.s.}}{\to} X, \quad \text{then } \{X_n\} \overset{P}{\to} X.$$

The converse of this, however, is not true.

CONVERGENCE IN DISTRIBUTION. Let $F_{X_n}(\cdot)$ be the distribution function of the random variable X_n, and $F_X(\cdot)$ be the distribution function of the random variable X.

The sequence $\{X_n\}$, $n = 1, 2, \ldots$, *converges in distribution* (or *law*) to the random variable X if, at every two points x and y, $x < y$, at which $F_X(\cdot)$ is continuous, as n tends to ∞,

$$P[x < X_n \leqslant y] = F_{X_n}(y) - F_{X_n}(x) \to F_X(y) - F_X(x)$$

$$= P[x < X \leqslant y].$$

This type of convergence is often denoted as

$$\{X_n\} \overset{L}{\to} X \qquad \text{or} \qquad F_{X_n}(\cdot) \overset{L}{\to} F_X(\cdot).$$

Alternatively,

$$\{X_n\} \overset{L}{\to} X \quad \text{if, as } n \to \infty, \ F_{X_n}(t) \overset{L}{\to} F_X(t)$$

for all points t at which $F_X(t)$ is continuous.

3.7.1 Chebychev's Inequality

This inequality, named after the Russian probabilist Chebychev, is a frequently used tool in statistics. Though Chebychev's inequality does not fall precisely into the general category of limit theorems, it is presented here because it is often used to prove limit theorems. In effect, this inequality formally states the fact that a small variance implies that large deviations from the mean are improbable.

Let X be a random variable (discrete or continuous) with a finite mean $E(X) = \mu$, and a finite variance $E(X - \mu)^2 = \sigma^2$. Then, for any $t > 0$,

$$P\{|X - \mu| > t\sigma\} \leqslant \frac{1}{t^2}.$$

3.7.2. The Weak Law of Large Numbers

Let $X_1, X_2, \ldots, X_n$ be a sequence of random variables, independent but not necessarily identically distributed, each having a finite mean $\mu_i = E(X_i)$, $i = 1, 2, \ldots, n$, and each having a variance σ_i^2, where

$$\frac{1}{n} \sum_{i=1}^{n} \sigma_i^2 \to 0, \quad \text{as } n \to \infty.$$

Then, according to the weak law of large numbers, for any $\epsilon > 0$,

$$\lim_{n \to \infty} P\left\{ \left| \frac{1}{n} \sum_{i=1}^{n} X_i - \frac{1}{n} \sum_{i=1}^{n} \mu_i \right| > \epsilon \right\} = 0.$$

This result can be proved using Chebychev's inequality, and this law of large numbers is also known as the Chebychev form of the law of large numbers.

3.7.3 The Strong Law of Large Numbers

Let the random variables $X_1, X_2, \ldots, X_n$ be independent and identically distributed with a finite mean $\mu = E(X_i)$, $i = 1, 2, \ldots, n$. Then, according to the strong law of large numbers,

$$P\left\{ \lim_{n \to \infty} \frac{1}{n} \sum_{i=1}^{n} X_i = \mu \right\} = 1.$$

The strong law of large numbers was first formulated by Cantelli, after Borel had discussed certain special cases.

3.7.4 The Central Limit Theorem

The laws of large numbers stated above establish an approximation of sums of independent random variables. Often, in practice, one needs an indication as to how frequently the sum in question deviates from a certain constant by a specified amount. The answer to this question is provided by the *central limit theorem*. In effect, the central limit theorem refers to the convergence in distribution of the sum of independent random variables.

The following version of the central limit theorem for independent, identically distributed random variables is a special case of a very general theorem that was proved by Lindberg (see Feller, 1965, p. 256).

Let $\{X_k\}$ be a sequence of mutually independent, identically distributed random variables with a finite common mean, $\mu = E(X_k)$, and a finite common variance, $\sigma^2 = \mathrm{Var}(X_k)$. Then, for every fixed β, as $n \to \infty$,

$$P\left\{ \frac{S_n - n\mu}{n^{1/2}\sigma} < \beta \right\} \to \frac{1}{\sqrt{2\pi}} \int_{-\infty}^{\beta} e^{-x^2/2} \, dx,$$

where

$$S_n = X_1 + X_2 + \cdots X_n.$$

If the members of the sequence of random variables $\{X_k\}$ are independent but not identically distributed, satisfaction of the Lyapunov condition provides a similar result. In practice, it is easier to apply Lyapunov's condition than Lindberg's condition, which is stated in the general form of Lindberg's theorem. It is for this reason that Lyapunov's condition for the convergence in distribution of independent but not identically distributed random variables is given here.

Let $\{X_n\}$ be a sequence of mutually independent random variables, each having a finite mean $E(X_n)$, and the finite $(2+\delta)$th central moment $\mu(2 \pm \delta; n)$, for some $\delta > 0$.

Then Lyapunov's condition requires that

$$\lim_{n\to\infty} \frac{1}{\sigma^{2+\delta}[S_n]} \sum_{k=1}^{n} \mu(2+\delta;k) = 0,$$

where

$$\mu(2+\delta;k) = E\big[|X_k - E(X_k)|^{2+\delta}\big]$$

and

$$\sigma^2[S_n] = \sum_{k=1}^{n} \text{Var}(X_k).$$

If Lyapunov's condition is satisfied, then, for every fixed β, as $n\to\infty$,

$$P\left\{ \frac{S_n - n\mu}{n^{1/2}\sigma} < \beta \right\} \to \frac{1}{\sqrt{2\pi}} \int_{-\infty}^{\beta} e^{-x^2/2}\,dx.$$

For many practical problems, satisfaction of Lyapunov's condition is often checked by choosing $\delta = 1$. The Lyapunov condition is often needed in reliability work when one wishes to ascertain the asymptotic normality of certain estimators. For example, this condition is used in Chapter 9 on accelerated life testing.

3.8 EXTREME-VALUE THEORY

In this section, some elements of the theory of extreme values are presented. The relevance of this theory to the failure of materials due to fracture or fatique, to the breakdown of dielectrics, and to the corrosion of metals has been explored by Epstein (1960) in a very comprehensive paper. In effect, an application of this theory attempts to explain in a systematic way why certain distributions may be expected to occur in connection with the fracture of materials. For a better appreciation of the theory of extreme values in all its facets, the reader is referred to Gumbel (1958) or to the classic expository paper on this subject by Gnedenko (1943).

3.8.1 The Distribution of Smallest Values

In Section 3.5.1, the exact distribution of the first order statistic was derived. Specifically, it was shown that

$$F_{X_{(1)}}(x) = 1 - [1 - F_X(x)]^n,$$

and this result was applied to the case of an exponential distribution (see the chain model in Section 3.5.1). Often the function $F_X(x)$ is not of a simple form; for example, powers of the normal distribution function are not easy to work with, and hence the determination of $F_{X_{(1)}}(x)$ is an involved problem. One can overcome this problem by using a technique given by Cramér (1946, p. 371), which applies when it is reasonable to assume that n is large. In fracture problems, this assumption is reasonable since n, the number of flaws, is large.

Define a random variable η_n as

$$\eta_n = nF_X(X_{(1)}).$$

Let

$$\Gamma_n(u) = P(\eta_n \leqslant u), \quad 0 \leqslant u \leqslant n.$$

$$\therefore \quad \Gamma_n(u) = P\left[X_{(1)} \leqslant F_X^{-1}\left(\frac{u}{n}\right)\right] = F_{X_{(1)}}\left[F_X^{-1}\left(\frac{u}{n}\right)\right].$$

Hence

$$\Gamma_n(u) = 1 - \left(1 - \frac{u}{n}\right)^n,$$

And this is true for any $F_X(x)$.

Let

$$\Gamma(u) = \lim_{n \to \infty} \Gamma_n(u) = \lim_{n \to \infty} \left[1 - \left(1 - \frac{u}{n}\right)\right]^n = 1 - e^{-u}, \quad u \geqslant 0.$$

It follows from the above result that, since the sequence of distribution functions $\Gamma_n(u)$ converges to $1 - e^{-u}$, the sequence of random variables η_n converges in distribution to a random variable, say η.

The probability density function of η is obtained by taking the derivative of $1 - e^{-u}$; thus

$$f_\eta(u) = \Gamma'(u) = e^{-u}, \quad u \geqslant 0.$$

Since $\eta_n = nF_X(X_{(1)})$, it is clear that the sequence of random variables $X_{(1)}$ converges in distribution to a random variable Y, where

$$Y = F_X^{-1}\left(\frac{\eta}{n}\right),$$

with the random variable η as defined before.

Thus the limiting distribution of the first order statistic $X_{(1)}$ is given by the distribution of Y. The following examples, chosen from Epstein (1960), are illustrative of the foregoing concepts

Example 1. Let

$$f_X(x) = \frac{1}{A}, \quad 0 \leqslant x \leqslant A,$$

$$= 0, \quad \text{otherwise.}$$

Also let

$$\eta_n = nF_X(X_{(1)}) = \frac{nX_{(1)}}{A}$$

or

$$X_{(1)} = \frac{A}{n}\eta_n.$$

Since the limiting distribution of η_n is the distribution of η,

$$X_{(1)} \xrightarrow{L} \frac{A}{n}\eta.$$

Since the probability density function of η is

$$f_\eta(x) = e^{-x},$$

it follows by a transformation of variables that the limiting density function of $X_{(1)}$ is

$$f_{X_{(1)}}(x) = \frac{n}{A}e^{-nx/A}, \quad x \geqslant 0,$$

$$= 0, \quad \text{otherwise,}$$

and that the limiting distribution of $X_{(1)}$ is

$$F_{X_{(1)}}(x) = 1 - e^{-(n/A)x}, \quad x \geqslant 0.$$

Example 2. Let

$$f_X(x) = \lambda e^{-\lambda x}, \quad \lambda > 0, \quad x \geqslant 0,$$

$$= 0, \quad \text{otherwise.}$$

As before, let $\eta_n = nF_X(X_{(1)})$. Hence

$$\eta_n = n(1 - e^{-\lambda X_{(1)}})$$

or

$$X_{(1)} = \frac{1}{\lambda}\log_e\left[\frac{1}{1-(\eta_n/n)}\right] = \frac{1}{\lambda}\left[\frac{\eta_n}{n} + \left(\frac{\eta_n}{n}\right)^2 + \left(\frac{\eta_n}{n}\right)^3 + \cdots\right].$$

As $n\to\infty$, $\eta_n \xrightarrow{L} \eta$, and, ignoring all powers greater than 1, one has

$$X_{(1)} \xrightarrow{L} \frac{\eta}{n\lambda}.$$

Since the asymptotic probability density function of η is of the form e^{-x}, it follows by a transformation of variables that

$$f_{X_{(1)}}(x) = n\lambda e^{-n\lambda x}, \quad x \geqslant 0,$$

$$= 0, \quad \text{elsewhere,}$$

and

$$F_{X_{(1)}}(x) = 1 - e^{-n\lambda x}, \quad x \geqslant 0.$$

This result is identical to that obtained in the example of Section 3.5.1, where the exact distribution of $X_{(1)}$ was obtained.

Example 3. Let

$$f_X(x) = \frac{\beta(x-\gamma)^{\beta-1}}{\alpha^\beta}, \quad \gamma \leqslant x \leqslant \alpha+\gamma; \ \alpha,\beta>0; \ \gamma \geqslant 0,$$

$$= 0, \quad \text{otherwise.}$$

It is easy to verify that for this density

$$F_X(x) = \left(\frac{x-\gamma}{\alpha}\right)^\beta, \quad \gamma \leqslant x \leqslant \alpha+\gamma.$$

The substitution $\eta_n = nF_X(X_{(1)})$ gives

$$X_{(1)} = \alpha\left(\frac{\eta_n}{n}\right)^{1/\beta} + \gamma,$$

and for large n it follows that

$$f_{X_{(1)}}(x) = n\beta \left(\frac{x-\gamma}{\alpha} \right)^{\beta-1} \exp\left[-n\left(\frac{x-\gamma}{\alpha} \right)^{\beta} \right]$$

and

$$F_{X_{(1)}}(x) = 1 - \exp\left[-n\left(\frac{x-\gamma}{\alpha} \right)^{\beta} \right], \quad x \geqslant \gamma.$$

The above form of the distribution of $X_{(1)}$ is known as the Weibull distribution and arises in the statistical theory of strength (Weibull, 1939).

Alternatively, the Weibull distribution can also be derived from other considerations. For example, the exact distribution of the first order statistic from a Weibull distribution is also a Weibull. Let

$$F_X(x) = 1 - \exp\left[-\left(\frac{x-\gamma}{\alpha} \right)^{\beta} \right], \quad x \geqslant \gamma;\ \alpha, \beta > 0;\ \gamma \geqslant 0,$$

$$= 0, \quad \text{otherwise.}$$

Clearly, $F_X(x)$ is a Weibull distribution. Since

$$F_{X_{(1)}}(x) = 1 - [1 - F_X(x)]^n,$$

it follows that

$$F_{X_{(1)}}(x) = 1 - \exp\left[-n\left(\frac{x-\gamma}{\alpha} \right)^{\beta} \right],$$

which is again a Weibull distribution.

Another derivation of the Weibull distribution is obtained by means of the hazard-rate concept. This is discussed in Section 4.4.

Gumbel (1958) calls the limiting distribution obtained in the above three examples the *Type III asymptotic distribution of the smallest extreme*. These distributions arise when two conditions are met.

1. The range over which the underlying density function is defined is bounded from below, that is, $F_X(x) \equiv 0$, for $x \leqslant \gamma$, for some finite γ.

2. $F_X(x)$ behaves like $\alpha(x-\gamma)^{\beta}$, for some $\alpha, \beta > 0$, as $x \to \gamma$.

If the range of the density is unlimited from below, and if, for some $\alpha, \beta > 0$,

$$\lim_{x \to \infty} (-x)^\alpha F_X(x) = \beta,$$

the limiting distribution of the smallest order statistic is referred to by Gumbel (1958) as the *Type II asymptotic distribution of the smallest extreme*. Such asymptotic distributions are not useful in failure studies, however, since $X_{(1)}$ is defined in the negative domain (see Problem 3.2.2).

If the underlying density function $f_X(x)$ is of such a form that it tends to zero exponentially as $x \to -\infty$, the limiting distribution of the first order statistic is referred to by Gumbel as the *Type I asymptotic distribution of the smallest extreme*.

For example, if

$$f_X(x) = \frac{1}{\sqrt{2\pi}} e^{-x^2/2}, \quad -\infty < x < \infty,$$

it has been shown by Cramér (1946, pp. 374–375) that the limiting distribution of the first order statistic $X_{(1)}$ is of the form

$$F_{X_{(1)}}(x) = 1 - \exp\left[-\exp\left(\frac{x-\gamma}{\alpha} \right) \right],$$

where γ and $\alpha > 0$ are constants, and $-\infty < x < \infty$.

The Type I asymptotic distribution was used by Gumbel extensively in his study of extremal phenomena. For this reason this family of distributions is also known as the *Gumbel distribution* (see Chapter 4).

To summarize this section, it can be stated that there are only three possible asymptotic types of distributions for the smallest order statistics (Fisher and Tippett, 1928). They are of the following forms:

Type I.

$$F_{X_{(1)}}(x) = 1 - \exp\left[-\exp\left(\frac{x-\gamma}{\alpha} \right) \right], \quad -\infty < x < \infty, \quad \alpha > 0$$

Type II.

$$F_{X_{(1)}}(x) = 1 - \exp\left[-\left(-\frac{x-\gamma}{\alpha} \right)^{-\beta} \right], \quad -\infty < x \leqslant \gamma, \quad \alpha, \beta > 0$$

Type III.

$$F_{X_{(1)}}(x) = 1 - \exp\left[-\left(\frac{x-\gamma}{\alpha}\right)^{\beta}\right], \quad \gamma \leqslant x < \infty, \; \alpha, \beta > 0.$$

These three types of distributions possess a *self-locking property*. In other words, if the underlying parent distribution function $F_X(x)$ is already in one of the forms given above, the distribution of the smallest order statistic $X_{(1)}$, for any n, is also of the same form, but with an appropriate change in the parameters. The proof of this result is left as an exercise to the reader.

3.8.2. The Distribution of Largest Values

In Section 3.5.1, the exact distribution of the largest order statistic was derived. Specifically, it was shown that

$$F_{X_{(n)}}(x) = [F_X(x)]^n.$$

In this section, procedures are developed for determining the asymptotic distribution of the largest order statistic. Here, again, a technique suggested by Cramér (1946, p. 371) is employed.

Define a random variable ξ_n as

$$\xi_n = n[1 - F_X(X_{(n)})].$$

Let

$$\Lambda_n(u) = P(\xi_n \leqslant u), \quad 0 \leqslant u \leqslant n.$$

$$\therefore \quad \Lambda_n(u) = P\{n[1 - F_X(X_{(n)})] \leqslant u\} = P\left[F_X(X_{(n)}) \geqslant 1 - \frac{u}{n}\right]$$

$$= P\left[X_{(n)} \geqslant F^{-1}\left(1 - \frac{u}{n}\right)\right]$$

or

$$\Lambda_n(u) = 1 - P\left[X_{(n)} \leqslant F^{-1}\left(1 - \frac{u}{n}\right)\right] = 1 - \left(1 - \frac{u}{n}\right)^n;$$

this result is true for any $F_X(x)$. Let

$$\Lambda(u) = \lim_{n\to\infty} \Lambda_n(u) = 1 - e^{-u}, \quad u \geqslant 0.$$

It follows from this result that, since the sequence of distribution functions $\Gamma_n(u)$ converges to $1 - e^{-u}$, the sequence of random variables ξ_n converges in distribution to a random variable ξ.

The probability density function of ξ is obtained by taking the derivative of $1 - e^{-u}$; thus

$$f_\xi(u) = e^{-u}, \quad u \geqslant 0.$$

Since $\xi_n = n(1 - F_{X_{(n)}})$, it is clear that the sequence of random variables $X_{(n)}$ converges in distribution to a random variable Y, where

$$Y = F_X^{-1}\left(1 - \frac{\xi}{n}\right).$$

The following examples illustrate the foregoing ideas.

Example 1. Let

$$f_X(x) = \frac{1}{A}, \quad 0 \leqslant x \leqslant A,$$

$$= 0, \quad \text{otherwise.}$$

Let

$$\xi_n = n[1 - F_X(X_{(n)})] = n\left(1 - \frac{X_{(n)}}{A}\right)$$

or

$$X_{(n)} = A - \frac{A\xi_n}{n}.$$

As $n \to \infty$, $\xi_n \overset{L}{\to} \xi$, and hence

$$X_{(n)} \overset{L}{\to} A - \frac{A\xi}{n}.$$

Since the limiting density function of ξ_n is e^{-u}, it can be shown by a simple transformation of variables that

$$f_{X_{(n)}}(x) = \frac{n}{A} \exp\left[\frac{-n(A-x)}{A}\right], \quad 0 \leqslant x \leqslant A,$$

$$= 0, \quad \text{elsewhere,}$$

and

$$F_{X_{(n)}}(x) = \exp\left[\frac{-n(A-x)}{A}\right], \quad 0 \leqslant x \leqslant A.$$

Gumbel calls the limiting distribution of the form obtained above the *Type III asymptotic distribution of the largest extreme.* Such distributions arise when two conditions are met.

1. The range over which the underlying density function is defined is bounded from above, that is, $F_X(x) \equiv 1$, for $x \geqslant \alpha$.
2. For some finite γ, $1 - F_X(x)$ behaves like $\alpha(2-x)^\beta$, for some $\alpha, \beta > 0$, as $x \to 2$.

If the range of the density is unlimited from above, and if, for some $\alpha, \beta > 0$,

$$\lim_{x \to \infty} x^\alpha [1 - F_X(x)] = \beta,$$

the limiting distribution of the largest-order statistic is referred to by Gumbel as the *Type II asymptotic distribution of the largest extreme* (see Problem 3.22).

If the underlying density function $f_X(x)$ is of such a form that it tends to zero exponentially as $x \to \infty$, the limiting distribution of the largest-order statistic is referred to by Gumbel as the *Type I asymptotic distribution of the largest extreme.*

Example 2. If

$$f_X(x) = \lambda e^{-\lambda x}, \quad x \geqslant 0, \lambda > 0,$$

$$= 0, \quad \text{otherwise,}$$

then

$$\xi_n = n[1 - F_X(X_{(n)})] = ne^{-\lambda X_{(n)}}$$

or

$$X_{(n)} \overset{L}{\to} \frac{\log n}{\lambda} - \frac{\log \xi}{\lambda}.$$

Since the asymptotic probability density function of ξ is e^{-u}, it follows that

$$f_{X_{(n)}}(x) = n\lambda e^{-\lambda x - n\exp(-\lambda x)}, \quad x \geqslant 0,$$

$$= 0, \quad \text{otherwise,}$$

and

$$F_{X_{(n)}}(x) = e^{-n\exp(-\lambda x)}, \quad x \geqslant 0.$$

This section can be summarized by noting that the limiting distributions of the largest-order statistic are exhausted by the following types (Fisher and Tippett, 1928):

Type I:

$$F_{X_{(n)}}(x) = \exp\left\{ -\exp\left[-\left(\frac{x-\gamma}{\alpha} \right) \right] \right\}, \quad -\infty < x < \infty, \quad \alpha > 0$$

Type II:

$$F_{X_{(n)}}(x) = \exp\left[-\left(\frac{x-\gamma}{\alpha} \right)^{-\beta} \right], \quad x \geqslant \gamma, \quad \alpha, \beta > 0$$

Type III:

$$F_{X_{(n)}}(x) = \exp\left[-\left(-\frac{x-\gamma}{\alpha} \right)^{\beta} \right], \quad x \leqslant \gamma, \quad \alpha, \beta > 0$$

Here, again, it can be shown that these distributions possess the self-locking (or *closure*) property discussed at the end of Section 3.8.1. For a rigorous treatment of the theory of extreme values the reader is referred to the work of Gnedenko (1943).

PROBLEMS

3.1. A random variable X has a Poisson probability mass function with a parameter μ. Find the mean and the variance of X. What is the generating function of X? Compute the mean and variance, using the generating function.

3.2. Let X_1 and X_2 be independently distributed Poisson random variables with parameters μ_1 and μ_2, respectively. What are the probability mass function and the moment generating function of $Y = X_1 + X_2$? If $Y = X_1 - X_2$, is the mass function of Y given by the Poisson law?

3.3.   A random variable X is said to have a *gamma probability density* function with parameters λ and β if

$$f_X(t) = \frac{e^{-\lambda t}(\lambda t)^{\beta-1}\lambda}{\Gamma(\beta)}, \quad t \geqslant 0; \lambda, \beta > 0.$$

What is the moment generating function of X? If $\beta = 1$, X is said to have an exponential p.d.f.

3.4.   What is the n-fold convolution of an exponential density with itself? Is this of a recognizable form?

3.5.   Under what conditions will the convolution of n independent random variables, each having a gamma density with parameters λ_i and β_i $(i = 1, 2, \dots, n)$, be of the gamma form? What are the parameters of the resulting gamma density?

3.6.   Let X have a binomial probability mass function with parameters n and p. Using the continuity theorem of the moment generating function, show that, for small p and large n, the mass function of X can be approximated by a Poisson-type mass, whereas for moderate p and large n it can be approximated by a normal density.

3.7.   A random variable X has a probability density function given by

$$f_X(x) = \tfrac{1}{2} \quad -\tfrac{1}{2} \leqslant x \leqslant \tfrac{3}{2}.$$

If $Y = X^2$, find the p.d.f. of Y.

3.8.   If a random variable X has a gamma probability density function with parameters λ and β (see Problem 3.3), show that the p.d.f. of $Y = 1/X$ is given by

$$f_Y(y) = \frac{e^{-\lambda/y}}{\Gamma(\beta)}\lambda^\beta \left(\frac{1}{y}\right)^{\beta+1}, \quad y > 0.$$

The above p.d.f. is often referred to as an *inverted gamma* probability density function. Show that the mean and the variance of Y are given by

$$\frac{\lambda}{\beta-1} \quad \text{for } \beta > 1 \quad \text{and} \quad \frac{\lambda^2}{(\beta-1)^2(\beta-2)} \quad \text{for } \beta > 2,$$

respectively.

3.9.   Show that the mean and the variance of Y, where Y has a beta probability density function

$$f_Y(y) = \frac{\Gamma(\alpha+\beta)}{\Gamma(\alpha)\Gamma(\beta)} y^{\alpha-1}(1-y)^{b-1}, \quad 0 < y < 1; \alpha, \beta > 0,$$

are

$$\frac{\alpha}{\alpha+\beta} \quad \text{and} \quad \frac{\alpha\beta}{(\alpha+\beta)^2(\alpha+\beta+1)},$$

respectively.

3.10.   Let X_1 and X_2 be independent random variables, each having a gamma probability density function with parameters λ_1, β_1 and λ_1, β_2, respectively, where

$$f_{X_i}(x) = \frac{e^{-\lambda_1 x}(\lambda_1 x)^{\beta_i-1}\lambda_1}{\Gamma(\beta_i)}, \quad i = 1, 2; x \geqslant 0; \lambda_1, \beta_i > 0.$$

Using the methods of Section 3.3.2, find the p.d.f. of $Y = X_1 + X_2$. Verify your result, using the moment generating function. Can this result be generalized?

3.11. Solve Problem 3.10 for X_1 and X_2 normally distributed with parameters μ_1, σ_1 and μ_2, σ_2, respectively.

3.12. Let X and Y be two random variables whose joint probability density function is given by

$$f_{X,Y}(x,y) = 24y(1-x), \quad 0 \leqslant x \leqslant 1, 0 \leqslant y \leqslant x.$$

Find the p.d.f. of $Z = Y/X$.

3.13. Find the expected value of $Z = XY$, where X and Y are jointly distributed as

$$f_{XY}(x,y) = xe^{-x(1+y)}, \quad x,y \geqslant 0.$$

Find the moment generating function of X, and verify your answer.

3.14. A random variable X is normally distributed with a mean μ and a variance σ^2, both of which are unknown. A random sample of size n is chosen from this density, the observable values being $X_1, X_2, \ldots, X_n$. The following two quantities can be computed:

$$\bar{X} = \frac{1}{n} \left(\sum_{i=1}^{n} X_i \right) \quad \text{and} \quad s^2 = \frac{1}{n-1} \sum_{i=1}^{n} (X_i - \bar{X})^2.$$

Find the probability density function of the random variable s^2.

3.15. For Problem 3.14, find the probability density function of the quantity

$$U = \frac{(\bar{X} - \mu)\sqrt{n}}{s}.$$

What would be the p.d.f. of U if s^2 were defined as

$$s^2 = \frac{\sum_{i=1}^{n} (X_i - \bar{X})^2}{n}?$$

3.16. What are the moment estimators of the parameters λ and β for the probability density function given in Problem 3.3?

3.17. *a.* For Problem 3.14, show that the maximum-likelihood estimator of σ^2 is biased. What is the unbiased estimator of σ^2?

b. Find the maximum-likelihood estimators of the parameters in the following two probability density functions:

$$f_X(x) = \binom{n}{x} p^x (1-p)^{n-x}, \quad x = 0, 1, 2, \ldots, n; n \text{ is known};$$

$$f_X(x) = \frac{e^{-\mu}\mu^x}{x!}, \quad x = 0, 1, 2, \ldots.$$

3.18. Find the maximum-likelihood estimators of the parameters in the following three probability density functions:

(a)

$$f_X(x) = \frac{1}{\eta}, \quad \gamma - \frac{\eta}{2} \leqslant x \leqslant \gamma + \frac{\eta}{2},$$

$$= 0, \quad \text{otherwise};$$

(b)

$$f_X(x) = \frac{1}{\theta} e^{-(x-a)/\theta}, \quad x \geqslant a, \theta > 0,$$

$$= 0, \text{ otherwise;}$$

(c)

$$f_X(x) = \frac{1}{2a} e^{-|x-b|/a}, \quad a > 0, b \geqslant 0,$$

$$= 0, \text{ otherwise.}$$

3.19. Find the 95% confidence limits for the parameter μ described in Problem 3.14.

3.20. For the probability density functions given in Problem 3.18 find the large-sample variances and convariances of the maximum-likelihood estimators of the parameters.

3.21. Find the distribution and the probability density function of the first and the last order statistics, respectively, from the following p.d.f.:

$$f_X(x) = \frac{1}{\eta}, \quad 0 \leqslant x \leqslant \eta,$$

$$= 0, \quad \text{elsewhere.}$$

3.22. Find the limiting (asymptotic) distribution of the smallest order statistic $X_{(1)}$ from the following probability density function:

$$f_X(x) = \frac{1}{x^2}, \quad x \leqslant -1,$$

$$= 0, \quad \text{otherwise.}$$

3.23. Find the limiting distribution of the largest order statistic $X_{(n)}$ from the following probability density function:

$$f_X(x) = 0, \quad x < 1,$$

$$= \frac{\alpha}{x^{\alpha+1}}, \quad x \geqslant 1, \alpha > 0.$$

REFERENCES

Aitken, A. C. (1935), On Least Squares and Linear Combinations of Observations, *Proceedings of the Royal Society of Edinburgh*, Vol. 55, p. 42.

Biometrika (1941), Tables of Percentage Points of the Incomplete Beta Function and of the Chi-Squared Distribution, Vol. 32.

Biometrika (1943), Tables of Percentage Points of the Inverted Beta Distribution, Vol. 33.

Cramér, H. (1946), *Mathematical Methods of Statistics*, Princeton University Press, Princeton, New Jersey.

Epstein, B. (1960), Elements of the Theory of Extreme Values, *Technometrics*, Vol. 2, No. 1, pp. 27–41.

Feller, W. (1965), *An Introduction to Probability Theory and Its Applications*, Vol. II, John Wiley and Sons, New York.

Ferguson, T. S. (1967), *Mathematical Statistics, A Decision Theoretic Approach*, Academic Press, New York.

Fisher, R. A. and L. H. C. Tippett (1928), Limiting Forms of the Frequency Distribution of the Largest or Smallest Member of a Sample, *Proceedings of the Cambridge Philosophical Society*, Vol. V, No. XXIV, Part II, pp. 180–190.

Gnedenko, B. V. (1943), Sur la distribution limite du terme maximum d'une serie aléatoire, *Ann. Math.*, Vol. 44, p. 423.

Gumbel, E. (1958), *Statistics of Extremes*, Columbia University Press, New York.

Hsu, P. L. (1938), On the Best Quadratic Estimate of the Variance, *Statistical Research Memoirs*, Vol. 2, pp. 91–104.

Kao, J. H. K. (1961), The Beta Distribution in Reliability and Quality Control, *Proceedings of the Seventh National Symposium on Reliability and Quality Control*, pp. 496–511.

Lieblein, J. (1954), A New Method of Analyzing Extreme Value Data, Technical Note 3053, National Advisory Committee for Aeronautics.

Mann, N. R. (1969), Optimum Estimators for Linear Functions of Location and Scale Parameters, *Annals of Mathematical Statistics*, Vol. 40, No. 6, pp. 2149–2155.

Mood, A. M. and F. Graybill (1963), *Introduction to the Theory of Statistics*, McGraw-Hill Book Company, New York.

Parzen, E. (1960), *Modern Probability Theory and Its Applications*, John Wiley and Sons, New York.

Rao, C. R. (1949), Sufficient Statistics and Minimum Variance Estimates, Proceedings of the Cambridge Philisophical Society, Vol. 45, pp. 213–217.

Weibull, W. (1939), A Statistical Theory of Strength of Materials, *Ingeniors Vetenskaps Akademien Handlingar*, No. 151; The Phenomenon of Rupture in Solids, *ibid.*, No. 153.

Widder, D. V. (1941), *The Laplace Transform*, Princeton University Press, Princeton, New Jersey.

CHAPTER 4

Statistical Failure Models

4.1. THE HAZARD-RATE CONCEPT

The measure of an equipment's reliability is the infrequency with which failures occur in time. A failure distribution represents an attempt to describe mathematically the length of the life of a material, a structure, or a device. There are many physical causes that individually or collectively may be responsible for the failure of a device at any particular instant. At present it is not possible to isolate these physical causes and mathematically account for all of them, and therefore, the choice of a failure distribution is still an art. If one tries to rely on actual observations of time to failure to distinguish among the various nonsymmetrical probability functions, he is still faced with a problem because nonsymmetric distributions are importantly different at the tails and actual observations are sparse, particularly at the right-hand tail, because of limited sample size.

In view of these difficulties, it is necessary to appeal to a concept that makes it possible to distinguish between the different distribution functions on the basis of a physical consideration. Such a concept is based on the failure-rate function, which is known in the literature of reliability as the *hazard rate*. In actuarial statistics the *hazard rate* goes under the name of *force of mortality*, in extreme-value theory it is called the *intensity function*, and in economics its reciprocal is called *Mill's Ratio*.

Let $F(x)$ be the distribution function of the time-to-failure random variable X, and let $f(x)$ be its probability density function. Then the hazard rate, $h(x)$, is defined as

$$h(x) = \frac{f(x)}{1 - F(x)}. \tag{4.1}$$

Here $1 - F(x)$ is called the reliability at time x and will be denoted by either $R(x)$ or $\overline{F}(x)$. The hazard rate, which is a function of time, has a

116

probabilistic interpretation; namely, $h(x)\,dx$ represents the probability that a device of age x will fail in the interval $(x, x + dx)$, or

$$h(x) = \lim_{\Delta x \to 0} \left[\frac{P\{\text{a device of age } x \text{ will fail in the interval } (x, x + \Delta x) \mid \text{it has survived up to } x\}}{\Delta x} \right].$$

On the basis of physical considerations, one is at liberty to choose the functional form of $h(x)$ for a particular device. Once this is done, a differential equation in $h(x)$ is obtained, from which $f(x)$ and $F(x)$ can be recovered.

To assist the choice of $h(x)$, three types of failures generally have been recognized as having a time characteristic. The first one, called the *initial failure*, manifests itself shortly after time $x = 0$ and gradually begins to decrease during the initial period of operation. A good example of this type is seen in the standard human mortality table, in which it is assumed that up to the age of 10 years a child can die of hereditary defects but, having lived past this age, is almost certainly free of such defects. The second one, called the *chance failure*, occurs during the period in which a device exhibits a constant failure rate, generally lower than that prevailing during the initial period. The cause of this failure is attributed to unusually severe and unpredictable environmental conditions occurring during the operating time of the device. For example, in human mortality tables, it is assumed that deaths between the ages of 10 and 30 years are generally due to accidents. The third type, called the *wear-out failure*, is associated with the gradual depletion of a material or with an accumulation of shocks, fatigue, and so on. Again, in human mortality tables, after the age of 30 years an increasing proportion of deaths are attributed to old age. The three types of failures have been classically represented by the *bath-tub curve* (see Figure 4.1), wherein each of the three segments of the curve represents one of the three time periods: initial, chance, and wear-out.

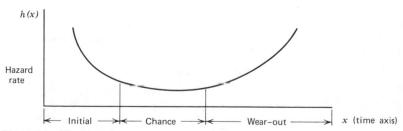

Figure 4.1. The bathtub curve.

It was stated above that, given the functional form of $h(x)$, the $f(x)$ and the $F(x)$ could be easily determined. Equations (4.2) and (4.3) below show this. It is assumed that $F(0^-) = 0$, and that $F(+\infty) = 1$. Since

$$\int_0^x f(s)\, ds = F(x),$$

$$\frac{d}{dx} F(x) = f(x).$$

Now, it follows from Equation (4.1) that

$$h(x)\, dx = \frac{dF(x)}{1 - F(x)}$$

or

$$\int_0^t h(x)\, dx = -\log[1 - F(x)]\big|_0^t.$$

Thus,

$$\log \frac{1 - F(t)}{1 - F(0)} = -\int_0^t h(x)\, d(x)$$

or

$$1 - F(t) = \exp\left[-\int_0^t h(x)\, dx \right]. \tag{4.2}$$

Taking derivatives, one obtains

$$f(t) = h(t)\exp\left[-\int_0^t h(x)\, dx \right]. \tag{4.3}$$

In the sections to follow, this technique will be used to develop the commonly used failure distributions. The hazard rates for some of these distributions are given in Figure 4.2; their moments and other properties are summarized in Table 4.1 at the end of this chapter. Methods of estimating the parameters of most of these distributions are discussed in Chapter 5.

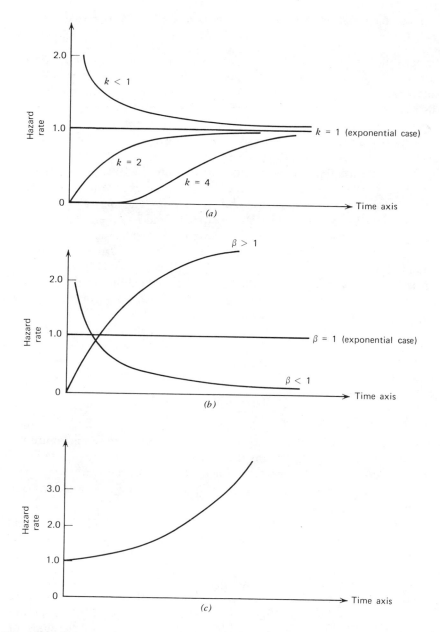

Figure 4.2 Hazard-rate curves for some commonly used failure distributions. (*a*) Gamma distribution for $\lambda = 1$. (*b*) Weibull distribution for $\gamma = 0$, $\eta = 1$. (*c*) Type I extreme-value distribution (smallest values), $\gamma = 0$, $\eta = 1$.

119

4.2. THE POISSON PROCESS AND THE EXPONENTIAL DISTRIBUTION

In reliability studies, the exponential distribution plays a role of importance analogous to that of the normal distribution in other areas of statistics. An acceptable justification for the assumption of an exponential distribution to life studies was initially discussed by Epstein (1953) and by Davis (1952). More recently, a mathematical argument has been advanced to support the plausibility of the exponential distribution as the failure law of complex equipment (Barlow and Proschan, 1965, p. 18). Although many life random variables cannot be adequately described by the exponential distribution, an understanding of the theory in the exponential case facilitates the treatment of more general situations. The desirability of the exponential distribution is due to its simplicity and its inherent association with the well-developed theory of Poisson processes. Also, many times certain quantities computed from the exponential distribution serve as bounds for similar quantities that need to be computed from other, less tractable distributions (see Section 7.3.2). The applicability of the exponential distribution is limited, however, because of its *lack of memory* property; this property requires that previous use does not affect future life, and the exponential distribution is the only distribution with this property (Feller, 1957, p. 413).

To indicate why the exponential distribution may occur, and what implications the assumption of an exponential distribution carries, it is necessary to discuss the Poisson process.

4.2.1 The Poisson Process

The exponential distribution corresponds to a purely random failure pattern; mathematically this means that, whatever be the cause of a failure, it occurs according to the postulates of a Poisson process with some parameter, λ. The Poisson probability law will now be briefly derived by means of differential equations.

Consider a system (a unit) subjected to instantaneous changes due to the occurrence of random events (shocks). All the random events are assumed to be of the same kind, and one is interested in their total number.

Let $P_m(t)$ be the probability that exactly m random events occur during a time interval of length t. Assume that, as $t \to 0$,

$$\frac{1 - P_0(t)}{t} = \lambda,$$

where λ is a positive constant, and $1 - P_0(t)$ is the probability of one or more occurrences of the random event. Thus, for a small interval of length

h, the probability of one or more occurrences of the random event is $1 - P_0(h) = \lambda h + o(h)$, where $o(h)$ represents a function $g(h)$, defined for $h > 0$ with the property that

$$\lim_{h \to 0} \frac{g(h)}{h} = 0.$$

The physical process that induces the occurrence of the random events is characterized by two postulates.

1. The process is time homogeneous, and future occurrences of the random event are independent of its past occurrences. Thus the probability of at least one event occurring in a time period of duration h is

$$p(h) = \lambda h + o(h), \quad \lambda > 0, \ h \to 0.$$

2. The probability of two or more occurrences during an interval h is $o(h)$.

These postulates lead to a system of differential equations for $P_m(h)$. Now $p(h) = P_1(h) + P_2(h) + P_3(h) + \cdots$ can be written as

$$p(h) = P_1(h) + \sum_{m=2}^{\infty} P_m(h).$$

But, by postulate 2 above,

$$\sum_{m=2}^{\infty} P_m(h) = o(h).$$

Now $P_m(h)$ will be calculated for every m. It is easy to see that

$$P_m(t+h) = P_m(t)P_0(h) + P_{m-1}(t)P_1(h) + \sum_{k=2}^{m} P_{m-k}(t)P_k(h).$$

But $P_0(h) = 1 - p(h)$, and $P_1(h) = p(h) - o(h)$. Also

$$\sum_{k=2}^{m} P_{m-k}(t)P_k(h) \leqslant \sum_{k=2}^{m} P_k(h) = o(h).$$

$$\therefore \quad P_m(t+h) = P_m(t)[1 - p(h)] + P_{m-1}(t)P_1(h) + o(h)$$

$$= P_m(t)[1 - \lambda h - o(h)] + P_{m-1}(t)P_1(h) + o(h),$$

using postulate 1. Or

$$\frac{P_m(t+h) - P_m(t)}{h} = -\lambda P_m(t) + \lambda P_{m-1}(t) + \frac{o(h)}{h}.$$

But

$$P'_m(t) = \lim_{h \to 0} \frac{P_m(t+h) - P_m(t)}{h} = -\lambda P_m(t) + \lambda P_{m-1}(t), \qquad m = 1, 2, \ldots .$$

$$(4.4)$$

For the case $m = 0$, one has

$$P_0(t+h) = P_0(t) P_0(h) = P_0(t)[1 - p(h)],$$

and therefore

$$\lim_{h \to 0} \frac{P_0(t+h) - P_0(t)}{h} = - \lim_{h \to 0} P_0(t) \frac{p(h)}{h}.$$

But, by postulate 1,

$$\lim_{h \to 0} \frac{p(h)}{h} = \lambda,$$

and therefore

$$P'_0(t) = -\lambda P_0(t).$$

$$(4.5)$$

Equations (4.4) and (4.5) are differential equations whose solutions may be obtained using Laplace transforms. Define the Laplace transform of $P(t)$, $\tilde{P}(s)$, as

$$\tilde{P}(s) = L[P(t)] = \int_0^\infty e^{-st} P(t) \, dt$$

and

$$\tilde{P}'(s) = L[P'(t)] = s\tilde{P}(s) - P(0).$$

Applying the definition of the Laplace transform for Equation (4.5), one has

$$s\tilde{P}_0(s) - P_0(0) = -\lambda \tilde{P}_0(s).$$

The initial conditions require that $P_0(0)=1$, and $P_m(0)=0$, $m=1,2,\ldots$.

$$\therefore \qquad \tilde{P}_0(s) = \frac{1}{s+\lambda}.$$

To find $P_0(t)$ one must take the inverse Laplace transform of $\tilde{P}_0(s)$. From a table of inverse Laplace transforms (Widder, 1941),

$$L^{-1}\left\{\tilde{P}_0(s)\right\} = e^{-\lambda t}.$$

Now $P'_m(t) = -\lambda P_m(t) + \lambda P_{m-1}(t)$, by Equation (4.4). Its Laplace transform is $s\tilde{P}_m(s) - P_m(0) = -\lambda\tilde{P}_m(s) + \lambda\tilde{P}_{m-1}(s)$. This can be solved recursively to give

$$\tilde{P}_m(s) = \frac{\lambda^m}{(s+\lambda)^{m+1}},$$

whose inverse Laplace transform gives us

$$P_m(t) = \frac{e^{-\lambda t}(\lambda t)^m}{m!}, \qquad m=0,1,2,\ldots. \qquad (4.6)$$

4.2.2 The Exponential Distribution

The probability density function (p.d.f.) of the exponential distribution can be obtained either from the hazard-rate concept or by considering the waiting time between arrivals in a Poisson process. The latter situation will be considered first.

Consider a situation wherein the device under consideration is subjected to an environment E, which is some sort of random process. Suppose that this process generates shocks, which are distributed according to the Poisson distribution, with a parameter λ. The device will fail only if a shock occurs and will not fail otherwise. One is interested in the random variable X, where X denotes the time interval between the successive occurrences of shocks. In the situation being discussed, X represents the time to failure of the device.

Let

$$R(x) = P(X \geqslant x) = P[\text{no shocks occurred during } (0,x)]$$

and let $x=0$ denote the time when the most recent shock occurred.

Since the shocks occur according to the postulates of a Poisson process, the probability that 0 shock occurs can be obtained from Equation (4.6) as $R(x) = e^{-\lambda x}$. Thus

$$P(X \leqslant x) = 1 - e^{-\lambda x},$$

and the p.d.f. of X is given by

$$f_X(x) = \lambda e^{-\lambda x}.$$

The same expression for the p.d.f. of X could be obtained from the hazard-rate concept, since the assumption of a random occurrence of shocks with a parameter λ implies a constant hazard rate, $h(x) = \lambda$, for $x \geqslant 0$. Now $f_X(x)$ can be obtained from Equation (4.3) as

$$f_X(x) = h(x) \exp\left[-\int_0^x h(s)\, ds \right]$$

or

$$f_X(x) = \lambda e^{-\lambda x},$$

and Equation (4.2) gives

$$F_X(x) = 1 - e^{-\lambda x}.$$

A more general form for the exponential distribution can be obtained if

$$h(x) = 0, \quad 0 \leqslant x < A,$$
$$= \lambda, \quad x \geqslant A.$$

Then

$$f_X(x) = \lambda e^{-\lambda(x-A)}, \quad x \geqslant A,$$
$$= 0, \quad x < A,$$

and

$$F_X(x) = 1 - e^{-\lambda(x-A)}, \quad x \geqslant A,$$
$$= 0, \quad x < A.$$

Often A is referred to as the *threshold* or the *shift* parameter. This section is concluded by emphasizing the fact that the exponential distribution can be chosen as a failure distribution if and only if the assumption of a constant hazard rate can be justified. This assumption implies that the failure of a device is due, not to its deterioration as a result of wear, but to random shocks that occur according to the postulates of a Poisson process. Conversely, it is also to be noted that the interarrival time between the epochs (shocks) of a Poisson process has an exponential distribution.

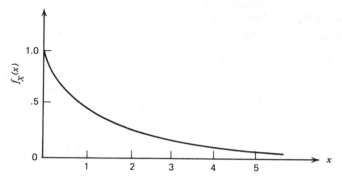

Figure 4.3 Unit exponential density function.

An exponential distribution with $A = 0$ and $\lambda = 1$, the *unit exponential density*, is illustrated in Figure 4.3. Its moments and other properties are summarized in Table 4.1 at the end of this chapter.

4.3 THE GAMMA DISTRIBUTION

The gamma distribution is a natural extension of the exponential distribution and has sometimes been considered as a model in life-test problems (Gupta and Groll, 1961). It can be derived by considering the time to the kth successive arrival in a Poisson process or, equivalently, by considering the k-fold convolution (see Section 3.1.3) of an exponential distribution. The gamma distribution is the continuous analog of the negative binomial distribution (Section 3.2.1), which can also be obtained by considering the sum of k variables with a common geometric distribution.

Consider a situation in which the unit under consideration operates in an environment where shocks are generated according to a Poisson distribution, with a parameter λ. Further suppose that the unit will fail only if exactly k shocks occur and will not fail until then. One is interested in the random variable $X^{(k)}$, where $X^{(k)}$ denotes the time for the occurrence of the kth shock. In the situation being considered, $X^{(k)}$ represents the time to failure of the unit.

To obtain the p.d.f. of $X^{(k)}$, $f_{X^{(k)}}(x)$, it is to be noted that

$$P[x < X^{(k)} < x + \Delta x] = P[\text{exactly } k - 1 \text{ shocks occur in } (0, x)$$

$$\text{and exactly 1 shock occurs in } (x, x + \Delta x)].$$

Since the number of shocks that occur in $[0, x]$ is given by the Poisson mass function, equation (4.6),

$$f(x)\,\Delta x = \lim_{\Delta x \to 0} P(x < X^{(k)} < x + \Delta X) = \frac{e^{-\lambda x}(\lambda x)^{(k-1)}}{(k-1)!}\lambda\,\Delta x.$$

Hence

$$f_{X^{(k)}}(x) = \frac{e^{-\lambda x}(\lambda x)^{k-1}\lambda}{\Gamma(k)}, \quad k \geqslant 1,\, x \geqslant 0, \tag{4.7}$$

where $\Gamma(k) = (k-1)!$ is the *gamma function*.
The gamma function is defined by

$$\Gamma(t) = \int_0^\infty e^{-s} s^{t-1}\,ds.$$

It generalizes the factorials in the sense that $\Gamma(k+1) = k!$, for $k = 0, 1, 2, \ldots$.
The distribution function of $X^{(k)}$, $F_{X^{(k)}}(x)$, can be obtained as follows:

$$1 - F_{X^{(k)}}(x) = P(X^{(k)} > x) = P[k-1 \text{ or fewer shocks in } (0, x)]$$

$$= \sum_{j=0}^{k-1} \frac{e^{-\lambda x}(\lambda x)^j}{j!}.$$

$$\therefore \qquad F_{X^{(k)}}(x) = \sum_{j=k}^\infty \frac{e^{-\lambda x}(\lambda x)^j}{j!}. \tag{4.8}$$

Since $X^{(k)}$ is the time to the kth shock, and since the time between the ith and the $(i-1)$th shock, X_i, has an exponential distribution with a parameter λ, one has

$$X^{(k)} = X_1 + X_2 \cdots + X_i \cdots + X_k.$$

The p.d.f. of $X^{(k)}$ is defined by the k-fold convolution of the X_i's (Feller, 1965, p. 46) as

$$f_{X^{(k)}}(x) = \left[f_{X_i}(x) \right]^{k*} = \int_0^x \left[f_{X_i}(x-s) \right]^{(k-1)*} f_{X_i}(s)\,ds,$$

where

$$f_{X_i}(x) = \lambda e^{-\lambda x}.$$

Accordingly,

$$\left[f_{X_i}(x)\right]^{2*} = \int_0^x f_{X_i}(x-s)f_{X_i}(s)\,ds = \int_0^x \lambda^2 e^{-(x-s)\lambda}e^{-\lambda s}\,ds$$

$$= \frac{\lambda^2 x e^{-\lambda x}}{\Gamma(2)}$$

In general, it can be deduced that

$$f_{X^{(k)}}(x) = \left[f_{X_i}(x)\right]^{k*} = \frac{\lambda^k x^{k-1}e^{-\lambda x}}{\Gamma(k)},$$

which is the *gamma probability density function* given by Equation (4.7).

Here $\lambda^{-1} > 0$ is the trivial scale parameter, but $k > 0$, the *shape parameter*, is essential. For integer values of k, the gamma p.d.f. is also known as the *Erlangian probability density function*; and, if $k = 1$, the gamma density reduces to an exponential density.

The gamma distribution with $\lambda = 1$ is illustrated in Figure 4.4 for different values of the shape parameter k. Its moments and other properties are summarized in Table 4.1.

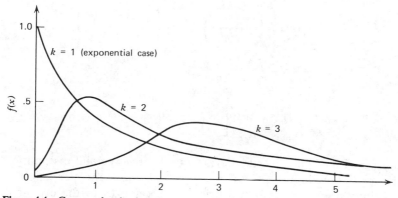

Figure 4.4 Gamma density function.

4.4 THE WEIBULL DISTRIBUTION

Recently the Weibull distribution has emerged as the most popular parametric family of failure distributions. Its applicability to a wide variety of failure situations was discussed by Weibull (1951); it has been used to describe vacuum-tube failures (Kao, 1958) and ball-bearing failures

(Lieblein and Zelen, 1956). Whereas the applicability of the exponential distribution is limited because of the assumption of a constant hazard rate, the family of Weibull distributions can be written to include increasing and decreasing hazard rates as well. Since many failures encountered in practice, especially those pertaining to nonelectronic parts, show an increasing failure rate (i.e., due to deterioration or wear), the Weibull distribution is useful in describing failure patterns of this type.

The Weibull distribution can be derived either from the hazard rate concept or as the asymptotic distribution of the smallest order statistic from a specified probability distribution function. The Weibull distribution was introduced in Section 3.8.1; in this section a derivation of this distribution, using an appropriate hazard rate, is illustrated.

If the hazard rate $h(x)$ is some power function of x, say

$$h(x) = \frac{\beta}{\eta} \left(\frac{x - \gamma}{\eta} \right)^{\beta - 1}, \quad \beta, \eta > 0; \ \gamma \geqslant 0; \ x \geqslant \gamma,$$

$$= 0, \quad x < \gamma.$$

equations (4.3) and (4.2) give

$$f_X(x) = \frac{\beta}{\eta} \left(\frac{x - \gamma}{\eta} \right)^{\beta - 1} \exp\left[-\left(\frac{x - \gamma}{\eta} \right)^{\beta} \right], \quad x \geqslant \gamma$$

and

$$F_X(x) = 1 - \exp\left[-\left(\frac{x - \gamma}{\eta} \right)^{\beta} \right], \quad x \geqslant \gamma.$$

As illustrated in Figure 4.2, the hazard rate for the Weibull distribution is decreasing (increasing) in $x - \gamma$ if $\beta < 1$ ($\beta > 1$), and is independent of x if $\beta = 1$. When $\beta = 1$ the Weibull distribution specializes to the exponential distribution, and when $\beta = 2$ the resulting distribution is known as the *Rayleigh distribution*; β is also known as the *shape parameter* and γ is the *location parameter*.

The derivation of the Weibull distribution as an asymptotic distribution of the smallest values from a distribution function $F(x)$, where

$$F(x) = \left(\frac{x - \gamma}{\eta} \right)^{\beta}, \quad \gamma \leqslant x \leqslant \eta + \gamma; \ \eta, \beta > 0; \ \gamma \geqslant 0,$$

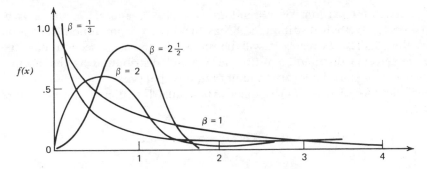

Figure 4.5 Weibull density function.

was shown in Section 3.8.1. It is clear that the Weibull distribution belongs to the family of Type III asymptotic distributions of the smallest extreme discussed in Section 3.8. Since the asymptotic distributions of extreme values possess a self-locking property (Section 3.8), it is true that the Weibull distribution can also be derived as an exact distribution of the smallest extreme from a Weibull distribution (or, equivalently, from a Type III asymptotic distribution of the smallest extreme).

A Weibull distribution with $\gamma = 0$ and $\eta = 1$ is illustrated for different values of the shape parameter β in Figure 4.5. Moments and other properties are summarized in Table 4.1.

4.5 THE GUMBEL (EXTREME-VALUE) DISTRIBUTION

The Type I asymptotic distribution of the smallest or the largest extreme is also known as the Gumbel distribution and is often used as a failure model for series and parallel systems, as well as in cases where failure is due to corrosive processes. In general, its applicability can be justified whenever the phenomenon causing failure depends on the smallest or the largest value of a variable, the underlying distribution of which is of the exponential type.[†] It is called the Gumbel distribution after Gumbel (1958), who used it extensively in the study of floods, aeronautics, meteorology, breaking strengths, geology, and naval engineering. Its application to air pollution problems is discussed by Singpurwalla (1972), and by Barlow and Singpurwalla (1973).

The Gumbel distribution for the smallest extreme can be derived either from the hazard-rate concept or by the methods of Section 3.8 by taking $F(x)$ to be some kind of an exponential function. It is also true that, if X is

[†]For a discussion of exponential type distributions see Gumbel (1958).

the natural logarithm of a random variable T, and if T has a two-parameter Weibull distribution, X has the Type I asymptotic distribution of the smallest extreme. It will be shown in the next section that the extreme-value distribution is to the Weibull distribution as the normal distribution is to the logarithmic normal distribution.

If the hazard rate $h(x)$ is some exponential function of x (see Figure 4.2), say

$$h(x) = e^x, \quad -\infty < x < \infty,$$

Equations (4.3) and (4.2) give

$$f_X(x) = \exp(x)\exp(-e^x), \quad -\infty < x < \infty, \tag{4.9}$$

and

$$F_X(x) = 1 - \exp[-\exp(x)], \quad -\infty < x < \infty. \tag{4.10}$$

A standard form for the Type I asymptotic distribution of the smallest extreme is given by Gumbel (1958, p. 159) as

$$F_X(x) = 1 - \exp\left[-\exp\left(\frac{x - \gamma}{\eta}\right)\right], \quad -\infty < x < \infty. \tag{4.11}$$

This is also the form given in Section 3.8.1.

The Gumbel distribution for the largest extreme can be derived from more fundamental arguments by considering a corrosion process (Lloyd and Lipow, 1962, p. 140).

Consider a surface that has microscopic pits of varying depths. Assume that the depths of these pits increase because of chemical corrosion, and that a failure occurs whenever any one pit depth penetrates the thickness of the surface. Assume also that the initial pit depths have an exponential distribution, and that the time to penetration is proportional to the difference between the thickness of the surface and the initial depth of the pit. Determining the time to failure distribution of the surface is the problem of interest.

Let T be the thickness of the surface, and let t_i be the depth of the ith pit at time zero. Also let $\tau_i = k(T - t_i)$ be the time to failure of the ith pit, $i = 1, 2, \ldots, N$.

Since t_i cannot be greater than T, the probability density function of t_i is a truncated exponential with parameters λ and T, and consequently

$$P(t_i \geqslant t) = 1 - F_{t_i}(t) = \frac{e^{-\lambda t} - e^{-\lambda T}}{1 - e^{-\lambda T}}, \quad 0 \leqslant t \leqslant T.$$

Now

$$P(\tau_i \leqslant \tau) = P\left(t_i \geqslant T - \frac{\tau}{k}\right) = \frac{e^{\lambda \tau / k} - 1}{e^{\lambda T} - 1}, \quad 0 \leqslant \tau \leqslant kT$$

$$= G(\tau), \text{ say.}$$

Let X be the time to failure of the surface; then

$$X = \min_i (\tau_i), \quad i = 1, 2, \ldots, N.$$

Since X is the first order statistic from a sample of N τ_i's,

$$F_X(x) = P(X \leqslant x) = 1 - [1 - G(x)]^N.$$

If N, the number of pits, is very large, as is usually the case, then, following Section 3.8.1, it can be shown that

$$F_X(x) = 1 - \exp\left[-\frac{N}{e^{\lambda x} - 1} (e^{\lambda x / k} - 1) \right],$$

and this can be transformed into the form of the Type I asymptotic distribution of the largest extreme given in Section 3.8.2.

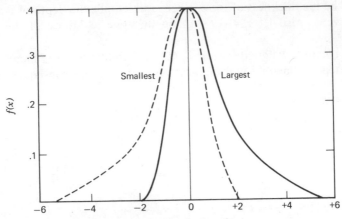

Figure 4.6   Type I extreme-value density function.

The Type I asymptotic distributions for the largest and the smallest extremes are illustrated in Figure 4.6 for the case $\gamma = 0$ and $\eta = 1$. Note that each density function is the mirror image of the other.

4.6   THE LOGARITHMIC NORMAL DISTRIBUTION

Until recently, the logarithmic normal distribution received relatively minor attention in the statistical literature, mainly because its applicability was limited to some rare situations in small-particle statistics, economics, and biology (Goldthwaite, 1961). Of late, however, because of its wide applicability to reliability problems, especially in the area of maintainability, and to certain fracture problems, its use has become more widespread. For semiconductor devices its plausibility was empirically demonstrated by Howard and Dodson (1961) and by Peck (1961). Its acceptability as a failure distribution is indicated by the life-test sampling plans developed for it (Gupta, 1962).

The logarithmic normal distribution implies that the logarithms of the lifetimes are normally distributed; hence it can be easily derived by a simple logarithmic transformation. The hazard rate of the logarithmic normal distribution as a function of time is an increasing function followed by a decreasing function, and can be shown to approach zero for large lifetimes and at the initial time (Goldthwaite, 1961). For this reason the derivation of the logarithmic normal distribution from the hazard rate is difficult. Its derivation by a logarithmic transformation is shown below.

Let X be the time-to-failure random variable of a device, and let $T = \log_e X$ be distributed normally with parameters μ and σ. Thus

$$f_T(t) = \frac{1}{\sqrt{2\pi}\,\sigma} \exp\left[-\frac{1}{2}\left(\frac{t-\mu}{\sigma}\right)^2 \right], \quad -\infty < t < \infty.$$

It follows from the above that the p.d.f. of X, $g_X(x)$, is given by

$$g_X(x) = \frac{1}{\sqrt{2\pi}\,\sigma x} \exp\left[-\frac{1}{2}\left(\frac{\log x - \mu}{\sigma}\right)^2 \right], \quad x > 0,$$

$$= 0, \quad \text{elsewhere.} \tag{4.12}$$

This is the logarithmic normal p.d.f. for X and is customarily written as follows: $X \sim \Lambda(\mu, \sigma^2)$.

The logarithmic normal distribution can be derived more fundamentally by considering a physical process wherein failure is due to fatigue cracks. It is because of this derivation that the plausibility of the logarithmic normal distribution for failure problems seems acceptable.

Let $X_1 < X_2 < \cdots < X_n$ be a sequence of random variables that denote the sizes of a fatigue crack at successive stages of its growth. A "proportional effect model" is assumed for the growth of these cracks (Kao, 1965). This implies that the crack growth at stage i, $X_i - X_{i-1}$, is randomly proportional to the size of the crack, X_{i-1}, and that the item fails when the crack size reaches X_n.

Let $X_i - X_{i-1} = \pi_i X_{i-1}$, $i = 1, 2, \ldots, n$, where π_i, the constant of proportionality, is a random variable. The initial size of the crack X_0, can be interpreted as the size of minute flaws, voids, and the like in the item. The π_i are assumed to be independently distributed random variables that need not have a common distribution for all i's.

Thus

$$\frac{X_i - X_{i-1}}{X_{i-1}} = \pi_i, \quad i = 1, 2, \ldots, n,$$

or

$$\sum_{i=1}^{n} \pi_i = \sum_{i=1}^{n} \frac{X_i - X_{i-1}}{X_{i-1}}.$$

If the change, $X_i - X_{i-1}$, is small at each step,

$$\sum_{i=1}^{n} \pi_i = \sum_{i=1}^{n} \frac{\Delta X_{i-1}}{X_{i-1}},$$

where

$$\Delta x_{i-1} = X_i - X_{i-1}.$$

In the limit, as $\Delta X_{i-1} \rightarrow 0$, and n becomes large,

$$\sum_{i=1}^{n} \pi_i = \int_{X_0}^{X_n} \frac{1}{X}\, dX = \log X_n - \log X_0$$

or

$$\log X_n = \sum_{n=1}^{n} \pi_i + \log x_0.$$

Since the π_i, by assumption, are independently distributed random variables, by the central limit theorem (see Section 3.7.4), it follows that they converge in distribution to a normal distribution. Thus $\log X_n$, the life length of the item, for large n, is asymptotically normally distributed, and hence X_n has a logarithmic normal distribution.

Other approaches to the derivation of the distribution of the time to failure due to fatigue are discussed in Section 4.11.

4.7   A GENERAL DISTRIBUTION OF TIME TO FAILURE

The failure models discussed in the preceding sections comprised a single distribution characterizing either a random (time-independent) or a wear-out (time-dependent) failure. In practice, a unit can suffer either one of these failures, and a statistical model characterizing the failure process should provide for both eventualities. For specific cases, models of this nature have been discussed by several authors (Kao, 1959; Cohen, 1965) under the name of *mixed* or *composite distributions*. In this section, a more general time-to-failure distribution is derived using a pure death process model, and the plausibility of such a distribution as a failure model is discussed.

Consider a unit that can be in any one of a finite number of states at time t. Let $E_1, E_2, \ldots, E_{n-1}$ represent the operative states, and let E_n denote the failed state. The state E_n is an *absorbing state*, in the sense that, once the unit reaches this state, it remains there. The failure process is such that a transition can occur only from a lower state E_i to the next higher state E_{i+1}, $i = 1, \ldots, n-1$, and also from any lower state E_i, $i = 1, 2, \ldots, n-1$, to the absorbing state E_n; no transitions from a higher to a lower state are possible.

Let $P_n(t)$ denote the probability that the device is in state E_n at time t.

Two postulates for this process are assumed.

1. The process is time homogeneous and state independent, in the sense that transitions from one state to another are independent of the time of transition and the states of transition. The probability of a transition from state E_i to state E_{i+1}, $i = 1, 2, \ldots, n-2$, in a time interval $(t, t+h)$ is $\lambda h + o(h)$. Also, the probability of a transition from any state E_i, $i = 1, 2, \ldots$, $n-2$, to the absorbing state E_n in a time interval $(t, t+h)$ is $\mu h + o(h)$. When the unit has reached state E_{n-1}, the probability of a transition from this state to the absorbing state E_n, in the time interval $(t, t+h)$, is $(\lambda + \mu)h + o(h)$.

2. The probability of two or more transitions in the time interval $(t, t+h)$ is $o(h)$.

Let X denote the time-to-failure random variable of the unit under consideration and let $f_X(\cdot)$ be its p.d.f. If X is the time for the unit to become absorbed into the state E_n,

$$f_X(x)\Delta x = P[\text{time to become absorbed falls in the interval } (x, x+\Delta x)].$$

Following the arguments used in Section 4.2.1, one can easily see that

$$f_X(x)\Delta x = P_{n-1}(x)[(\lambda + \mu)\Delta x + o(\Delta x)]$$

$$+ \sum_{i=1}^{n-2} P_i(x)\{[1 - \lambda(\Delta x) - o(\Delta x)][\mu(\Delta x) + o(\Delta x)]\}$$

$$= P_{n-1}(x)[(\lambda + \mu)\Delta x + o(\Delta x)]$$

$$+ \sum_{i=1}^{n-2} P_i(x)\mu\Delta x + o(\Delta x), \quad \text{by postulate 2.}$$

Dividing throughout by Δx and taking the limit as $\Delta x \to 0$ gives

$$f_X(x) = P_{n-1}(x)\lambda + \mu \sum_{i=1}^{n-1} P_i(x). \tag{4.13}$$

The probabilities $P_i(x), i = 1, \ldots, n-1$, may be obtained by considering the following equations:

$$P_i(t+h) = P_i(t)[1 - \lambda h - \mu h - o(h)]$$

$$+ P_{i-1}(t)[\lambda h + o(h)], \quad i = 2, 3, \ldots, n-1,$$

and

$$P_1(t+h) = P_1(t)[1 - \lambda h - \mu h - o(h)].$$

Dividing throughout by h, and taking the limit as $h \to 0$, one has, for $i = 2, 3, \ldots, n-1$,

$$P_i'(t) = -(\lambda + \mu)P_i(t) + \lambda P_{i-1}(t) \qquad (4.14)$$

and

$$P_1'(t) = -(\lambda + \mu)P_1(t). \qquad (4.15)$$

If it is assumed that the device is in state E_1 at time $t = 0$, the initial conditions are

$$P_1(0) = 1 \quad \text{and} \quad P_i(0) = 0, \, 2 \leqslant i \leqslant n - 1.$$

Equations (4.14) and (4.15) are differential equations in $P_i(t)$ whose solutions can be obtained by using Laplace transforms and the initial conditions stated above. Following the notation used in Section 4.2.1, one has for the Laplace transform of Equation (4.15)

$$s\tilde{P}_1(s) - P_1(0) = -(\lambda + \mu)\tilde{P}_1(s)$$

or

$$\tilde{P}_1(s) = \frac{1}{s + \lambda + \mu}.$$

The inverse Laplace transform of this gives $P_1(t) = e^{-(\lambda + \mu)t}$.

Applying the Laplace transform to Equation (4.14), one has

$$s\tilde{P}_i(s) - P_i(0) = -(\lambda + \mu)\tilde{P}_i(s) + \lambda\tilde{P}_{i-1}(s), \quad i = 2, 3, \ldots, n - 1.$$

This equation when solved recursively gives

$$\tilde{P}_i(s) = \frac{\lambda^{i-1}}{(s + \lambda + \mu)^i}, \quad i = 2, 3, \ldots, n - 1.$$

The inverse Laplace transform of $\tilde{P}_i(s)$ gives

$$P_i(t) = \frac{e^{-(\lambda + \mu)t}(\lambda t)^{i-1}}{(i-1)!}, \quad i = 2, 3, \ldots, n - 1.$$

Substitution into Equation (4.13) gives

$$f_X(x) = e^{-\mu x}\left\{\left[\frac{e^{-\lambda x}(\lambda x)^{n-2}\lambda}{(n-2)!}\right] + \mu\sum_{i=1}^{n-1}\left[\frac{e^{-\lambda x}(\lambda x)^{i-1}}{(i-1)!}\right]\right\}.$$

Using the techniques of Section 4.3, it is easy to verify that the incomplete gamma function can be expressed as the summation of an infinite series, that is,

$$\sum_{j=n}^{\infty}\frac{e^{-\lambda t}(\lambda t)^j}{j!} = \int_0^{\lambda t}\frac{u^{n-1}e^{-u}du}{\Gamma(n)}.$$

Therefore

$$f_X(x) = e^{-\mu x}\left\{\frac{\lambda e^{-\lambda x}(\lambda x)^{n-2}}{(n-2)!} + \mu\left[1 - \int_0^{\lambda x}\frac{e^{-u}u^{n-1}du}{\Gamma(n-1)}\right]\right\}.$$

Changing the variable of integration by putting $u = \lambda x$, one can obtain the distribution function of X as

$$F_X(t) = 1 - e^{-\mu t}\left[1 - \int_0^t\frac{\lambda^{n-1}x^{n-2}e^{-\lambda x}}{\Gamma(n-1)}\,dx\right].$$

The survival to time t is given by

$$R_X(t) = e^{-\mu t}\left[1 - \int_0^t\frac{\lambda^{n-1}x^{n-2}e^{-\lambda x}}{\Gamma(n-1)}\,dx\right]. \qquad (4.16)$$

This equation shows that the probabilities of the events "no chance failure to time t" and "no wear-out failure before time t" multiply to give the probability of no failure before time t.

4.8   MIXED, COMPOSITE, AND COMPETING-RISK MODELS

In Section 4.7 a general distribution of time to failure was derived by essentially considering transitions from a higher state to a lower state under some simplifying postulates. In this section some other failure models that can also be viewed as general failure distributions are derived under certain physical assumptions. The main reason for presenting these distributions is that they allow flexibility in fitting and explaining failure data—a capability that may be useful.

4.8.1 The Mixed Distribution Model

The mixed distribution model can be considered as a special case of the general time-to-failure distribution discussed in Section 4.7. Such models have been considered by several workers (Kao, 1959; Cohen, 1965; Harris and Singpurwalla, 1968), although in different contexts. Instead of a single distribution, a mixture of two or more distributions, each representing a type of failure, is postulated. It has been a common practice to assume that distributions that are mixed belong to the same family but differ in the values of their parameters. The mixed normal distribution considered by Cohen (1965) is claimed to be applicable to the study of wind velocities and physical dimensions of mass-produced items. The mixed Weibull distribution discussed by Kao (1959) is useful in reliability studies, especially those involving electron tubes.

From a theoretical point of view, mixtures of distributions present interesting problems. For example, Barlow and Proschan (1965) prove that the DHR property (see Section 7.3) is closed under mixtures. In this section a mixed distribution is defined and its suitability as a failure model briefly discussed.

Let $F_{X_i}(x)$ be the cumulative distribution function (c.d.f.) of a random variable $X_i, i = 1, 2, \ldots, k$. Then a k-fold mixed c.d.f. is defined as

$$F_X(x) = \sum_{i=1}^{k} p_i F_{X_i}(x), \quad 0 \leqslant p_i \leqslant 1, \text{ and } \sum_{i=1}^{k} p_i = 1.$$

Often $F_{X_i}(x)$ is referred to as the *ith subpopulation in c.d.f. form*, and the p_i are called the *mix parameters*. A physical interpretation for p_i is given in Section 4.8.3.

Alternatively, a k-fold mixed probability density function is given by

$$f_X(x) = \sum_{i=1}^{k} p_i f_{X_i}(x),$$

where $f_{X_i}(x)$ is the *ith subpopulation in p.d.f. form*.

A possible motivation for the mixed Weibull distribution is provided by the following example.

Consider devices whose failures can be classified as types: (a) catastrophic or sudden, or (b) wear-out or delayed. Catastrophic failures occur as soon as the device is exposed to risk, whereas wear-out or delayed failures occur after a specific time period has elapsed. It is generally believed that wear-out failures are due to an aging or a decaying of the material or the device. For most applications, since it is usually the poor-quality devices that fail prematurely, the hazard rate of the population of devices decreases with time. Thus a Weibull distribution with a location parameter 0 and a shape parameter β_1 less than 1 would be an appropriate model for describing catastrophic failures. Since wear-out commences after a particular time period has elapsed, a Weibull distribution with a nonzero location parameter γ and a shape parameter β_2 greater than 1 would be an appropriate model for describing such failures. Combining the two failure models yields a twofold mixed Weibull distribution as the underlying failure distribution of the device. Thus the failure model of the device is

$$F_X(x) = p\left[1 - \exp\left(\frac{x}{\eta_1}\right)^{\beta_1}\right] + (1-p)\left\{1 - \exp\left[-\left(\frac{x-\gamma}{\eta_2}\right)^{\beta_2}\right]\right\},$$

for $\eta_1, \eta_2 > 0$, $0 < \beta_1 < 1$, $\beta_2 > 1$, $\gamma > 0$, and $0 \leqslant p \leqslant 1$. The probability density function of this mixed Weibull is easily seen to be

$$f_X(x) = \frac{p\beta_1}{\eta_1}\left(\frac{x}{\eta_1}\right)^{\beta_1 - 1}\exp\left(-\frac{x}{\eta_1}\right)^{\beta_1}$$

$$+ \frac{(1-p)\beta_2}{\eta_2}\left(\frac{x-\gamma}{\eta_2}\right)^{\beta_2 - 1}\exp\left[-\left(\frac{x-\gamma}{\eta_2}\right)^{\beta_2 - 1}\right].$$

The above equations state that, for a given application, the lifetime of a device is a mixed chance variable depending on the wear-out quality of the device and its ability to withstand the severity of the causes of catastrophic failures. The parameters $\eta_1, \eta_2, p, \beta_1, \beta_2$, and γ have to be estimated from failure data. Several methods for estimating the parameters of mixed models have been proposed, although none of them has the properties of the methods discussed in Chapter 5. For a graphical estimation of the parameters of the mixed Weibull distribution the reader is referred to Kao (1959).

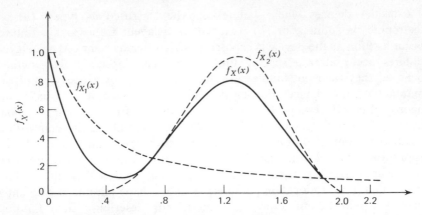

Figure 4.7 A two-fold mixed Weibull density function.

The p.d.f. of a mixed Weibull distribution with $\eta_1 = \eta_2 = 1$, $\beta_1 = \frac{1}{2}$, $\beta_2 = 3$, $\gamma_1 = 0$, and $\gamma_2 = 0.4$ is illustrated in Figure 4.7 for $p = 0.2$. Here $f_{X_1}(x)$ [$f_{X_2}(x)$] denotes the Weibull with parameters $\eta_1, \beta_1, \gamma_1 [\eta_2, \beta_2, \gamma_2]$.

The kth moment of the mixed Weibull distribution, μ_k, is given as (cf. Kao, 1959)

$$\mu_k = p\eta_1{}^k \Gamma\left(\frac{k}{\beta} + 1\right) + q\gamma_2{}^k + qk \sum_{i=0}^{k-1} \binom{k-1}{i} \gamma_2{}^i \eta_2{}^{k-i} \frac{1}{\beta_2} \Gamma\left(\frac{k-i}{\beta_2}\right)$$

or

$$\mu_1 = p\eta_1 \Gamma\left(\frac{1}{\beta} + 1\right) + q\left[\gamma + \eta_2 \Gamma\left(\frac{1}{\beta_2} + 1\right)\right],$$

which gives us the mean. Here $q = 1 - p$.

4.8.2 THE COMPOSITE DISTRIBUTION MODEL

The main reason for considering a composite distribution model is that it can sometimes provide flexibility in fitting and explaining failure data. It will be shown later that this distribution can also be used as an approximation to the twofold mixed Weibull distribution discussed in Section 4.8.1 if p is small and γ is large. Since estimation of the parameters of a mixed Weibull distribution can be an involved procedure, such an approximation can be very useful from a practical point of view.

An r-component composite c.d.f. is defined as

$$F_X(x) = F_j(x), \quad S_j \leqslant x \leqslant S_{j+1}, j = 0, 1, 2, \ldots, r.$$

Here $F_j(x)$ is called the ith component of a composite distribution in c.d.f. form. The parameters S_j are termed the *partition parameters*. It is clear than an r-component composite Weibull distribution has $r-1$ partition parameters. An r-component composite p.d.f. is simply

$$f_X(x) = f_j(x), \quad S_j \leqslant x \leqslant S_{j+1}, \quad j = 0, 1, 2, \ldots, r,$$

where $f_j(x)$ is the derivative of $F_j(x)$.

A two-component composite model for $\eta_1 = \eta_2 = 1$, $\beta_1 = \frac{1}{2}$, $\beta_2 = 3$, $\gamma_1 = 0$, $\gamma_2 = 0.4$, and $S = 1$ is shown in Figure 4.8 where $f_1(x)$ $[f_2(x)]$ denotes the Weibull distribution with parameters $\eta_1, \beta_1, \gamma_1 (\eta_2, \beta_2, \gamma_2)$.

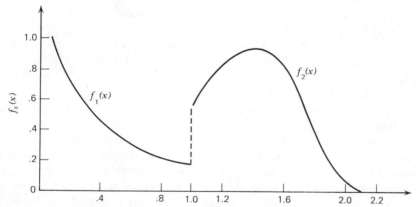

Figure 4.8 A two-component composite Weibull density function.

For small values of p and large values of γ, the twofold mixed Weibull model of Section 4.8.1 can be approximated by a two-component composite failure model. Thus

$$F_X(x) = 1 - \exp\left(-\frac{x}{\eta_1}\right)^{\beta_1}, \quad 0 \leqslant x \leqslant \delta,$$

$$= 1 - \exp\left(-\frac{x}{\eta_2}\right)^{\beta_2}, \quad \delta \leqslant x < \infty.$$

The partition parameter δ can be expressed in terms of the other parameters since, at $x = \delta$,

$$1 - \exp\left(-\frac{\delta}{\eta_1} \right)^{\beta_1} = 1 - \exp\left(-\frac{\delta}{\eta_2} \right)^{\beta_2}$$

or

$$\delta = \left(\frac{\eta_1^{\beta_1}}{\eta_2^{\beta_2}} \right)^{1/(\beta_1 - \beta_2)}.$$

4.8.3 THE COMPETING-RISK MODEL

The competing-risk model, as identified by Makeham (1873) and restated by Neyman (1950), Cornfield (1957), and Berkson and Elveback (1960), was concerned primarily with the analysis of human and animal mortality data. More recently, its potential use in the areas of life testing and reliability analysis has been explored by Anello (1968). Like the mixed distribution model, the competing-risk model allows flexibility in explaining failure data when there are multiple modes of failure. The inherent difference between the competing-risk model and the mixed distribution model is pointed out below.

Suppose that a device exhibits k modes (risks) of failure, $m_1, m_2, \ldots, m_k$, and that a random lifetime on this item occurs as follows: When the device begins operation, each failure mode simultaneously generates a random lifetime that is independent of the other modes. Thus, in effect, k lifetimes, denoted by $X_1, X_2, \ldots, X_k$, simultaneously begin; lifetime X_i corresponds to the ith mode of failure. Failure of the device occurs as soon as any one of the lifetimes, say X_i, is realized. In effect, if the life length of the device is denoted by a random variable X, then

$$X = \min (X_1, X_2, \ldots, X_k) \equiv X_{(1)}.$$

If $F_{X_i}(x)$ is the cumulative distribution function of X_i, the c.d.f. of $X, F_X(x)$, is given by

$$F_X(x) = 1 - \prod_{i=1}^{k} \left[1 - F_{X_i}(x) \right]. \tag{4.17}$$

The above derivation of the competing-risk model not only is independent of the functional form of the $F_{X_i}(x)$, but also allows for the $F_{X_i}(x)$ to be all different. In effect, this means that each failure mode can have any failure distribution and that all the failure distributions need not be alike. However, the derivation does require that all the k modes operate independently of each other. If the modes do not operate independently, that is, if one mode influences the behavior of the other modes, Equation 4.17 for the competing-risk model will have to be amended to account for any relationship that may exist between the modes of failure.

The conceptual difference between the competing-risk model and the mixed-population model is that in the former all the k modes begin to generate random lifetimes simultaneously, whereas in the latter only one of the k possible modes generates a random lifetime that causes part failure. The selection of this mode is governed by the parameter p_i, where p_i can be interpreted as the probability of randomly selecting a device that is predestined to fail because of mode i.

Some very general results for the competing-risk model that help to explain the behavior of certain kinds of failure data are now derived.

Consider a device on which k risks (modes) are jointly but independently operating. If only the ith risk, with a *risk-specific hazard rate* $h_i(S)$ were effective, the probability that the device will survive to time x is

$$1 - F_{X_i}(x) = \exp\left[-\int_0^x h_i(S)\,dS \right],$$

where $\int_0^x h_i(S)\,dS$ is the cumulative hazard due to risk i, at time x. From Equation (4.17) it is clear that the probability that the device will survive all the k risks is

$$1 - F_X(x) = \prod_{i=1}^k \exp\left[-\int_0^x h_i(S)\,dS \right] = \exp\left[-\int_0^x \sum_{i=1}^k h_i(S)\,dS \right].$$

This equation leads to the inference that the total hazard to the device at time S, say $h(S)$, is the sum of the k independent risk-specific hazards at time S, that is

$$h(S) = h_1(S) + h_2(S) + \cdots h_k(S).$$

The particular form of the competing-risk model depends on an additional assumption concerning the behavior of $h_i(S)$ with time. If it is assumed that the risk-specific hazard rates are constant over the period of observation, as is done by Berkson and Elveback (1960), the competing-risk model takes a simple and tractable form. If on the contrary the risk-specific hazard rates are increasing functions of time, as is assumed by Anello (1968) the competing-risk model may take a less manageable form.

More specifically, if it is assumed that $h_i(S) = h_i$, for all i and for all $S > 0$, then

$$h(S) = h_1 + h_2 + \cdots h_k = h, \text{ say.}$$

The probability that the device survives to time x in the case of a constant risk-specific hazard rate is given by

$$1 - F_X(x) = \exp(-hx). \tag{4.18}$$

This equation provides a convenient explanation of the often observed result that part failure times are exponential, even though several modes of failure are observed. In effect, it means that, if each of the k failure modes has an exponential failure distribution with a failure rate of h_i, the device exhibits an exponential failure distribution with a failure rate of $h = \sum_{i=1}^{k} h_i$.

To consider the case in which the risk-specific hazard rate is an increasing function of time, let

$$h_i(S) = \beta_i \eta_i^{\beta_i} S^{\beta_i - 1}, \quad \text{for all } S > 0; \ \eta_i, \beta_i > 0; \ i = 1, 2, \ldots k.$$

Recall that the above form of the hazard rate corresponds to the hazard rate of the Weibull distribution given in Section 4.4.

The total hazard at time S is given by

$$h(S) = \sum_{i=1}^{k} \beta_i \eta_i^{\beta_i} S^{\beta_i - 1},$$

and the probability that the device survives to time x is

$$1 - F_X(x) = \exp\left(-\int_0^x \sum_{i=1}^{k} \beta_i \eta_i^{\beta_i} S^{\beta_i - 1} \, dS\right). \tag{4.19}$$

If $\beta_i = \beta$, for $i = 1, 2, \ldots, k$, Equation (4.19) becomes

$$1 - F_X(x) = \exp\left(-x^\beta \sum_{i=1}^{k} \eta_i^\beta\right),$$

and this implies that, if each of the k failure modes has a Weibull failure distribution with the same shape parameter β and a scale parameter η_i, the device exhibits a Weibull failure distribution with a shape parameter β and a scale parameter $\sum_{i=1}^{k} \eta_i^\beta$.

4.9   STOCHASTIC FAILURE RATE MODELS

The failure models discussed in the preceding sections could be derived from their corresponding failure-rate functions under the assumption that any parameters involved were not random variables. If it is assumed that the parameters are random variables with known probability distributions, called *prior* distributions (see Chapter 8), the failure-rate function becomes a random function. An unconditional failure distribution can then be derived from elementary considerations, and this section is devoted to such distributions.

Before discussing the conceptual justification of random parameters for reliability problems, certain definitions are in order.

Suppose that a parameter λ of the hazard rate is assumed to be a random variable with a prior distribution $G(\lambda)$. Then, analogously to Equation (4.1), the *conditional hazard rate* is defined as

$$h(x|\lambda) = \frac{f(x|\lambda)}{1 - F(x|\lambda)}.$$

Following Equation (4.2), the *conditional failure distribution* is given as

$$F(t|\lambda) = 1 - \exp\left[-\int_0^t h(x|\lambda)\, dx\right],$$

from which it follows that

$$F(t) = \int_{-\infty}^{\infty} F(t|\lambda)\, dG(\lambda)$$

is the *unconditional time-to-failure distribution*.

4.9.1   Conceptual Justification of Random Parameters

The effect of an acceptance sampling plan is to modify the distribution of lot quality. In other words, the density function for an incoming lot quality is multiplied by a factor, the probability of acceptance, and then renormalized, resulting in a distribution of outgoing lot quality. Since the probability of acceptance is high for all sufficiently good lots, a bin containing all accepted lots will contain components having, not the same λ, but rather a range of values of λ. A prior distribution on λ can be specified if there is some knowledge of how the values of λ are distributed. For the situation described here, a gamma prior distribution for λ is quite conceivable because it is skewed to the right, implying a low probability of acceptance for large values of λ. Of course, other skewed distribution or perhaps a truncated normal distribution could also fit the situation described. A uniform prior distribution for λ results from a situation wherein each lot is homogeneous within itself, but different lots may have different values of λ and any one value of λ appears as often as any other.

A two-point prior distribution occurs in a situation wherein λ may take only two values, with probabilities p and $1-p$, respectively.

In any case the question of concern is, given such a bin of components, what is the lifetime distribution of a component picked at random from this bin?

4.9.2 The Exponential Conditional Failure Distribution

Suppose that the failure distribution is exponential with a scale parameter Λ that is random. Thus

$$f_\tau(t|\Lambda=\lambda)=\lambda\exp(-\lambda t), \quad \lambda>0, t\geqslant 0.$$

Assume that the prior distribution $G(\lambda)$ is a gamma distribution with parameters α and β; that is,

$$dG(\lambda;\alpha,\beta)=\frac{\lambda^\alpha\exp(-\lambda/\beta)}{\beta^{\alpha+1}\Gamma(\alpha+1)}d\lambda, \quad \alpha>1, \beta>0.$$

It follows that the unconditional distribution of time to failure $F_\tau(t)$ is given by

$$F_\tau(t)=\int_0^\infty (1-e^{-\lambda t})\,dG(\lambda)$$

$$=1-(\beta t+1)^{-(\alpha+1)}.$$

By a simple transformation, $X=\beta\tau+1$, $a=\alpha+1$, and $b=1$, it is easy to see that

$$F_X(x)=1-(b|x)^a, \quad x>b,$$

which is also known as the *Pareto* distribution.

The failure-rate function of $F_\tau(t)$ given above is easily seen to be

$$h^*(t) = \frac{(dF(t)/dt)}{R(t)}$$

$$= \frac{(\alpha+1)^\beta}{\beta t + 1}.$$

Clearly $h^*(t)$ decreases in t, implying that decreasing failure rates can also arise from situation of the type discussed here.

If a two-point prior distribution is assumed on Λ such that

$$P(\Lambda = \lambda_1) = p$$

and

$$P(\Lambda = \lambda_2) = 1 - p,$$

then

$$\bar{F}_\tau(t) = 1 - pe^{-\lambda_1 t} - (1-p)e^{\lambda_2 t}.$$

It is easy to see that $F_\tau(t)$ is a special case of the mixed Weibull model discussed in Section 4.8.1 when $\gamma = 0$ and $\beta = 1$.

If a uniform prior distribution is assumed on Λ such that

$$G(\lambda) = \begin{cases} 0, & \lambda < a, \\ \dfrac{\lambda - a}{b - a}, & a \leqslant \lambda \leqslant b, \\ 1 & \lambda > b, \end{cases}$$

then

$$F(t) = 1 - \frac{e^{-at} - e^{-bt}}{t(b-a)}.$$

A similar approach can be taken if it is assumed that the conditional failure distribution is a Weibull; for the details the reader is referred to Harris and Singpurwalla (1968). The estimation of the parameters of such failure models is treated in Harris and Singpurwalla (1969).

4.10   A BIVARIATE EXPONENTIAL DISTRIBUTION

A bivariate exponential distribution is a special case of a multivariate exponential distribution and can be used to describe the failure behavior of two-unit systems in which the lifetimes of the two units in service are not independent but depend on one another in a particular way. Specifically, a multivariate exponential distribution is a multivariate distribution with exponential marginals and can be derived by considering the occurrence of random events (shocks) in a Poisson process (Marshall and Olkin, 1967). It can also be derived by considering a multivariate extension of a central property of the exponential distribution: that the distribution of residual life is independent of age. Applications of a bivariate exponential distribution to reliability problems are discussed by Harris (1968). The following derivation of the bivariate exponential distribution is due to Marshall and Olkin (1967).

Suppose that the components in any two-component system (series or parallel) fail after receiving a shock that could be fatal. Three independent Poisson processes, $P_1(\lambda_1)$, $P_2(\lambda_2)$, and $P_3(\lambda_{12})$, govern the occurrence of the shocks with parameters λ_1, λ_2, and λ_{12}, respectively. Events in the process $P_1(\lambda_1)$ are shocks applicable only to the first component, which may cause it to fail with a probability p_1, and events in the process $P_2(\lambda_2)$ are shocks applicable only to the second component, which may cause it to fail with a probability p_2. However, events in the process $P_3(\lambda_{12})$ are shocks applicable to both components, which may cause either one or both or none to survive.

Let p_{01} (p_{10}) be the probability that a shock due to $P_3(\lambda_{12})$ causes the first (second) component to fail and the second (first) to survive. Also, let $p_{00}(p_{11})$ be the probability that a shock due to $P_3(\lambda_{12})$ causes both the components to fail (survive).

Let X and Y denote the random lifetimes of the first and the second components, respectively. Then

$$\overline{F}_{XY}(x,y) = P(X \geqslant x, Y \geqslant y) \quad \text{for } 0 \leqslant x \leqslant y.$$

Let event $E \equiv (X \geqslant x, Y \geqslant y)$. Clearly,
$E \equiv j$ nonfatal shocks due to $P_1(\lambda_1)$ in time $(0,x)$, and
  j nonfatal shocks due to $P_2(\lambda_2)$ in time $(0,y)$, and
  k non fatal shocks due to $P_3(\lambda_{12})$ in time $(0,x)$, and
  l shocks due to $P_3(\lambda_{12})$ in time $(y - x)$ such that all the l shocks are not fatal to the second component.

Then

$$P(X \geqslant x, Y \geqslant y) = \left\{ \sum_{j=0}^{\infty} \frac{e^{-\lambda_1 x}(\lambda_1 x)^j}{j!}(1-p_1)^j \right\} \cdot \left\{ \sum_{j=0}^{\infty} \frac{e^{-\lambda_2 y}(\lambda_2 y)^j}{j!}(1-p_2)^j \right\}$$

$$\left\{ \sum_{l=0}^{\infty} \sum_{k=0}^{\infty} \left[\frac{e^{-\lambda_{12}x}(\lambda_{12}x)^k}{k!}(p_{11})^k \right] \left[\frac{e^{-\lambda_{12}(y-x)}(\lambda_{12}(y-x))^l}{l!}(p_{11}+p_{01})^l \right] \right\}$$

With minor modifications, each of the terms within the curly braces can be written as a sum of Poisson random variables, and hence

$$P(X \geqslant x, Y \geqslant y) = \exp\{-x(\lambda_1 p_1 + \lambda_{12} p_{01}) - y[\lambda_2 p_2 + \lambda_{12}(p_{00} + p_{10})]\}$$

since $p_{00} + p_{10} + p_{01} + p_{11} = 1$.
Similarly, for $0 \leqslant y \leqslant x$,

$$P(X \geqslant x, y \geqslant y) = \exp\{-x[\lambda_1 p_1 + \lambda_{12}(p_{00} + p_{01})] - y(\lambda_2 p_2 + \lambda_{12} p_{10})\}.$$

Therefore, in general, for either $0 \leqslant x \leqslant y$ or $0 \leqslant y \leqslant x$,

$$P(X \geqslant x, Y \geqslant y) = \exp[-\delta_1 x - \delta_2 y - \delta_{12} \max(x,y)], \qquad (4.20)$$

where $\delta_1 = \lambda_1 p_1 + \lambda_{12} p_{01}$, $\delta_2 = \lambda_2 p_2 + \lambda_{12} p_{10}$, and $\delta_{12} = \lambda_{12} p_{00}$.
If $p_1 = p_2 = p_{00} = 1$, every occurrence of a shock is fatal, and so

$$P(X \geqslant x, Y \geqslant y) = \exp[-\lambda_1 x - \lambda_2 y - \lambda_{12} \max(x,y)]. \qquad (4.21)$$

The model given by Equation (4.20) is known as the *nonfatal shock model*, whereas that of Equation (4.21) is termed the *fatal shock model*. The bivariate (multivariate) exponential distribution is abbreviated as BVE (MVE).

4.10.1 Some Properties and Generalizations of the BVE Distribution

Marshall and Olkin (1967) have investigated several interesting properties of the BVE distribution, a few of which are highlighted in this section.

1. The BVE distribution [Equation (4.21)] has exponential marginals given by

$$P(X \geqslant x) = \exp[-(\lambda_1 + \lambda_{12})x]$$

and

$$P(Y \geqslant y) = \exp[-(\lambda_2 + \lambda_{12})y].$$

2. The Laplace transform of the BVE distribution exists and is given by

$$\psi(s,t) = \frac{(\lambda + s + t)(\lambda_1 + \lambda_{12})(\lambda_2 + \lambda_{12}) + st\lambda_{12}}{(\lambda + s + t)(\lambda_1 + \lambda_{12} + s)(\lambda_2 + \lambda_{12} + t)},$$

where $\lambda = \lambda_1 + \lambda_2 + \lambda_{12}$.

The moments of the BVE distribution obtained from $\psi(s,t)$ are

$$E(X) = (\lambda_1 + \lambda_{12})^{-1}, \qquad \text{Var}(X) = (\lambda_1 + \lambda_{12})^{-2},$$

$$E(Y) = (\lambda_2 + \lambda_{12})^{-1}, \qquad \text{Var}(Y) = (\lambda_2 + \lambda_{12})^{-2},$$

and

$$E(XY) = \lambda^{-1}\left[(\lambda_1 + \lambda_{12})^{-1} + (\lambda_2 + \lambda_{12})^{-1}\right].$$

3. The BVE distribution [equation (4.21)] can be generalized to the multivariate case to give the MVE distribution with parameters $\lambda_i, \lambda_{ij}, \ldots,$ $\lambda_{12\ldots n}$ as

$$P(X_1 \geqslant x_1, X_2 \geqslant x_2, \ldots, X_n \geqslant x_n) = \exp\left[-\sum_{i=1}^{n} \lambda_i x_i - \sum_{i<j} \lambda_{ij} \max(x_i, x_j)\right.$$

$$\left. -\sum_{i<j<k} \lambda_{ijk} \max(x_i, x_j, x_k) \ldots - \lambda_{12\ldots n} \max(x_i, x_2, \ldots, x_n)\right].$$

For example, if $n = 3$ one has a three-component system for which the fatal shock model is given by

$$P(X_1 \geqslant x_1, X_2 \geqslant x_2, X_3 \geqslant x_3) = \exp[-\lambda_1 x_1 - \lambda_2 x_2 - \lambda_3 x_3 - \lambda_{12} \max(x_1, x_2)$$

$$-\lambda_{13} \max(x_1, x_3) - \lambda_{23} \max(x_2, x_3) - \lambda_{123} \max(x_1, x_2, x_3)].$$

4.11 THE FATIGUE-LIFE MODEL (THE BIRNBAUM-SAUNDERS DISTRIBUTION)

To characterize failures due to fatigue crack extension, Birnbaum and Saunders (1969) proposed a life distribution based on two parameters. This distribution, for a nonnegative random variable, is derived, using considerations from renewal theory (see Section 10.1.2), via an idealization of the number of cycles necessary to force a fatigue crack to grow past a critical value. A distribution such as this, which is obtained from considerations of the basic characteristics of the fatigue process, is more persuasive in its implications than any other distribution chosen for ad hoc reasons, such as goodness-of-fit tests and the like. The reasonableness of the approach used to derive this model is indicated by its ability to offer a probabilistic interpretation of Miner's (1945) rule, a deterministic rule that attempts to predict the fatigue life of a specimen under repeated cyclic loading (Birnbaum and Saunders, 1968). Freudenthal and Shinozuka (1961), on the basis of heuristic engineering considerations, have presented a distributional form akin to the model given by Birnbaum and Saunders, and have also substantiated the validity of their model using several sets of fatigue data. It is for these reasons that a presentation of the basic approach used by Birnbaum and Saunders in deriving their model is believed to be useful.

Consider a material specimen that is subjected to a sequence of m loads, $\{l_i\}$, $i = 1, 2, \ldots, m$, each of which causes a deflection of the specimen, thereby imposing a stress on it. Also assume that the loading is *cyclic*, so that in each cycle the same sequence of loads (and in the same order) is applied to the specimen as in the previous cycle. The loading scheme can be depicted as follows:

$$\underbrace{l_1, l_2, \ldots, l_m}_{\text{Cycle 1}} \quad \underbrace{l_{m+1}, l_{m+2}, \ldots, l_{2m},}_{\text{Cycle 2}} \quad \ldots, \quad \underbrace{l_{jm+1}, l_{jm+2}, \ldots, l_{jm+m}}_{\text{Cycle } (j+1)}$$

By assumption, $l_{jm+i} = l_{km+i}$ for all $j \neq k$. Birnbaum and Saunders have also assumed that the loading is continuous; in order to make this statement more precise, the load is visualized as a continuous unimodal function defined on the unit interval. The value of this function at any time gives the stress imposed by the deflection of the specimen.

Continuous loading implies that, for all $i = 1, 2, \ldots, m$,

$$l_{i+1}(0) = l_{i(1)} = l_{i-1}(0).$$

Physically, it is reasonable to conceive that fatigue failure is due to the initiation, the growth, and the ultimate extension of a dominant crack past some critical length for the first time. At the imposition of each load l_i, the crack is extended by a random amount, the randomness being due to variations in the material, the magnitude of the stress, the number of prior loads, and other such factors.

Having established the preceding physical framework, it is necessary to make two assumptions of a statistical nature.

Assumption 1. The incremental crack extension X_i due to a load l_i in cycle j is a random variable whose distribution is governed by all the loads $l_j, j < i$, and the actual crack extensions that have preceded it in cycle j alone.

Assumption 2. The total crack extension Y_j due to the jth cycle is a random variable with a mean μ and a variance σ^2, for $j = 1, 2, \ldots$.

Notice that the total crack extension due to the $(j+1)$st cycle of loads is

$$Y_{j+1} = X_{jm+1} + X_{jm+2} + \cdots X_{jm+m'}, \quad j = 0, 1, 2, \ldots .$$

Assumption 1 is rather restrictive and may not be valid for certain applications. This assumption ensures that, regardless of the dependence among the successive random extensions due to the loads in a particular cycle, the total random crack extensions are independent from cycle to cycle. The plausibility of this assumption in aeronautical fatigue studies is briefly stated by Birnbaum and Saunders (1969).

Under a repeated application of n cycles of loads, the total extension of a dominant crack in a specimen can be written as

$$W_n = \sum_{j=1}^{n} Y_j.$$

Let C be an integer-valued, nonnegative random variable denoting the smallest number of cycles at which W_n exceeds a critical value ω, that is, failure of the specimen occurs. Clearly, $P(C \leqslant n) = P(W_n \geqslant \omega)$, and this implies that

$$P(C \leqslant n) = 1 - P\left(\sum_{j=1}^{n} Y_j \leqslant \omega \right).$$

By assumption 2 the Y_j's have a mean μ and a variance σ^2; thus the Y_j's can be standardized to give

$$P(C \leqslant n) = 1 - P\left(\sum_{j=1}^{n} \frac{Y_j - \mu}{\sigma \sqrt{n}} \leqslant \frac{\omega - n\mu}{\sigma \sqrt{n}} \right).$$

Assumption 1 ensures the independence of the Y_j's; and, if n is large (a criterion easy to satisfy in fatigue studies), the central limit theorem applies. Hence, by Section 3.7 and the symmetry of the normal distribution,

$$P(C \leqslant n) = 1 - \Phi\left(\frac{\omega - n\mu}{\sigma \sqrt{n}} \right) = \Phi\left(\frac{n\mu}{\sigma \sqrt{n}} - \frac{\omega}{\sigma \sqrt{n}} \right),$$

where

$$\Phi(x) = \int_{-\infty}^{x} \frac{1}{\sqrt{2\pi}} e^{-s^2/2} \, ds.$$

It should be remarked here that the convergence in distribution to normality used above is only an approximation, which improves as n becomes large (see Section 3.7). However, for many applications, especially those pertaining to fatigue studies, this approximation may be satisfactory.

The above derivation, which involved a nonnegative integer-valued random variable C, can be naturally extended to continuous variables. Let T, a continuous nonnegative random variable, denote the time to failure of the material specimen with a distribution function $F_T(t)$. If T is viewed as the continuous analog of C, and t as the continuous analog of n,

$$F_T(t) = P(T \leqslant t) = P(C \leqslant n) = \Phi\left(\frac{\mu\sqrt{n}}{\sigma} - \frac{\omega}{\sigma\sqrt{n}} \right).$$

Replacing n by t, one has

$$F_T(t) = \Phi\left(\frac{\mu}{\sigma} t^{1/2} - \frac{\omega}{\sigma} t^{-1/2} \right).$$

If

$$\alpha = \frac{\sigma}{\sqrt{\mu\omega}} \quad \text{and} \quad \beta = \frac{\omega}{\mu},$$

then

$$F_T(t; \alpha, \beta) = \Phi\left\{ \frac{1}{\alpha}\left[\left(\frac{t}{\beta}\right)^{1/2} - \left(\frac{t}{\beta}\right)^{-1/2} \right] \right\} \tag{4.22}$$

is a two-parameter failure distribution for the fatigue life of a device with parameters $\alpha, \beta > 0$. The parameters α and β can be interpreted as shape and scale, respectively.

The two-parameter distribution derived here is a plausible model for the distribution of fatigue life in the same manner as the two-parameter logarithmic normal distribution derived in Section 4.6. Of course the choice of a failure distribution for fatigue life has to be based not only on physical considerations (as was done in this section and Section 4.6), but also on the ability of these distributions to conform to the observed fatigue data. Some properties of the fatigue-life model, such as its density function, its moments, and its failure rate, are briefly discussed below.

4.11.1 Some Properties of the Fatigue-Life Model

Since

$$F_T(t;\alpha,\beta) = \Phi\left\{\frac{1}{\alpha}\left[\left(\frac{t}{\beta}\right)^{1/2} - \left(\frac{\beta}{t}\right)^{1/2}\right]\right\}, \quad \alpha,\beta > 0,$$

it follows that

$$Z = \frac{1}{\alpha}\left[\left(\frac{T}{\beta}\right)^{1/2} - \left(\frac{\beta}{T}\right)^{1/2}\right]$$

is distributed normally with a mean 0 and a variance 1. It therefore follows that the probability density function of $T, f_T(t;\alpha,\beta)$ is

$$f_T(t;\alpha,\beta) = \frac{1}{2\sqrt{2\pi}\ \alpha^2\beta t^2}\ \frac{t^2-\beta^2}{(t/\beta)^{1/2}-(\beta/t)^{1/2}}\exp\left[-\frac{1}{2\alpha^2}\left(\frac{t}{\beta}+\frac{\beta}{t}-2\right)\right],$$

$$t>0; \alpha,\beta>0. \tag{4.23}$$

For $\alpha=\beta=1, f_T(t;\alpha,\beta)$ takes a form that is suitable for graphing the probability density function. This is illustrated in Figure 4.9. Thus

$$f_T(t;1,1) = \frac{t^{1/2}+t^{-1/2}}{2t\sqrt{2\pi}}\exp\left(-\frac{t}{2}-\frac{1}{2t}+1\right), \quad t>0. \tag{4.24}$$

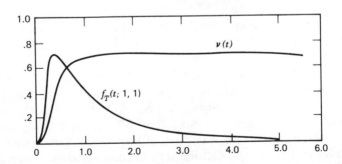

Figure 4.9 The probability density function and average hazard rate for the fatigue-life distribution.

The mean and the variance of T can be found in the usual manner by integration. For ease of computation, however, the following alternative approach is preferable.

Let a random variable X be distributed normally with a mean 0 and a variance $\alpha^2/4$. It follows that $2X$ is also distributed normally with a mean 0, but a variance α^2. Moreover, since Z has a unit normal distribution, αZ is distributed normally with a mean 0 and a variance α^2. Thus, in distribution, $2X$ and αZ are alike, so that

$$2X = \left(\frac{T}{\beta}\right)^{1/2} - \left(\frac{\beta}{T}\right)^{1/2}$$

or

Hence

$$T = \beta(1 + 2X^2 + 2X\sqrt{1+X^2}).$$

and

$$E(T) = \beta\left(1 + \frac{\alpha^2}{2}\right)$$

$$V(T) = (\alpha\beta)^2\left(1 + \frac{5\alpha^2}{4}\right). \tag{4.25}$$

The hazard rate of the fatigue-life model has also been investigated by Birnbaum and Saunders (1969), who established that, although the hazard rate is not increasing, the average hazard rate (see Section 9.4.1) is nearly nondecreasing. Specifically, they have shown that in the limit, as $t \to \infty, (1/t)\int_0^t h(s)\,ds$ approaches a constant; here

$$h(t)\,(\text{the hazard rate}) = \frac{f_T(t; 1, 1)}{1 - F_T(t; 1, 1)}.$$

The probability density function, $f_T(t; 1, 1)$, and the average hazard rate, $v(t) = (1/t)\int_0^t h(s)\,ds$, are shown in Figure 4.9.

4.12 SUMMARY

In this chapter, several distributions that are reasonable models for the life distribution of a device are presented. Practically all these distributions have been motivated by considering either a physical failure process or the wear-out and aging properties of the device. Since these distributions may not be appropriate for several situations arising in practice, the objective

Table 4.1 Some failure distributions

Failure Distribution	Probability Density or Distribution Function	Moment Generating Functions	Mean	Variance
The exponential with parameter $\lambda>0$	$f(x;\lambda)=\lambda e^{-\lambda x}, \quad x>0$	$\left(1-\dfrac{t}{\lambda}\right)^{-1}$	$\dfrac{1}{\lambda}$	$\dfrac{1}{\lambda^2}$
The gamma with parameters $\lambda>0$, and $k\geqslant 1$	$f(x;\lambda,k)=\dfrac{e^{-\lambda x}(\lambda x)^{k-1}\lambda}{\Gamma(k)}, \quad x\geqslant 0$	$\left(1-\dfrac{t}{\lambda}\right)^{-k}$	$\dfrac{k}{\lambda}$	$\dfrac{k}{\lambda^2}$
The Weibull with parameters $\gamma\geqslant 0$ and $\beta,\eta>0$	$f(x;\gamma,\eta,\beta)=\left\{\dfrac{\beta}{\eta}\left(\dfrac{x-\gamma}{\eta}\right)^{\beta-1}\exp\left[-\left(\left(\dfrac{x-\gamma}{\eta}\right)^{\beta}\right)\right]\right\}, \quad x>\gamma$	—	$\gamma+\eta\Gamma\left(\dfrac{\beta+1}{\beta}\right)$	$\eta^2\left[\Gamma\left(\dfrac{\beta+2}{\beta}\right)-\Gamma^2\left(\dfrac{\beta+1}{\beta}\right)\right]$
Bivariate exponential fatal shock model with parameters $\lambda_1,\lambda_2,\lambda_{12}>0$	$F_{XY}(x,y;\lambda_1,\lambda_2,\lambda_{12})=$ $\exp(-\lambda_1 x-\lambda_2 y-\lambda_{12}\max(x,y))$	—	$E(X)=(\lambda_1+\lambda_{12})^{-1}$ $E(Y)=(\lambda_2+\lambda_{12})^{-1}$	$\mathrm{Var}(X)=(\lambda_1+\lambda_{12})^{-2}$ $\mathrm{Var}(Y)=(\lambda_2+\lambda_{12})^{-2}$
Fatigue-life model (Birnbaum-Saunders Distribution) with parameters $\alpha,\beta>0$	$f_T(t;\alpha,\beta)=\dfrac{1}{2\sqrt{2\pi}\,\alpha^2\beta t^2}\left\{\dfrac{t^2-\beta^2}{(t/\beta)^{1/2}-(\beta/t)^{1/2}}\exp\left[-\dfrac{1}{2\alpha^2}\left(\dfrac{t}{\beta}+\dfrac{\beta}{t}-2\right)\right]\right\}$, $x>0$	—	$\beta\left(1+\dfrac{\alpha^2}{2}\right)$	$\left(1+\dfrac{5\alpha^2}{4}\right)(\alpha\beta)^2$

here has been to illustrate the various techniques, approaches, assumptions, and arguments that can be used to justify the choice of a failure model. The reader who does not wish to use any of the distributions given here can proceed to derive his own model for failure to suit his particular situation, perhaps by applying some of the ideas presented in this chapter.

As stated before, the final choice of a failure distribution should be based not only on physical considerations, but also on the ability of a plausible distribution to explain the observed failure data. What has to be avoided, of course, is to choose a failure distribution based purely on failure data, which are sparse in many situations. Essentially, the failure data and an understanding of the physical process causing failure should complement each other in the process in which one arrives at a suitable failure distribution.

For convenience, some of the failure distributions discussed in this chapter are summarized in Table 4.1.

PROBLEMS

4.1. Find the moment generating function of a gamma distribution, with parameters α and β, where β is the shape parameter. Verify your answer with that given in Table 4.1.

4.2. A series system of n components is such that each component has a failure distribution given by

$$F_X(x) = 1 - \exp\left[-\left(\frac{x}{\alpha} \right)^{1/b} \right], \quad x \geqslant 0.$$

If the components operate independently, and if n is large, what is the failure distribution of the system?

4.3. If the components in Problem 4.2 are connected in parallel, what is the failure distribution of the system?

4.4. Consider the corrosion process discussed in Section 4.5. What is the failure distribution of the time-to-corrode random variable if it is assumed that the initial pit depths have a Rayleigh distribution with parameters λ and $\gamma = 0$.

4.5. Verify Equations (4.23) and (4.24).

REFERENCES

Anello, C. (1968), On the Maximum-Likelihood Estimation of Failure Probabilities in the Presence of Competing Risks, *Technical Paper*, RAC-TP-291, Research Analysis Corporation, McLean, Virginia.

Barlow, R. E. and F. Proschan, (1965), *Mathematical Theory of Reliability*, John Wiley and Sons, New York.

Barlow, R. E. and Singpurwalla, N. D. (1973), Averaging Time and Maxima for Dependent Observations, *Proceedings of the Symposium on Statistical Aspects of Air Quality Data*, Triangle Universities Consortium on Air Pollution, Chapel Hill, N. C.

Berkson, J. and L. Elveback, (1960), Competing Exponential Risks with Particular Reference to the Study of Smoking and Lung Cancer, *Journal of the American Statistical Association*, Vol. 55, pp. 415-428.

Birnbaum, Z. W. and S. C. Saunders, (1968), A Probabilistic Interpretation of Miner's Rule, *SIAM Journal of Applied Mathematics*, Vol. 16, pp. 637-652.

Birnbaum, Z. W. and S. C. Saunders, (1969), A New Family of Life Distributions, *Journal of Applied Probability*, Vol. 6, pp. 319-327.

Cohen, A. C. (1965), Estimation of Mixtures of Poisson and Mixtures of Exponential Distributions, *NASA Technical Memorandum* NASA TMX 53245.

Cornfield, J. (1957), The Estimation of the Probability of Developing a Disease in the Presence of Competing Risks, *American Journal of Public Health*, Vol 47, pp. 601-607.

Davis, D. J. (1952), An Analysis of Some Failure Data, *Journal of the American Statistical Association*, Vol. 47, pp. 113-150.

Epstein, B. (1958), The Exponential Distribution and Its Role in Life-Testing, *Industrial Quality Control*, Vol. 15, No. 6, pp. 2-7.

Feller, W. (1957), *An Introduction to Probability Theory and Its Applications*, Vol. I, 2nd Ed., John Wiley and Sons, New York.

Feller, W. (1965), *An Introduction to Probability Theory and Its Applications*, Vol. II, John Wiley and Sons, New York.

Freudenthal, A. M. and M. Shinozuka, (1961), Structural Safety under Conditions of Ultimate Load-Failure and Fatigue, *WADD Technical Report* 61-77.

Goldthwaite, L. (1961), Failure Rate Study for the Lognormal Lifetime Model, *Proceedings of the Seventh National Symposium on Reliability and Quality Control*, pp. 208-213.

Gumbel, E. J. (1958), *Statistics of Extremes*, Columbia University Press, New York.

Gupta, S. (1962), Order Statistics from the Gamma Distribution, *Technometrics*, Vol. 2, pp. 243-262.

Gupta, S. and P. Groll, (1961), Gamma Distribution in Acceptance Sampling Based on Life Tests, *Journal of the American Statistical Association*, Vol. 56, pp. 943-970.

Harris, C. M. and N. D. Singpurwalla, (1968), Life Distributions Derived from Stochastic Hazard Rates, *IEEE Transactions on Reliability*, Vol. R-17, No. 2, pp. 70-79.

Harris, C. M. and N. D. Singpurwalla, (1969), On Estimation in Weibull Distribution with Random Scale Parameters, *Naval Research Logistics Quarterly*, Vol. 16, No. 3, pp. 405-410.

Harris, R. (1968), Reliability Applications of a Bivariate Exponential Distribution, *Journal of the Operations Research Society of America*, Vol. 6, No. 1, pp. 18-27.

Howard, B. T. and G. A. Dodson, (1961), High Stress Aging to Failure of Semiconductor Devices, *Proceedings of the Seventh National Symposium on Reliability and Quality Control.*

Kao, J. H. K. (1958), Computer Methods for Estimating Weibull Parameters in Reliability Studies, *I.R.E. Transactions on Reliability and Quality Control*, PGRQC No. 13, pp. 15-22.

Kao, J. H. K. (1959), A Graphical Estimation of Mixed Weibull Parameters in Life-Testing of Electron Tubes, *Technometrics*, Vol. 1, No. 4, pp. 389-407.

Kao, J. H. K. (1965), Statistical Models in Mechanical Reliability, *Proceedings of the Eleventh National Symposium on Reliability and Quality Control*, pp. 240–247.

Lieblein, J., and M. Zelen, (1956), Statistical Investigation of the Fatigue Life of Deep-Groove Ball Bearings, *Journal of Research, National Bureau of Standards*, Vol. 57, pp. 273-316.

Lloyd, D. K. and M. Lipow, (1962), *Reliability: Management, Methods, and Mathematics*, Prentice-Hall, Englewood Cliffs, New Jersey.

Makeham, W. M. (1873), On an Application of the Theory of the Composition of Decremental Forces, *Institute of Actuaries Journal*, Vol. 18, pp. 317-322.

Marshall, A. W. and I. Olkin, (1967), A Multivariate Exponential Distribution, *Journal of the American Statistical Association*, Vol. 62, pp. 30-44.

Miner, M. A. (1945), Cumulative Damage in Fatigue, *Journal of Applied Mechanics*, Vol. 12, pp. A159–A164.

Neyman, J. (1950), *First Course in Probability and Statistics*, Henry Holt Company, New York, pp. 68-69.

Peck, D. S. (1961), Semiconductor Reliability Predictions from Life Distribution Data, *Proceedings of the AGET Conference on Reliability of Semiconductor Devices*.

Singpurwalla, N. D. (1972), Extreme Values from a Lognormal Law with Applications to Air Pollution Problems, *Technometrics*, Vol. 14, No. 3, pp. 703-711.

Weibull, W. (1951), A Statistical Distribution Function of Wide Applicability, *Journal of Applied Mechanics*, Vol. 18, pp. 293-297.

Widder, D. V. (1941), *The Laplace Transform*, Princeton University Press, Princeton, New Jersey.

CHAPTER 5

Point and Interval Estimation Procedures for Lifetime Distributions (Failure Models)

In this chapter, various procedures for obtaining point and interval estimates for parameters of several lifetime distributions are discussed and illustrated. The parameters of interest are sometimes measures of central tendency, such as the distribution mean or mode, and sometimes the parameters that appear in the expressions giving the density and distribution functions. More often for lifetime distributions, however, the quantity of interest is a distribution percentile, also known as the *reliable life* of the item to be tested, corresponding to some specified population survival proportion; or it is the population proportion surviving at least until a specified time, t_m. The reliability at time t_m or *probability of survival* at least until a time t_m, symbolically $P[T \geqslant t_m]$ where T is failure time, is often denoted as $R(t_m; \boldsymbol{\theta})$. If a survival proportion, γ, is specified, then, by solving $\gamma = R(t_{1-\gamma}; \boldsymbol{\theta})$ for the $100(1-\gamma)$th distribution percentile, $t_{1-\gamma}$, an expression for $t_{1-\gamma}$ is obtained. In order to determine a point estimate of $R(t_m; \boldsymbol{\theta})$ or of $t_{1-\gamma}(\boldsymbol{\theta})$, when the form of $F_T(t; \boldsymbol{\theta})$ is known and is either expressible in closed form or tabulated for specific values of the elements of $\boldsymbol{\theta}$, one need only estimate the parameter vector $\boldsymbol{\theta}$. Unless the vector $\boldsymbol{\theta}$ consists of a single element, however, obtaining confidence intervals for $R(t_m)$ or $t_{1-\gamma}$ requires more than simple point or interval estimation of the unknown parameter vector $\boldsymbol{\theta}$. Examples of obtaining confidence bounds for reliability and reliable life when two parameters are unknown are given in Sections 5.1.2 and 5.2.3.

The data from which one determines the point and interval estimates of reliability, reliable life, and other lifetime distribution parameters of interest are very often obtained from what are called *life tests*, that is, a sample of n items from a population of interest is put into an environment as similar as possible to the one which the items will experience in actual use, and one or more stresses of usually constant severity are applied. If the life test is terminated at a specified time t_0 before all n items have failed, one speaks of *Type I censoring* of the life test. *Type II censoring* occurs when the life test is terminated at the time of a particular (the rth, say, $r \leqslant n$) failure. In Type I censoring the number of failures and all the failure times are random variates, whereas in Type II censoring the number of failures is considered fixed, and the only random variates are the failure times.

Censoring may also be conducted progressively, that is, items may be removed from life test throughout the duration of the test. Such a model has been treated to some extent for the two-parameter Weibull distribution.

Because of the "lack of memory" of the exponential distribution, one can use the exponential model when conducting a life test *with replacement*. This means that, at the time of failure of any item, the item is replaced by an essentially new one, and the life test is then continued. Section 5.1.3 deals with this situation.

In Chapter 6, testing of hypotheses, reliability demonstrations, selection of sample sizes and censoring numbers, and so forth, are discussed for some of the distributions considered in this chapter, most particularly the exponential distribution.

5.1 THE EXPONENTIAL DISTRIBUTION

The exponential distribution as a failure model was derived in Section 4.2.2. The probability density function of an exponential variate X is given by

$$f_X(x) = \begin{cases} \lambda \exp[-\lambda(x-\mu)], & x \geqslant \mu, \\ 0, & x < \mu; \lambda > 0, \mu \geqslant 0. \end{cases} \tag{5.1}$$

The cumulative distribution function, that is, $P(X \leqslant x)$, is

$$F_X(x) = \begin{cases} 1 - \exp[-\lambda(x-\mu)], & x \geqslant \mu, \\ 0, & x < \mu; \lambda > 0, \mu \geqslant 0. \end{cases} \tag{5.2}$$

The density and cumulative distribution functions are sometimes expressed in terms of the scale parameter, $\theta = \lambda^{-1}$.

The expected value of $W = X - \mu$ is given by

$$E(W) = \int_0^\infty \frac{w}{\theta} \exp\left(\frac{-w}{\theta}\right) dw = \int_0^\infty \theta z \exp(-z) dy.$$

From results demonstrated in Section 4.3 relating to the gamma distribution, one sees that $E(W)$ (or the mean of the distribution of $X - \mu$) is equal to $\theta \Gamma(2) = \theta 1! = \theta$. The mean of the distribution of X is therefore $\mu + \theta$. If μ is known to be zero, then, of course, the mean of X is simply θ. Similarly,

$$E(X^2) = \int_\mu^\infty \frac{x^2}{\theta} \exp\left[\frac{-(x - \mu)}{\theta} \right] dx = \int_0^\infty (\theta z + \mu)^2 \exp(-z)\, dy$$

$$= \theta^2 \Gamma(3) + 2\mu\theta\Gamma(2) + \mu^2 = 2\theta^2 + 2\mu\theta + \mu^2.$$

Thus the variance of both W and X is equal to $E(X^2) - (EX)^2 = \theta^2$. If X represents failure time, λ is referred to as the hazard rate, and in this case, if $\mu = 0$, θ is called the mean time to failure.

As discussed in Section 4.2, the exponential distribution has been and continues to be used extensively as a model for failure-time distributions. It will be shown in Section 7.3 that the exponential distribution provides bounds on the reliability functions of distributions from certain restricted classes.

Zelen and Dannemiller (1961) have demonstrated that the use of a one-parameter exponential model for failure analysis when, in fact, the correct distribution is a two-parameter Weibull distribution with decreasing hazard rate (see Section 5.2) tends to lead to incorrect conclusions concerning the population from which the life-test sample is selected. In statistical parlance, the exponential assumption is not *robust*, or cannot be used indiscriminantly when the underlying model is other than exponential.

An exponential model can be used in certain situations in which hazard rate is not constant if an appropriate transformation of the failure-time data is made. Suppose that the ordered failure times $t_{(1)}, \ldots, t_{(r)}$ from a sample of size n are observed. If it is known that the observations are from a population of Weibull variates with known shape and location parameters β and μ, respectively, and unknown scale parameter δ [see Equation (5.6), Section 5.2], then $x_{(1)} = (t_{(1)} - \mu)^\beta, \ldots, x_{(r)} = (t_{(r)} - \mu)^\beta$ can be considered ordered observations from an exponential distribution with unknown scale parameter (mean) $\theta = \delta^\beta$. To make such a transformation, it is necessary to have life-tested the item of interest extensively enough to have established that the Weibull shape parameter is independent of the level of stress applied and that β and μ can be assumed to be known for given environmental conditions. Such a situation is analogous to one in which a normal distribution has a fixed known standard deviation, but an unknown mean.

Saunders (1968) discusses situations in which the scatter parameter (such as the Weibull shape parameter) is fixed for a particular material or fabric with respect to the distribution over time of fatigue failures. Lieblein and Zelen (1956) demonstrate that the Weibull location and shape parameters for failure-time distributions of deep-groove ball bearings are 0 and between 1 and 2, respectively. Methods of testing whether or not a particular set of data (or transformed data) is from an exponential distributic า are given in Chapter 7.

Point and interval estimation procedures for the exponential distribution vary accordingly as the location parameter μ is or is not 0. The parameter μ, incidentally, is sometimes called a *guarantee* or *threshold parameter*, since the probability of a failure occurring before time μ is zero. The choice of estimation procedures also depends on whether a life-testing procedure is terminated at a specified time (Type I censoring) or at the time of an observed failure (Type II censoring) and whether or not testing is conducted with replacement. Testing without replacement and the case in which μ is known (i.e., the one-parameter exponential distribution) will be considered first.

5.1.1 Point and Interval Estimation Procedures for the One-Parameter Exponential Distribution; Testing without Replacement

If the location parameter μ of an exponential distribution is 0, or is known and can be set equal to 0 by an appropriate transformation of the data, point and interval estimation procedures are simplified significantly. For a random variate X chosen from such a population, the probability density function and the cumulative distribution function are given by Equations (5.1) and (5.2), respectively, with $\mu \equiv 0$. It is assumed here that testing is conducted without replacement.

(a) **Estimation for Type II Censoring; μ Known.** Suppose that a sample of size n with failure-time distribution given by (5.2) with $\mu \equiv 0$ has been subjected to life testing and that the test is terminated at the time of the rth failure, $r \leqslant n$. More generally, one might assume that a sample of size n has been randomly selected from a one-parameter exponential population, that the observations from the sample become available in order, and that experimentation is terminated at the time that the rth observation becomes available.

For $\mu \equiv 0$, the joint density function of the ordered observations $X_{(1)}, \ldots, X_{(r)}$, $X_{(i)} \leqslant X_{(i+1)}$, $i = 1, \ldots, r-1$ and $r = 1, \ldots, n$, is given by

$$f_{X_{(1)}, \ldots, X_{(r)}}(x_1, \ldots, x_r) = \frac{n!}{(n-r)! \theta^r} \exp\left\{ -\frac{\left[\sum_{i=1}^{r} x_i + (n-r)x_r \right]}{\theta} \right\},$$

$$0 \leqslant x_1 \leqslant \ldots \leqslant x_r. \tag{5.3}$$

Clearly, if the right-hand side of (5.3) is maximized with respect to θ, one obtains, as the maximum-likelihood estimator of θ,

$$\hat{\theta} = \frac{\sum_{i=1}^{r} X_{(i)} + (n-r)X_{(r)}}{r}.$$

[The maximization of the right side of (5.3) is facilitated by first taking its logarithm.] One may also observe from (5.3), by using Neyman's criterion given in Section 3.4, that $\hat{\theta}$ is a sufficient statistic for θ.

To show that the density of $\hat{\theta}$ is complete and that $\hat{\theta}$ is unbiased, it is necessary to know the distribution of $\hat{\theta}$. This can be obtained by considering an alternative form of the density function of $X_{(1)}, \ldots, X_{(r)}$,

$$f_{X_{(1)}, \ldots, X_{(r)}}(x_1, \ldots, x_r) = \prod_{i=1}^{r} \left\{ \frac{(n-i+1)}{\theta} \exp\left[\frac{-(n-i+1)(x_i - x_{i-1})}{\theta} \right] \right\},$$

$$0 \leqslant x_1 \leqslant \ldots \leqslant x_r, \tag{5.4}$$

with x_0 defined as zero. From (5.4), each $(n-i+1)(X_{(i)} - X_{(i-1)})/\theta$, $i = 1, \ldots, r$ (with $x_0 \equiv 0$), can be seen to have an independent exponential distribution with mean unity. Thus

$$\frac{r\hat{\theta}}{\theta} = \frac{\sum_{i=1}^{r} (n-i+1)(X_{(i)} - X_{(i-1)})}{\theta}$$

is a sum of independent exponentials with mean 1. Therefore, from results of Section 4.3 relating to a k-fold convolution of independent exponentials, $Y = r\hat{\theta}/\theta$ has a gamma distribution with scale parameter 1 and shape parameter r. Finally,

$$E\left(\frac{r\hat{\theta}}{\theta}\right) = \int_0^{\infty} \frac{xx^{r-1}}{\Gamma(r)} \exp(-x)\, dx = \frac{\Gamma(r+1)}{\Gamma(r)} = \frac{r!}{(r-1)!} = r,$$

so that $E(\hat{\theta}) = \theta$.

The fact that the density of $\hat{\theta}$ is complete will now be demonstrated. Consider the arbitrary function $u(\hat{\theta})$. The expectation of u is given by

$$E[u(\hat{\theta})] = \int_0^{\infty} \frac{u(t)(t)^{r-1} \exp(-rt/\theta)\, dt}{(\theta/r)^r \Gamma(r)}, \quad r = 1, 2, \ldots,$$

which is proportional to the Laplace transform of $\hat{\theta}^{r-1}u(\hat{\theta})$. Hence, by the unicity of the Laplace transform, for all r/θ greater than 0, if $E[u(\hat{\theta})]=0$, then $u(\hat{\theta})\equiv 0$. This completes the proof.

Because the statistic $\hat{\theta}$ is based on a complete sufficient statistic, it is, by the Lehmann-Scheffé-Blackwell theorem discussed in Section 3.4.1, the unique minimum-variance estimator among unbiased estimators of θ. The statistic $\hat{\theta}$ is also efficient, that is, its variance is equal to the Cramér-Rao bound for regular unbiased estimators of θ (see Problem 5.1 and Section 3.4.1). The expression for the variance of $\hat{\theta}$ can be determined by first finding the expectation of $Y^2 = (r\hat{\theta}/\theta)^2$. From results above concerning the bias of $\hat{\theta}$, one sees that $E(Y^2)=\Gamma(r+2)/\Gamma(r)=(r+1)r$. Then, $\text{Var}(Y) = E(Y^2)-[E(Y)]^2 = r^2+r-r^2 = r$, so that $\text{Var}(\hat{\theta})=\theta^2/r$.

Since θ is a scale parameter, the best invariant estimator of θ, defined in Section 3.4 under "Completeness," is

$$\tilde{\theta} = \frac{\hat{\theta}}{1+1/r} = \frac{r\hat{\theta}}{r+1}.$$

The mean-squared error of $\tilde{\theta}$ is

$$\frac{\theta^2/r}{1+1/r} = \frac{\theta^2}{r+1}.$$

The estimators $\hat{\theta}$ and $\tilde{\theta}$ clearly approach equivalence as r approaches infinity. They are both consistent and asymptotically normal, and the mean-squared error of $\tilde{\theta}$ is equal to the Cramér-Rao bound for regular invariant estimators of the scale parameter θ (see Section 3.4.1 and Problem 5.1).

It was observed earlier that in Equation (5.4) each $S_i=(n-i+1)(X_{(i)}-X_{(i-1)})/\theta, i=1,\ldots,r$, has an independent exponential distribution with scale parameter equal to 1. By referring to the density function of a chi-square variate given in Section 3.3 one can assure himself that each $2S_i$, $i=1,\ldots,r$, has an independent chi-square distribution with 2 degrees of freedom. Then, from Section 3.3, one can easily infer that $\sum_{i=1}^{r}2S_i=2r\hat{\theta}/\theta$ (being the sum of r independent chi squares) has a chi-square distribution with $2r$ degrees of freedom. A lower confidence bound on θ at confidence level $1-\alpha$ is therefore given by $2r\hat{\theta}/\chi^2_{1-\alpha}(2r)$, where $\chi^2_\gamma(k)$ is the 100γth percentile of a chi-square distribution with k degrees of freedom. The interval $[2r\hat{\theta}/\chi^2_{1-\alpha/2}(2r), 2r\hat{\theta}/\chi^2_{\alpha/2}(2r)]$ is a two-sided confidence interval for θ at confidence level $1-\alpha$. To obtain such intervals one simply substitutes calculated values for $\hat{\theta}$ and tabulated values of chi-square percentiles. (An example is given on p. 166 after the discussion of estimation of reliability and reliable life.) The one-sided confidence bound for θ

is *UMA* (uniformly most accurate; see Section 3.4 and Problem 5.2), and two-sided confidence intervals are *UMA* among unbiased intervals. If a mission time t_m is specified, one can obtain a point estimate for the reliability function $R(t_m) = 1 - F(t_m)$ by substituting either $\hat{\theta}$ or $\tilde{\theta}$ for θ in $\exp(-t_m/\theta)$. A lower confidence bound for $R(t_m)$ at level $1-\alpha$ can be obtained by substituting $2r\hat{\theta}/\chi^2_{1-\alpha}(2r)$ for θ in the same expression. Two-sided confidence intervals are obtained similarly.

The minimum-variance unbiased point estimator of $R(t_m)$ [derived for $r = n$ by Pugh (1963) and Basu (1964)] is given by

$$R^*(t_m) = \begin{cases} \left(1 - \dfrac{t_m}{n\hat{\theta}}\right)^{n-1}, & n\hat{\theta} > t_m, \\ 0, & n\hat{\theta} \leqslant t_m. \end{cases}$$

Graphical results displayed by Zacks and Even (1966) demonstrate, however, that, except for $t_m/n\hat{\theta}$ relatively large or close to zero, the maximum-likelihood estimator (MLE) has smaller mean-squared error than the minimum-variance unbiased estimator of $R(t_m)$. As one might expect, the discrepancy in mean-squared error between the two estimators decreases with n. For $r = n$ as small as 4, the MLE has considerably smaller mean-squared error over the range $(\frac{1}{2}, 2)$ for $t_m/n\hat{\theta}$. For $t_m/n\hat{\theta}$ greater than about 3.7, $R^*(t_m)$ has less mean-squared error, but for both estimators mean-squared error is quite small in this case.

If a survival proportion γ is specified, x_P, the $100(1-\gamma)$th percentile of the distribution, is defined by $\gamma \equiv 1 - P = \exp(-x_P/\theta)$ as

$$x_P = \theta \ln\left(\frac{1}{\gamma}\right) = \theta \ln\left(\frac{1}{1-P}\right).$$

Point or interval estimators of x_P (the reliable life corresponding to the survival proportion γ) are obtained by substituting for θ the point or interval estimators given above for this parameter.

Example. If the observed exponential failure times in days from a censored sample of size 24 are 6.2, 9.2, and 16.9, that is, $r = 3, n = 24$, then $\hat{\theta} = [6.2 + 9.2 + 22 \ (16.9)]/3 = 129.1$ days and $\tilde{\theta} = 96.8$ days with mean-squared errors $\theta^2/3$ and $\theta^2/4$, respectively. Point estimates of reliable life $x_{.10}$ corresponding to a survival proportion of .90 are $\hat{X}_{.10} = \hat{\theta}\ln(1/ .90) = 129.1 \ (.10536) = 13.6$ days and $\tilde{X}_{.10} = \tilde{\theta}\ln(1/.90) = 10.2$ days. An 80% lower confidence bound for $X_{.10}$ is $6(13.6)/\chi^2_{.80} \ (6) = 81.6/8.558 = 9.5$ days. If a mission time of 10 days is specified, an 80% lower confidence bound on reliability is given by $\exp[-(10 \times 8.558)/(6 \times 129.1)] \cong .896$. Examples showing the use of the statistic $r\hat{\theta}$ for solutions of problems involved with reliability demonstration are given in Section 6.2.4.

DECREASING LIFE-TEST TIME WHEN OBTAINING ESTIMATES UNDER TYPE II CENSORING. Epstein and Sobel (1953) have pointed out that the mean-squared error in estimating θ for the fixed-failure-time model is a function of θ and r only, that is, θ^2/r or $\theta^2/(r+1)$, and thus is independent of sample size n. Since failures tend to occur more rapidly as n increases, one can decrease the waiting time for the observance of r failures by increasing n for fixed $r \leqslant n$. It would be desirable to do this, however, only if any increased expense of the life-test procedure was small compared with the advantage gained by saving time.

Epstein and Sobel give a table of ratios of expected waiting times, $\{E(X_{(r)})\}$, to observe the rth failure in samples of sizes n and r, respectively, where $E(X_{(r)}) = \theta \sum_{j=1}^{r}(n-j+1)^{-1}$ (see Problem 6.27 in Chapter 6). Essentially the same tabulated values can be obtained by using ratios of median values rather than expected values of the rth observation. The latter method eliminates most of the calculation involved for large r. One uses the fact (derived in Section 7.4.2) that $1 - F_{X_{(r)}}(x_r)$, for F any continuous distribution and the unordered X's independently and identically distributed, has a beta distribution with parameters $n - r + 1$ and r. Thus $v_{.50}(n-r+1,r)$, with $v_{\gamma}(a,b)$ the 100γth percentile of a beta distribution with parameters a and b, is the median of $\exp(-X_{(r)}/\theta)$, and $-\theta \ln v_{.50}(n-r+1,r)$ is the median of $X_{(r)}$. The ratio $\tau_{n,r}$ of median waiting times to observe the rth failures in samples of sizes n and r, respectively, is therefore given by $\ln v_{.50}(n-r+1,r)/\ln v_{.50}(1,r)$. One can also easily determine, by use of the beta percentiles, values of $\rho_{r,n} = \ln v_{.50}(n-r+1,r)/\ln v_{.50}(1,n)$, the ratio for samples of size n of median waiting times to observe the rth and nth failures, respectively.

Values of 50th percentiles $v_{.50}(n-r+1,r)$ of appropriate beta distributions can be found easily from tabulations of Harter (1964) and of Bracken (1966) or can be approximated by $2^{-1/n} - [(r-1)/(n-1)] \cdot (2^{1-1/n} - 1)$ (see Johnson, 1963). Therefore the median weighting time for 4 failures out of 4 test items is approximately $-\theta \ln[2^{-1/4} - (2^{3/4} - 1)] \cong -\theta \ln(.159) \cong 1.84\theta$, while the median weighting time for 4 failures out of 20 test items is $-\theta \ln[.966 - (3/19) \cdot (.932)] \cong .20\theta$. The ratio of the latter value to the former, .108, is very slightly larger than the ratio .104 based on expected, rather than median, waiting times. Selection of r and n for a given life test is discussed in some detail in Section 6.2.4. A few values of $\tau_{r,n}$ (actually based on expected waiting times) and $\rho_{r,n}$ (based on median waiting times) are given in Tables 5.1 and 5.2. The value for $\rho_{2,4}$ shown in Table 5.2 is calculated as $\ln v_{.50}(3,2)/\ln v_{.50}(1,4)$, where $\ln v_{.50}(1,4)$ from calculations above is approximately 1.84. Since $\ln v_{.50}(3,2)$ is approximately equal to $\ln[2^{-1/4} - \frac{1}{3}(2^{3/4} - 1)] \cong .48$, $\rho_{2,4} \cong .26$.

Table 5.1   Some values of $\tau_{r,n}$

r \ n	1	2	3	4	5	10	15	20
1	1	.50	.33	.25	.20	.10	.067	.050
2		1	.56	.39	.30	.14	.092	.068
3			1	.59	.43	.18	.12	.087
4				1	.62	.23	.14	.104
5					1	.28	.18	.125
10						1	.35	.23

Table 5.2   Some values of $\rho_{r,n}$

n \ r	1	2	3	4	5
1	1				
2	.28	1			
3	.14	.44	1		
4	.094	.26	.52	1	
5	.068	.18	.33	.56	1

One should note that values of $\rho_{r,n}$ do not apply to estimators of θ with the same mean-squared error, that is, sample size n is fixed but not censoring number r, which determines the mean-squared error of the estimator of θ. Tables 5.1 and 5.2 can be made to apply to Weibull failure-time populations by raising each entry to the power $\xi = \beta^{-1}$, with β defined in (5.6), as long as the entries are considered to be ratios of median waiting times.

PREDICTION INTERVALS FOR SAMPLE OBSERVATIONS. In statistical methods generally, and particularly in this chapter, confidence intervals refer to interval estimates for the unknown parameter(s) of a probability distribution. In life testing, however, a problem arises that is not adequately covered by the methods of confidence intervals. Suppose that n units are placed on test and that the test is censored (without replacement) at $r \leqslant n$. The failure times then occur in order, and it may be required on the basis of the first $k < r$ failure times, $X_{(1)}, X_{(2)}, \ldots, X_{(k)}$, to predict the rth ordered observation, $X_{(r)}$. A prediction for $X_{(r)}$ will give an estimate of when the test will be completed, which is $X_{(r)}$, or, similarly, it will predict the additional testing time required, namely, $X_{(r)} - X_{(k)}$.

Lawless (1971) has treated the case of interval estimation of $X_{(r)}$ when lifetime follows the exponential distribution, and his results are described in this section. If S_k is defined as

$$S_k = \sum_{i=1}^{k} X_{(i)} + (n-k)X_{(k)},$$

a point estimate of $E(X_{(r)})$ is given by

$$\hat{\theta} \sum_{j=1}^{r} (n-j+1)^{-1} \quad \text{with} \quad \hat{\theta} = \frac{S_k}{k}.$$

Lawless showed that the random variable $U = (X_{(r)} - X_{(k)})/S_k$ has probability distribution

$$f_U(u) = \frac{(n-k)!}{(r-k-1)!(n-r)!} \sum_{i=0}^{r-k-1} \binom{r-k-1}{i} (-1)^i [1 + (n-r+i+1)u],$$

$$u > 0.$$

The cumulative distribution function of U is

$$P(U < t) = 1 - \frac{(n-k)!}{(r-k-1)!(n-r)!}$$

$$\times \sum_{i=0}^{r-k-1} \frac{\binom{r-k-1}{i}(-1)^i}{(n-r+i+1)[1+(n-r+i+1)t]^k}.$$

This distribution function is easily programmed on a computer. Suppose, then, that t_γ is chosen so that $P(U < t_\gamma) = \gamma$. Then clearly,

$$P\left(\frac{X_{(r)} - X_{(k)}}{S_k} < t_\gamma\right) = \gamma,$$

so that

$$P(X_{(r)} < X_{(k)} + t_\gamma S_k) = \gamma$$

and $X_{(k)} + t_\gamma S_k$ is an upper γ *prediction* limit for $X_{(r)}$. The word confidence is avoided because of possible confusion with parameter estimation. It should be noted that always $X_{(r)} \geqslant X_{(k)}$.

As pointed out by Lawless, two cases of special interest exist. First, if $r = n$, the distribution function of U simplifies to

$$P(U < t) = \sum_{i=0}^{n-k} \binom{n-k}{i} (-1)^i (1 + it)^{-k}.$$

Second, if $k = r - 1$, that is, if all but the last failure time have been observed, $(r - 1)(n - r + 1)U$ has an F distribution with $(2, 2r - 2)$ degrees of freedom and the required prediction limit can be obtained directly.

In the replacement case (with either type of censoring), if $X_{(k)}$ is the elapsed time from the start of testing until the kth failure, the random variable $k(X_{(r)} - X_{(k)})/(r - k)X_{(k)}$ has an F distribution with $(2r - 2k, 2k)$ degrees of freedom, and again a prediction interval for $X_{(r)}$ can be obtained directly.

A problem related to this topic, that is, deciding whether or not to continue a test in the face of early failures, is discussed in Brown and Rutemiller (1971).

Example. Suppose that $n = 4$ units are placed on test (without replacement), the test is scheduled to stop when all 4 units have failed, and the part lifetimes have identical exponential failure distributions. Suppose further that at the second failure $X_{(1)} = 71.5$ and $X_{(2)} = 84.7$. This is the special case $r = n$ with $r = n = 4$ and $k = 2$. Then

$$P(U < t = 3) = \sum_{i=0}^{n-k} \binom{n-k}{i} (-1)^i (1 + it)^{-k}$$

$$= \sum_{i=0}^{2} \binom{2}{i} (-1)^i (1 + 3i)^{-2}$$

$$\cong .90.$$

Since $S_2 = 325.6$, then $X_{(2)} + t_{.90} S_2 = 84.7 + 3(325.6) = 1061.5$, and 1061.5 is an upper one-sided .90 prediction interval for $X_{(4)}$.

An alternative approach allows for the direct (noniterative) determination of approximate closed-form prediction intervals for $X_{(r)}$ when $P(U < t)$ is given. It is derived in Mann and Grubbs (1973). Note that

$$(X_{(r)} - X_{(k)}) = \sum_{i=k+1}^{r} (X_{(i)} - X_{(i-1)})$$

$$= \sum_{i=k+1}^{r} [2(n-1+1)]^{-1} 2(n-i+1)(X_{(i)} - X_{(i-1)}).$$

Thus $Q = (X_{(r)} - X_{(k)})/\theta$ is a sum of weighted independent chi squares with mean

$$\sum_{i=k+1}^{r} \frac{1}{n-i+1} \quad \text{and} \quad \text{variance} \quad \sum_{i=k+1}^{r} \frac{1}{(n-i+1)^2}.$$

One can now use the approach of Grubbs (1971), discussed in Section 5.1.2 and applying to a sum Q of weighted independent chi squares with mean m and variance v. Grubbs uses the chi-square approximation, that is, $2mQ/v$ is approximately chi square with $2m^2/v$ degrees of freedom, which Patnaik (1949) applies when Q is a weighted noncentral chi square, a weighted sum of independent noncentral chi squares.

Also, it is known that

$$\frac{2k\hat{\theta}}{\theta} = \frac{2 \sum_{i=1}^{k} (n-i+1)(X_{(i)} - X_{(i-1)})}{\theta}.$$

has a chi-square distribution with $2k$ degrees of freedom, and that Q and $\hat{\theta}$ are independent since they involve different independent exponential variates. Therefore, from Section 3.3.5(c),

$$F_r = \frac{\left[\sum_{i=k+1}^{r} 1/(n-i+1) \right]^{-1} (X_{(r)} - X_{(k)})}{\hat{\theta}}$$

has approximately the Fisher-Snedecor F distribution with

$$\frac{2\left[\sum_{i=k+1}^{r} 1/(n-i+1)\right]^2}{\sum_{i=k+1}^{r} 1/(n-i+1)^2}$$

and $2k$ degrees of freedom. In agreement with Lawless' result, if $k = r - 1$, then $(n - r + 1)(X_{(r)} - X_{(k)})/\hat{\theta}$ has exactly an F distribution with 2 and $2k = 2(r - 1)$ degrees of freedom.

For the numerical example given earlier, in which $n = r = 4, k = 2$, the variate F_r has degrees of freedom $2(1 + \frac{1}{2})^2/(1 + \frac{1}{4}) = 3.6$ and 4. Thus, for $F_\gamma(2a, 2b)$ the 100γth percentile of an F distribution with $2a$ and $2b$ degrees of freedom, $X_{(r)}$ is less than or equal to $84.7 + F_{.90}(3.6, 4)$ $(325.6/2) \times (1.5)$ with probability approximately equal to .90.

In general, for noninteger degrees of freedom, one can use the relationship

$$F(2a, 2b) = \frac{(b/a) V(a,b)}{1 - V(a,b)},$$

where $V(a,b)$ is a beta variate with parameters a and b. A method is given in Section 10.4.1 for estimating percentiles of $-\ln V$ or $-\ln(1 - V)$, shown in Section 10.4.1 to be also sums of weighted chi-squares. To apply the weighted chi-square approximation for $-\ln(1 - V)$ to this example, one first calculates the mean m and variance v of $-\ln(1 - V)$ for V with parameters 3.6/2 and 4/2:

$$m = \psi(2 + 1.8) - \psi(2)$$

and

$$v = -\psi'(3.8) + \psi'(2),$$

where $\psi(z)$ and $\psi'(z)$ are the first and second derivatives, respectively, of $\ln \Gamma(z)$. One approximates $\psi(z)$ by $\ln(z - .5)$ and $\psi'(z)$ by $1/(z - .5)$, so that $m \cong 1.1939 - .4055 = .7884$ and $v \cong -.3030 + .6666 = .3636$. Therefore $2(.7884)[-\ln(1 - V)]/.3636 = -4.336\ln(1 - V)$ is approximately chi square with 3.419 degrees of freedom. The 90th percentile of $\chi^2(3.419)$ is approximately 6.87, so that the 90th percentile of $-\ln(1 - V)$ is approximately 1.58, and the 90th percentiles of V and F are approximately $1 - \exp(-1.58) = .794$ and $(4/3.6)(.794)/.206 = 4.28$, respectively. The value 1.58 can be obtained by use of the Wilson-Hilferty transformation of chi square to normality, discussed in Section 5.1.2. For χ_γ^2, the 100γth percentile of a chi-square variate with ν degrees of freedom

$$\chi_\gamma^2 \cong \nu \left(1 - \frac{2}{9\nu} + \frac{\phi_\gamma \sqrt{2}}{3\sqrt{\nu}} \right)^3$$

where ϕ_γ is the 100γth percentile of a standard normal variate. Hence the approximate upper prediction limit for $X_{(4)}$ is $84.7 + 4.28(244.2) = 1129.9$.

Clearly, for the given data, the prediction interval corresponding to a specified probability based on the F approximation involves more calculation than the approach of Lawless for determining the probability associated with a given interval. In the general case, however, the F approximation may be considerably simpler than even a single iterative calculation of the exact expression of Lawless, and it has the further advantage that no iteration is necessary to calculate the interval corresponding to a specified probability.

At the end of the material on "Estimation for Type II Censoring; μ Not Known" in Section 5.1.2(a), the approximate prediction-interval technique of Mann and Grubbs (1973) is applied to the case involving an unknown exponential location parameter.

(b) Estimation for Type I Censoring; μ Known. If a life test is terminated at a specified time t_0 and the exponential location parameter is 0, the joint probability density function of $X_{(1)}, \ldots, X_{(r)}$ is given by

$$f_{X_{(1)}, \ldots, X_{(r)}}(x_1, \ldots, x_r) = \frac{n!}{(n-r)! \theta^r} \exp \left\{ - \left[\sum_{i=1}^{r} x_i + (n-r)t_0 \right] \right\},$$

$$0 \leqslant x_1 \leqslant \ldots \leqslant x_r < t_0.$$

The maximum-likelihood estimator of θ is easily seen to be

$$\hat{\theta} = \frac{\sum_{i=1}^{r} X_{(i)} + (n-r)t_0}{r}, \quad r \neq 0.$$

This estimator is biased for small samples but has all the desirable asymptotic properties associated with the MLE of θ under Type II censoring.

To obtain a confidence interval for θ, one may consider the probability that the number of failures, K, in a sample of size n is equal to r at time t_0:

$$P(K = r) = \frac{n!}{(n-r)! r!} [R(t_0)]^{n-r} [1 - R(t_0)]^r.$$

Then a conservative (because K is discrete and the failure times are not used) lower confidence bound ρ at level $1-\alpha$ for $R(t_0)$ is given by the solution of

$$\sum_{i=n-r}^{n} \binom{n}{i} \rho^i (1-\rho)^{n-i} = \alpha = \int_0^\rho \frac{\Gamma(n+1)}{\Gamma(n-r)\Gamma(r+1)} x^{n-r-1}(1-x)^r \, dx.$$

(Such a confidence bound, based on the binomial distribution of K, is discussed at greater length in Section 7.4.3.) From the relationship between the binomial and the beta distributions demonstrated in Section 7.4.3, it can be seen that ρ is equal to $1-v_{1-\alpha}(r+1,n-r)=v_\alpha(n-r,r+1)$, the 100αth percentile of the beta distribution with parameters $n-r$ and $r+1$. From this, one can determine, from $R(t_0)=\exp(-t_0/\theta)$, a lower confidence bound for θ at level $1-\alpha$ as $t_0/\ln(1/\rho)$. If t_0 is 24 days, the sample size is 15 and the number of failures is 2; then a 90% lower confidence bound for θ is given by $24/\ln(1/.6827) \cong 63$ days. Examples of calculations related to tests of hypotheses for this model are given in Section 6.2.

Percentiles of the beta distribution can be evaluated by using tables of percentiles of the F distribution and the transformation

$$V(\nu_1,\nu_2) = \frac{(\nu_1/\nu_2)F(2\nu_1,2\nu_2)}{1+(\nu_1/\nu_2)F(2\nu_1,2\nu_2)}$$

where $V(a,b)$ is a beta variate with parameters a and b, and $F(2a,2b)$ is an F-distributed variate with degrees of freedom $2a$ and $2b$.

Bartholomew (1963) discusses methods for obtaining exact confidence bounds for this model. He shows that a confidence bound based only on the number of failures (considered above) is extremely efficient, for $R(t_0) \geqslant .5$, compared with exact confidence bounds based on $\hat\theta$, which are very difficult to obtain. Bartholomew also gives expressions for the mean and the second, third, and fourth moments about the mean of $\hat\theta/\theta$.

5.1.2 Point and Interval Estimation Procedures for the Two-Parameter Exponential Distribution; Testing without Replacement

It is assumed here that μ is not zero. Moreover, μ is assumed to be unknown, so that it cannot be set equal to zero by an appropriate transformation of the data. Testing is without replacement. Estimation procedures for Type II censoring are considered first.

(a) **Estimation for Type II Censoring; μ Not Known.** For μ unknown and Type II censoring, the joint density of $X_{(1)},\dots,X_{(r)}$ is given by

$$f_{X_{(1)},\ldots,X_{(r)}}(x_1,\ldots,x_r) = \frac{n!}{(n-r)!\theta^r}\exp\left\{\frac{-\left[\sum_{i=1}^{r}(x_i-\mu)+(n-r)(x_r-\mu)\right]}{\theta}\right\},$$

$$0 \leqslant x_{(1)} \leqslant \ldots \leqslant x_{(r)}.$$

Results relating to point and interval estimation of θ and μ for this model were first derived by Epstein and Sobel (1954). The reader is urged to satisfy himself that the maximum-likelihood estimators of μ and θ are

$$\hat{\mu} = X_{(1)} \qquad \text{and} \qquad \hat{\theta} = \frac{1}{r}\left[\sum_{i=1}^{r}X_{(i)}+(n-r)X_{(r)}-nX_{(1)}\right],$$

respectively. The MLE's $X_{(1)}$ and $\hat{\theta}$ represent joint sufficient statistics for μ and θ, and it can be shown that these statistics are complete (see Sections 3.4 and 5.1.1 and Problem 5.4). Therefore, by the Lehmann-Scheffé-Blackwell theorem, unique minimum-variance unbiased estimators θ^* and μ^* of θ and μ, respectively, exist and are functions of $\hat{\theta}$ and $X_{(1)}$. Specifically, they are $\theta^* = [r/(r-1)]\hat{\theta}$ and $\mu^* = X_{(1)} - \theta^*/n$.

The variances of the minimum-variance unbiased estimators of the two parameters are given by

$$\text{Var}(\theta^*) = \frac{\theta^2}{r-1} \qquad \text{and} \qquad \text{Var}(\mu^*) = \frac{\theta^2 r}{n^2(r-1)}$$

and their covariance is

$$\text{Cov}(\theta^*,\mu^*) = -\frac{\theta^2}{n(r-1)}.$$

(see Problem 5.6). From the expressions for θ^* and μ^*, the variance of θ^*, and the covariance of θ^* and μ^*, expressions for the unique best invariant estimators $\tilde{\theta}$ and $\tilde{\mu}$ of the two parameters can be calculated. [The best invariant estimators are those with smallest mean-squared error among estimators $\bar{\theta}$ and $\bar{\mu}$ such that $E(\bar{\theta}-\theta)/\theta$ and $E(\bar{\mu}-\mu)/\theta$ are independent of both θ and μ (see Section 3.4 and Mann, 1969c).] These are

$$\tilde{\theta} = \hat{\theta} \qquad \text{and} \qquad \tilde{\mu} = X_{(1)} - \frac{\theta^*}{n} + \frac{\tilde{\theta}}{n(r-1)} = X_{(1)} - \frac{\hat{\theta}}{n}.$$

The mean-squared errors of these estimators are given by

$$\text{MSE}(\tilde{\theta}) = \frac{\theta^2}{r} \quad \text{and} \quad \text{MSE}(\tilde{\mu}) = \frac{r\theta^2}{n^2(r-1)} - \frac{\theta^2}{n^2(r-1)r}$$

$$= \frac{r\theta^2}{n^2(r-1)}\left(\frac{r^2-1}{r^2}\right) = \frac{(r+1)\theta^2}{n^2 r}.$$

One would ordinarily prefer to use minimum-variance unbiased estimators rather than best invariant estimators *only* if a series of estimates was to be averaged.

One can determine confidence intervals for μ by using the fact that $2n(X_{(1)} - \mu)/\theta$ and $2(r-1)\theta^*/\theta$ are independent and each is distributed as chi square with 2 and $2(r-1)$ degrees of freedom, respectively (see Problem 5.5). The ratio $W = n(X_{(1)} - \mu)/\theta^*$ has an F distribution with degrees of freedom 2 and $2(r-1)$, respectively (see Section 3.3). Hence a lower confidence bound for μ at confidence level $1 - \alpha$ is given by an observed value of $X_{(1)} - \theta^* F_{1-\alpha}[2, 2(r-1)]/n$, where $F_{1-\alpha}[2, 2(r-1)]$ is the $100(1-\alpha)$th percentile of the F distribution with 2 and $2(r-1)$ degrees of freedom. An upper bound for μ (with probability 1 attached) is, of course, given by the smallest observed value of X, namely, $x_{(1)}$.

One can eliminate the need for tables of the F distribution in obtaining a confidence interval for μ by using the fact (derived by Epstein and Sobel, 1954) that the 100γth percentile $F_\gamma[2, 2(r-1)]$ of the F distribution with 2 and $2r-2$ degrees of freedom is given by $(r-1)[(1-\gamma)^{-1/(r-1)} - 1]$ (see Problem 5.7). Thus, for two-parameter exponential data, the 4 smallest observations from a sample of size $5, x_{(1)} = 36.0, x_{(2)} = 57.5, x_{(3)} = 81.5, x_{(4)} = 127.2$, one obtains

$$\hat{\mu} = 36.0 \quad \text{and} \quad \theta^* = \frac{36.0 + 57.5 + 81.5 + 2(127.2) - 5(36.0)}{3} = 83.1.$$

Then, since $F_{.50}(2,6) = .78$, a 50% lower confidence bound for μ is given by $36.0 - .78(83.1)/5 = 23.1$.

Since $2(r-1)\theta^*/\theta$ is distributed as chi square with $2(r-1)$ degrees of freedom, a lower confidence bound for θ at confidence level $1 - \alpha$ can be obtained as $2(r-1)\theta^*/\chi^2_{1-\alpha}(2r-2)$, with $\chi^2_{1-\alpha}(2r-2)$ the $100(1-\alpha)$th percentile of χ^2 with $2r-2$ degrees of freedom. For the given data, a 50% lower confidence bound for θ is $6(83.1)/5.348 = 93.2$.

CONFIDENCE INTERVALS FOR $R(t_m)$ AND PREDICTION INTERVALS FOR SAMPLE OBSERVATIONS. Grubbs (1971) has suggested a method by which one can, for a fixed mission time t_m, determine an approximate confidence bound for $R(t_m) \equiv \exp[-(t_m - \mu)/\theta]$, the population survival proportion, or reliability, at time t_m. Note that

$$\ln\left[\frac{1}{R(t_m)}\right] = \frac{t_m - \mu}{\theta} \equiv \frac{X_{(1)} - \mu}{\theta} + \left[\frac{t_m - X_{(1)}}{(r-1)\theta^*}\right]\left[\frac{(r-1)\theta^*}{\theta}\right].$$

Thus

$$P\left\{\ln\left[\frac{1}{R(t_m)}\right] > \eta\right\} = P\left\{\frac{X_{(1)} - \mu}{\theta} + \frac{Z(r-1)\theta^*}{\theta} > \eta\right\},$$

where $Z = (t_m - X_{(1)})/[(r-1)\theta^*]$. Using the fact that $2n(X_{(1)} - \mu)/\theta$ and $2(r-1)\theta^*/\theta$ are independent chi-square variates with 2 and $2r-2$ degrees of freedom, respectively, one obtains

$$Q \equiv \ln\left[\frac{1}{R(t_m)}\right] = \frac{1}{2n}\chi^2(2) + \frac{Z}{2}\chi^2(2r-2).$$

If Z is observed as z and is considered to be fixed, the distribution of Q can be obtained as a sum of weighted chi-square variates. The parameter Q, expressed as the variate $c_1\chi^2(2) + c_2\chi^2(2r-2)$ with $c_1 = 1/2n$ and $c_2 = z/2$, can, on the basis of results of Patnaik (1949), be approximated by a single central chi-square variate. To accomplish this, it is necessary to determine the mean and variance of Q, given $Z = z$. One obtains, since $E[\chi^2(\nu)] = \nu$,

$$E(Q|z) = \frac{1}{n} + \frac{t_m - x_{(1)}}{\theta^*} \equiv m,$$

say. Also, since $\text{Var}[\chi^2(\nu)] = 2\nu$,

$$\text{Var}(Q|z) = \frac{1}{n^2} + \frac{(t_m - x_{(1)})^2}{(r-1)(\theta^*)^2} \equiv v, \text{ say.}$$

Then, by Patnaik's result, $2mQ/v$ is approximately chi square with $2m^2/v$ degrees of freedom. Because one would not expect $2m^2/v$ to be an integer, in general, the Wilson-Hilferty (1931) transformation of chi square to normality can be used. This transformed variate, $\{[\chi^2(\nu)/\nu]^{1/3} + 2/9\nu - 1\} \times (9\nu/2)^{1/2}$, is approximately normally distributed with mean 0 and variance 1.

Thus one obtains

$$P\left\{R(t_m) > \exp\left\{-m\left[1 - \frac{v}{(3m)^2} + \frac{\phi_{1-\alpha}v^{1/2}}{3m}\right]^3\right\}\right\} = 1 - \alpha,$$

where $\phi_{1-\alpha}$ is the $100(1-\alpha)$th percentile of a standard normal distribution, and m and v are defined as above.

Note that here Grubbs has used what has been termed a *fiducial* approach, since the data are assumed to be fixed and the parametric function $Q = (t_m - \mu)/\theta$ is considered to be a random variate whose posterior distribution (see Section 2.8) is approximated. The fiducial approach approximates optimum confidence intervals very well in this case, but must be modified for cases considered in Section 10.4 in which prior density functions are implicitly specified for more than a single parametric function.

For the two-parameter exponential data given earlier, that is, $n = 5$, $r = 4$, with $x_{(1)} = 36.0$, $x_{(2)} = 57.5$, $x_{(3)} = 81.5$, and $x_{(4)} = 127.2$, one obtains, for $t_m = 30.0$,

$$m = \frac{1}{5} + \frac{30.0 - 36.0}{83.1} = .1278 \quad \text{and} \quad v = \frac{1}{25} + \frac{(30.0 - 36.0)^2}{3(83.1^2)} = .0417.$$

Then an approximate 90% lower confidence bound for $R(30.0)$ is

$$\exp\left\{-(.1278)\left[1 - \frac{.0417}{.1470} + 1.282\sqrt{.0417/.1470}\right]^3\right\} \cong .71$$

To calculate a 50% lower confidence bound for $R(30)$ one substitutes $\phi_{.50} = 0$ for $\phi_{.90} = 1.282$. An alternative and much more difficult procedure would be to determine a distribution percentile ω of $-Z_\gamma$, using Monte Carlo simulation methods or numerical integration. Then, for $P = 1 - \gamma$ and $Y_1 = (X_{(1)} - \mu)/\theta$, values of $\omega_{1-\alpha}$ are defined implicitly by

$$P\left\{\frac{Y_1 - \ln[1/R(x_P)]}{(r-1)(\theta^*/\theta)} < \omega_{1-\alpha}\right\} = 1 - \alpha.$$

One could determine a lower confidence bound at level $1 - \alpha$ for the value of x_P, the distribution percentile corresponding to the γth survival proportion with $R(x_P) = \gamma$, and hence $\ln[1/R(x_P)] = \ln(1/\gamma)$. Such a bound would be given by $x_{(1)} - \omega_{1-\alpha}(r-1)\theta^*$. Pierce (1972) shows that the approximation of Grubbs works well in comparison with an exact classical confidence bound obtained by numerical integration.

The weighted chi-square approximation, in conjunction with the Wilson-Hilferty transformation, gives surprisingly good approximations to percentiles of sums of weighted independent chi-squares, such as Q. When the mean m and the variance v of Q are known exactly, the combined approximation to the $100p$th percentile of Q agrees with the corresponding exact percentile to within about 3 units in the third significant figure for $.01 \leqslant p \leqslant .99$ and $2m^2/v \geqslant 3$. Examples of the use of this approximation and comparisons with exact values are given in Sections 5.1.1, 5.2.3, 5.2.4 and 10.4. and in references cited therein.

The approximate prediction-interval technique of Mann and Grubbs (1973), discussed in Section 5.1.1 for the case in which μ is known, can be applied here also. The approximate chi-square variate,

$$2\frac{\left[\left(\sum_{i=r+1}^{m} 1/(n-i+1)\right)/\sum_{i=r+1}^{m} 1/(n-i+1)^2\right](X_{(m)}-X_{(r)})}{\theta}, \quad m>r$$

with

$$X_{(m)}-X_{(r)} = \sum_{i=r+1}^{m}(X_i - X_{(i-1)})$$

has a distribution independent of μ and independent of the chi-square variate,

$$\frac{2(r-1)\theta^*}{\theta} = \frac{2\sum_{i=2}^{r}(n-i+1)(X_{(i)}-X_{(r)})}{\theta}.$$

Thus a prediction interval for $X_{(m)}$ based on the first r observations of an exponential variate can be obtained from the fact that

$$F_m = \frac{\left[\sum_{i=r+1}^{m} 1/(n-i+1)\right]^{-1}(X_{(m)}-X_{(r)})}{\theta^*}$$

has approximately an F distribution with

$$\frac{2\left[\sum_{i=r+1}^{m} 1/(n-i+1)\right]^2}{\sum_{i=r+1}^{m} 1/(n-i+1)^2}$$

and $2(r-1)$ degrees of freedom.

For the data given above with $n = 5$, $r = 4$, $\theta^* = 83.1$, $x_{(4)} = 127.2$, we obtain, as an exact (because $r = m - 1$) 75% upper prediction interval for $X_{(5)}$,

$$127.2 + F_{.75}(2,6)(83.1) = 273.6.$$

(b) Estimation for Type I Censoring; μ Not Known. If a life test, instead of being terminated at the time of an observed failure, is terminated at some specified time t_0, then, for testing without replacement, the joint probability density function, or the likelihood function, of the r life-test ordered failure times $X_{(1)}, \ldots, X_{(r)}$ observable before time t_0 is

$$f_{X_{(1)}, \ldots, X_{(r)}}(x, \ldots, x_r) = \frac{n!}{(n-r)! \theta^r}$$

$$\cdot \exp \left\{ - \frac{\left[\sum_{i=1}^{r} (x_i - \mu) + (n-r)(t_0 - \mu) \right]}{\theta} \right\}, \quad 0 < x_1 \leqslant \ldots \leqslant x_r < t_0.$$

For Type I censoring, the maximum-likelihood estimators of μ and θ are

$$\hat{\mu} = X_{(1)} \quad \text{and} \quad \hat{\theta} = \frac{1}{r} \left[\sum_{i=1}^{r} X_i + (n-r) t_0 - n X_{(1)} \right], \quad r \neq 0,$$

respectively. One cannot, however, obtain estimators of μ as functions of $\hat{\mu}$ having mean-squared error independent of μ.

5.1.3   Point and Interval Estimation Procedures for Testing with Replacement

The following applies to a life-test situation such that any item, at the time of its failure, is repaired and replaced on life test with essentially zero time assumed for repair. Here we consider Type I censoring first.

(a) Estimation for Type I Censoring with Replacement. For $\mu = 0$, one observes a Poisson process in which r failures are observed in time $t_r = n t_0$, where t_0 is the time of termination of the life test and n is the number of items in the sample subjected to life test. For this model, the number of observed failures, K, is a random variate with

$$P(K = r) = \left(\frac{1}{r} \right)! (\lambda t_r)^r \exp(-\lambda t_r)$$

with $\lambda = 1/\theta$. If μ is known to be 0, the maximum-likelihood estimator $\hat{\lambda}$ of $\lambda = \theta^{-1}$ is $r/t_r = r/nt_0$, and $\hat{\theta} = \hat{\lambda}^{-1} = nt_0/r$. Then

$$P(K \leqslant r) = \sum_{k=0}^{r} \frac{1}{k!} (\lambda t_r)^k \exp(-\lambda t_r) = \int_{\lambda t_r}^{\infty} \frac{z^r}{r!} \exp(-z)\, dz,$$

an incomplete gamma function with shape parameter $r+1$ (see Section 4.3). Therefore $\mathrm{Prob}(K \leqslant r) = \mathrm{Prob}[\chi^2(2r+2) > 2\lambda t_r]$. Thus, if one observes r failures, then, with probability at least $1 - \alpha$ (we say *at least* because K is discrete, as in the case discussed in Section 7.4.3), $2\lambda r\hat{\theta} < \chi^2_{1-\alpha}(2r+2)$, or $2nt_0/\chi^2_{1-\alpha}(2r+2)$ is a conservative lower confidence bound for θ at confidence level $1 - \alpha$. If 20 electron tubes are life-tested with replacement for 10 days, and 3 failures are observed, a 90% lower confidence bound for the mean time to failure is $40(10)/\chi^2_{.90}(8) = 400/13.36 = 29.9$ days. A 90% lower confidence bound for the probability of survival until time $t_m = 5$ days is $\exp(-5/29.9) = .85$. An assumption of exponential failure time for electronic components is often used.

Similarly, it can be shown that $\mathrm{Prob}(K > r) = \mathrm{Prob}[\chi^2(2r) < 2\lambda t_r]$. Hence a conservative upper confidence bound for θ at level $1 - \alpha$ is given by $2r\hat{\theta}/\chi^2_{\alpha}(2r)$, and a conservative two-sided confidence interval at level $1 - \alpha$ is

$$\left[\frac{2r\hat{\theta}}{\chi^2_{1-\alpha/2}(2r+2)}, \ \frac{2r\hat{\theta}}{\chi^2_{\alpha/2}(2r)} \right].$$

Cox (1953) discusses the interval $[(2r\hat{\theta}/\chi^2_{1-\alpha/2}(2r+1),\ 2r\hat{\theta}/\chi^2_{\alpha/2}(2r+1)]$, which he states is slightly narrower than that given above, but sometimes has a true confidence coefficient less than $1 - \alpha$. For μ not known, the probability density function of $X_{(1)}, \ldots, X_{(r)}$ is

$$f_{X_{(1)}, \ldots, X_{(r)}}(x_1, \ldots, x_r) = (n\lambda)^r \exp[-n\lambda(t_0 - \mu)], \quad 0 < x < \ldots < x_r < t_0.$$

The maximum-likelihood estimators of μ and λ for this model are $\hat{\mu} = X_{(1)}$ and $\hat{\lambda} = r/[n(t_0 - X_{(1)})], r \neq 0$, respectively. As in the case of Type I censoring without replacement, one cannot obtain confidence intervals for μ or reliability parameters that are invariant under translations of the data.

(b) Estimation for Type II Censoring with Replacement. If a life test is performed with replacement and testing is terminated at the time of the rth observed failure, then, for μ known to be 0, the joint probability density function of the ordered failure times $X_{(1)}, \ldots, X_{(r)}$ is given by

$$f_{X_{(1)}, \ldots, X_{(r)}}(x_1, \ldots, x_r) = (n\lambda)^r \exp(-n\lambda x_r), \quad 0 < x_1 \leqslant \ldots \leqslant x_r.$$

From this expression, one can obtain the maximum-likelihood estimator of λ as $(r/X_{(r)})/n$ and of θ as $(nX_{(r)})/r$. For this model, one can also write the joint probability density of $X_{(1)}, \ldots, X_{(r)}$ as

$$(n\lambda)^r \exp\left[-n\lambda \sum_{i=1}^r (x_i - x_{i-1})\right], \quad x_0 \equiv 0 < x_1 \leqslant \ldots \leqslant x_r. \quad (5.5)$$

In other words, each $S_i = X_i - X_{i-1}$, $i = 1, \ldots, r$, with $X \equiv 0$, has an independent exponential distribution with scale parameter $(\lambda n)^{-1}$. Since $X_{(r)} = \sum_{i=1}^r S_i$, the density of $X_{(r)}$ for testing with replacement is given by

$$f_{X_{(r)}}(x_r) = \frac{(n\lambda)^r x_r^{r-1}}{(r-1)!} \exp(-n\lambda x_r).$$

Consequently, $2n\lambda X_{(r)}$ has a chi-square distribution with $2r$ degrees of freedom, and confidence bounds for λ, θ, reliable life, or reliability can be obtained as functions of observed values of $X_{(r)}$ and values to be obtained from tables of the chi-square distribution function. If μ is not known to be equal to 0, the likelihood function of $X_{(1)}, \ldots, X_{(r)}$ is

$$f_{X_{(1)}, \ldots, X_{(r)}}(x_1, \ldots, x_r) = (n\lambda)^r \exp[-n\lambda(x_r - \mu)], \quad x_1 < \ldots < x_r,$$

and the maximum-likelihood estimators of μ and λ are

$$\hat{\mu} = X_{(1)} \quad \text{and} \quad \hat{\lambda} = \frac{r}{n(X_{(r)} - X_{(1)})}.$$

By writing the joint density function of the ordered X's as in (5.5), one can see that

$$2n\lambda(X_{(1)} - \mu) \quad \text{and} \quad 2n\lambda(X_{(r)} - X_{(1)}) = 2n\lambda \sum_{i=2}^r [X_{(i)} - \mu - (X_{(i-1)} - \mu)]$$

are independent chi-square variates with 2 and $2(r-1)$ degrees of freedom, respectively. Thus the methods described in Section 5.2.1 for obtaining confidence intervals for the two parameters can be applied in this *testing-with-replacement* model.

Also, since the expectation of a chi-square variate is equal to its degrees of freedom,

$$E(\hat{\theta}) = E\left[\frac{n(X_{(r)} - X_{(1)})}{r}\right] = \frac{(r-1)\theta}{r} \quad \text{and} \quad E(\hat{\mu}) = \mu + \frac{\theta}{n}.$$

The estimators $X_{(1)}$ and $\hat{\theta}$ can be shown, exactly as for the case of testing without replacement (see Problem 5.4), to be complete sufficient statistics. Therefore $\theta^* = r\hat{\theta}/(r-1)$ and $\mu^* = X_{(1)} - \theta^*/n$ are minimum-variance unbiased estimators of θ and μ, respectively. Since the same variance-covariance relationships apply here as in the nonreplacement model, the best invariant estimators can be defined similarly, that is, $\tilde{\theta} = \hat{\theta}$ and $\tilde{\mu} = X_{(1)} - \hat{\theta}/n$.

For a sample (with replacement) of size 5 from an exponential population for which the first observed failure time is 56 hours and the fourth and last observed failure time, occurring at the termination of the life test, is 200 hours, the maximum-likelihood estimate of θ for μ known is $5(200)/4 = 250$ hours, and an 80% lower confidence bound for θ is $10(200)/\chi^2_{.80}(8) \cong 181$ hours. If μ is not assumed to be known, $\hat{\mu} = 56$ hours and $\hat{\theta} = 5(200 - 56)/4 = 180$ hours. An 80% lower confidence bound for θ is then $10(200 - 56)/\chi^2_{.80}(6) \cong 168$ hours, and a 50% lower confidence bound for μ is $56 - F_{.50}(2,6)(\frac{1}{3})(144) = 18$ hours, since $F_{.50}(2,6) = .78$. Unbiased estimators of θ and μ are $\theta^* = (4/3)\hat{\theta} = 240$ hours and $\mu^* = \hat{\mu} - \theta^*/5 = 56 - 48 = 8$ hours, and the estimators with smallest mean-squared error are $\tilde{\theta} = \hat{\theta} = 180$ hours, and $\tilde{\mu} = \hat{\mu} - \hat{\theta}/5 = 56 - 180/5 = 20$ hours.

Reliability demonstration tests relating to the exponential models considered in this section are discussed and illustrated in Chapter 6, Sections 6.2.4 and 6.3.1. Examples and problems relating to statistical tests of hypotheses concerning functions of θ and the selection of sample size n and censoring number r are also given there.

PROBLEMS FOR SECTION 5.1

Problems 5.1 through 5.7 apply to Type II censoring and testing without replacement.

5.1. For the one-parameter exponential distribution with unknown scale parameter θ, show that the maximum-likelihood estimator $\hat{\theta}$ is efficient. Also, show that the mean-squared error of the best invariant estimator $\tilde{\theta}$ of θ is equal to the Cramér-Rao bound for regular invariant estimators of θ.

5.2. A lower confidence bound L, a function of $T(\mathbf{x})$ for a parameter θ, is a uniformly most accurate bound L_{min} at level $1 - \alpha$ (has, for any $\theta' < \theta$ and all θ, the minimum probability among all lower confidence bounds at level $1 - \alpha$ of being less than θ') if $\mathbf{X}$ has density $f_{\mathbf{X}}(\mathbf{x}, \theta) = C(\theta) \exp[Q(\theta)T(\mathbf{x})]h(\mathbf{x})$, where Q is strictly monotone. Show that $2r\hat{\theta}/\chi^2_{1-\alpha}(2r)$ is a uniformly most accurate lower confidence bound for the exponential scale parameter θ at level $1 - \alpha$ when μ is known.

5.3. Calculate, for $n=12$, $r=6$, $\tau_{n,r}$ and $\rho_{r,n}$, the ratios of median waiting times to observe, respectively, the rth failure in samples of sizes n and r and, for samples of size n, the median waiting times to observe the rth and nth failures. Use the approximation $2^{-1/n} - [(r-1)/(n-1)](2^{1-1/n}-1)$ for the median of a beta distribution with parameters $n-r+1$ and r.

5.4.*a*. Show that the maximum-likelihood estimators $X_{(1)}$ and $\hat{\theta}$ are sufficient statistics for an exponential distribution with location and scale parameters μ and θ, respectively, when censoring is Type II.

b. Prove that these statistics have densities that are complete.

5.5. For a two-parameter exponential distribution with location parameter μ and scale parameter θ, demonstrate that $2n(X_{(1)}-\mu)/\theta$ and $2(r-1)\theta^*/\theta$ are independent and that each is distributed as chi square with 2 and $2(r-1)$ degrees of freedom, respectively, when censoring is Type II. *Hint*: See Equation (5.4), which gives an alternative method of writing the exponential density function when μ is zero.

5.6.*a* Show that the variances of the minimum-variance unbiased estimators of the exponential distribution parameters μ and θ are $\theta^2 r/(n^2 r - n^2)$ and $\theta^2/(r-1)$. (Use the hint given in Problem 5.5.)

b. Using the fact that $X_{(1)}$ and $\hat{\theta}$ are independent, show that the covariance of the minimum-variance unbiased estimators of μ and θ is $-\theta^2/(nr-n)$.

5.7. Using the fact that a beta-distributed variate V with parameters ν_1 and ν_2 is equal to $(\nu_1/\nu_2)F/[1+(\nu_1/\nu_2)F]$ when F has Snedecor's F distribution with $2\nu_1$ and $2\nu_2$ degrees of freedom, show that $F_\gamma(2,2r-2)=(r-1)[1-\gamma]^{-1/(r-1)}-1]$. The density function of the beta variate V is given by

$$f_V(v) = \frac{(\nu_1+\nu_2-1)!}{(\nu_1-1)!(\nu_2-1)!} v^{\nu_1-1}(1-v)^{\nu_2-1}, \quad 0<v\leqslant 1; \quad \nu_1,\nu_2>0.$$

5.8. Show that, for a two-parameter exponential distribution with scale parameter θ, the best invariant estimator of the location parameter μ for Type II censoring and testing with replacement is $X_{(1)}-\hat{\theta}/n$ with mean-squared error $(r+1)\theta^2/n^2 r$.

5.9. Show that, for $\mu=0$ and testing with replacement, a conservative upper confidence bound for θ at level $1-\alpha$ is given by $r\hat{\theta}/\chi_\alpha^2(2r)$ when censoring is Type I.

5.2 THE WEIBULL DISTRIBUTION

Two derivations of the Weibull distribution as a distribution of failure times are given in Section 4.4. The probability density function of a random variate T having the three-parameter Weibull distribution is given by

$$f_T(t) = \begin{cases} \dfrac{\beta}{\delta}\left(\dfrac{t-\mu}{\delta}\right)^{\beta-1} \exp\left[-\left(\dfrac{t-\mu}{\delta}\right)^\beta\right], & t \geqslant \mu, \\[4mm] 0, & t<\mu;\ \beta,\delta>0; \mu\geqslant 0. \end{cases} \qquad (5.6)$$

The parameter μ for the Weibull distribution, as for the exponential distribution, is sometimes called a *guarantee* or *threshold parameter* since, if T is failure time, a failure occurs before time μ with probability 0. The parameter δ, sometimes called the characteristic life, is a scale parameter specifying the $100[1 - \exp(-1)]$th distribution percentile of $T - \mu$. The parameter β (or β^{-1}) determines the shape of the distribution. If $\beta > 1$, the distribution is unimodal and the hazard function $(\beta/\delta)[(T - \mu)/\delta]^{\beta-1}$ is increasing with T. If $\beta \leqslant 1$, there is no mode and the density function decreases monotonically with T. If $\beta < 1$ or $\beta = 1$, the hazard function is increasing or constant, respectively. If β tends to infinity, the distribution of $T - \mu$ tends toward a degenerate distribution at δ. It should be noted here, as in Section 5.1, that if μ and β are known one can consider the random variate $(T - \mu)^{\beta}$ as an exponentially distributed variate with scale parameter $\theta = \delta^{\beta}$. In such a case, the methods described in Section 5.1 can be used for point and interval estimation of functions of θ.

The mean of the three-parameter Weibull distribution is given by $\mu + \delta\Gamma(1 + 1/\beta)$ and the variance by $\delta^2[\Gamma(1 + 2/\beta) - \Gamma^2(1 + 1/\beta)]$, where the definition of $\Gamma(\cdot)$ is given in Section 4.3. The distribution median is $\mu + \delta(\ln 2)^{1/\beta}$, and the distribution mode (for $\beta > 1$) is $\mu + \delta(1 - 1/\beta)^{1/\beta}$. The reliability function is $R(t) = \exp[-(t/\delta)^{\beta}]$, and reliable life corresponding to a survival proportion $\gamma = 1 - P$ is $t_P = \delta[\ln(1/\gamma)]^{1/\beta}$.

The derivation, given in Section 4.4, of the Weibull distribution as an asymptotic distribution of a smallest extreme suggests the type of failure phenomenon for which the Weibull model is appropriate. If the item or component of interest is of a type in which a large number of flaws exist, if the underlying distribution of flaw severity is of the necessary form, and if the failure of any item depends on its severest flaw (or weakest spot), the Weibull distribution is applicable for failure analysis. This distribution has been used, for example, in the analysis of failure in electronic components, ball bearings, semiconductor devices, photoconductive cells, motors and capacitors, and various biological organisms and psychological test situations, as well as for the study of breaking strength and fatigue in textiles. It has also been used to analyze corrosion resistance, leakage failure of dry batteries, return of goods after shipment, marketing life expectancy of drugs, number of downtimes per shift, and particle-size data.

5.2.1 Point Estimation for the Three-Parameter Weibull Distribution

Analytical estimates of the three distribution parameters, when all are unknown, should in general be obtained iteratively. Very often the method of maximum likelihood is used. It is to be noted that, if censoring of a life test is at time t_0 rather than at the time of the rth failure $T_{(r)}$, the procedure described below can be used with t_0 substituted for $T_{(r)}$.

The likelihood function L of $(r - r_0)$ order statistics, $T_{(r_0+1)} \leqslant T_{(r_0+2)}$ $\leqslant \cdots \leqslant T_{(r)}$, is given by

$$L = f_{T_{(r_0+1)}, \ldots, T_{(r)}}(t_{r_0+1}, \ldots, t_r)$$

or

$$
L = \frac{n!}{(n-r)! r_0!} \left(\frac{\beta}{\delta} \right)^{r-r_0} \prod_{i=r_0+1}^{r} \left(\frac{t_i - \mu}{\delta} \right)^{\beta-1}
$$

$$
\cdot \exp\left[-\sum_{i=r_0+1}^{r} \left(\frac{t_i - \mu}{\delta} \right)^{\beta} - (n-r)\left(\frac{t_r - \mu}{\delta} \right)^{\beta} \right] \left\{ 1 - \exp\left[-\left(\frac{t_{r_0+1} - \mu}{\delta} \right)^{\beta} \right] \right\}^{r_0},
$$

$$(5.7)$$

so that

$$
\frac{\partial \ln L}{\partial \delta} = -\frac{(r-r_0)\beta}{\delta} + \beta \frac{\sum_{i=r_0+1}^{r} (t_i - \mu)^{\beta}}{\delta^{\beta+1}} + \frac{\beta(n-r)(t_r - \mu)^{\beta}}{\delta^{\beta+1}}
$$

$$
- \frac{\beta(t_{r_0+1} - \mu)^{\beta}}{\delta^{\beta+1}} \cdot \frac{r_0 \exp\left[-(t_{r_0+1} - \mu)^{\beta}/\delta^{\beta} \right]}{1 - \exp\left[-(t_{r_0+1} - \mu)^{\beta}/\delta^{\beta} \right]}
$$

$$(5.8)$$

$$
\frac{\partial \ln L}{\partial \beta} = (r - r_0)\left(\frac{1}{\beta} - \ln\delta \right) + \sum_{i=r_0+1}^{r} \ln(t_i - \mu) - \sum_{i=r_0+1}^{r} \left(\frac{t_i - \mu}{\delta} \right)^{\beta} \ln\left(\frac{t_i - \mu}{\delta} \right)
$$

$$
- (n-r)\left(\frac{t_r - \mu}{\delta} \right)^{\beta} \ln\left(\frac{t_r - \mu}{\delta} \right) + r_0(t_{r_0+1} - \mu)^{\beta} \ln\left(\frac{t_{r_0+1} - \mu}{\delta} \right)
$$

$$
\cdot \frac{\exp\left\{ -\left[(t_{r_0+1} - \mu)/\delta \right]^{\beta} \right\}/\delta^{\beta}}{1 - \exp\left[-(t_{r_0+1} - \mu)^{\beta}/\delta^{\beta} \right]},
$$

$$(5.9)$$

and

$$\frac{\partial \ln L}{\partial \mu} = (1-\beta) \sum_{i=r_0+1}^{r} (t_i-\mu)^{-1} + \frac{\beta}{\delta^\beta} \sum_{i=r_0+1}^{r} (t_i-\mu)^{\beta-1}$$

$$+ (n-r)\frac{\beta}{\delta^\beta}(t_r-\mu)^{\beta-1} - \beta r_0 (t_{r_0+1}-\mu)^{\beta-1}$$

$$\times \cdot \frac{\exp\left[-(t_{r_0+1}-\mu)^\beta/\delta^\beta\right]/\delta^\beta}{1-\exp\left[-(t_{r_0+1}-\mu)^\beta/\delta^\beta\right]}. \tag{5.10}$$

Lemon (1974) has recently modified the likelihood equations so that one need iteratively solve only two equations for estimates of μ and β, which together then specify an estimate of δ. Here, the procedure based on the iterative solution of (5.8), (5.9) and (5.10) is discussed. A computer program listing for this procedure can be obtained from Dr. H. L. Harter, Aerospace Research Laboratories, Wright-Patterson Air Force Base, Ohio.

Assuming that $t_1 \leqslant \cdots \leqslant t_r$ are the smallest ordered observations in a sample of size n, one begins iterative estimation of the three parameters, one at a time, in the cyclic order δ, β, and μ after initial estimates are selected. Initially r_0 is assumed to be 0. Any parameters assumed to be known can, of course, be eliminated from the estimation procedure. If μ is known, the maximum-likelihood procedure can be simplified significantly, as discussed in Section 5.2.2. [If β is known but μ is not known, the tables of Harter and Dubey (1967) can be used to test hypotheses concerning the distribution mean and variance.]

Harter and Moore (1965) suggest the rule of false position (iterative linear interpolation) for determining the value, if any, of the parameter, being estimated at any given step, that satisfies the appropriate likelihood equation into which the latest estimates (or known values) of the other two parameters have been substituted. Positive values of the estimates $\hat{\delta}$ of δ and $\hat{\beta}$ of β can always be found in this way. It may be, however, that no value in the interval $(0,t_1)$ satisfies the likelihood equation (5.10). In such a case, the likelihood function in $(0,t_1)$ is either decreasing monotonically, so that $\hat{\mu}=0$, or increasing monotonically, so that $\hat{\mu}=t_1$. The latter situation occurs when $\hat{\beta}$ is less than or equal to 1, since then the right-hand side of Equation (5.10), for $r_0=0$, contains only positive terms. Once this has occurred, it is impossible to continue the iterative estimation procedure without making some modification. Harter and Moore suggest censoring the smallest observation at this point and any others equal to it (r_0

observations in all). Iteration then continues until the results of successive steps agree to within some assigned tolerance.

Dubey (1966) proposes an alternative solution of the problem of $\hat{\mu} = t_1$. He points out that $\mu^* = t_1 - \delta\Gamma(1 + 1/\beta)/n^{1/\beta}$ is an unbiased estimate of μ. Of course, since δ and β are not known, this estimate cannot be used directly but can be employed with estimates substituted for δ and β. One can continue the iterative estimation procedure by using such an estimate for μ in place of t_1.

The graphical procedure described in Section 5.2.2 provides a convenient method for obtaining initial estimates of μ, δ^β, and β. In the graphing process, an iterative procedure is necessary if the initial estimate of μ is not zero, that is, if the observed failure times do not plot into a straight line without modification of the data. The graphical method is also used alone for estimation of the parameters.

Another method of obtaining quick initial estimates of the three Weibull parameters is suggested by Dubey (1966). He proposes, for μ, a simple estimator of the form

$$\bar{\mu} = \frac{T_{(1)}T_{(k)} - T_{(j)}^2}{T_{(1)} + T_{(k)} - 2T_{(j)}},$$

where $T_{(k)}$ is any ordered observable variate such that $T_{(1)} < T_{(j)} < (T_{(1)}T_{(k)})^{1/2}$.

A slight modification of the two-order-statistic estimator for β (or $\xi = 1/\beta$) for μ known, which is discussed in Section 5.2.2, is then used as an initial estimator of β:

$$\bar{\beta} = \frac{2.99}{\ln(T_{([.9737n]+1)} - \bar{\mu}) - \ln(T_{([.1673n]+1)} - \bar{\mu})}$$

with $[x]$ indicating the largest integer less than x.

Then δ is estimated by

$$\bar{\delta} = \frac{T_{([.1673n]+1)} - \bar{\mu}}{.183^{1/\bar{\beta}}}.$$

For a size-40 sample of semiconductor devices, life-tested until all fail, if the times of the 1st, 4th, and 30th ordered failure times are 2, 3, and 198 hours, $[2(198) - 3^2]/[2 + 198 - 2(3)] = 387/194 = 1.93$ is an estimate, $\bar{\mu}$, of μ. Use of this estimate to estimate β gives $\bar{\beta} = 2.99/[\ln(2256 - 1.93) - \ln(6 - 1.93)] = 1.21$, where 6 hours and 2256 hours are the 7th and 39th ordered failure times, respectively. Finally, one obtains $\bar{\delta} = (6 - 1.93)/.183^{1/1.21} = 16.5$ hours.

For Type II censoring, alternative estimators of β and δ can be obtained by subtracting $\bar{\mu}$ from each failure time and then using estimators of $\xi = \beta^{-1}$ and $\eta = \ln\delta$, given in Section 5.2.2, which are linear functions of the logarithms of the ordered failure times. Any of these estimators of β and δ can also be used in an iterative procedure by evaluating $\mu^* = t_1 - \delta\Gamma(1 + 1/\beta)/n^{1/\beta}$ with the values specified for β and δ by the linear estimators of $1/\beta$ and $\ln\delta$ substituted. In fact, unless one has available a working computer program for evaluating the maximum-likelihood estimates, use of the linear estimation procedures is probably more practical. This is particularly true for small to medium sample sizes. The iterative process terminates when the change in the estimate μ^* is sufficiently small from one iteration to the next. A procedure for obtaining, from the iterative solution of a single equation, a *median unbiased* estimate of μ, a 50% lower (or upper) confidence bound, is described in Section 5.2.3(d).

Rockette, Antle, and Klimko (1973) conjecture that there are never more than two solutions to the likelihood equations. They demonstrate that, if there exists a solution of the likelihood equations that is a local maximum, there is a second solution that is a saddle point. They also show that, even if a solution $(\hat{\mu}, \hat{\beta}, \hat{\delta})$ is a local maximum, $L(\hat{\mu}, \hat{\beta}, \hat{\delta})$ may be less than $L(\mu_0, \beta_0, \delta_0)$, where

$$
\mu_0 = t_1, \qquad \beta_0 = 1, \quad \text{and} \quad \delta_0 = \frac{\sum_{j=1}^{r} (t_j - t_1) + (n - r)(t_r - t_1)}{r}.
$$

5.2.2 Point Estimation for the Two-Parameter Weibull Distribution

For the two-parameter Weibull distribution ($\mu = 0$), there are several methods by which one can obtain good point estimates of the two unknown parameters, β and δ. The methods include the iterative solution of the maximum-likelihood equations and several types of linear estimation techniques. Moment estimators (based on moments of the distribution of $X = \ln T$) have been used for this two-parameter distribution but have been shown (see Mann, 1968) to be less efficient than estimators based on only a few ordered observations. For this reason, and because of the fact that considerable effort is required in calculating moment estimates for this distribution, their use is not recommended. In the following, maximum-likelihood and various linear estimators of functions of the parameters β and δ are discussed.

(a) Maximum-Likelihood Estimation for μ Known. When the Weibull location parameter μ is known to be 0 or has been set equal to 0 by an appropriate transformation of the data, then, for a sample of size n, the likelihood function $L = f_{T_{(1)}, \ldots, T_{(r)}}(t_1, \ldots, t_r)$ can be expressed as

$$L = \frac{n!}{(n-r)!} \prod_{i=1}^{r} \left\{ \frac{\beta}{\delta} \left(\frac{t_i}{\delta} \right)^{\beta-1} \exp\left[-\left(\frac{t_i}{\delta} \right)^{\beta} \right] \right\} \cdot [1 - F(t_s)]^{n-r},$$

where for Type I censoring $t_s = t_0$, a specified test termination time, and for Type II censoring $t_s = t_{(r)}$, the observed time of the rth failure, at which time testing is terminated. The equations from which the maximum-likelihood estimates are obtained (see Section 5.2.1) can, for μ known, be put into forms that greatly facilitate their solution, namely,

$$\left[\frac{\sum_{i=1}^{r} t_i^{\beta} \ln t_i + (n-r) t_s^{\beta} \ln t_s}{\sum_{i=1}^{r} t_i^{\beta} + (n-r) t_s^{\beta}} - \frac{1}{\beta} \right] - \frac{1}{r} \sum_{i=1}^{r} \ln t_i = 0 \qquad (5.11)$$

and

$$\delta^{\beta} = \frac{1}{r} \left[\sum_{i=1}^{r} t_i^{\beta} + (n-r) t_s^{\beta} \right]. \qquad (5.12)$$

There are several methods by which one can obtain fairly good approximations of $\hat{\beta}$ and $\hat{\delta}$ to use as first guesses in the maximum-likelihood iterative procedure. About the simplest, and probably as good as any, are the two-order statistic estimators (with $\hat{\mu} = 0$) suggested by Dubey and given in Section 5.2.1. One can also use graphical plotting techniques [discussed in Section 5.2.2(b)]. A rather efficient iterative technique for obtaining $\hat{\beta}$, the solution of $f(\hat{\beta}) = 0$, Equation (5.11), is that of Newton-Raphson, wherein the $(j+1)$st successive approximation, $\hat{\beta}_{j+1}$ to $\hat{\beta}$, is given by $\hat{\beta}_{j+1} = \hat{\beta}_j - f(\hat{\beta}_j)/f'(\hat{\beta}_j)$.

The maximum-likelihood method has the advantage that it can be applied to a life-test model in which censoring is progressive, that is, a model in which a portion of the survivors is withdrawn from life test several times during the test. If survivors are removed from life test only at (or approximately at) the time of a failure, the progressive censoring is of Type II, and, for sufficiently small sample sizes, estimates and confidence bounds can also be obtained by certain of the linear techniques described in the following section (see also Mann, 1969a, 1971). If censoring is not strictly of Type II, however, maximum-likelihood procedures should be used.

The equations to be solved for obtaining maximum-likelihood estimates for β and δ are given by Cohen (1965). The equations are similar to (5.11) and (5.12), except that the r_l observations censored at time t_l are assigned the value of t_l, where t_l may or may not be the time of a failure. Unless either censoring is of Type II exclusively (i.e., each t_l *is* the time of a failure), or all parameters except δ are known, it is impossible to obtain confidence bounds for reliability or reliable life. The Bayesian technique of Bogdanoff and Pierce (1973) can be used for cases in which tables are not available for obtaining confidence bounds from Type II progressively censored data.

For the Weibull distribution, the order statistics are the sufficient statistics. Thus, unless $r=2$ the sufficient statistics are not complete (see Section 3.4), and no small-sample optimality properties can be claimed for the maximum-likelihood estimators. The MLE's of δ and β are, however, asymtotically unbiased as well as asymptotically normal and asymptotically efficient (see Table 5.4).

(b) Linear Estimation Techniques. Suppose that one considers the random variate X, the natural logarithm of T, where T is taken to be failure time. If T is from a population of failure times having a two-parameter Weibull distribution with scale parameter δ and shape parameter β, then X has a Type I asymptotic distribution of smallest (extreme) values (see Section 3.8), given by

$$F_X(x) = 1 - \exp\left[-\exp\left(\frac{x-\eta}{\xi} \right) \right], \quad \xi > 0, \tag{5.13}$$

where $\eta = \ln\delta$ and $\xi = 1/\beta$. The parameter η is a location parameter, the mode of the distribution of X, and ξ is a scale parameter, with $\pi\xi/\sqrt{6}$ equal to the standard deviation of X. The parameter x_P, the $100P\%$ point of the distribution of X (which is often called simply the extreme-value distribution) is equal to $\eta + \ln\{\ln[1/(1-P)]\}\xi \equiv \eta + y_P\xi$.

Because ξ is a scale parameter and η is a location parameter [i.e.,

$$F_X(x) = G\left(\frac{x-\eta}{\xi} \right)$$

for some G], various methods of linear estimation of linear functions of these parameters can be used effectively if the model is one of Type II censoring or approximately that. Some of these methods make use of all the observations, whereas others are based on only a few of the ordered observed values of X. All of the linear estimation methods are convenient in that one needs only the sample data and a table of weights in order to obtain the estimates in the form of weighted sums of the observations. Also, these methods are easily adaptable to computer usage by having the necessary weights available on IBM cards or stored on magnetic tape. Graphical estimation procedures for obtaining estimates usually involve even less effort than do the linear estimation rules that they are approximating.

1. BEST LINEAR UNBIASED AND BEST LINEAR INVARIANT ESTIMATORS. The generalized Gauss-Markov theorem, discussed in Section 3.6, specifies least-squares estimators as the unique estimators with minimum variance among unbiased linear functions of the observable variates when these variates have expectations that are linear functions with known coefficients of unknown parameters and covariance matrix $\sigma^2 \mathbf{B}$ with $\mathbf{B}$ known (see Problem 5.16). This theorem has been applied to the estimation of linear functions of location and scale parameters such as η and ξ (see Section 3.6.2). Of particular interest is the linear function defining the $100P$th distribution percentile, $x_P = \eta + y_P \xi$, of the extreme-value distribution. Lieblein (1954) and Lieblein and Zelen (1956) were the first to apply the generalized Gauss-Markov theorem to the estimation of the extreme-value parameters η, ξ, and x_P, using complete and censored samples, respectively.

In cases in which a minimum-variance unbiased estimator of a location parameter is known to exist and the location parameter and a scale parameter are both unknown, the estimator is linear in the observations and hence coincides with the minimum-variance (best) linear unbiased estimator of the specified location parameter. Also, for distributions for which best unbiased estimators of scale parameters exist, linear approximations to these estimators in the form of best linear unbiased estimators are shown to be highly efficient relative to the best unbiased estimator. For example, one can estimate the Gaussian scale parameter, the standard deviation, by means of a weighted sum of the ordered observations with an efficiency, relative to the best unbiased estimator (obtainable for complete samples only), that ranges between .988 and 1.0 as sample size varies. Efficiency here is defined as the ratio of the reciprocals of the variances of two corresponding estimators.

If one is determining a single point estimate and does not intend to average a series of estimates obtained over a period of time, he need not restrict himself to the use of unbiased estimators (see Section 3.6.3). In other words, one can, unless averaging of estimates of η, ξ, and/or x_P is required, use linear estimators of these parameters that have uniformly smaller mean-squared errors than the corresponding best linear unbiased estimators.

Such estimation functions are the best linear invariant estimators discussed in Section 3.6.3. These estimators are defined by sets of weights (functions of censoring number r and sample size n) for multiplying the ordered observations. The weights $A(n,r,i)$ and $C(n,r,i)$ defining the best linear invariant estimators,

$$\tilde{\eta} = \sum_{i=1}^{r} A(n,r,i) X_{(i)}, \qquad \tilde{\xi} = \sum_{i=1}^{r} C(n,r,i) X_{(i)}, \qquad \text{and}$$

$$\tilde{X}_P = \tilde{\eta} + \tilde{\xi} \ln \left[\ln \left(\frac{1}{1-P} \right) \right]$$

of η, ξ, and x_P, respectively, are linear functions of the best linear unbiased estimator weights (see Section 3.6.2). Values of $A(n,r,i)$ and $C(n,r,i)$ are given in Table 5.3 for $r=2(1)n, n=2(1)13$. Values proportional to the mean-squared errors, $E(L\eta)\xi^2$ and $E(L\xi)\xi^2$, of $\tilde{\eta}$ and $\tilde{\xi}$, respectively, and $E(CP) = E[(\tilde{\xi}-\xi)(\tilde{\eta}-\eta)]/\xi^2$ are also given, and from these values one can, if necessary, convert the best linear invariant estimates to best linear unbiased estimates. This is explained in Section 5.2.2(b)-2, which follows. Corresponding tables of sample sizes 2–15 and 2–25 can be found in publications of Mann (1967b and 1967a, respectively).

As an example of the use of the tables, consider a situation in which a sample of 10 items is subjected to life test and the test is terminated at the time of the third failure. The three observed failure times are 304, 460, and 612 hours. Then, from Table 5.3, $\tilde{\eta} = -.4086$ (ln 304) $-.3404$ (ln 460) $+1.749$ (ln 612) $= 6.797$ and $\tilde{\xi} = -.3213$ (ln 304) $-.2979$ (ln 460) $+.6191$ (ln 612) $= .308$. The logarithm of reliable life corresponding to a survival probability of .90 (or the 10th percentile of the extreme-value distribution), $x_{.10} = \eta + \xi \ln [\ln (1/.90)]$, is estimated by $\tilde{X}_{.10} = \tilde{\eta} + \tilde{\xi} \ln [\ln (1/.90)] = 6.797 + .308 (-2.25037) = 6.104$. One can then form the estimates $\exp(\tilde{X}_{.10}) = 446$ hours and $\exp(\tilde{\eta}) = 897$ hours of $t_{.10} = \exp(x_{.10})$, the 10th percentile of the distribution of Weibull variates, and of $\delta = \exp(\eta)$, the Weibull characteristic life. A procedure for converting these estimates to best linear unbiased estimates (in case the estimates are to be averaged) is given in the discussion in Section 5.2.2(b)-2, "Combining Best Linear Unbiased Estimators from Subsamples."

Monte Carlo investigations of maximum-likelihood estimation procedures by Harter and Moore (1965) indicate that the maximum-likelihood estimators and the best linear invariant estimators of ξ and x_P for P small have nearly equal mean-squared errors. One can, therefore, use either of the methods and obtain point estimates that are equally close, on the average, to any of these parameters. (The linear estimation procedures are,

Table 5.3 Weights for obtaining best linear invariant estimates of parameters of the extreme-value distribution

$A(n, r, i)$ – WEIGHT FOR ESTIMATING η
$C(n, r, i)$ – WEIGHT FOR ESTIMATING ξ
$\xi^2\, E(LU) = \xi^2\, E(L\eta)$ – MEAN SQUARED ERROR FOR $\tilde{\eta}$
$\xi^2\, E(CP) = E[(\tilde{\eta}-\eta)(\tilde{\xi}-\xi)]$
$\xi^2\, E(LB) = \xi^2\, E(L\xi)$ – MEAN SQUARED ERROR FOR $\tilde{\xi}$

n	r	i	A(n,r,i)	C(n,r,i)
2	2	1	0.110731	-0.421383
		2	0.889269	0.421383
E(LU)			0.65712995	
E(CP)				0.03757418
E(LB)				0.41583918
3	2	1	-0.166001	-0.452110
		2	1.166001	0.452110
E(LU)			0.79546061	
E(CP)				0.25750956
E(LB)				0.45005549
3	3	1	0.081063	-0.278666
		2	0.251001	-0.190239
		3	0.667936	0.468904
E(LU)			0.40240741	

n	r	i	A(n,r,i)	C(n,r,i)
5	2	1	-0.481434	-0.472962
		2	1.481434	0.472962
E(LU)			1.24921018	
E(CP)				0.53379141
E(LB)				0.47230837
5	3	1	-0.137958	-0.306562
		2	-0.025510	-0.257087
		3	1.163468	0.563650
E(LU)			0.49029288	
E(CP)				0.16612899
E(LB)				0.29419192
5	4	1	-0.006983	-0.217766
		2	0.059652	-0.199351
		3	0.156664	-0.118927

194

```
E(CP)                          -0.01842169
E(LB)                           0.25634620

4   2   1   -0.346974   -0.465455
        2    1.346974    0.465455
E(LU)        1.01477788
E(CP)                    0.41350875
E(LB)                    0.46438768

4   3   1   -0.044975   -0.297651
        2    0.080057   -0.234054
        3    0.956918    0.531705
E(LU)        0.42315147
E(CP)                    0.08477554
E(LB)                    0.28172930

4   4   1    0.064336    0.203052
        2    0.147340   -0.182749
        3    0.261510   -0.070109
        4    0.526813    0.455910
E(LU)        0.29247651
E(CP)                   -0.02831210
E(LB)                    0.18386193
```

```
4            0.790668    0.536044
             0.29062766
             0.03076329
                         0.20241894
E(CP)
E(LB)

5   5   1    0.052975   -0.158131
        2    0.103531   -0.155707
        3    0.163808   -0.111820
        4    0.246092   -0.005600
        5    0.433593    0.431259
E(LU)        0.23040495
E(CP)       -0.02913523
E(LB)                    0.14284288

6   2   1   -0.588298   -0.477782
        2    1.588298    0.477782
             1.48102383
             0.63148980
                         0.47734078
E(LU)
E(CP)
E(LB)

6   3   1   -0.211474   -0.311847
        2   -0.112994   -0.271381
        3    1.324468    0.583229
             0.57539484
             0.23269670
                         0.30173252
E(LU)
E(CP)
E(LB)
```

Table 5.3 (continued)

n	r	i	A(n,r,i)	C(n,r,i)
6	4	1	-0.063569	-0.225141
		2	-0.006726	-0.209083
		3	0.079802	-0.146386
		4	0.990412	0.580610
E(LU)			0.31552097	
E(CP)			0.08035062	
E(LB)				0.21242254
6	5	1	0.007521	-0.169920
		2	0.048328	-0.166319
		3	0.101608	-0.129510
		4	0.172859	-0.054453
		5	0.669685	0.520201
E(LU)			0.22351297	
E(CP)			0.0088019	
E(LB)				0.15690540
6	6	1	0.044826	-0.128810
		2	0.079377	-0.132102
		3	0.117541	-0.111951
		4	0.163591	-0.064666
		5	0.226486	0.031796
		6	0.368179	0.405733
E(LU)			0.19030430	
E(CP)			-0.02771574	
E(LB)				0.11657671
7	2	1	-0.676894	-0.481140
		2	1.676894	0.481140
7	6	1	0.013524	-0.138436
		2	0.041588	-0.140342
		3	0.075499	-0.121821
		4	0.117461	-0.082938
		5	0.172092	-0.015394
		6	0.579835	0.498931
E(LU)			0.18269947	
E(CP)			-0.00130057	
E(LB)				0.12760617
7	7	1	0.038743	-0.108323
		2	0.064086	-0.113479
		3	0.090785	-0.103569
		4	0.120971	-0.078748
		5	0.157657	-0.032632
		6	0.207825	0.054727
		7	0.319934	0.382022
E(LU)			0.16219070	
E(CP)			-0.02578937	
E(LB)				0.09836496
8	2	1	-0.752513	-0.483616
		2	1.752513	0.483616
E(LU)			1.91861540	
E(CP)			0.78453314	
E(LB)				0.48337662
8	3	1	-0.323875	-0.317890
		2	-0.243808	-0.288231

E(LU)
E(CP)
E(LB)

7 3
E(LU) 1.70468001
E(CP)   0.71366553
E(LB)   0.48082310

     1 -0.272195   -0.315369
     2 -0.184061   -0.281139
     3   1.456255   0.596507
E(LU)
E(CP)   0.66758707
E(LB)   0.28885432   0.30681307

7 4
     1 -0.110274   -0.229691
     2 -0.060226   -0.215613
     3   0.018671   -0.164168
     4   1.151829   0.609472
E(LU)
E(CP)   0.35340223   0.12260834
E(LB)   0.21884662

7 5
     1 -0.030368   -0.176203
     2   0.004333   -0.172399
     3   0.052957   -0.141218
     4   0.117599   -0.082820
     5   0.855480   0.572640
E(LU)
E(CP)   0.23316740   0.04212562
E(LB)   0.16497315

E(LU)
E(CP)
E(LB)

3
E(LU) 1.567683   0.606120
E(CP) 0.76198737
E(LB)   0.33734068   0.31047652

8 4
     1 -0.149973   -0.232805
     2 -0.105015   -0.220324
     3 -0.032257   -0.176675
     4   1.287245   0.629805
E(LU)
E(CP)   0.39805551   0.15928131
E(LB)   0.22335819

8 5
     1 -0.062656   -0.180231
     2 -0.032248   -0.176510
     3   0.012767   -0.149566
     4   0.072446   -0.101642
     5   1.009691   0.607948
E(LU)
E(CP)   0.25192092   0.07129172
E(LB)   0.17037848

8 6
     1 -0.013509   -0.143834
     2   0.010292   -0.145006
     3   0.041357   -0.128393
     4   0.080475   -0.095696
     5   0.130327   -0.043280
     6   0.751058   0.556209
E(LU)
E(CP)   0.18599844   0.02247163
E(LB)   0.13422386

Table 5.3 (continued)

n	r	i	A(n,r,i)	C(n,r,i)
8	7	1	0.015973	-0.116317
		2	0.036729	-0.120331
		3	0.060439	-0.110582
		4	0.088239	-0.088450
		5	0.122062	-0.050995
		6	0.165529	0.009700
		7	0.511030	0.476975

E(LU)   0.15505149
E(CP) -0.00641304
E(LB)   0.10726405

n	r	i	A(n,r,i)	C(n,r,i)
8	8	1	0.034052	-0.093270
		2	0.053552	-0.098886
		3	0.073452	-0.093994
		4	0.095062	-0.079752
		5	0.119768	-0.053918
		6	0.149934	-0.010179
		7	0.191236	0.069325
		8	0.282943	0.360675

E(LU)   0.14136026
E(CP) -0.02386561
E(LB)   0.08501680

n	r	i	A(n,r,i)	C(n,r,i)
9	2	1	-0.818444	-0.485517
		2	1.818444	0.485517

E(LU)   2.12272209
E(CP)   0.84680378
E(LB)   0.48532951

n	r	i	A(n,r,i)	C(n,r,i)
9	3	1	-0.368833	-0.319786

n	r	i	A(n,r,i)	C(n,r,i)
9	7	1	-0.004220	-0.120988
		2	0.013386	-0.124245
		3	0.035068	-0.115091
		4	0.061198	-0.095508
		5	0.093013	-0.064162
		6	0.132740	-0.017187
		7	0.668815	0.537180

E(LU)   0.15547192
E(CP)   0.01139509
E(LB)   0.11278822

n	r	i	A(n,r,i)	C(n,r,i)
9	8	1	0.016797	-0.100011
		2	0.032919	-0.104750
		3	0.050582	-0.099608
		4	0.070497	-0.086226
		5	0.093635	-0.063541
		6	0.121560	-0.028346
		7	0.157175	0.026525
		8	0.456836	0.455956

E(LU)   0.13496842
E(CP) -0.00906894
E(LB)   0.09236358

n	r	i	A(n,r,i)	C(n,r,i)
9	9	1	0.030338	-0.081777
		2	0.045872	-0.087308
		3	0.061368	-0.085084
		4	0.077742	-0.076470
		5	0.095769	-0.060667
		6	0.116517	-0.035136
		7	0.141932	0.006001

9 2

2	-0.295280	-0.293621
3	1.664113	0.613407

E(LU)
E(CP) 0.85621748
E(LB) 0.37995861   0.31324611

9 4

1	-0.184461	-0.235080
2	-0.143505	-0.223891
3	-0.075815	-0.185970
4	1.403781	0.644941

E(LU)
E(CP) 0.44625568
E(LB) 0.19160927   0.22671251

9 5

1	-0.090726	-0.183061
2	-0.063541	-0.179515
3	-0.021495	-0.155825
4	-0.034159	-0.115133
5	1.141604	0.633534

E(LU)
E(CP) 0.27605014
E(LB) 0.09715351   0.17429417

9 6

1	-0.037118	-0.147411
2	-0.016377	-0.148150
3	0.012499	-0.133219
4	0.049305	-0.105060
5	0.095614	-0.062073
6	0.896078	0.595913

E(LU)
E(CP) 0.19579592
E(LB) 0.04378261   0.13880129

8 9

8	0.176764	0.078828
9	0.253697	0.341614

E(LU)
E(CP) 0.12529518
E(LB) -0.02209438   0.07482425

10 2

1	-0.876869	-0.487022
2	1.876869	0.487022
	2.31744054	

E(LU)
E(CP) 0.90232208
E(LB) 0.48687150

10 3

1	-0.408602	-0.321265
2	-0.340443	-0.297858
3	1.749045	0.619124

E(LU)
E(CP) 0.94907551
E(LB) 0.41795081   0.31541467

10 4

1	-0.214930	-0.236817
2	-0.177223	-0.226688
3	-0.113820	-0.193159
4	1.505973	0.656663

E(LU)
E(CP) 0.49619736
E(LB) 0.22047816   0.22930885

Table 5.3 (continued)

n	r	i	A(n,r,i)	C(n,r,i)
10	5	1	-0.115524	-0.185169
		2	-0.090868	-0.181821
		3	-0.051341	-0.160697
		4	0.000925	-0.125311
		5	1.256809	0.652997
E(LU)			0.30344549	
E(CP)			0.12033056	
E(LB)			0.17727542	
10	6	1	-0.058017	-0.149985
		2	-0.039595	-0.150451
		3	-0.012513	-0.136941
		4	0.022314	-0.112224
		5	0.065750	-0.075721
		6	1.022062	0.625321
E(LU)			0.20973843	
E(CP)			0.06299841	
E(LB)			0.14219828	
10	7	1	-0.022198	-0.124170
		2	-0.006909	-0.126894
		3	0.013224	-0.118392
		4	0.037994	-0.100924
		5	0.068153	-0.073988
		6	0.105164	-0.035501
		7	0.804572	0.579868
E(LU)			0.16066059	
E(CP)			0.02762724	
E(LB)			0.11670571	
10	8	1	0.001179	-0.104082

n	r	i	A(n,r,i)	C(n,r,i)
10	10	1	0.027331	-0.072734
		2	0.040034	-0.077971
		3	0.052496	-0.077242
		4	0.065408	-0.071876
		5	0.079263	-0.061652
		6	0.094638	-0.045420
		7	0.112414	-0.020698
		8	0.134239	0.017927
		9	0.164178	0.085070
		10	0.230001	0.324597
E(LU)			0.11252220	
E(CP)			-0.02050852	
E(LB)			0.06679250	
11	2	1	-0.929310	-0.488243
		2	1.929310	0.488243
E(LU)			2.50340024	
E(CP)			0.95239887	
E(LB)			0.48812000	
11	3	1	-0.444245	-0.322452
		2	-0.380642	-0.301277
		3	1.824887	0.623729
E(LU)			1.03995578	
E(CP)			0.45220741	
E(LB)			0.31715930	
11	4	1	-0.242206	-0.238188
		2	-0.207204	-0.228941
		3	-0.147490	-0.198888
		4	1.596900	0.666017

2   0.014889   -0.108163
3   0.030998   -0.103119
4   0.049734   -0.090835
5   0.071745   -0.070902
6   0.098114   -0.041560
7   0.130649   0.000799
8   0.602692   0.517864

E(LU) 0.13403554
E(CP) 0.00474963
E(LB) 0.09704810

10 9

1   0.016841   -0.087538
2   0.029807   -0.092405
3   0.043570   -0.089839
4   0.058640   -0.081428
5   0.075576   -0.066855
6   0.095169   -0.044670
7   0.118707   -0.011816
8   0.148575   0.038159
9   0.413116   0.436394

E(LU) 0.11965747
E(CP) -0.01043859
E(LB) 0.08100409

11 5

E(LU) 0.54681985
E(CP) 0.24653583
E(LB) 0.23138012

1   -0.137718   -0.186803
2   -0.115110   -0.183651
3   -0.077762   -0.164597
4   -0.028411   -0.133278
5   1.359000   0.668329

E(LU) 0.33282848
E(CP) 0.14129911
E(LB) 0.17962678

11 6

1   -0.076739   -0.151936
2   -0.060142   -0.152221
3   -0.034581   -0.139907
4   -0.001490   -0.117886
5   0.039518   -0.086131
6   1.133434   0.648081

E(LU) 0.22640907
E(CP) 0.08045010
E(LB) 0.14483423

11 7

1   -0.038349   -0.126507
2   -0.024842   -0.128838
3   -0.005964   -0.120951
4   0.017632   -0.105219
5   0.046354   -0.081602
6   0.081182   -0.048929
7   0.923987   0.612047

E(LU) 0.16905710
E(CP) 0.04246025
E(LB) 0.11966982

Table 5.3 (continued)

n	r	i	A(n,r,i)	C(n,r,i)
11	8	1	-0.012943	-0.106922
		2	-0.001050	-0.110498
		3	0.013869	-0.105662
		4	0.031661	-0.094405
		5	0.052723	-0.076693
		6	0.077815	-0.051525
		7	0.108161	-0.016860
		8	0.729765	0.562564
E(LU)			0.13669382	
E(CP)				0.01751192
E(LB)				0.10043756
11	9	1	0.004425	-0.091115
		2	0.015498	-0.095437
		3	0.028023	-0.092780
		4	0.042178	-0.084833
		5	0.058340	-0.071581
		6	0.077093	-0.052182
		7	0.099349	-0.024880
		8	0.126592	0.013606
		9	0.548502	0.499201
E(LU)			0.11809425	
E(CP)				0.00058414
E(LB)				0.08503131
11	10	1	0.016502	-0.077717
		2	0.027205	-0.082449
		3	0.038291	-0.081388

n	r	i	A(n,r,i)	C(n,r,i)
12	3	1	-0.476530	-0.323426
		2	-0.416836	-0.304093
		3	1.893367	0.627519
E(LU)			1.12857097	
E(CP)			0.48338667	
E(LB)				0.31859354
12	4	1	-0.266888	-0.239300
		2	-0.234180	-0.230796
		3	-0.177681	-0.203562
		4	1.678749	0.673657
E(LU)			0.59748043	
E(CP)				0.27026774
E(LB)				0.23307201
12	5	1	-0.157792	-0.188109
		2	-0.136884	-0.185142
		3	-0.101445	-0.167790
		4	-0.054640	-0.139693
		5	1.450761	0.680734
E(LU)			0.36338878	
E(CP)				0.16042600
E(LB)				0.18153147
12	6	1	-0.093679	-0.153471
		2	-0.078561	-0.153632
		3	-0.054320	-0.142329
		4	-0.022769	-0.122474

```
                                              5    0.016136   -0.094355
                                              6    1.233193    0.666261
                                        E(LU)      0.24490094
                                        E(CP)      0.09641022
                                        E(LB)      0.14694548

                                     12  7
                                              1   -0.052987   -0.128308
                                              2   -0.040893   -0.130339
                                              3   -0.023072   -0.123007
                                              4   -0.000515   -0.108712
                                              5    0.026930   -0.087681
                                              6    0.059918   -0.059256
                                              7    1.030620    0.637304
                                        E(LU)      0.17967935
                                        E(CP)      0.05607919
                                        E(LB)      0.12200601

                                     12  8
                                              1   -0.025785   -0.109045
                                              2   -0.015312   -0.112224
                                              3   -0.001353   -0.107627
                                              4    0.015634   -0.097276
                                              5    0.035853   -0.081361
                                              6    0.059835   -0.059315
                                              7    0.088444   -0.029900
                                              8    0.842684    0.596748
                                        E(LU)      0.14186580
                                        E(CP)      0.02930146
                                        E(LB)      0.10304331


              4    0.050160   -0.075977
              5    0.063170   -0.066222
              6    0.077772   -0.051429
              7    0.094625   -0.030120
              8    0.114811    0.000537
              9    0.140333    0.046381
             10    0.377130    0.418384
        E(LU)      0.10756449
        E(CP)     -0.01109747
        E(LB)      0.07207183

     11  11
              1    0.024850   -0.065444
              2    0.035456   -0.070318
              3    0.045727   -0.070456
              4    0.056215   -0.067076
              5    0.067261   -0.060207
              6    0.092560   -0.049300
              7                -0.033156
              8    0.108034   -0.009427
              9    0.127068    0.026879
             10    0.153197    0.089148
             11    0.210412    0.309357
        E(LU)      0.10212039
        E(CP)     -0.01910164
        E(LB)      0.06030372

     12  2
              1   -0.976872   -0.489254
              2    1.976872    0.489254
        E(LU)      2.68127021
        E(CP)      0.99799849
        E(LB)      0.48915157
```

Table 5.3 (continued)

n	r	i	A(n,r,i)	C(n,r,i)
12	9	1	-0.006944	-0.093658
		2	-0.002669	-0.097540
		3	0.014239	-0.094893
		4	0.027669	-0.087448
		5	0.043189	-0.075371
		6	0.061225	-0.058180
		7	0.082441	-0.034802
		8	0.107856	-0.003342
		9	0.667655	0.545234

E(LU) 0.11929957
E(CP) 0.01087297
E(LB) 0.08799386

n	r	i	A(n,r,i)	C(n,r,i)
12	10	1	0.006411	-0.080881
		2	0.015598	-0.085171
		3	0.025675	-0.083952
		4	0.036799	-0.078714
		5	0.049211	-0.069610
		6	0.063256	-0.056237
		7	0.079438	-0.037675
		8	0.098522	-0.012272
		9	0.121752	0.022956
		10	0.503338	0.481555

E(LU) 0.10573191
E(CP) -0.00210755
E(LB) 0.07557509

n	r	i	A(n,r,i)	C(n,r,i)
12	11	1	0.015982	-0.069798
		2	0.024997	-0.074285
		3	0.034156	-0.074131
		4	0.043790	-0.070617

E(LU)

n	r	i	A(n,r,i)	C(n,r,i)
13	3	1	-0.506031	-0.324239
		2	-0.449735	-0.306454
		3	1.955765	0.630694

E(LU) 1.21480934
E(CP) 0.51198847
E(LB) 0.31979363

n	r	i	A(n,r,i)	C(n,r,i)
13	4	1	-0.289420	-0.240219
		2	-0.258687	-0.232349
		3	-0.205024	-0.207450
		4	1.753131	0.680018

E(LU) 0.64778295
E(CP) 0.29204583
E(LB) 0.23448055

n	r	i	A(n,r,i)	C(n,r,i)
13	5	1	-0.176109	-0.189177
		2	-0.156637	-0.186381
		3	-0.122893	-0.170454
		4	-0.078337	-0.144971
		5	1.533976	0.690983

E(LU) 0.39459617
E(CP) 0.17799724
E(LB) 0.18310709

n	r	i	A(n,r,i)	C(n,r,i)
13	6	1	-0.109140	-0.154711
		2	-0.095246	-0.154785
		3	-0.072165	-0.144347
		4	-0.041997	-0.126268
		5	-0.004940	-0.101028
		6	1.323488	0.681140

E(LU) 0.26460952

```
                                                          E(CP)   0.11109896
                                                          E(LB)   0.14867755

                                                 13  7    1  -0.066358  -0.129743
                                                          2  -0.055414  -0.131538
                                                          3  -0.038503  -0.124701
                                                          4  -0.016879  -0.111609
                                                          5   0.009416  -0.092649
                                                          6   0.040810  -0.067475
                                                          7   1.126930   0.657714
                                                          E(LU)   0.19187273
                                                          E(CP)         0.06864731
                                                          E(LB)         0.12390133

          5    0.054149   -0.063891
          6    0.065515   -0.053621
          7    0.078264   -0.039034
          8    0.092958   -0.018715
          9    0.110521    0.009948
         10    0.132666    0.052280
         11    0.347003    0.401864
   E(LU)   0.09775217
   E(CP)        -0.01134890
   E(LB)         0.06487266


                                                 13  8    1  -0.037540  -0.110704
                                                          2  -0.028206  -0.113563
                                                          3  -0.015049  -0.109206
                                                          4   0.001231  -0.099644
                                                          5   0.020686  -0.085204
                                                          6   0.043677  -0.065581
                                                          7   0.070830  -0.039995
                                                          8   0.944372   0.623896
                                                          E(LU)   0.14885020
                                                          E(CP)         0.04022462
                                                          E(LB)         0.10512398

  12 12   1    0.022771   -0.059449
          2    0.031776   -0.063952
          3    0.040408   -0.064601
          4    0.049122   -0.062489
          5    0.058175   -0.057754
          6    0.067800   -0.050137
          7    0.078281   -0.039010
          8    0.090017   -0.023199
          9    0.103664   -0.000505
         10    0.120475    0.033696
         11    0.143566    0.091751
         12    0.193947    0.295648
   E(LU)   0.09348388
   E(CP)        -0.01785537
   E(LB)         0.05495436


  13  2   1   -1.020378   -0.490105
          2    2.020377    0.490105
   E(LU)   2.85169694
   E(CP)         1.03985071
   E(LB)         0.49001823
```

Table 5.3 (continued)

n	r	i	A(n,r,i)	C(n,r,i)
13	9	1	-0.017389	-0.095590
		2	-0.008934	-0.099109
		3	0.001863	-0.096521
		4	0.014684	-0.089554
		5	0.029637	-0.078490
		6	0.047027	-0.063068
		7	0.067346	-0.042607
		8	0.091328	-0.015928
		9	0.774437	0.580865

E(LU) 0.12250342
E(CP) 0.02046326
E(LB) 0.09030201

n	r	i	A(n,r,i)	C(n,r,i)
13	10	1	-0.002927	-0.083170
		2	-0.005067	-0.087085
		3	0.014356	-0.085792
		4	0.024891	-0.080789
		5	0.036816	-0.072325
		6	0.050389	-0.060181
		7	0.065995	-0.043768
		8	0.084201	-0.022048
		9	0.105863	0.006715
		10	0.615348	0.528441

E(LU) 0.10607774
E(CP) 0.00635741
E(LB) 0.07818835

n	r	i	A(n,r,i)	C(n,r,i)
13	12	1	0.015382	-0.063288
		2	0.023100	-0.067492
		3	0.030818	-0.067892
		4	0.038824	-0.065622
		5	0.047302	-0.060887
		6	0.056444	-0.053540
		7	0.066482	-0.043158
		8	0.077739	-0.028970
		9	0.090699	-0.009644
		10	0.106166	0.017233
		11	0.125627	0.056547
		12	0.321416	0.386713

E(LU) 0.08961947
E(CP) -0.01136145
E(LB) 0.05895232

13 11

1	0.007628	-0.072617
2	0.015408	-0.076746
3	0.023732	-0.076418
4	0.032743	-0.072938
5	0.042611	-0.066531
6	0.053556	-0.057014
7	0.065876	-0.043886
8	0.080005	-0.026244
9	0.096594	-0.002552
10	0.116703	0.029910
11	0.465143	0.465037
	-0.00388188	0.09583611

E(LU)
E(CP)
E(LB)   0.06795140

13 13

1	0.021005	-0.054436
2	0.028757	-0.058585
3	0.036127	-0.059535
4	0.043501	-0.058259
5	0.051078	-0.054942
6	0.059028	-0.049472
7	0.067533	-0.041504
8	0.076831	-0.030398
9	0.087274	-0.015037
10	0.099441	0.006644
11	0.114446	0.038943
12	0.135068	0.093324
13	0.179913	0.283257
	0.08619744	-0.01674914

E(LU)
E(CP)
E(LB)   0.05046988

207

of course, simpler to implement.) For estimating η the two methods are also comparable except when n is very small, in which case the linear estimator is very slightly preferable. Comparisons of mean-squared error for these two estimation methods and best linear unbiased estimators are given in Table 5.4.

Also shown in Table 5.4 are $A(\eta)$ and $A(\xi)$ and $\tilde{A}(\eta)$ and $\tilde{A}(\xi)$, the Cramér-Rao lower bounds (see Section 3.4) for the mean-squared errors of regular unbiased and regular invariant estimators, respectively, of η and ξ. The bounds for censored unbiased estimators are based on results of Harter and Moore (1968), and those for invariant estimators depend on results of Mann (1969b).

One can observe that the maximum-likelihood, best linear invariant, and best linear unbiased estimators all appear to have mean-squared errors that in general approach their respective bounds as n increases for fixed r/n. It can be shown, in fact, that all of these estimators are asymptotically efficient and asymptotically normal (see Section 3.4 and Mann, 1969b) and hence are asymptotically equal to their respective Cramér-Rao bounds. These asymptotic properties also apply to estimators of $\delta = \exp(\eta)$ and $t_P = \exp(x_P)$, obtained by transforming appropriately any of the estimators of η and x_P discussed above. One should note, however, that any estimator of δ and t_P that is a function of unbiased estimators of η and ξ will be unbiased only asymptotically and not for small n.

The estimation of η and ξ by the use of best linear invariant estimator weights has been applied to Type II progressively censored samples, that is, to life-test situations in which any number of items may be removed from test at the time of any failure. Weights for obtaining the estimates for this model for sample sizes 2–9 are available in publications of Mann (1970b, 1971). Critical values for obtaining 95% tolerance bounds (confidence bounds on reliable life) at a 90% confidence level for all possible censorings of samples of sizes 2–6 from three ordered observations are given by Mann (1969a, 1968). See Bogdanoff and Pierce (1973) for treatment of cases for which available tabulations do not apply.

2. APPROXIMATIONS TO THE BEST LINEAR UNBIASED AND BEST LINEAR INVARIANT ESTIMATORS FOR USE WITH CENSORED SAMPLES LARGER THAN 25. For censored samples of size 25 and smaller, the best linear unbiased and best linear invariant estimators described above are recommended, accordingly as unbiased estimates are or are not required. For samples of sizes 30, 50, and 100 and certain censorings, one can use weights yielding approximations to the best linear invariant estimators of η and ξ derived by Johns and Lieberman (1966). What is needed, however, is an efficient estimation method that can be easily applied to censored samples of moderate to large size without the necessity of a great number of tables.

Table 5.4 Comparison of expected losses of various estimators of η and ξ under type II censoring from above when loss is squared error divided by ξ^2

Sample Size	Censoring Number	Best Linear Unbiased η^*	Maximum Likelihood $\hat{\eta}$	Best Linear Invariant $\tilde{\eta}$	C-R Bound Unbiased $A(\eta)$	C-R Bound Invariant $\tilde{A}(\eta)$	Best Linear Unbiased ξ^*	Maximum Likelihood $\hat{\xi}$	Best Linear Invariant $\tilde{\xi}$	C-R Bound Unbiased $A(\xi)$	C-R Bound Invariant $\tilde{A}(\xi)$
10	2	3.904	2.359	2.317	1.648	1.279	.949	.487	.487	.474	.322
10	3	1.204	.999[a]	.949	.719	.629	.461	.313[a]	.315	.307	.235
10	4	.559	.515[a]	.496	.393	.367	.298	.233[a]	.229	.222	.182
10	5	.321	.315[a]	.303	.251	.244	.215	.172[a]	.177	.172	.146
10	6	.214	.226[a]	.210	.181	.179	.166	.141[a]	.142	.137	.121
10	7	.162	.166[a]	.161	.145	.145	.132	.113[a]	.117	.112	.101
10	8	.134	.137[a]	.134	.125	.125	.107	.094[a]	.094	.093	.085
10	9	.120	.122[a]	.120	.115	.115	.088	.077[a]	.081	.077	.071
10	10	.113	.114[a]	.113	.111	.110	.072	.063[a]	.067	.061	.057
20	2	7.033	3.919	3.880	3.026	2.198	.975	.494	.494	.487	.327
20	4	1.197	1.055[a]	.978	.824	.714	.316	.247[a]	.240	.237	.192
20	6	.456	.428[a]	.413	.360	.334	.184	.160[a]	.155	.153	.133
20	8	.232	.228[a]	.221	.197	.189	.127	.117[a]	.133	.111	.100
20	10	.141	.143[a]	.138	.126	.123	.096	.090[a]	.087	.086	.079
20	12	.098	.099[a]	.097	.091	.090	.075	.074[a]	.070	.069	.064
20	14	.076	.078[a]	.076	.072	.072	.061	.060[a]	.057	.056	.053
20	16	.065	.066[a]	.065	.063	.063	.050	.050[a]	.047	.046	.044
20	18	.059	.060[a]	.059	.058	.058	.041	.042[a]	.039	.038	.037
20	20	.056	.056[a]	.056	.055	.055	.033	.033[a]	.032	.030	.030

[a]Computed from empirical sampling results of Harter and Moore (1968).

The following procedure satisfies such criteria.

Simplified Linear Estimators. An unbiased linear estimator of ξ, based on the first r of n extreme-value order statistics, was suggested by Bain (1972) and modified by Englehardt and Bain (1973). It is of the form

$$\xi^{**} = \frac{\sum_{i=1}^{r} |X_{(s)} - X_{(i)}|}{nk_{r,n}}, \qquad (5.14)$$

with $k_{r,n} = (1/n)\sum_{i=1}^{r} E|Z_s - Z_i|$, where $Z_i = (X_{(i)} - \eta)/\xi$, and

$s = r$ for $r \leqslant .9n$,

$s = n$ for $r = n, n = \leqslant 15$,

$s = n - 1$ for $r = n, 16 \leqslant n \leqslant 24$,

and

$s = [.892n] + 1$ for $r = n, n \geqslant 25$, with $[y]$ the largest integer less than y.

Since $E(Z_r - Z_i) = E(Z_r) - E(Z_i)$, and since

$$E(Z_i) = \sum_{k=1}^{i} (-1)^{k-1} n \binom{n-1}{i-1} \binom{i-1}{k-1} \frac{-\gamma - \ln(n-i+k)}{n-i+k} \qquad (5.15)$$

(with γ equal to Euler's constant, approximately .5772157), an appropriate value of $k_{r,n}$ can be calculated (with the use of a computer) for almost any combination of r and n. Values of $E(Z_i)$ have been tabulated by White (1967) for samples of size n, $n = 1(1)50(5)100$. Values of $k_{r,n}$ generated by Monte Carlo methods with Monte Carlo sample size of 20,000 are tabulated in Section 5.2.3, Table 5.11, for selected values of r and n. In Table 5.11 the value of s is r for $r < n$ and is $[.892n]$ rather than $[.892n] + 1$ for $r = n$. Bain (1972) gives exact values of $k_{r,n}$ for $n = 5$, 15, 20, 30, 60, and 100 and $r/n = .1(.1).9$ for integer r. He also tabulates asymptotic values of $k_{r,n}$ for these censorings.

Since ξ is a scale parameter and ξ^{**} is unbiased, the estimation of ξ can be improved considerably in terms of the mean-squared error for extensive censoring if the variance is known. Bain notes that, for r/n about .5 or less, $\text{Var}(\xi^{**}) \cong (nk_{r,n})^{-1}\xi^2$. Thus (from Section 3.4)

$$\tilde{\tilde{\xi}} = \frac{\xi^{**}}{1 + (nk_{r,n})^{-1}} = \frac{\sum_{i=1}^{r} (X_{(r)} - X_{(i)})}{1 + nk_{r,n}}$$

has smaller mean-squared error than ξ^{**} for r/n less than .5. Also, we have computed and displayed in Table 5.11 values from which $l_{r,n}$ with

$l_{r,n}\xi^2$ equal to the variance of ξ^{**} can be determined. Therefore

$$\tilde{\xi} = \frac{\xi^{**}}{1+l_{r,n}} = \frac{\sum_{i-1}^{r} |X_{(s)} - X_{(i)}|}{nk_{r,n}(1+l_{r,n})}$$

has mean-squared error $l_{r,n}\xi^2/(1+l_{r,n})$.

For $r=2$, as n approaches infinity, the mean-squared errors of ξ^{**} and $\tilde{\xi}$ approach ξ^2 and $.5\xi^2$, respectively.

For $r/n = .7$ or less, the asymptotic efficiency of ξ^{**} with respect to the Cramér-Rao bound for regular unbiased estimators of ξ is at least .977. For $n=20$, the efficiency of ξ^{**} with respect to ξ^*, the best linear unbiased estimator of ξ, is at least .93 for $r \leqslant .9n$. For $r=n$, the efficiency of ξ^{**} with respect to ξ^* is .808 for $n=25$ and .760 for n very large.

Now, in order to make use of ξ^{**} in estimating η or x_P, it should be noted that $E(X_{(s)})$ is given by $\eta + E(Z_s)\xi$, with $E(Z_s)$ defined in Equation (5.15). Thus linear unbiased estimators of η and x_P, respectively, are given by

$$\eta^{**} = X_{(s)} - E(Z_s)\xi^{**}$$

and

$$X_P^{**} = \eta^{**} + \xi^{**} \ln\left[\ln\left(\frac{1}{1-P} \right) \right].$$

(Values of $E(Z_s)$ are displayed in Table 5.11.) For r/n less than .5 these estimators are similar to their respective best linear unbiased estimators.

One can derive the expression for η^{**} for $r=2$, in which case η^{**} is the best (linear) unbiased estimator of η, in the following manner:

$$E(X_{(1)}) = \eta + \xi \int_0^\infty n \ln x \exp(-nx)\,dx = \eta + \xi \int_0^\infty (\ln v - \ln n)\exp(-v)\,dv$$

$$= \eta - \xi(\gamma + \ln n),$$

where γ is Euler's constant, approximately .5772157. Therefore, for $r=2$,

$$\eta^{**} = \eta^* = X_{(1)} + (\gamma + \ln n)\xi^{**}.$$

To determine an expression for ξ^{**} for $r=2$, we note that, for $\ln W = X_{(1)} - X_{(2)}$,

$$f_W(w) = \frac{n(n-1)w^{(1/\xi)-1}}{\xi(n-1+w^{1/\xi})^2}, \quad 0 < w \leqslant 1. \tag{5.16}$$

Then, expanding the denominator and determining $E(-\ln W)$, one obtains

$$\xi^{**} = \frac{X_{(2)} - X_{(1)}}{n \ln[1 + 1/(n-1)]}$$

so that

$$\eta^{**} = X_{(2)} + \left\{ -1 + \frac{\gamma + \ln n}{n \ln[1 + 1/(n-1)]} \right\} (X_{(2)} - X_{(1)}).$$

For $r = 2$ and large n, the mean-squared error of the estimator of η can be cut approximately in half by using, instead of η^{**}, $\tilde{\tilde{\eta}} = \eta^{**} - B_{r,n} \tilde{\tilde{\xi}}$, where $B_{r,n} \xi^2$ is the covariance of η^{**} and ξ^{**}. Values of $nB_{r,n}$ are also given in Table 5.11.

For $r = 3$, $n = 30$,

$$\xi^{**} = \frac{2X_{(3)} - X_{(1)} - X_{(2)}}{30(.069)} \quad \text{and} \quad \eta^{**} = X_{(3)} + 2.44\xi^{**},$$

where $k_{3,30} = .069$ and $E(Z_3) = -2.44$ are given in Table 5.11. Then

$$\tilde{\tilde{\xi}} = \frac{\xi^{**}}{1 + l_{3,30}} = \frac{\xi^{**}}{1 + 14.64/30} \quad \text{and} \quad \tilde{\tilde{\eta}} = \eta^{**} - \left(\frac{36.03}{30} \right) \tilde{\tilde{\xi}},$$

where $36.03 = nB_{3,30}$. The mean-squared error of $\tilde{\tilde{\xi}}$ is $[(14.64/30)/(1 + 14.64/30)]\xi^2$. For the variance of η^{**} equal to $A\xi^2$, the mean-squared error of $\tilde{\tilde{\eta}}$ is $[A - B_{3,30}^2/(1 + l_{3,30})]\xi^2 = (A - 1.2^2/1.49)\xi^2$. From Table 5.12 in Section 5.2.3, one finds for $r = 3$, $n = 30$, $A = 3.35$.

For 4.45, 5.87, 7.15, 8.29, 9.26, and 11.04, the first six observations from a size-60 sample of Weibull failure times, one obtains, using Table 5.11,

$$\xi^{**} = \frac{5\ln 11.04 - \ln[(4.45)(5.87)(7.15)(8.29)(9.26)]}{60(.085)} = .478$$

and

$$\tilde{\tilde{\xi}} = \frac{.478}{1 + .194} = .400.$$

Also, one finds

$$\eta^{**} = 2.401 + 2.34(.478) = 3.52 \quad \text{and} \quad \tilde{\tilde{\eta}} = 3.52 - .461(.400) = 3.34.$$

Estimates of δ and $t_{.10}$, reliable life corresponding to a survival probability of 90%, are given by $\exp(\tilde{\tilde{\eta}}) = 28.2$ and $\exp\{\tilde{\tilde{\eta}} + \tilde{\tilde{\xi}}\ln[\ln(1/.90)]\} = \exp(2.44) = 11.5$.

Combining Best Linear Unbiased Estimators from Subsamples. For samples of moderately large size, another approach is to divide the unordered sample randomly into subsamples of size less than or equal to 25. One then obtains best linear unbiased estimates for each subsample and calculates an optimum weighted average (discussed below) of these estimates to approximate a best linear unbiased estimate for the entire sample. If there is not a great deal of censoring, it is unnecessary to convert this unbiased estimate to an approximation of the best linear invariant estimate. (If censoring is extensive, however, it is considerably simpler to use the process based on ξ^{**} and η^{**}, just described.) This technique of combining subsample estimates can also be applied to the estimation of ξ from two samples having different values for η but the same value for ξ.

The best linear unbiased estimators ξ_j^* and η_j^* of ξ and η, respectively, for the jth subsample are equal to $\tilde{\xi}_j/[1 - E_j(L\xi)]$ and $\tilde{\eta}_j + E_j(CP)\xi_j^*$, respectively, where $E_j(L\xi)\xi^2$ and $E_j(CP)\xi^2$ are the values of $E[(\tilde{\xi} - \xi)^2]$ and $E[(\tilde{\xi} - \xi)(\tilde{\eta} - \eta)]$ associated with the combination of r and n corresponding to the jth subsample. Consider any specified linear combination $\phi = l_1\eta + l_2\xi$ of η and ξ, where l_1 or l_2 can be 0. Let k be the total number of subsamples, and let Q_j be equal to the variance of the best linear unbiased estimator, $\phi_j^* = l_1\eta_j^* + l_2\xi_j^*$, of ϕ for the jth subsample, $j = 1, 2, \ldots, k$. The variance Q_j is thus equal to $(l_1^2 A_j + 2l_1 l_2 B_j + l_2^2 C_j)\xi^2$, with C_j equal to $E_j(L\xi)/[1 - E_j(L\xi)]$, B_j equal to $E_j(CP)/[1 - E_j(L\xi)]$, A_j equal to $E_j(L\eta) + [E_j(CP)]/[1 - E(L\xi)]$, and $E(L\eta)$ equal to $E[\tilde{\eta} - \eta)^2]$. The optimum linear combination of the unbiased subsample estimates of ϕ is, then,

$$\bar{\phi} = \frac{\sum_{j=1}^{k} (1/Q_j)(l_1\eta_j^* + l_2\xi_j^*)}{\sum_{i=1}^{k} (1/Q_i)}.$$

The variance of $\bar{\phi}$ is equal to $\xi^2/\Sigma_{j=1}^{k}(1/Q_j)$.

If one has failure data applying to 43 capacitors that have been life-tested, the sample of size 43 can be randomly divided into 2 subsamples of sizes 22 and 21. If, in the first subsample, 2 items did not fail, one considers $n_1 = 22, r_1 = 20$. If all of the items in the second subsample failed, $n_2 = r_2 = 21$. Then, for estimating $\xi, Q_j = E_j(L\xi)/[1 - E_j(L\xi)], j = 1, 2$. From tables of Mann (1967a), one obtains

$$Q_1\xi^2 = \text{Var}(\xi_1^*) = \frac{\xi^2(.03514)}{1 - .03514} = .03642\xi^2$$

and

$$Q_2\xi^2 = \text{Var}(\xi_2^*) = \frac{\xi^2(.03047)}{1 - .03047} = .03143\xi^2,$$

so that

$$\bar{\xi} = \frac{\xi_1^* / .03642 + \xi_2^* / .03143}{27.46 + 31.82}.$$

If ξ_1^* and ξ_2^* are equal to .623 and .592, respectively, $\bar{\xi}$ is equal to .606. The variance of $\bar{\xi}$ is equal to $\xi^2 / (27.46 + 31.82) = .017\xi^2$. Note that here the values of η in the subsamples are not required to be equal.

3. ESTIMATION BY GRAPHICAL PLOTTING TECHNIQUES. Among the most easily obtained point estimates of the parameters of the Weibull distribution are graphical approximations to the best linear unbiased estimates. In obtaining such estimates, one effectively plots each ordered observation versus an estimate of the central tendency of the reduced observation and then visually fits the least-squares line. The parameter estimates are then obtained as functions of values of specified properties of the visually fitted line. This method has been fully explored in publications of Kao (1959, 1960), which give explicit details for its implementation.

If Weibull probability paper is used for failure analyses, one plots the ith ordered observed failure time $t_{(i)}$ [which is converted by the probability paper to the natural logarithm $x_{(i)}$ of $t_{(i)}$] versus some estimate, $p_{i,n}$, of $F(t_{(i)})$, $i = 1, 2, \ldots, r$. The probability estimate $p_{i,n}$ is converted by probability paper to $v_{i,n} = \ln\{\ln[1/(1 - p_{i,n})]\}$. On most probability paper one plots the observations on the horizontal axis so that the sum of horizontal rather than vertical squared deviations should be minimized. This linear estimation procedure results from the fact that, since $1 - F(t) = \exp[-(t/\delta)^\beta]$,

$$\ln\left\{ \ln\left[\frac{1}{1 - F(t)} \right] \right\} = \beta \ln t - \beta \ln \delta, \quad \ln \delta = \eta.$$

The slope of this line provides an estimate of $\beta = 1/\xi$, and the p intercept of the plot with the principal ordinate gives an estimate of $-\beta\eta$. The t intercept of the plot with the principal abscissa is an estimate of η.

If probability paper is not used, one can plot the ith ordered natural logarithm of observed failure time $X_{(i)}$ directly versus v_i, some characteristic value of the ith reduced order statistic, $Z_i = (X_{(i)} - \eta)/\xi$. The characteristic value selected is often the mean, median, or mode of Z_i. [Expected values of Z_i have been published by Mann, 1968a, for $n = 1(1)25$ and by White, 1967, for $n = 1(1)50(5)100$.] Suppose that a plot is made, and a curve that is concave downward results. Then often one can select by trial and error a value of $\hat{\mu} < t_{(1)}$, which when subtracted from each $t_{(i)}$, $i = 1, 2, \ldots, r$, will convert the curve to a line. The value $\hat{\mu}$ provides an estimate of μ, the location parameter in the three-parameter form of the Weibull distribution. The trial-and-error method of selecting $\hat{\mu}$ consists of decreasing the estimate if the curve plotted is concave downward and increasing it if the curve is concave upward.

In Figure 1 below the data apply to six aluminum specimens of the same type that were placed on test in a corrosive environment with additional stress. Failures were observed at times 3.45, 5.00, 5.90, 7.30, 8.60, and 10.20 days. In the plot labeled ①, the raw failure times are plotted against $[100(i-\frac{1}{2})/6]\%, i = 1, \dots, 6$. In the second plot (labeled ②), 3.00 was subtracted from each observed failure time before the plot was made; in the plot labeled ③, 1.00 was subtracted from each observed failure time. The line drawn through the points of the third plot has a slope that gives an estimate of $\beta = 1/\xi$. To obtain this estimate, approximately 2.6 here, one can draw a line perpendicular to the plot and originating at the circle at the top of the page. The value given by the intersection of this line with the value in the left-hand margin is the estimate. The estimate $-\beta\eta$ (approximately -5.0) is obtained by reading the value in the right-hand margin corresponding to the point where the plotted line intersects the

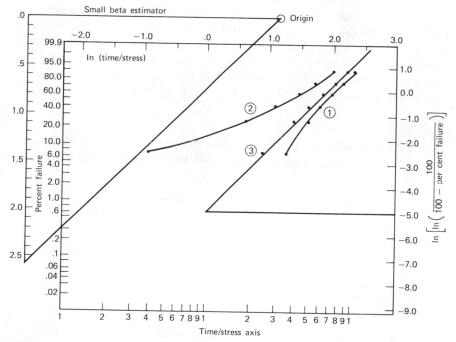

Figure 5.1 Weibull probability plot.

ordinate labeled 0.0. Also, the point (approximately 1.9) where the plotted line intersects the line labeled 0.0 in the right-hand margin is an estimate of η.

A graphical plot provides a representation of the data that is easily understood by the layman. It is extremely useful in detecting outliers, in estimating a nonzero Weibull location parameter μ, and in helping with a decision as to whether observed failure times are from a Weibull distribution. One would not expect, however, that parameter estimators defined by graphical procedures, which are necessarily subjective, would be as efficient for estimation as the linear estimators they approximate.

Table 5.5 gives, for μ known, the biases and mean-squared errors for estimating ξ, assuming no loss due to subjective errors caused by visual fitting for a complete sample of size 6. Most of the original calculations for this table were made by Kimball (1960). For the true Gauss-Markov least-squares estimators, v_i is equal to $E[(X_{(i)}-\eta)/\xi]$, as in row 4. The difference in the estimator weights and the theoretical efficiency of the two estimators, however, is accounted for by the fact that, in graphically approximating the true least-squares line, one tacitly and incorrectly assumes that the covariance matrix of the vector of order statistics is equal to $\xi^2 I$, where I is the identity matrix.

It appears, if generalizations can be made from Table 5.5, that one can do best in terms of mean-squared error by using $p_{i,n}=(i-\frac{1}{2})/n$, as in the example above. The convention based on $p_{i,n}=i/(n+1)$ is often used, possibly because $i/(n+1)$ is the expected value of $F(X_{(i)})$. The convention based on $p_{i,n}$ equal to the median rank of the ith reduced order statistic [$v_{i,n}$ equal to the median of $Z_i=(X_{(i)}-\eta)/\xi$ as in row 5] is also used

Table 5.5   Bias and expected squared error in linearly estimating the parameter ξ from complete samples of size 6, using various plotting and estimation conventions

Basis of Plotting or Estimation Convention	Bias	Mean-Squared Error
1. $v_{i,n}=\ln\{\ln[1/(1-p_{i,n})]\}, p_{i,n}=i/(n+1)$	.2284 ξ	.3011 ξ^2
2. $v_{i,n}=\ln\{\ln[1/(1-p_{i,n})]\}, p_{i,n}=(i-\frac{3}{8})/(n+\frac{1}{4})$	.0220 ξ	.1758 ξ^2
3. $v_{i,n}=\ln\{\ln[1/(1-p_{i,n})]\}, p_{i,n}=(i-\frac{1}{2})/n$	$-.0547\,\xi$	.1543 ξ^2
4. $v_{i,n}=E[(X_{i,i}n-u)/b]$	0.0	.1686 ξ^2
5. $v_{i,n}=$ median of $(X_{i,n}-\eta)/\xi$	.0632 ξ	.1928 ξ^2
6. $v_{i,n}=$ mode of $(X_{i,n}-\eta)/\xi$	.1916 ξ	.2692 ξ^2
7. Gauss-Markov least squares	0.0	.1320 ξ^2

extensively, together with the procedure described by Johnson (1951), wherein the approximate least-squares line is drawn so that half of the plotted values lie above the line and half below. This gives an estimate that is median unbiased (has positive error with probability .5 and negative error with probability .5). Note that all the plotting conventions given in Table 5.5 are functions of i and n only and do not vary with censoring number r. Graphical plotting, in fact, is not strictly applicable to censored samples without modification.

Weibull probability paper can be ordered from Technical and Engineering Aids for Management (TEAM), Box 25, Tamworth, New Hampshire 03886.

4. LINEAR ESTIMATORS BASED ON A FEW ORDERED OBSERVATIONS. For small to moderate sample sizes, procedures based on two or three ordered observations can be used to define point estimators that are nearly as efficient (i.e., have mean-squared error nearly as small) as maximum-likelihood and best linear invariant estimators. For very large uncensored samples, there exist linear estimators based on as few as 10 ordered observable variates with extremely high asymptotic efficiencies relative to the Cramér-Rao bounds for regular invariant estimators (see Section 3.4.1).

Since best linear invariant and best linear unbiased estimates are relatively easy to obtain for samples of size 25 or less, values for obtaining two- and three-order–statistic point estimates of η, ξ, and x_P are not given here. They are available for samples of sizes 2–15 in Mann (1970c) and for certain larger sample sizes in Mann (1970a).

If sample size is nearly complete and is as large as 80 or so, the two-order-statistic estimator of ξ, determined iteratively by Lieblein (1954) and Dubey (1967), has asymptotically smallest mean-squared error among invariant two-order-statistic estimators of ξ. It is asymptotically unbiased and asymptotically normal, and its asymptotic efficiency with respect to the Cramér-Rao bound for regular unbiased estimators is .66. It is of the form

$$.33[X_{(.9737n)} - X_{(.1673n)}], \qquad (5.17)$$

where the value .33 is defined by $[\ln(-\ln.0263) - \ln(-\ln.8327)]^{-1}$, and $\eta + \xi\ln[-\ln(1-\lambda)]$ gives the asymptotic expectation of $X_{(\lambda n)}$, and where, if the order numbers of the two order statistics are not integers, they should be replaced by $[.9737]+1$ and $[.1673n]+1$, with $[x]$ indicating the largest integer less than x. Thus, from a sample of size 147, one can estimate ξ by $.33(X_{(144)} - X_{(25)})$. The asymptotic variance of this asymptotically unbiased estimator is $\xi^2(.608/147)/.66 = .0063\xi^2$ (see Kimball, 1949), where $.608\xi^2/n$ is the Cramér-Rao lower bound for regular unbiased estimators of ξ.

The asymptotically most efficient two-order-statistic estimator of η is that determined by Dubey (1967), namely,

$$.446X_{(.40n)} + .554X_{(.82n)}, \tag{5.18}$$

with $.40n$ and $.82n$ equal to $[.40n]+1$ and $[.82n]+1$, respectively, if they are not integers. This estimator is asymptotically unbiased, and its asymptotic efficiency with respect to the Cramér-Rao bound $(1.109\xi^2/n)$ for unbiased estimation of η is .82. Derivations of the coefficients and order numbers of the ordered observations that give the asymptotically best two-order-statistic estimators of ξ and η depend on the results of Mosteller (1946), which demonstrate that $X_{(\lambda n)}$, the λnth order statistic of a sample of size n from $f_X(x)$, is asymptotically normal with mean x_λ and variance $\lambda(1-\lambda)/\{n[f(x_\lambda)]^2\}$, and that the covariance of the λnth and δnth observations, $1 \leqslant \lambda n \leqslant \delta n \leqslant n$, is asymptotically equal to $\lambda(1-\delta)/\{nf(x_\lambda)f(x_\delta)\}$ (see Problem 5.19). Note that x_λ for extreme-value variates is equal to $\eta + \xi\ln\{\ln[1/(1-\lambda)]\}$, so that $f(x_\lambda)$ is equal to $(1-\lambda)\ln[1/(1-\lambda)]/\xi$.

The asymptotic efficiencies of the estimators of $x_{.10}$ and $x_{.05}$, determined by Mann (1968b) by hand calculations extending tabulated results from small to moderately large sample sizes, are .82 and .81, respectively. These estimators, which are asymptotically unbiased, are of the form

$$X_{([.40n]+1)} + C_x(X_{([.98n]+1)} - X_{([.13n]+1)}), \tag{5.19}$$

where, for estimating $x_{.10}$, $C_x = -.473$ and, for estimating $x_{.05}$, $C_x = -.689$. The Cramér-Rao lower bounds for regular unbiased estimators of $x_{.10}$ and $x_{.05}$ are $5.345\xi^2/n$ and $8.000\xi^2/n$, respectively.

From complete (or nearly complete) samples of sufficiently large size, one can obtain estimates of η and ξ with efficiencies of .9720 and .9442, respectively, by forming the appropriate linear combination of 10 observed ordered X's. This has been demonstrated by Chan and Kabir (1969), who present asymptotic efficiencies of specified linear combinations of from 2 to 10 ordered observations. Given below are the spacings and coefficients of Chan and Kabir for estimating η, ξ, and x_P from 10 ordered observations chosen from nearly complete large samples.

The estimators η, ξ, and x_P, respectively, are of the forms

$$\eta' = \sum_{i=1}^{10} a_i' X_{([\kappa_i n]+1)}, \qquad \xi' = \sum_{i=1}^{10} b_i' X_{([\lambda_i n]+1)}$$

and

$$X_P' = \eta' + \xi'\ln\left[\ln\left(\frac{1}{1-P}\right)\right],$$

where $\kappa_i = (i - .5)/10$ and λ_i, a_i', and b_i' are as shown in Table 5.6. Note that

$$\sum_{i=1}^{10} a_i' = 1 \quad \text{and} \quad \sum_{i=1}^{10} b_i' = 0.$$

This ensures that the estimators are invariant, that is, have mean-squared error proportional to ξ^2.

In immersion tests of stress-corroded aluminum, the time for noticeable cracking to appear was recorded for every item of a sample of size 100. The logarithms of the 1st, 3rd, 7th, 15th, 66th, 81st, 91st, 97th, and 100th times to failure (cracking) were respectively, $-.301$, .262, .663, 1.08, 2.05, 2.25, 2.42, 2.58, and 2.77 log weeks. In addition, the logarithms of the 6th, 16th, 26th, 36th, 46th,...,96th failure times were .520, 1.21, 1.41, 1.56 1.66, 1.83, 2.05, 2.10, 2.30, and 2.58 log weeks.

Under assumptions of a two-parameter Weibull distribution for failure time, one estimates ξ by

$\xi' = -.0284(-.301) - .0616(-.301) - .0934(.262) - .1130(.663) -$

$.0652(1.08) + .1653(2.05) + .0950(2.25) + .0596(2.42) + .0309(2.58) +$

$$.0107(2.77) = .663,$$

and η by

$\eta' = .1742(.520) + .1792(1.21) + .1416(1.41) + .1186(1.56) + .1005(1.66)$

$+ .0850(1.83) + .0710(2.05) + .0582(2.10) + .0467(2.30) + .0250(2.58)$

$= 1.45$ log weeks.

Table 5.6 Values for obtaining ten-order-statistic estimates of extreme-value parameters

i	λ_i	a_i'	b_i'
1	.0014	.1742	$-.0284$
2	.0078	.1792	$-.0616$
3	.0249	.1416	$-.0934$
4	.0624	.1186	$-.1130$
5	.1412	.1005	$-.0652$
6	.6501	.0850	.1653
7	.8002	.0710	.0950
8	.9002	.0582	.0596
9	.9616	.0467	.0309
10	.9918	.0250	.0107

An estimate of reliable life corresponding to a survival proportion of 90% is then $\exp\{\eta' + \xi' \ln[\ln(1/.90)]\} = \exp[1.45 - 2.251(.663)] \cong 1$ week.

Indications are that samples of size about 80 will yield approximately the specified asymptotic efficiencies for the linear estimators based on only a few ordered observations.

5.2.3 Confidence Intervals under Type II Censoring for Weibull Parameters

For a two-parameter Weibull distribution under Type II censoring, confidence intervals for and tests of hypotheses concerning the parameters can be based on many of the point estimators discussed in the preceding section. For this model, the statistics that tend to yield tests with highest power (highest probability of rejecting a false hypothesis) and confidence intervals with greatest accuracy [lowest probability of including a value greater than (less than) any $g'(\eta,\xi) > (<)g(\eta,\xi)$ in an upper (a lower) confidence interval for $g(\eta,\xi)$] are the functions that estimate the parameters η and ξ with smallest mean-squared errors. These are, as discussed earlier, the maximum-likelihood and best linear invariant estimators. Bogdanoff and Pierce (1973) and Lawless (1973) show that use of a Bayesian approach in which one conditions on ancillary statistics, $(X_{(i)} - \hat\eta)/\hat\xi$, $i = 1,\ldots,r$, offers no improvement over these classical methods in obtaining confidence bounds for this distribution. Following descriptions of the various methods for obtaining confidence intervals for functions of δ and β with μ known, a procedure for calculating a lower confidence bound for μ is described.

(a) Intervals for Censored Samples of Size Less than 26. Unfortunately, the distributions of functions of neither the maximum-likelihood nor the best linear invariant estimator can be found analytically for small samples. For samples as large as about 80, the statistics $\tilde\xi/\xi$, $\hat\xi/\xi$, $(\tilde X_P - x_P)/\tilde\xi$ and $(\hat X_P - x_P)/\hat\xi$, $0 < P < 1$ (where $\hat\xi$ and $\hat X_P$ are maximum-likelihood estimators), are approximately normally distributed. For small to moderately large sample sizes, distribution percentiles of invariant functions of best linear invariant and maximum-likelihood estimators have been computed and tabulated through the use of Monte Carlo simulation procedures (see Thoman, Bain, and Antle, 1969, 1970; Billman, Antle, and Bain 1971; and Mann and Fertig, 1973). The tabulations applying to functions of MLE's pertain to uncensored samples or selected proportional censorings, and are given for samples of size n, with n as large as 120 in some cases. The smallest n considered is 5. The tabulations of distribution percentiles of functions of best linear invariant estimators of η and ξ apply to all Type II censorings for samples of size n, $n = 3(1)25$. The latter tabulations are given for samples of sizes 3–13 in Tables 5.7, 5.8, 5.9, and 5.10. Table 5.10 also gives percentiles applying to $n = 22, 23$. Methods are explained in Section 5.2.3(c) for approximating percentiles of $\tilde{\xi}/\xi$ not given in Table 5.7 and percentiles of $(\tilde{\eta} - x_P)/\tilde{\xi}$ not shown in Tables 5.8, 5.9, and 5.10.

Table 5.7 gives values of w_ζ, with w_ζ the 100ζth distribution percentile of $W = \tilde{\xi}/\xi$ (where $\tilde{\xi}$, as defined earlier, is the best linear invariant estimator of ξ). Since $\text{Prob}(\tilde{\xi}/\xi > w_\alpha) = 1 - \alpha$ and $\text{Prob}(\tilde{\xi}/\xi < w_{1-\alpha}) = 1 - \alpha$, then $\xi \geq \tilde{\xi}/w_{1-\alpha}$ with probability $1 - \alpha$ and $\xi \leq \tilde{\xi}/w_\alpha$ with probability $1 - \alpha$. Thus $\tilde{\xi}/w_\alpha$ and $\tilde{\xi}/w_{1-\alpha}$ provide, respectively, an upper and a lower confidence bound for ξ at confidence level $1 - \alpha$.

A lower confidence bound at level $1 - \alpha$ for η or $x_P \equiv \eta + \xi \ln\{\ln[1/(1 - P)]\}$ can be based similarly on $v_{1-P,1-\alpha}$, the $100(1 - \alpha)$th percentile of $V_{1-P} = (\tilde{\eta} - x_P)/\xi/(\tilde{\xi}/\xi)$, since the distribution of V_{1-P} can be shown to be independent of functions of η and ξ.

Since $(\tilde{\eta} - x_P)/\tilde{\xi} < v_{1-P,1-\alpha}$ with probability $1 - \alpha$, $\text{Prob}(x_P > \tilde{\eta} - \tilde{\xi}v_{1-P,1-\alpha}) = 1 - \alpha$, or $\tilde{\eta} - \tilde{\xi}v_{1-P,1-\alpha}$ is a lower confidence bound for x_P at confidence level $1 - \alpha$. A $(1 - \alpha)$-level lower confidence bound for $t_P = \exp(x_P)$, reliable life corresponding to a survival proportion of $1 - P$, is therefore given by $\exp(\tilde{\eta} - \tilde{\xi}v_{1-P,1-\alpha})$. If $P = 1 - \exp(-1)$, then $V_{1-P} = (\tilde{\eta} - \eta)/\tilde{\xi}$. Thus Table 5.8 corresponding to $P = 1 - \exp(-1)$ gives values for obtaining confidence bounds for the extreme-value location parameter η or the Weibull scale parameter $\delta = \exp(\eta)$. Tables 5.9 and 5.10 give values of $v_{1-P,\zeta}$ for $P = .10$ and $.05$, respectively, and values of ζ ranging from $.02$ to $.95$. Use of these various tables to test hypotheses concerning ξ, η, $x_{.10}$, and $x_{.05}$ is discussed in Chapter 6.

As an example of the use of the tables to obtain a confidence bound for ξ and η or a distribution percentage point, consider the following. The seven shortest times to failure during a life test of eight specimens of stress-corroded metal are 50.3, 64.1, 75.2, 85.9, 96.1, 107.0, and 120.0 days. Experience has shown that, under the conditions of the life test, lives of stress-corroded metal specimens such as those tested tend to have a two-parameter Weibull distribution. Thus one can use Table 5.3 to determine estimates of ξ and η, obtaining $\tilde{\xi} = .268$ and $\tilde{\eta} = 4.65$. Then a 75% lower confidence bound for the 10th percentile of the distribution of failure times is $\exp[\tilde{\eta} - \tilde{\xi}(3.46)] = 41$ days. A 75% upper confidence bound for ξ is $\tilde{\xi}/.67 = .40$. The respective confidence bounds are calculated by the use of Table 5.9 and Table 5.7.

(b) Intervals for Large, Complete Samples. For nearly complete samples of size about 80, linear estimators and maximum-likelihood estimators attain approximately their asymptotic properties. Therefore tables of standard normal variates can be used with functions of the estimators to obtain confidence intervals.

Table 5.7 Percentiles of the distribution of W

n	r	0.02	0.05	0.10	0.25	0.40	0.50	0.60	0.75	0.90	0.95	0.98
3	3	0.11	0.17	0.25	0.42	0.57	0.67	0.78	0.99	1.33	1.56	1.86
4	3	0.10	0.15	0.22	0.39	0.53	0.64	0.75	0.96	1.32	1.56	1.90
	4	0.20	0.28	0.37	0.54	0.68	0.77	0.86	1.05	1.33	1.53	1.77
5	3	0.09	0.14	0.21	0.37	0.51	0.61	0.73	0.94	1.32	1.59	1.93
	4	0.18	0.26	0.34	0.50	0.64	0.74	0.84	1.03	1.35	1.55	1.87
	5	0.28	0.36	0.44	0.60	0.73	0.82	0.91	1.07	1.33	1.50	1.70
6	3	0.09	0.14	0.21	0.36	0.50	0.61	0.72	0.93	1.32	1.59	1.92
	4	0.18	0.25	0.32	0.49	0.62	0.72	0.82	1.01	1.33	1.55	1.84
	5	0.25	0.33	0.42	0.58	0.71	0.79	0.89	1.05	1.33	1.51	1.73
	6	0.33	0.41	0.50	0.65	0.77	0.85	0.93	1.07	1.31	1.46	1.64

7

3	1.92	1.56	1.30	0.92	0.71	0.59	0.49	0.35	0.20	0.14	0.08
4	1.82	1.54	1.32	1.01	0.81	0.71	0.62	0.48	0.31	0.24	0.17
5	1.75	1.52	1.33	1.05	0.88	0.78	0.70	0.56	0.40	0.32	0.25
6	1.67	1.48	1.32	1.07	0.92	0.84	0.75	0.63	0.47	0.39	0.32
7	1.60	1.43	1.30	1.08	0.95	0.87	0.80	0.69	0.54	0.46	0.38

8

3	1.95	1.58	1.31	0.92	0.70	0.59	0.49	0.35	0.19	0.13	0.08
4	1.83	1.55	1.33	1.00	0.81	0.70	0.61	0.47	0.31	0.23	0.16
5	1.76	1.52	1.33	1.05	0.87	0.77	0.68	0.55	0.39	0.31	0.23
6	1.69	1.49	1.32	1.06	0.91	0.82	0.74	0.62	0.46	0.38	0.30
7	1.62	1.45	1.30	1.08	0.94	0.86	0.78	0.67	0.52	0.44	0.36
8	1.56	1.41	1.28	1.09	0.96	0.89	0.82	0.71	0.58	0.50	0.42

9

3	1.92	1.58	1.31	0.92	0.70	0.59	0.49	0.34	0.19	0.13	0.08
4	1.84	1.55	1.33	1.00	0.80	0.70	0.60	0.47	0.31	0.23	0.16
5	1.76	1.52	1.33	1.04	0.86	0.77	0.68	0.54	0.39	0.31	0.23
6	1.70	1.48	1.31	1.06	0.90	0.81	0.73	0.60	0.45	0.38	0.30
7	1.65	1.46	1.30	1.07	0.93	0.85	0.77	0.66	0.50	0.43	0.35
8	1.59	1.42	1.28	1.08	0.95	0.88	0.81	0.70	0.55	0.48	0.40
9	1.53	1.39	1.27	1.08	0.97	0.90	0.84	0.74	0.60	0.53	0.45

Table 5.7 (continued)

n	r	0.02	0.05	0.10	0.25	0.40	0.50	0.60	0.75	0.90	0.95	0.98
10	3	0.08	0.13	0.19	0.34	0.48	0.59	0.71	0.93	1.31	1.59	1.92
	4	0.16	0.23	0.30	0.46	0.60	0.70	0.80	1.00	1.33	1.57	1.86
	5	0.23	0.30	0.38	0.54	0.68	0.77	0.86	1.04	1.33	1.53	1.77
	6	0.29	0.37	0.45	0.60	0.73	0.81	0.90	1.06	1.32	1.49	1.71
	7	0.34	0.42	0.50	0.65	0.77	0.84	0.92	1.07	1.31	1.46	1.66
	8	0.39	0.47	0.54	0.69	0.80	0.87	0.95	1.08	1.29	1.43	1.60
	9	0.43	0.51	0.59	0.73	0.83	0.89	0.96	1.08	1.28	1.40	1.55
	10	0.48	0.55	0.62	0.76	0.85	0.91	0.98	1.09	1.26	1.38	1.51
11	3	0.08	0.13	0.19	0.34	0.48	0.59	0.71	0.97	1.31	1.60	1.97
	4	0.15	0.22	0.30	0.46	0.60	0.70	0.80	1.00	1.34	1.58	1.87
	5	0.22	0.30	0.38	0.54	0.67	0.76	0.86	1.04	1.34	1.54	1.82
	6	0.28	0.36	0.44	0.60	0.73	0.81	0.90	1.07	1.33	1.52	1.73
	7	0.33	0.41	0.49	0.65	0.76	0.84	0.92	1.08	1.32	1.48	1.67
	8	0.38	0.46	0.54	0.68	0.80	0.87	0.95	1.08	1.31	1.45	1.62
	9	0.42	0.50	0.57	0.71	0.82	0.89	0.96	1.09	1.29	1.42	1.58
	10	0.46	0.54	0.61	0.74	0.85	0.91	0.98	1.09	1.27	1.38	1.53
	11	0.50	0.57	0.64	0.77	0.87	0.93	0.99	1.09	1.25	1.36	1.49

12

n											
3	0.08	0.13	0.19	0.34	0.48	0.58	0.70	0.92	1.30	1.56	1.87
4	0.16	0.22	0.30	0.46	0.60	0.70	0.80	1.00	1.33	1.55	1.82
5	0.23	0.30	0.38	0.54	0.67	0.76	0.86	1.04	1.33	1.53	1.78
6	0.29	0.36	0.44	0.60	0.72	0.81	0.90	1.06	1.33	1.49	1.72
7	0.34	0.41	0.50	0.65	0.76	0.84	0.93	1.08	1.31	1.47	1.66
8	0.38	0.46	0.54	0.68	0.79	0.87	0.95	1.08	1.30	1.45	1.61
9	0.42	0.50	0.57	0.71	0.82	0.89	0.96	1.09	1.29	1.43	1.58
10	0.45	0.53	0.61	0.74	0.84	0.90	0.97	1.09	1.28	1.40	1.55
11	0.49	0.56	0.64	0.76	0.86	0.92	0.98	1.09	1.27	1.37	1.51
12	0.53	0.60	0.66	0.78	0.87	0.93	0.99	1.09	1.24	1.35	1.46

13

n											
3	0.08	0.13	0.19	0.33	0.48	0.58	0.69	0.91	1.30	1.58	1.95
4	0.15	0.22	0.29	0.45	0.59	0.69	0.79	0.99	1.33	1.57	1.86
5	0.22	0.30	0.37	0.53	0.67	0.75	0.85	1.03	1.34	1.55	1.79
6	0.28	0.36	0.43	0.59	0.72	0.80	0.89	1.06	1.33	1.51	1.72
7	0.33	0.40	0.48	0.64	0.76	0.84	0.92	1.07	1.32	1.48	1.67
8	0.37	0.45	0.53	0.67	0.79	0.86	0.94	1.08	1.30	1.45	1.62
9	0.42	0.49	0.56	0.70	0.81	0.88	0.95	1.08	1.29	1.42	1.58
10	0.44	0.52	0.60	0.73	0.83	0.89	0.96	1.08	1.28	1.40	1.55
11	0.48	0.55	0.62	0.75	0.85	0.91	0.97	1.08	1.26	1.38	1.51
12	0.51	0.58	0.65	0.77	0.86	0.92	0.98	1.08	1.25	1.36	1.47
13	0.54	0.61	0.68	0.79	0.88	0.93	0.99	1.09	1.24	1.33	1.44

Table 5.8 Percentiles of the distribution of $(\tilde{u}-u)/\tilde{b}$

n	r	0.02	0.05	0.10	0.25	0.40	0.50	0.60	0.75	0.90	0.95	0.98
3	3	-4.47	-2.54	-1.49	-0.52	-0.10	0.10	0.31	0.69	1.46	2.12	3.39
4	3	-6.92	-3.85	-2.32	-0.84	-0.29	-0.04	0.18	0.50	1.06	1.55	2.43
	4	-2.37	-1.50	-0.96	-0.37	-0.08	0.09	0.75	0.55	1.07	1.49	2.15
5	3	-9.35	-5.22	-3.04	-1.22	-0.50	-0.19	0.06	0.40	0.86	1.20	1.76
	4	-3.13	-1.94	-1.24	-0.50	-0.16	0.02	0.18	0.45	0.88	1.72	1.74
	5	-1.63	-1.08	-0.73	-0.31	-0.06	0.08	0.22	0.47	0.89	1.20	1.64
6	3	-10.54	-6.12	-3.72	-1.56	-0.69	-0.32	-0.04	0.33	0.75	1.02	1.39
	4	-3.69	-2.39	-1.59	-0.67	-0.25	-0.05	0.12	0.38	0.76	1.03	1.42
	5	-2.05	-1.36	-0.91	-0.38	-0.11	0.04	0.17	0.40	0.77	1.04	1.41
	6	-1.29	-0.91	-0.64	-0.28	-0.06	0.07	0.19	0.41	0.77	1.04	1.39

7											
3	-13.00	-7.39	-4.45	-1.87	-0.89	-0.48	-0.16	0.26	0.68	0.90	1.20
4	-4.67	-2.95	-1.94	-0.84	-0.36	-0.13	0.05	0.32	0.66	0.89	1.20
5	-2.48	-1.59	-1.10	-0.48	-0.17	-0.02	0.12	0.34	0.66	0.89	1.21
6	-1.54	-1.04	-0.73	-0.32	-0.10	0.03	0.15	0.35	0.67	0.90	1.20
7	-1.09	-0.79	-0.56	-0.26	-0.06	0.05	0.17	0.36	0.68	0.90	1.18

8											
3	-14.36	-8.15	-5.01	-2.14	-1.04	-0.58	-0.21	0.24	0.67	0.88	1.12
4	-5.34	-3.30	-2.18	-0.99	-0.43	-0.19	0.02	0.30	0.64	0.83	1.07
5	-2.78	-1.86	-1.25	-0.56	-0.22	-0.05	0.10	0.32	0.62	0.82	1.07
6	-1.80	-1.20	-0.83	-0.36	-0.12	0.01	0.13	0.33	0.63	0.82	1.08
7	-1.28	-0.88	-0.61	-0.27	-0.07	0.04	0.15	0.33	0.63	0.82	1.08
8	-0.97	-0.70	-0.50	-0.22	-0.05	0.06	0.16	0.34	0.63	0.82	1.07

9											
3	-15.68	-9.12	-5.64	-2.38	-1.17	-0.66	-0.28	0.20	0.66	0.86	1.06
4	-6.31	-3.78	-2.47	-1.08	-0.50	-0.24	-0.01	0.28	0.61	0.79	1.00
5	-3.19	-2.10	-1.40	-0.63	-0.26	-0.08	0.08	0.30	0.58	0.76	0.98
6	-2.01	-1.38	-0.94	-0.41	-0.15	-0.01	0.11	0.30	0.57	0.76	0.99
7	-1.43	-0.99	-0.70	-0.31	-0.10	0.02	0.13	0.31	0.57	0.76	0.99
8	-1.08	-0.76	-0.55	-0.25	-0.07	0.04	0.14	0.31	0.58	0.76	0.99
9	-0.87	-0.64	-0.47	-0.21	-0.05	0.05	0.15	0.32	0.58	0.76	0.98

Table 5.8 (continued)

n	r	0.98	0.95	0.90	0.75	0.60	0.50	0.40	0.25	0.10	0.05	0.02
10	3	1.07	0.87	0.66	0.17	-0.34	-0.76	-1.29	-2.58	-6.05	-9.98	-17.45
	4	0.96	0.77	0.60	0.77	-0.04	-0.28	-0.58	-1.22	-2.70	-4.17	-6.54
	5	0.93	0.72	0.56	0.28	0.05	-0.12	-0.31	-0.73	-1.56	-2.37	-3.56
	6	0.92	0.71	0.54	0.28	0.09	-0.04	-0.19	-0.48	-1.03	-1.51	-2.21
	7	0.93	0.70	0.54	0.28	0.11	-0.00	-0.12	-0.35	-0.77	-1.08	-1.56
	8	0.93	0.71	0.53	0.28	0.12	0.02	-0.08	-0.27	-0.62	-0.86	-1.20
	9	0.93	0.71	0.54	0.29	0.13	0.04	-0.06	-0.23	-0.50	-0.70	-0.97
	10	0.92	0.71	0.54	0.29	0.14	0.04	-0.04	-0.20	-0.44	-0.60	-0.80
11	3	1.07	0.87	0.65	0.13	-0.42	-0.85	-1.41	-2.76	-6.42	-10.68	-18.52
	4	0.92	0.75	0.58	0.24	-0.10	-0.36	-0.66	-1.37	-2.95	-4.57	-7.26
	5	0.88	0.69	0.54	0.26	0.01	-0.16	-0.37	-0.81	-1.75	-2.58	-4.00
	6	0.85	0.66	0.52	0.26	0.06	-0.07	-0.22	-0.53	-1.16	-1.67	-2.45
	7	0.86	0.65	0.50	0.26	0.09	-0.02	-0.15	-0.40	-0.85	-1.21	-1.70
	8	0.86	0.65	0.50	0.26	0.10	0.00	-0.11	-0.30	-0.66	-0.92	-1.30
	9	0.86	0.65	0.50	0.26	0.11	0.02	-0.08	-0.25	-0.54	-0.76	-1.06
	10	0.86	0.65	0.50	0.27	0.12	0.03	-0.06	-0.21	-0.46	-0.63	-0.87
	11	0.85	0.65	0.50	0.27	0.12	0.03	-0.05	-0.19	-0.42	-0.55	-0.75

12

row											
3	1.10	0.88	0.64	0.10	-0.49	-0.97	-1.58	-3.03	-6.92	-11.23	-19.08
4	0.92	0.75	0.58	0.21	-0.14	-0.40	-0.74	-1.47	-3.17	-4.81	-7.44
5	0.84	0.68	0.53	0.24	-0.01	-0.20	-0.42	-0.89	-1.88	-2.72	-4.17
6	0.81	0.64	0.50	0.25	0.05	-0.10	-0.26	-0.60	-1.27	-1.83	-2.63
7	0.80	0.62	0.48	0.25	0.08	-0.04	-0.17	-0.42	-0.92	-1.32	-1.91
8	0.79	0.62	0.48	0.25	0.09	-0.01	-0.12	-0.33	-0.71	-1.00	-1.41
9	0.80	0.62	0.47	0.25	0.10	-0.01	-0.09	-0.27	-0.58	-0.80	-1.15
10	0.80	0.62	0.47	0.25	0.11	0.02	-0.07	-0.23	-0.48	-0.67	-0.91
11	0.80	0.62	0.47	0.25	0.11	0.03	-0.06	-0.20	-0.43	-0.58	-0.78
12	0.79	0.62	0.47	0.25	0.11	0.03	-0.05	-0.19	-0.39	-0.53	-0.69

13

row											
3	1.09	0.88	0.65	0.08	-0.54	-1.02	-1.64	-3.21	-7.41	-11.66	-19.77
4	0.93	0.76	0.59	0.20	-0.19	-0.48	-0.82	-1.60	-3.37	-5.21	-8.22
5	0.84	0.68	0.54	0.24	-0.04	-0.24	-0.47	-0.96	-1.99	-2.95	-4.44
6	0.79	0.64	0.51	0.25	0.03	-0.13	-0.31	-0.66	-1.35	-1.94	-2.86
7	0.77	0.61	0.47	0.25	0.06	-0.06	-0.19	-0.46	-0.98	-1.40	-2.04
8	0.75	0.59	0.46	0.24	0.08	-0.02	-0.14	-0.36	-0.77	-1.06	-1.52
9	0.74	0.58	0.45	0.24	0.09	-0.00	-0.10	-0.29	-0.61	-0.86	-1.18
10	0.74	0.58	0.45	0.24	0.10	-0.01	-0.08	-0.24	-0.52	-0.72	-1.00
11	0.75	0.58	0.45	0.24	0.11	0.02	-0.06	-0.21	-0.45	-0.63	-0.85
12	0.75	0.59	0.45	0.25	0.11	0.03	-0.05	-0.19	-0.41	-0.56	-0.74
13	0.75	0.59	0.45	0.25	0.11	0.04	-0.05	-0.18	-0.38	-0.51	-0.67

Table 5.9 Percentiles of the distribution of $V_{.90}$

n	r	0.02	0.05	0.10	0.25	0.40	0.50	0.60	0.75	0.90	0.95	0.98
3	3	0.75	1.10	1.43	2.18	2.88	3.40	4.06	5.50	8.99	13.16	20.93
4	3	0.78	1.16	1.49	2.18	2.82	3.33	3.96	5.38	9.03	13.07	20.23
	4	0.87	1.16	1.46	2.06	2.60	2.99	3.45	4.40	6.47	8.39	11.66
5	3	0.78	1.18	1.51	2.17	2.79	3.27	3.87	5.74	8.78	12.58	20.38
	4	0.97	1.23	1.51	2.09	2.61	2.99	3.44	4.40	6.49	8.48	11.73
	5	0.97	1.23	1.49	2.02	2.49	2.82	3.20	3.93	5.48	6.73	8.66
6	3	0.73	1.18	1.53	2.15	2.73	3.18	3.74	4.98	8.74	11.74	18.65
	4	1.00	1.28	1.55	2.10	2.60	2.98	3.41	4.30	6.33	8.18	11.39
	5	1.02	1.29	1.54	2.05	2.50	2.82	3.21	3.94	5.42	6.73	8.89
	6	1.02	1.27	1.53	2.01	2.42	2.70	3.04	3.67	4.86	5.83	7.31

7	3	17.54	11.12	7.80	4.79	3.60	3.08	2.66	2.13	1.53	1.18	0.64
	4	10.90	7.89	6.16	4.21	3.33	2.91	2.57	2.10	1.58	1.31	1.04
	5	8.44	6.68	5.36	3.87	3.15	2.80	2.49	2.06	1.57	1.33	1.08
	6	7.23	5.82	4.86	3.63	3.01	2.70	2.42	2.03	1.56	1.32	1.08
	7	6.37	5.25	4.46	3.44	2.90	2.62	2.37	2.00	1.55	1.32	1.08

8	3	16.36	10.67	7.51	4.62	3.48	3.01	2.62	2.11	1.52	1.13	0.49
	4	10.75	7.79	5.96	4.10	3.77	2.88	2.56	2.10	1.60	1.33	1.04
	5	8.62	6.50	5.28	3.82	3.12	2.78	2.49	2.08	1.60	1.36	1.11
	6	7.18	5.83	4.83	3.62	3.02	2.71	2.43	2.05	1.59	1.36	1.13
	7	6.40	5.31	4.49	3.46	2.93	2.64	2.38	2.03	1.58	1.36	1.12
	8	5.84	4.90	4.21	3.32	2.83	2.57	2.34	2.01	1.58	1.36	1.12

9	3	15.61	10.21	7.14	4.43	3.40	2.95	2.57	2.09	1.51	1.12	0.42
	4	10.26	7.39	5.77	4.00	3.21	2.84	2.52	2.10	1.61	1.36	1.06
	5	8.13	6.34	5.13	3.76	3.08	2.76	2.47	2.08	1.63	1.41	1.17
	6	7.06	5.67	4.74	3.59	2.99	2.70	2.43	2.06	1.62	1.41	1.19
	7	6.46	5.28	4.48	3.45	2.91	2.64	2.39	2.04	1.62	1.41	1.19
	8	5.94	4.95	4.26	3.34	2.84	2.59	2.36	2.02	1.61	1.40	1.19
	9	5.50	4.66	4.04	3.22	2.78	2.55	2.33	2.00	1.60	1.40	1.19

Table 5.9 (continued)

n	r	0.02	0.05	0.10	0.25	0.40	0.50	0.60	0.75	0.90	0.95	0.98
10	3	0.09	0.99	1.46	2.05	2.51	2.84	3.27	4.25	6.75	9.36	14.88
	4	0.99	1.34	1.62	2.08	2.48	2.77	3.13	3.90	5.56	7.17	9.60
	5	1.17	1.42	1.64	2.07	2.45	2.71	3.02	3.67	5.00	6.13	8.02
	6	1.20	1.43	1.64	2.05	2.41	2.66	2.94	3.53	4.67	5.59	6.99
	7	1.21	1.43	1.64	2.04	2.38	2.62	2.88	3.41	4.41	5.18	6.29
	8	1.21	1.43	1.63	2.02	2.35	2.58	2.83	3.31	4.22	4.91	5.83
	9	1.21	1.42	1.63	2.01	2.32	2.54	2.77	3.22	4.03	4.63	5.51
	10	1.21	1.42	1.62	1.99	2.30	2.50	2.72	3.13	3.86	4.41	5.16
11	3	0.09	0.90	1.42	2.01	2.45	2.77	3.17	4.07	6.41	9.11	14.47
	4	0.97	1.35	1.61	2.06	2.44	2.73	3.06	3.79	5.46	7.04	9.98
	5	1.18	1.43	1.64	2.05	2.41	2.68	2.98	3.60	4.90	6.07	7.83
	6	1.24	1.45	1.64	2.04	2.38	2.63	2.91	3.46	4.58	5.52	6.96
	7	1.25	1.45	1.64	2.03	2.35	2.59	2.86	3.36	4.36	5.16	6.34
	8	1.25	1.45	1.64	2.01	2.33	2.56	2.80	3.28	4.15	4.87	5.82
	9	1.25	1.44	1.64	2.00	2.31	2.53	2.76	3.21	4.01	4.63	5.54
	10	1.25	1.44	1.64	1.99	2.29	2.49	2.71	3.14	3.87	4.44	5.23
	11	1.25	1.45	1.63	1.98	2.28	2.46	2.67	3.06	3.76	4.76	4.94

3	12.96	8.40	6.00	3.89	3.08	2.71	2.41	1.98	1.37	0.75	-0.38
4	9.07	6.60	5.17	3.67	3.00	2.69	2.42	2.05	1.60	1.34	0.95
5	7.35	5.79	4.72	3.52	2.93	2.65	2.40	2.05	1.66	1.44	1.20
6	6.61	5.31	4.41	3.39	2.88	2.62	2.38	2.04	1.67	1.46	1.26
7	6.09	4.98	4.21	3.30	2.82	2.58	2.36	2.03	1.67	1.47	1.28
8	5.71	4.75	4.06	3.22	2.78	2.54	2.34	2.02	1.66	1.47	1.28
9	5.40	4.53	3.94	3.16	2.74	2.52	2.31	2.01	1.66	1.46	1.27
10	5.11	4.37	3.87	3.11	2.70	2.49	2.30	2.00	1.65	1.47	1.27
11	4.88	4.23	3.72	3.05	2.67	2.47	2.28	2.00	1.64	1.46	1.27
12	4.68	4.07	3.62	3.00	2.63	2.44	2.27	1.99	1.64	1.47	1.28

3	12.45	8.16	5.88	3.85	3.04	2.69	2.40	1.99	1.34	0.72	-0.45
4	8.82	6.45	5.10	3.64	2.98	2.68	2.42	2.06	1.60	1.31	0.88
5	7.32	5.75	4.71	3.49	2.92	2.64	2.40	2.07	1.67	1.45	1.20
6	6.49	5.30	4.43	3.38	2.86	2.61	2.38	2.07	1.68	1.48	1.27
7	6.02	4.96	4.23	3.30	2.82	2.58	2.36	2.06	1.68	1.49	1.30
8	5.63	4.73	4.06	3.22	2.78	2.55	2.34	2.04	1.68	1.49	1.30
9	5.32	4.55	3.94	3.16	2.75	2.52	2.33	2.03	1.68	1.49	1.30
10	5.11	4.37	3.83	3.12	2.72	2.50	2.31	2.03	1.68	1.49	1.30
11	4.90	4.23	3.74	3.06	2.69	2.48	2.30	2.02	1.67	1.49	1.30
12	4.73	4.09	3.65	3.02	2.65	2.46	2.28	2.01	1.67	1.49	1.30
13	4.51	3.97	3.57	2.97	2.62	2.44	2.27	2.01		1.49	1.30

Table 5.10 Percentiles of the distribution of $V_{.95}$

n	r	0.02	0.05	0.10	0.25	0.40	0.50	0.60	0.75	0.90	0.95	0.98
3	3	1.26	1.64	2.04	2.94	3.83	4.49	5.33	7.20	11.75	17.21	27.32
4	3	1.38	1.73	2.11	2.98	3.82	4.47	5.30	7.19	12.17	17.55	27.59
	4	1.36	1.69	2.04	2.78	3.44	3.95	4.51	5.73	8.40	10.88	15.06
5	3	1.44	1.79	2.16	3.00	3.81	4.46	5.27	7.16	12.07	17.36	28.30
	4	1.45	1.76	2.10	2.82	3.49	3.98	4.56	5.80	8.56	11.14	15.51
	5	1.44	1.74	2.06	2.72	3.29	3.70	4.17	5.11	7.06	8.68	11.14
6	3	1.48	1.83	2.20	2.99	3.77	4.37	5.15	6.93	11.53	16.66	26.85
	4	1.52	1.83	2.15	2.84	3.49	3.98	4.55	5.75	8.47	10.95	15.32
	5	1.51	1.81	2.12	2.76	3.33	3.73	4.21	5.17	7.08	8.82	11.58
	6	1.50	1.80	2.10	2.70	3.20	3.56	3.97	4.77	6.27	7.53	9.39

7											
3	25.31	16.07	11.20	6.80	5.04	4.30	3.71	2.97	2.22	1.87	1.52
4	14.80	10.80	8.39	5.69	4.49	3.94	3.47	2.86	2.19	1.88	1.59
5	11.18	8.84	7.12	5.13	4.18	3.72	3.32	2.78	2.16	1.86	1.59
6	9.40	7.61	6.33	4.76	3.96	3.57	3.21	2.72	2.14	1.84	1.57
7	8.19	6.73	5.76	4.47	3.79	3.45	3.14	2.68	2.14	1.85	1.58

8											
3	24.57	15.76	11.02	6.61	4.95	4.22	3.67	2.96	2.24	1.90	1.53
4	15.22	10.74	8.19	5.60	4.44	3.91	3.46	2.87	2.22	1.91	1.63
5	11.57	8.78	7.07	5.09	4.18	3.71	3.32	2.79	2.20	1.90	1.63
6	9.43	7.67	6.35	4.77	3.99	3.58	3.24	2.75	2.18	1.89	1.62
7	8.38	6.91	5.83	4.57	3.83	3.48	3.16	2.71	2.17	1.89	1.62
8	7.50	6.29	5.44	4.31	3.71	3.38	3.10	2.69	2.17	1.89	1.62

9											
3	23.80	15.33	10.71	6.45	4.85	4.18	3.63	2.95	2.26	1.93	1.55
4	14.41	10.40	8.02	5.50	4.40	3.86	3.44	2.88	2.25	1.96	1.67
5	11.05	8.59	6.90	5.07	4.13	3.70	3.33	2.80	2.23	1.95	1.69
6	9.46	7.51	6.27	4.77	3.98	3.58	3.24	2.76	2.22	1.95	1.68
7	8.40	6.91	5.86	4.53	3.85	3.50	3.18	2.73	2.21	1.94	1.67
8	7.73	6.39	5.53	4.36	3.73	3.41	3.12	2.70	2.20	1.94	1.68
9	7.09	6.00	5.22	4.19	3.63	3.35	3.08	2.68	2.19	1.95	1.69

Table 5.10 (continued)

n	r	0.02	0.05	0.10	0.25	0.40	0.50	0.60	0.75	0.90	0.95	0.98
10	3	1.51	1.91	2.26	2.91	3.56	4.06	4.70	6.73	10.24	14.50	23.00
	4	1.70	1.98	2.27	2.86	3.39	3.81	4.29	5.42	7.81	10.12	13.69
	5	1.72	1.97	2.24	2.80	3.31	3.65	4.07	4.97	6.82	8.39	11.00
	6	1.71	1.97	2.23	2.76	3.22	3.55	3.93	4.70	6.24	7.50	9.42
	7	1.71	1.96	2.22	2.73	3.18	3.48	3.80	4.50	5.79	6.83	8.29
	8	1.71	1.96	2.21	2.71	3.13	3.41	3.72	4.33	5.52	6.40	7.61
	9	1.70	1.96	2.21	2.69	3.08	3.35	3.64	4.20	5.23	6.01	7.12
	10	1.72	1.96	2.21	2.67	3.04	3.29	3.56	4.08	4.98	5.67	6.65
11	3	1.48	1.93	2.25	2.88	3.49	3.98	4.59	6.07	9.89	14.11	22.60
	4	1.73	2.01	2.27	2.83	3.37	3.76	4.24	5.31	7.71	10.03	14.44
	5	1.75	2.01	2.26	2.78	3.27	3.62	4.03	4.89	6.72	8.34	10.93
	6	1.75	2.00	2.24	2.74	3.19	3.52	3.90	4.63	6.16	7.42	9.39
	7	1.75	1.99	2.23	2.72	3.14	3.45	3.79	4.45	5.79	6.83	8.42
	8	1.75	1.99	2.22	2.70	3.10	3.39	3.70	4.31	5.46	6.38	7.65
	9	1.75	1.98	2.22	2.68	3.07	3.34	3.63	4.21	5.23	6.04	7.23
	10	1.76	2.00	2.22	2.67	3.04	3.28	3.56	4.09	5.03	5.75	6.73
	11	1.77	2.00	2.22	2.66	3.01	3.25	3.50	3.98	4.85	5.49	6.35

12

3	21.39	13.40	9.41	5.85	4.50	3.91	3.47	2.87	2.26	1.91	1.47
4	13.27	9.56	7.42	5.16	4.18	3.72	3.34	2.84	2.28	2.02	1.74
5	10.40	8.08	6.54	4.82	3.99	3.61	3.26	2.78	2.28	2.03	1.79
6	9.00	7.22	5.97	4.56	3.86	3.52	3.20	2.75	2.27	2.02	1.79
7	8.08	6.66	5.63	4.40	3.76	3.43	3.14	2.73	2.26	2.02	1.79
8	7.49	6.27	5.36	4.26	3.68	3.38	3.11	2.71	2.25	2.01	1.79
9	7.06	5.95	5.16	4.16	3.62	3.33	3.07	2.69	2.24	2.00	1.78
10	6.63	5.67	4.99	4.07	3.56	3.29	3.04	2.67	2.23	2.01	1.77
11	6.29	5.47	4.84	3.98	3.51	3.25	3.02	2.67	2.24	2.01	1.78
12	6.00	5.26	4.68	3.91	3.46	3.22	3.00	2.66	2.24	2.02	1.80

13

3	20.76	13.11	9.23	5.84	4.47	3.90	3.45	2.88	2.77	1.92	1.44
4	13.09	9.47	7.38	5.18	4.18	3.73	3.35	2.84	2.31	2.05	1.78
5	10.25	8.04	6.57	4.80	3.98	3.60	3.26	2.81	2.30	2.06	1.82
6	8.89	7.24	6.03	4.58	3.86	3.50	3.21	2.78	2.29	2.05	1.82
7	8.10	6.68	5.65	4.41	3.75	3.44	3.15	2.75	2.28	2.05	1.82
8	7.43	6.29	5.40	4.28	3.69	3.39	3.12	2.74	2.27	2.04	1.81
9	6.99	6.00	5.17	4.16	3.63	3.35	3.10	2.72	2.27	2.04	1.81
10	6.66	5.70	5.01	4.09	3.58	3.31	3.07	2.70	2.27	2.04	1.81
11	6.36	5.50	4.87	4.02	3.53	3.27	3.05	2.70	2.27	2.04	1.82
12	6.12	5.30	4.73	3.95	3.48	3.24	3.02	2.69	2.27	2.04	1.82
13	5.80	5.12	4.61	3.87	3.43	3.21	3.00	2.68	2.27	2.05	1.83

Table 5.10 (continued)

n	r	0.02	0.05	0.10	0.25	0.40	0.50	0.60	0.75	0.90	0.95	0.98
22	3	0.58	1.60	2.15	2.74	3.18	3.52	3.93	4.86	7.37	9.93	15.15
	4	1.71	2.07	2.33	2.78	3.18	3.47	3.82	4.55	6.26	7.80	10.66
	5	1.92	2.17	2.37	2.78	3.15	3.42	3.73	4.38	5.75	6.85	8.68
	6	1.98	2.19	2.38	2.77	3.12	3.37	3.66	4.25	5.41	6.34	7.71
	7	1.99	2.19	2.38	2.76	3.10	3.34	3.61	4.16	5.22	6.03	7.20
	8	1.99	2.19	2.37	2.75	3.08	3.31	3.57	4.08	5.03	5.75	6.77
	9	1.99	2.19	2.37	2.74	3.06	3.29	3.53	4.01	4.88	5.53	6.44
	10	1.99	2.19	2.37	2.73	3.04	3.26	3.50	3.94	4.78	5.38	6.16
	11	1.99	2.18	2.36	2.72	3.03	3.24	3.47	3.91	4.67	5.24	5.98
	12	1.99	2.19	2.36	2.72	3.02	3.21	3.45	3.87	4.58	5.14	5.81
	13	2.00	2.19	2.36	2.71	3.00	3.20	3.42	3.82	4.52	5.04	5.66
	14	2.00	2.18	2.37	2.71	3.00	3.19	3.40	3.79	4.46	4.94	5.56
	15	2.00	2.19	2.36	2.71	2.99	3.17	3.38	3.75	4.39	4.84	5.44
	16	2.00	2.19	2.37	2.71	2.98	3.16	3.36	3.73	4.34	4.76	5.30
	17	2.00	2.19	2.36	2.71	2.97	3.15	3.34	3.70	4.29	4.69	5.23
	18	2.00	2.19	2.37	2.70	2.97	3.14	3.33	3.67	4.24	4.64	5.15
	19	2.02	2.19	2.37	2.70	2.96	3.14	3.32	3.65	4.20	4.59	5.06
	20	2.02	2.20	2.38	2.70	2.96	3.12	3.30	3.63	4.16	4.51	4.98
	21	2.02	2.21	2.38	2.70	2.95	3.11	3.29	3.60	4.11	4.45	4.89
	22	2.04	2.22	2.38	2.70	2.95	3.11	3.27	3.57	4.06	4.38	4.79

3	15.32	9.91	7.21	4.84	3.88	3.49	3.18	2.74	2.13	1.56	0.50
4	10.66	7.84	6.20	4.54	3.80	3.46	3.18	2.79	2.33	2.07	1.69
5	8.89	6.93	5.70	4.35	3.71	3.41	3.16	2.79	2.37	2.16	1.92
6	7.78	6.34	5.33	4.22	3.64	3.37	3.13	2.79	2.38	2.19	1.99
7	7.17	5.97	5.14	4.12	3.59	3.33	3.10	2.77	2.38	2.19	2.00
8	6.76	5.68	4.98	4.04	3.55	3.31	3.08	2.76	2.38	2.19	2.00
9	6.37	5.52	4.85	3.98	3.51	3.28	3.07	2.75	2.38	2.19	2.00
10	6.16	5.33	4.73	3.92	3.48	3.26	3.05	2.74	2.37	2.19	2.00
11	5.93	5.21	4.63	3.88	3.46	3.24	3.03	2.74	2.37	2.18	2.00
12	5.78	5.08	4.55	3.84	3.43	3.22	3.02	2.73	2.37	2.18	2.00
13	5.59	4.98	4.48	3.80	3.41	3.20	3.01	2.73	2.37	2.18	2.01
14	5.52	4.91	4.45	3.77	3.39	3.18	3.00	2.72	2.36	2.19	2.01
15	5.35	4.81	4.39	3.74	3.37	3.17	3.00	2.71	2.36	2.19	2.02
16	5.31	4.75	4.33	3.72	3.35	3.16	2.99	2.71	2.36	2.19	2.02
17	5.20	4.70	4.29	3.69	3.34	3.15	2.98	2.71	2.37	2.19	2.01
18	5.14	4.63	4.24	3.67	3.33	3.14	2.97	2.71	2.37	2.20	2.02
19	5.07	4.58	4.20	3.65	3.31	3.13	2.97	2.71	2.37	2.20	2.03
20	4.99	4.53	4.15	3.62	3.30	3.12	2.96	2.70	2.37	2.20	2.03
21	4.92	4.47	4.11	3.60	3.29	3.12	2.95	2.70	2.38	2.21	2.04
22	4.82	4.42	4.07	3.58	3.28	3.10	2.95	2.70	2.38	2.21	2.05
23	4.74	4.36	4.03	3.55	3.26	3.10	2.94	2.70	2.38	2.21	

23

For example, the three-order–statistic estimator of $x_{.10}$ given by (5.19),

$$X_{.10}^0 = X_{([.40n]+1)} - .473(X_{([.98n]+1)} - X_{([.13n]+1)}),$$

where $[x]$ indicates the largest integer less than x, is asymptotically normal with mean $x_{.10}$. The asymptotic efficiency of $X_{.10}^0$ is .82, and the asymptotic variance of the MLE of $x_{.10}$ (since the asymptotic covariance of MLE's of ξ and η is $-.257\xi^2$) is equal to

$$\frac{1.104\xi^2}{n} + .608\left\{\ln\left[\ln\left(\frac{1}{.90}\right)\right]\right\}^2 \frac{\xi^2}{n} - 2(.257)\ln\left[\ln\left(\frac{1}{.90}\right)\right]\frac{\xi^2}{n}$$

(see Section 5.2.1). Therefore, since $\ln[\ln(1/.90)] = -2.25037$, the asymptotic variance of $X_{.10}^0$ is $(5.345/.82)\xi^2/n$, and $(X_{.10}^0 - x_{.10})/(2.55\xi/\sqrt{n})$ is asymptotically standard normal (with mean 0 and variance 1). To obtain a confidence bound for $x_{.10}$, one can simply use the estimator $\xi^0 = .33(X_{([.937n]+1)} - X_{([.1673n]+1)})$ for ξ, since $(X_{.10}^0 - x_{.10})/(2.55\xi^0/\sqrt{n})$ is also asymptotically normal with mean 0 and standard deviation 1.

One can similarly form asymptotically normal estimators for η and $x_{.05}$ by using the fact that the asymptotic variances of

$$\eta^0 = .446 X_{([.40n]+1)} + .554 X_{([.82n]+1)}$$

and

$$X_{.05}^0 = X_{([.40n]+1)} - .689(X_{([.98n]+1)} - X_{([.13n]+1)})$$

are $(1.109/.82)\xi^2/n$ and $(8.000/.81)\xi^2/n$, respectively.

The asymptotic distribution of ξ^0/ξ is normal with mean 1 and variance $(.608/.66)/n$, so that asymptotic confidence bounds for ξ can also be calculated from $(1.042\sqrt{n})(\xi^0/\xi - 1)$ by using tables of standard normal variates. The ten-order-statistic asymptotic estimators, η' and ξ', of η and ξ given in Section 5.2.2 will provide more accurate confidence bounds, however, since the asymptotic efficiencies of η' and ξ' are .9720 and .9442, respectively. One can use these estimators of ξ and η to obtain confidence bounds on any distribution percentile by incorporating the assumption that the asymptotic covariance of η' and ξ' is approximately $-.257\xi^2/n$. Thus the asymptotic variance can be calculated for any estimator $X_P' = \eta' + \ln[\ln(1/(1-P)]\xi'$, and tables of standard normal deviates can be used to obtain confidence bounds for x_P.

As an illustration of the use of these asymptotic methods, consider the example given in Section 5.2.2(b)-4. The ten-order–statistic asymptotically unbiased estimators ξ' and η' of ξ and η corresponding to the stress-corrosion test data are .641 and 1.45. Since the efficiency of ξ' is .9724 and the Cramér-Rao lower bound for the variance of unbiased estimators of ξ is $.608\xi^2/n$, the variance of ξ' is $.625\xi^2/100 = .00625\xi^2$, and a 95% upper confidence bound $\bar{\xi}$ for ξ is therefore given implicitly by $(.00625)^{-1/2}$ $(.641/\bar{\xi} - 1) = -1.645$, or $\bar{\xi} = .641/.87 = .74$.

(c) Intervals for Moderate to Large Sample Sizes.

1. THE PARAMETER $\xi = \beta^{-1}$. The unbiased linear estimator ξ^{**}, described in Section 5.2.2, can be used to obtain confidence bounds for the parameter ξ. Bain (1972) notes that the mean and the variance of

$$\frac{nk_{r,n}\xi^{**}}{\xi} = \frac{\sum_{i=1}^{r} (X_{(r)} - X_{(i)})}{\xi}$$

are both approximately equal to $nk_{r,n}$ for n large and r/n about .5 or less. He therefore suggests approximating the distribution of $2nk_{r,n}\xi^{**}/\xi$ by chi square with $2nk_{r,n}$ degrees of freedom.

A better approximation, applicable for all r, can be obtained for $n \geqslant 20$, however, by using the observation of van Montfort (1970) that for $X_{(i)}$, the ith order statistic from an extreme-value distribution and $Z_i = (X_{(i)} - \eta)/\xi)$, each

$$H_i = \frac{(X_{(i+1)} - X_{(i)})}{\xi[E(Z_{i+1}) - E(Z_i)]}, \quad i = 1, \ldots, n-1$$

has approximately an exponential distribution with mean 1, variance approximately 1, and H_i and H_j, $i \neq j$, have very nearly 0 covariance. Although the approximation is quite good generally for $n \geqslant 3$ for all distribution percentiles between .5 and .95, it depends on an asymptotic result concerning differences of adjacent-order statistics given by Pyke (1965).

The approximation is used in the construction of the goodness-of-fit statistic described for the extreme-value distribution in Section 7.1. The fact that the goodness-of-fit test statistic has very nearly a beta distribution suggests that each $2H_i$ has approximately an independent chi-square distribution with 2 degrees of freedom. The agreement of distribution percentiles with beta percentiles increases with n. For $n \geqslant 20$, the agreement is within a unit for the number of significant figures tabulated for the test statistic for 95th and lower percentiles.

Now note that, for $r \leqslant .90n$, $2nk_{r,n}\xi^{**}/\xi$ is equal to

$$
W_{r,n} = \frac{2 \sum\limits_{i=1}^{r} (X_{(r)} - X_{(i)})}{\xi} = \sum\limits_{i=1}^{r-1} i[E(Z_{i+1}) - E(Z_i)]2H_i
$$

and is therefore approximately a sum of weighted independent chi-square variates. The results of Patnaik, used in Sections 5.1.1 and 5.1.2, can now be applied to approximate the weighted sum of weighted chi squares by a chi-square $2mW/v$ with $2m^2/v$ degrees of freedom, where m and v are the mean and the variance, respectively, of $W_{r,n}$. The value $nk_{r,n}$ is the mean of $W_{r,n}/2$ so that

$$
\frac{2mW}{v} = \frac{8n^2 k_{r,n}^2 \xi^{**}}{4n^2 k_{r,n}^2 l_{r,n} \xi} = \frac{2\xi^{**}}{l_{r,n} \xi},
$$

and all that is needed in addition is the variance $(2nk_{r,n})^2 l_{r,n}$ of W. For r/n small, $v/4 \cong nk_{r,n}$ and $2mW/v \cong W$. In this case, the number of degrees of freedom, $2m^2/v \cong 2nk_{r,n}$, is in agreement with Bain. In general, $2m^2/v = 2/l_{r,n}$. Englehardt and Bain (1973) use asymptotic arguments to infer that $2(\xi^{**}/\xi)/l_{r,n}$ is approximately a chi-square variate with $2l_{r,n}^{-1}$ degrees of freedom when n becomes large with r fixed; that is, their proof applies to deeply censored samples.

Values of $k_{r,n}$ and of $nl_{r,n}$ are given in Table 5.11, along with values for estimating η, for $n = 25(5)60$, $r/n = .1(.10)1.0$, for integer r. Note that the expectation and the variance of ξ^{**} are ξ and $l_{r,n}\xi^2$, respectively. These results also apply when $r = n$, in which case s (in Table 5.11) is $[.892n]$, and ξ^{**} and $W_{r,n}$ are defined in terms of sums of absolute values of $X_{(s)} - X_{(i)}$, $i = 1, \ldots, n$. Englehardt and Bain (1972) give, for $r/n = .3(.1).9$, values of $nl_{r,n}$ applying to $n = 10, 20$, and ∞. (The entries in Table 5.11 were computed by Mann and Fertig, 1974a; a discussion of the precision of the tabulated values is given by these authors.) Since the number of degrees of freedom, $2m^2/v$, would not be expected to be an integer, the Wilson-Hilferty approximation of chi square to normality (discussed in Sections 5.1.1 and 5.1.2) can be used to approximate $z_{r,n}(\zeta)$, the 100ζth distribution percentile of ξ^{**}/ξ. Since ξ^{**}/ξ is a chi square divided by its degrees of freedom,

$$
z_{r,n}(\zeta) \cong \left(1 - \frac{v}{9m^2} + \frac{\phi_\zeta \sqrt{v}}{3m} \right)^3 \tag{5.20}
$$

is approximately equal to $(1 - l_{r,n}/9 + \phi_\zeta \sqrt{l_{r,n}}/3)^3$, where ϕ_ζ is the 100ζth percentile of the standard normal distribution. Then an upper confidence bound for ξ at level $1-\zeta$ is given by $\xi^*/z_{r,n}(\zeta)$. For the data given in Section 5.2.2, for which an estimate $\xi^{**}=.478$ is obtained, an 80% upper confidence bound for ξ is given by

$$\frac{(.478)}{\left(1-.194/9-.84\sqrt{.194/9}\ \right)^3}=.765.$$

The values of .085 and .194 were obtained for $k_{r,n}$ and $l_{r,n}$ from Table 5.11, $n=60$, $r/n=.10$.

This method of approximating distribution percentiles can also be used to obtain confidence bounds on ξ based on the best linear invariant estimator, $\tilde{\xi}$, or the best linear unbiased estimator, ξ^*, for values of n not given in Table 5.7, since $\tilde{\xi}$ and ξ^* can also be expressed as sums of weighted H_i's. In order to use this method, however, one needs to know the weights defining ξ^* or $\tilde{\xi}=\xi^*/(1+C)$ and $C=\mathrm{Var}(\xi^*/\xi)=E(L\xi)/[1-E(L\xi)]$. Then, since $E(\xi^*/\xi)=1$, if C is known, the mean m and the variance v of ξ^*/ξ are both known and the distribution of $2(\xi^*/\xi)/C$ can be fitted by a chi-square distribution with $2/C$ degrees of freedom. Results obtained by this approximation for percentiles of $\tilde{\xi}/\xi=\xi^*/[\xi(1+C)]$ agree with the values shown in Table 5.7 to within a unit in the second significant figure. For $n\geqslant 20$, the approximate values agree with tabulated values in Mann, Scheuer and Fertig (1971) to within a unit in the second decimal place.

Because of the fact that sums of independent chi-squares are equal to a chi square, one can also use this approach to obtain confidence bounds for ξ when samples have been divided into subsamples or when samples are from different populations with the same value of ξ. In the latter case the value of η need not be the same in the populations considered, since the estimators to be combined are independent of η. In the discussion in Section 5.2.2(b)-2, "Combining Best Linear Unbiased Estimators from Subsamples," the estimator given for ξ, $\bar{\xi}=\Sigma_{j=1}^k C_j^{-1}\xi_j^*/\Sigma_{j=1}^k C_j^{-1}$, has expectation ξ and variance $\xi^2/\Sigma_{j=1}^k C_j^{-1}$. Therefore,

$$2\left(\frac{\bar{\xi}}{\xi}\right)\sum_{j=1}^k C_j^{-1}=2\sum_{j=1}^k C_j^{-1}\left(\frac{\xi_j^*}{\xi}\right)$$

is an approximate chi square with $2\Sigma_{j=1}^k C_j^{-1}$ degrees of freedom. To obtain an upper confidence bound for ξ at level $1-\alpha$, one divides $\bar{\xi}$ by $\left[1-\left(9\Sigma_{j=1}^k C_j^{-1}\right)^{-1}+\phi_\alpha\left(\Sigma_{j=1}^k C_j^{-1}\right)^{-1/2}/3\right]^3$; see Equation (5.20).

Table 5.11 Values for calculating estimates $\xi^{**} = \sum_{i=1}^{r} |x_{(s)} - x_{(i)}|/(nk_{r,n})$, $\tilde{\xi} = \xi^{**}/(1 + l_{r,n})$, $\eta^{**} = x_{(s)} - E(Z_s)\xi^{**}$ and $\tilde{\eta} = \eta^{**} - B_{r,n}\tilde{\xi}$ and the approximate chi-square variate, $(2\xi^{**}/\tilde{\xi})/l_{r,n}$, with $2/l_{r,n}$ degrees of freedom

r	$k_{r,n}$	$nl_{r,n}$	$E(Z_s)$	$nB_{r,n}$	r	$k_{r,n}$	$nl_{r,n}$	$E(Z_s)$	$nB_{r,n}$
		$n=25$					$n=35$		
5	.169	5.97	−1.63	10.04	7	.182	5.55	−1.60	9.16
10	.406	2.45	−.75	2.13	14	.420	2.41	−.73	2.07
15	.690	1.48	−.16	.57	21	.707	1.46	−.14	.56
20	1.066	1.01	.40	.04	28	1.091	1.01	.42	.03
25	1.321	.81	.65	−.08	35	1.367	.81	.70	−.08
		$n=30$					$n=40$		
3	.069	14.64	−2.44	36.03	4	.077	12.70	−2.39	30.75
6	.176	5.71	−1.61	9.44	8	.184	5.34	−1.58	8.72
9	.291	3.45	−1.11	4.11	12	.299	3.30	−1.09	3.89
12	.415	2.43	−.73	2.07	16	.423	2.33	−.72	1.98
15	.549	1.85	−.42	1.10	20	.558	1.78	−.41	1.04
18	.698	1.46	−.14	.54	24	.708	1.41	−.13	.52
21	.870	1.20	.13	.22	28	.882	1.17	.14	.21
24	1.077	1.01	.41	.02	32	1.092	.99	.43	.02
27	1.356	.87	.74	−.10	36	1.377	.86	.77	−.09
30	1.302	.82	.62	−.08	40	1.340	.80	.67	−.07
		$n=45$					$n=55$		
9	.188	5.37	−1.57	8.75	11	.192	5.30	−1.56	8.48
18	.427	2.36	−.71	2.02	22	.432	2.39	−.71	1.98
27	.712	1.42	−.13	.53	33	.719	1.45	−.12	.52
36	1.098	.98	.43	.02	44	1.105	1.00	.44	.02
45	1.386	.79	.73	−.07	55	1.402	.81	.75	−.08
		$n=50$					$n=60$		
5	.082	12.35	−2.37	29.30	6	.085	11.65	−2.34	27.65
10	.190	5.40	−1.56	8.65	12	.193	5.10	−1.55	8.17
15	.306	3.39	−1.08	3.94	18	.309	3.25	−1.07	3.75
20	.430	2.41	−.71	2.03	24	.433	2.34	−.70	1.96
25	.565	1.82	−.40	1.06	30	.568	1.78	−.40	1.02
30	.716	1.45	−.12	.53	36	.720	1.44	−.12	.52
35	.890	1.18	.15	.21	42	.894	1.19	.16	.21
40	1.103	1.00	.44	.02	48	1.106	.99	.44	.01
45	1.393	.87	.78	−.10	54	1.397	.88	.79	−.09
50	1.368	.81	.70	−.09	60	1.381	.83	.72	−.08

For the data given in the example showing point estimation for this model, it was found that

$$\bar{\xi}= \frac{\xi_1^*/C_1+\xi_2^*/C_2}{C_1^{-1}+C_2^{-1}}=.606,$$

with $C_1=.03642$ and $C_2=.03143$. A 90% upper confidence bound for ξ is then $.606/[1-\frac{1}{9}/59.28-1.282(.130)/3]^3=.724$.

One could also use the approximate F statistic, $\xi_1^*/\xi^{*}2$, with $2/C_1$ and $2/C_2$ degrees of freedom, to test that $\xi_1=\xi_2$. Methods for dealing with approximate F statistics with noninteger degrees of freedom are discussed later in this section with regard to obtaining confidence bounds for t_P.

2. THE PARAMETER $\eta=\ln\delta$: SAMPLES THAT ARE NOT HIGHLY CENSORED.

To obtain approximate confidence bounds for η (or δ), the approach of Mann and Fertig (1974a) can be used. (Confidence bounds on x_P will be discussed subsequently.) First, recall from Section 5.1 that, for $T_{(s)}=\exp(X_{(s)})$, $R(T_{(s)})=\exp[-(T_{(s)}/\delta)^\beta]$ has a beta distribution with parameters $n-s+1$ and s. Results of Section 10.4 show that, if R has a beta distribution with integer parameters $n-s+1$ and s, then $-\ln R$ has approximately a weighted chi-square distribution with mean

$$\psi(n+1)-\psi(n-s+1)=\frac{\sum_{i=1}^{s}1}{n-i+1}\cong\ln(n+.5)-\ln(n-s+.5)$$

and variance

$$-\psi'(n+1)+\psi'(n-s+1)=\frac{\sum_{i=1}^{s}1}{(n-i+1)^2}\cong\frac{-1}{n+.5}+\frac{1}{n-s+.5},$$

where $s=1,\ldots,n$. The expressions for the mean and the variance of the reduced exponential order statistic $(T_{(s)}/\delta)^\beta$ are in agreement with those derived by Epstein and Sobel (1953) and cited in Section 5.1 and Chapter 6.

An expression that does not depend on β and has a distribution independent of both parameters is given by $Q_r=(T_{(s)}/\delta)^{\beta\xi/\xi**}=\exp[(X_{(s)}-\eta)/\xi^{**}]$, since $\xi=\beta^{-1}$. Therefore, if the distribution of Q_r (an approximate noncentral chi square raised to a power given by the inverse of an approximate chi square over degrees of freedom) can be found, a confidence bound for δ or $\eta=\ln\delta$ can be obtained.

For r small relative to n, $\xi^* \cong \xi^{**}$ and $\eta^* \cong \eta^{**} = X_{(s)} - E(Z_s)\xi^{**}$; and, since $\tilde{\xi} = \xi^*/(1 + C)$ and $\tilde{\eta} = \eta^* - B\tilde{\xi}$, one can, for $r/n \leqslant .5$, convert distribution percentiles of $(\tilde{\eta} - \eta)/\tilde{\xi}$ in Table 5.8 to approximate percentiles of Q_r if B, C, and $E(Z_s)$ are known. From these facts the distribution of Q_r can be approximated for $r/n \leqslant .5$. Since $V_0 \cong (\tilde{\eta} - \eta)/\tilde{\xi}$ is approximately equal to

$$\frac{X_{(s)} - \eta - E(Z_s)\xi^* - B\xi^*/(1 + C)}{\xi^*/(1 + C)},$$

$\ln Q_r$ is approximately equal to

$$\frac{V_0 + E(Z_s)(1 + C) + B}{1 + C},$$

where $C\xi^2$ is the variance of ξ^*, and $B\xi^2$ is the covariance of ξ^* and η^*. From Table 5.3, one obtains, for $r = 5$, $n = 12$,

$$C = \frac{E(L\xi)}{1 - E(L\xi)} = \frac{.1815}{.8185} = .222 \quad \text{and} \quad B = \frac{E(CP)}{1 - E(L\xi)} = \frac{.1604}{.8185} = .196.$$

From Table 7.2, one can determine $E(Z_5) = -.778$ recursively by knowing that $E(Z_1)$ is $-.5772157 + \ln(1/n)$. Then, since the 50th percentile of V_0, from Table 5.8, is $-.20$, the 50th percentile of Q_5 is approximately exp $\{[-.20 - .778(1.222) + .196]/1.222\} \cong .46$. Similarly, one can approximate the 90th, 95th, and 98th percentiles of Q_5 as .83, .94, and 1.07. Examination of a table of chi-square percentiles reveals that the corresponding distribution percentiles of $18Q_5$ are approximately those of a chi-square distribution with about 9 degrees of freedom.

From the result of Patnaik applied above to weighted chi-square variates, if Q_r with mean m and variance v is approximately weighted chi square, $2mQ_r/v$ is approximately chi square with $2m^2/v$ degrees of freedom. Thus, since Q_5 for $n = 12$ is approximately a weighted chi-square variate, its mean and variance are approximately $9/18 = .50$ and $2(.50)/18 = .055$. Suppose that one approximates the mean m and variance v of Q_5 by the mean m_5 and variance v_5 of $Q_5^{\xi^{**}/\xi} = (T_{(5)}/\delta)^\beta$, that is, $\Sigma_{j=1}^5 1/(n - j + 1)$ and $\Sigma_{j=1}^5 1/(n - j + 1)^2$, respectively, and obtains $m \cong .51$ and $v \cong .053$. Also, one sees that the mean m_6 of $(T_{(6)}/\delta)^\beta$ is approximately $.51 + 1/7 = .65$ and its variance v_6 is approximately $.025 + 1/49 = .073$. Since in these cases $2m_k^2/v_k > 3$, the 100γth percentile of $(T_{(k)}/\delta)^\beta$ can be approximated well by use of the Wilson-Hilferty transformation [see Equation (5.20)] as:

$$m_k \left(1 - \frac{v_k/m_k^2}{9} + \frac{\phi_\gamma \sqrt{v_k}/m_k}{3}\right)^3, \tag{5.21}$$

where ϕ_γ is the 100γth percentile of a standard normal variate. For $n=12$ one obtains approximately .48, .82, and .94 and .61, 1.01, and 1.15 for the 50th, 90th, and 95th percentiles of $(T_{(5)}/\delta)^\beta$ and $(T_{(6)}/\delta)^\beta$, respectively. Use of the latter percentiles for the distribution of Q_5, rather than those obtained above, in determining lower confidence bounds for δ or η will give conservative results. Use of percentiles of $(T_{(5)}/\delta)^\beta$ will give nearly correct results. [Note that the weighted chi-square approximations for the percentiles of the distribution of the exponential order statistics, $(T_{(5)}/\delta)^\beta$ and $(T_{(6)}/\delta)^\beta$, agree to within about 3 units in the third decimal place with those obtainable by taking logarithms of appropriate beta percentiles.]

As r and n both increase with r/n fixed, the distributions of Q_r, $(T_{(s)}/\delta)^\beta$, and $(T_{(s+1)}/\delta)^\beta$ will become more nearly equivalent because (1) the proportions of increments added to m_s and v_s to form m_{s+1} and v_{s+1} will decrease with respect to the total values, and (2) the value of ξ^{**} will grow stochastically closer to the value of $\xi = \beta^{-1}$, while the covariance of ξ^{**} and $X_{(s)}$ tends to grow smaller. What should be investigated first, therefore, are distribution percentiles of Q_3 for n both small and large.

Distribution percentiles of Q_3 can be approximated in the same manner as for Q_5 above. The 50th, 90th, and 98th percentiles of Q_3 for $n=6$ are approximately .625, 1.39, and 2.18. Then $5.37Q_3$ is approximately chi square with 4 degrees of freedom. Thus the mean of Q_3 is approximately $4/5.4 = .74$, and the variance is approximately $2(.74)/5.4 = .274$. For $n=6$, the means and variances of $(T_{(i)}/\delta)^\beta$, $i=3,4$, are $\frac{1}{4}+\frac{1}{5}+\frac{1}{6} \cong .617$, $\frac{1}{16}+\frac{1}{25}+\frac{1}{36} \cong .130$, $.6167+.3333 \cong .950$, and $.1303+.1111 \cong .241$. The 50th, 90th, and 98th percentiles of $(T_{(4)}/\delta)^\beta$ are approximately .868, 1.60, and 2.20, while the 50th, 90th, and 98th percentiles of $(T_{(3)}/\delta)^\beta$ are approximately .549, 1.10, and 1.56.

For $n=20$, $m_4 = \frac{1}{20}+\frac{1}{19}+\frac{1}{18}+\frac{1}{17} = .217$ and $v_4 = .0118$; while, for Q_3, calculations using tabulated values in Mann (1967a, 1968a) and Mann, Fertig, and Scheuer (1971) reveal that m is approximately .22 and v is about .03. Therefore the percentiles of Q_3 of interest lie between corresponding percentiles of $(T_{(3)}/\delta)^\beta$ and $(T_{(4)}/\delta)^\beta$ for $n=6$ but not for $n=20$.

A Monte Carlo investigation reveals, in fact, that, if r/n is .4 or more and n is 15 or more, a percentile $q_r(\gamma)$ of Q_r approximated by substituting

$$m_s = \sum_{i=1}^{s} \frac{1}{n-i+1} \cong \ln(n+.5) - \ln(n-s+.5)$$

and

$$v_s = \sum_{i=1}^{s} \frac{1}{(n-i+1)^2} \cong \frac{-1}{n+.5} + \frac{1}{n-s+.5}$$

for m_k and v_k, respectively, in (5.21) will be very nearly correct over the range .5 to .95, that is, within about a unit in the second significant figure. Then an approximate lower confidence bound for δ at confidence level γ is $T_{(s)}/[q_r(\gamma)]^{\xi**}$.

As an example, consider the following 15 times of onset of combustion instability observed during testing of 25 thrust chambers of a specified rocket engine in which testing was terminated at approximately the time of the fifteenth failure: .438, 2.413, 3.073, 3.079, 3.137, 3.198, 3.918, 4.287, 4.508, 4.981, 5.115, 5.592, 5.848, 5.958, 6.013. The estimate $\tilde{\xi}$ is computed from $\Sigma_{i=1}^{15}(\ln T_{15} - \ln T_i)/[(25)(.690)]$ and is found to be .445. The estimate of η, $\hat{\eta}$, is given as $\ln T_{15} - E(Z_{15})b = \ln 6.013 - (-.156)\ (.445) = 1.863$. The corresponding estimate of δ is then $\hat{\delta} = \exp(1.863) = 6.44$. With $m_k = .8870$ and $v_k = .0560$, $q_r(\gamma = .8)$ is approximately equal to 1.0768. [Use of tables of the incomplete beta functions yields, as the 80th percentile of the natural logarithm of the reciprocal of a beta variate with parameters $25 - 15 + 1 = 11$ and 15, $\ln(.3408) \cong 1.0764$.] A lower confidence bound for δ at confidence level .80 is then found to be $(6.013)/(1.0768)^{.445} = 5.82$.

Since the asymptotic linear estimators of ξ based on a few ordered observations can also be represented as linear combinations of $\xi H_i = (X_{(i+1)} - X_i)/E(Z_{i+1} - Z_i)$, $i = 1,\ldots,r$, they can also be approximated by weighted chi-square variates when sample size is large. For n very large, the estimators are unbiased and their variances are known. A 95% upper confidence bound for ξ is given by $\xi'/(1 - .625/9n - 1.645\sqrt{.625/9n}\ )^3$, where ξ' is the ten-order-statistic estimator of ξ, and $.625\xi^2/n$ is its asymptotic variance. For the example given in Section 5.2.2(b), ξ' is equal to .641 and n is 100. Therefore the 95% upper confidence bound is $.641/.8755 = .73$, essentially the value that was obtained by assuming normality for ξ'. The normal and noncentral chi-square confidence bounds (practically) coincide because the sample size is quite large.

3. THE PARAMETER $\eta = \ln \delta$ AND THE PARAMETRIC FUNCTIONS $R(t_m)$ AND t_p. To obtain confidence bounds for η or for $R(t_m)$ from censored data and sample sizes between 40 and 100, one can use maximum-likelihood estimates of functions of η and ξ, together with tabulated results of Billman, Antle, and Bain (1971). If censoring is about 50%, $\tilde{\xi}$ and $\tilde{\eta}$ can be used in place of the corresponding maximum-likelihood estimators, $\hat{\xi}$ and $\hat{\eta}$. Billman et al. give percentiles of $\sqrt{n}\ (\hat{\eta} - \eta)/\hat{\xi}$ for 0, 25, and 50% censoring of samples of sizes 40, 60, 80, 100, and 120.

Linear interpolation can be used to approximate percentiles applying to other censorings. Other tables in the same article by Bain et al. apply to the same sample sizes and censorings and to 90, 95, 98, and 99% confidence bounds for $R(t_m)$ as functions of

$$\hat{R}(t_m) = \exp\left\{-\exp\left[\frac{\ln(t_m) - \hat{\eta}}{\hat{\xi}}\right]\right\}.$$

For obtaining confidence bounds on η (or δ) from highly censored samples or confidence bounds on reliable life, $t_P = \exp(x_P)$, from large samples, the following results can be used. It is known from earlier discussion that ξ^{**}/ξ has an approximate chi-square-over-degrees-of-freedom distribution with $2/l_{r,n}$ degrees of freedom. It can be shown that under certain conditions

$$Q = \frac{\eta^{**} - x_P - \xi^{**} B_{r,n}/l_{r,n}}{(-y_P - B_{r,n}/l_{r,n})\xi},$$

with $y_P = (x_P - \eta)/\xi = \ln[-\ln(1 - P)]$, also has an approximate chi-square-over-degrees-of-freedom distribution. Recall that $B_{r,n}$ and $l_{r,n}$, tabulated for selected values of r and n in Table 5.11, are the covariance of η^{**}/ξ and ξ^{**}/ξ and the variance of ξ^{**}/ξ, respectively.

The approximate chi-square-over-degrees-of-freedom distribution for Q applies generally for sample size n of 10 or larger when $\gamma = 1 - P$, the specified reliability, is equal to or greater than .90. If $\gamma = \exp(-1)$, so that $x_P = \eta$, then Q is approximately chi square over degrees of freedom for $n \geqslant 20$. Also note that the covariance of $\eta^{**} - x_P - \xi^{**} B_{r,n}/l_{r,n}$ and ξ^{**} is equal to $\xi^2 B_{r,n} - \xi^2 (B_{r,n}/l_{r,n})l_{r,n} = 0$. Thus Q and ξ^{**}/ξ are independent, and their ratio has an approximate F distribution. One would expect, from results discussed earlier, that the approximation should be quite good except in the extreme tails of the distribution. The degrees of freedom for the approximate F variate,

$$F_x = \frac{\eta^{**} - x_P - \xi^{**} B_{r,n}/l_{r,n}}{\xi^{**}(-y_P - B_{r,n}/l_{r,n})}, \tag{5.22}$$

are $2(y_P + B_{r,n}/l_{r,n})^2/(A_{r,n} - B_{r,n}^2/l_{r,n})$ and $2/l_{r,n}$, where $A_{r,n}$ is the variance of η^{**}/ξ. The approximate chi-square and F distributions also apply to any efficient unbiased linear estimators of ξ and η. It is left to the reader to verify, from the fact that the degrees of freedom for $2mQ/v$ are $2m^2/v$, that the expression for the number of degrees of freedom for the numerator of F_x is correct. To obtain a lower confidence bound for $R(t_m)$ when t_m is specified, one can substitute $x_m = \ln t_m$ for x_P in (5.22) and $\ln[\ln(1/R(t_m)]$ for y_P.

Given now are some examples based on

$$\xi^* = \frac{\tilde{\xi}}{1 - E(L\xi)} \equiv \tilde{\xi}(1 + l_{r,n}) \quad \text{and} \quad \eta^* = \tilde{\eta} + \frac{E(CP)\tilde{\xi}}{1 - E(L\xi)} \equiv \tilde{\eta} + \frac{B_{r,n}\xi^*}{1 + l_{r,n}},$$

the best linear unbiased estimators of ξ and η, respectively, for samples of size 20 with various censorings. Values of $E(CP)$ and $E(L\xi)$ for $n = 20$ can be found in an extension of Table 5.5 found in Mann (1967a). Comparisons are made with exact values based on the extensions of Tables 5.8, 5.9, and 5.10 of Mann and Fertig, given in Mann, Fertig, and Scheuer (1971).

From the extension of Table 5.9, one obtains 3.59 as the 90th percentile of $V_{.90} = (\tilde{\eta} - x_{.10})/\tilde{\xi}$ for $n = 20, r = 10$. Thus a 90 percent lower confidence bound on the 10th percentile of the population is given by $\tilde{\eta} - 3.59\tilde{\xi}$, or

$$\eta^* - \frac{(B_{r,n} + 3.59)\xi^*}{1 + l_{r,n}} = \eta^* - \frac{(.0584 + 3.59)\xi^*}{1.0955} = \eta^* - 3.33\xi^*,$$

where the fact that $B_{r,n} = E(CP)/[1 - E(L\xi)]$ and $l_{r,n} = E(L\xi)/[1 - E(L\xi)]$ has been applied.

Using the F approximation, one obtains, as a 90% lower confidence bound on x_P,

$$\eta^* + \left[\frac{-B_{r,n}(1 - F_{x,1-\alpha})}{l_{r,n}} + y_P F_{x,1-\alpha} \right] \xi^* \qquad (5.23)$$

with $\alpha = .10$.

Then, since $B_{r,n}/l_{r,n} = .0584/.0955 = .6115$, and $y_P = \ln[-\ln(.90)] = -2.25$, there remains only the problem of determining the 90th percentile of an approximate F distribution with $2(1.64)^2/(.1409 - .0357) \cong 51$ and $2/.0955 \cong 20$ degrees of freedom, where $.1409 = E(L\eta) + [E(CP)]^2/[1 - E(L\xi)] = A_{r,n}$ is the variance of η^*/ξ. Use of a table of percentiles of the F distribution gives 1.69 for the needed value, so that the approximate lower confidence bound on $x_{.10}$ is $\eta^* - [.6115(1 - 1.69) + 2.25(1.69)] \xi^* = \eta^* - 3.38\xi^*$. Clearly, the approximation works very well in this case, in comparison with the exact expression, $\eta^* - 3.33\xi^*$, obtained by Mann and Fertig using Monte Carlo procedures.

For $n = 20, r = 10$, $y_{.01} = \ln[-\ln(.99)] = -4.6$ and the variate F_x has degrees of freedom $2(-4.6 + .6115)^2/(.1052) \cong 300$ and approximately 20 and, therefore, a 90th percentile of approximately 1.607. Then, from (5.23), the 90% approximate lower confidence bound on the logarithm of reliable life corresponding to $\gamma = .99$ is $\eta^* - [.6115(1 - 1.607) + 4.6(1.607)] \xi^* = \eta^* - 7.02\xi^*$. Using tables in Mann, Fertig, and Scheuer (1971) and Mann (1967a), one obtains $\eta^* - (7.59 + .0584)\xi^*/1.0955 = \eta^* - 6.98\xi^*$, a result very little different from the approximate expression.

For $n = 20, r = 5$, one obtains, from the table of exact values of percentiles of $(\tilde{\eta} - \eta)/\tilde{\xi}$, $-.41$ as the 50th percentile. Therefore the 50% lower confidence bound on η, based on η^* and ξ^*, with covariance $.335\xi^2$ and with variance $.2336\xi^2$ for ξ^*, is $\eta^* - \xi^*(-.41 + .335)/1.2336 = \eta^* + .069\xi^*$. The approximate 50% confidence bound is, from (5.23), given by $\eta^* + (1 - F_{x,.50})(-.335/.2336)\xi^*$. The degrees of freedom for F_x are $2(1.424^2)/(.70308 - .48095) \cong 18.5$ and $2/.226 \cong 8.56$. Here, to estimate F percentiles for noninteger degrees of freedom, one can use the relationship between the F and the beta distributions and the weighted chi-square approximation for percentiles of the negative logarithm of a beta variate. These are used in obtaining approximate exponential prediction intervals in Section 5.1.1.

First, one calculates the mean m and variance v of the approximate weighted chi-square variate $-\ln V$, where V has a beta distribution with parameters $18.5/2$ and $8.56/2$., $m = \psi(9.25 + 4.28) - \psi(9.25) \cong \ln(13.03) - \ln(8.75) = 2.567 - 2.169 = .398$ and $v = -\psi'(9.25) + \psi'(4.28) \cong -13.03^{-1} + 8.75^{-1} = .0376$. Then use of the Wilson-Hilferty transformation gives, as the 50th percentile of V,

$$\exp\left[-m\left(1 - \frac{v}{9m^2}\right)^3 \right] = \exp\left[-.398(1 - .0264)^3 \right] \cong .693.$$

Finally, $F_{.50}$ $(18.5, 8.56) \cong (8.56/18.5).693/(1 - .693) = 1.044$, so that the 50% approximate confidence bound based on F_x is $\eta^* + .044(.335/.2336)\xi^* = \eta^* + .063\xi^*$.

The 90% approximate confidence bound on η is obtained by using the Wilson-Hilferty transformation (see Sections 5.1.1 and 10.4) to approximate the 90th percentile of $-\ln V$ and hence the 10th percentile of the approximate beta variate V with parameters 9.25 and 4.28. From this, one calculates .499 as the approximate 10th percentile of F. Thus the approximate 90% lower confidence bound for η is $\eta^* - 1.434(1 - .499)\xi^* = \eta^* - .718\xi^*$, while the exact expression obtained by the Monte Carlo procedure is $\eta^* - .725\xi^*$.

Note that, for calculating a lower confidence bound for η at confidence level $1 - \alpha$, the 100αth percentile of F_x is used. For x_P with $\gamma = 1 - P \geqslant .90$, the $100(1 - \alpha)$th percentile of F_x is used for obtaining lower confidence bounds at level $1 - \alpha$. Investigation of the reason for this discrepancy is left to the reader.

In Table 5.12 are given values of $A_{r,n}$, the variance of η^{**}/ξ, for combinations of values of r and n specified in Table 5.11. Values needed in obtaining the lower confidence bound on x_P can be calculated in terms of

Table 5.12   Values of $A_{r,n}$ for use in evaluating the numbers of degrees of freedom, $2(y_p + B_{r,n}/l_{r,n})^2/(A_{r,n} - B_{r,n}^2/l_{r,n})$ and $2/l_{r,n}$, for the approximate F variate $(\eta^{**} - x_p - \xi^{**}B_{r,n}/l_{r,n})/[\xi^{**}(-y_p - B_{r,n}/l_{r,n})]$

r	$A_{r,n}$	r	$A_{r,n}$	r	$A_{r,n}$	r	$A_{r,n}$
$n=25$		$n=35$		$n=45$		$n=55$	
5	.897	7	.585	9	.434	11	.342
10	.180	14	.116	18	.096	22	.077
15	.079	21	.057	27	.043	33	.034
20	.053	28	.038	36	.030	44	.025
25	.052	35	.037	45	.028	55	.023
$n=30$		$n=40$		$n=50$		$n=60$	
3	3.350	4	2.145	5	1.611	6	1.275
6	.702	8	.489	10	.382	12	.305
9	.281	12	.202	15	.161	18	.129
12	.146	16	.107	20	.086	24	.070
15	.091	20	.067	25	.054	30	.044
18	.065	24	.048	30	.038	36	.033
21	.052	28	.041	35	.033	42	.025
24	.044	32	.034	40	.027	48	.022
27	.042	36	.031	45	.025	54	.021
30	.044	40	.033	50	.026	60	.021

observed data and values obtained from Table 5.11. To obtain the numbers of the degrees of freedom for the approximate F variate F_x, however, one needs, in addition to the values tabulated in Table 5.11, the values of $A_{r,n}$ given in Table 5.12. One can also use tabulations appearing in Mann (1970b, 1971) to apply the results just described to progressively censored samples of size $n = 2(1)9$.

(d) Confidence Intervals for the Three-Parameter Weibull Distribution.
For the three-parameter Weibull distribution, knowledge concerning the location or threshold parameter μ is often of primary interest. A lower confidence bound for μ can be determined by using fairly simple iterative procedures combined with existing tables.

In Chapter 7, a new goodness-of-fit test for the two-parameter Weibull against a three-parameter Weibull alternative is described. The test statistic is

$$P = \frac{\sum\limits_{i=k+1}^{r-1} H_i}{\sum\limits_{i=1}^{r-1} H_i},$$

where the appropriate value of k can be found in a tabulation of Mann and Fertig (1974b) with $k \cong r/3$ for $r \geqslant 15$ and where $\xi H_i = (X_{(i+1)} - X_{(i)})/[E(Z_{i+1}) - E(Z_i)]$, with $Z_i = (X_{(i)} - \eta)/\xi$. The distribution of P (except for the extreme right tail—above .95) is very nearly beta with parameters $r - k - 1$ and k. The similarity of the distribution of P to a beta distribution also increases with n. Thus, for two-parameter Weibull observable data,

$$\frac{[k/(r-k-1)]P}{1-P} = \frac{k}{r-k-1} \frac{\displaystyle\sum_{i=k+1}^{r-1} H_i}{\displaystyle\sum_{i=1}^{k} H_i}$$

is distributed approximately as Snedecor's F with $2(r - k - 1)$ and $2k$ degrees of freedom. Therefore F tables can be used for percentiles of the transformed distribution. Expected values of $Z_i, i = 1, \ldots, n$, $n = 1(1)50(5)100$, are given by White (1967), and appropriate differences of expected values of order statistics, are provided by Mann, Scheuer, and Fertig (1973) for $n = 3(1)25$.

If T has the three-parameter Weibull distribution, $X_{(i)} = \ln(T_{(i)} - \mu)$, $i = 1, \ldots, r$, is an extreme-value order statistic, and

$$P^*(\mu) = \frac{\displaystyle\sum_{i=k+1}^{r-1} H_i^*}{\displaystyle\sum_{i=1}^{r-1} H_i^*},$$

with

$$H_i^* = \frac{\ln(T_{(i+1)} - \mu) - \ln(T_{(i)} - \mu)}{E(Z_{i+1}) - E(Z_i)}$$

has the distribution of P defined above. Mann and Fertig (1974b) show that P^* is monotonically decreasing in μ, with $0 \leqslant \mu \leqslant T_{(1)}$. Thus a lower confidence bound $\underline{\mu}$ for μ at level $1 - \alpha$ can be determined by finding $\underline{\mu}$ such that $P^*(\underline{\mu})$ is equal to the $100(1 - \alpha)$th percentile of P, for an appropriate combination of k and r, or the $100(1 - \alpha)$th percentile of the beta distribution with parameters $r - k - 1$ and k. For $n = 20, r = 15$, a 90% confidence bound $\underline{\mu}$ can be determined iteratively by setting P^* equal to the 90th percentile of a beta distribution with parameters 9 and 5, and solving iteratively for $\underline{\mu}$. The lower confidence bound can be found fairly easily using *bisection techniques* if one has access to an on-line computer terminal. Bisection techniques involve the use of $t_{(1)}/2$ as a first guess for $\underline{\mu}$. For the first iteration, then, $t_{(1)}/4$ or $3t_{(1)}/4$ is used accordingly as $P^*(T_{(1)}/2)$ is too small or too large, and so forth. An upper bound for μ

with probability 1 attached is, of course, given by an observed value of $T_{(1)}$. An example of the calculation of S, which has the general form of P, is given in Section 7.1. Note that one can obtain a median unbiased estimate of μ, one which is too large with 50% probability and too small with 50% probability, by using the iterative procedure described, with $\alpha = .50$.

5.2.4 Prediction Intervals and Warranty Periods for the Two-Parameter Weibull Distribution

Because of the asymptotic result of Pyke (1965), applying in particular to differences of adjacent ordered observations from the extreme-value distribution, the methods used in Section 5.1.1 in deriving approximate exponential prediction intervals can be extended to the two-parameter Weibull distribution. Thus, if one has observed the first r failure times, $r < n$, of a size-n sample from a Weibull distribution, it is possible to make a prediction of the time of the mth failure, $r < m \leqslant n$, corresponding to some specified probability.

Also, because it has been established that F_x, defined in Section (5.2.1), has approximately a classical Fisher-Snedecor F distribution, it is possible to demonstrate that a similar simple F approximation can be used to obtain Weibull warranty periods. A warranty period is the time, associated with a given probability (assurance level), before which no failures will occur in a lot of specified size to be manufactured in the future. Values for obtaining warranty periods for the Weibull distribution have been tabulated for a few special cases by Mann and Saunders (1969) and Mann (1970d). Except for the situation in which only the first two failures have been observed in a prior sample, however, the tabulated values are not based on optimum point estimators of the Weibull distribution parameters, and they are rather difficult to determine. A new method, proposed by Lawless (1973b), involves conditioning on ancillary statistics and calculation by the investigator of the warranty period, associated with the (conditional) assurance level, by means of a computer procedure involving numerical integration. The simple approximate procedures are discussed below.

(a) **Prediction Intervals.** It was shown in Section 5.2.3(c) that the unbiased linear estimator ξ^*, or any unbiased linear estimator of the scale parameter ξ based on the first r ordered observations, can be considered to be approximately a sum of weighted chi squares and hence ξ^*/ξ is approximately chi square over degrees of freedom. The estimator ξ^* or any other unbiased linear estimator of ξ based on $X_{(1)}, \ldots, X_{(r)}$, the extreme-value order statistics, involves $X_{(i)} - X_{(i-1)}$, $i = 2, \ldots, r$. The difference

$$X_{(m)} - X_{(r)} = \sum_{i=r+1}^{m} \left(X_{(i)} - X_{(i-1)} \right)$$

clearly involves different gaps (differences of ordered observations). Thus, by Pyke's (1965) result, $X_{(m)} - X_{(r)}$ is approximately independent of ξ^* or other unbiased linear estimators of ξ for sample size n sufficiently large. Therefore, as in approximating exponential prediction intervals in Section 5.1.1, one forms

$$F_m = \frac{X_{(m)} - X_{(r)}}{\xi^* [E(Z_m) - E(Z_r)]}, \qquad (5.24)$$

where ξ^* may be replaced by ξ^{**} for moderately large samples. The distribution of F_m is approximately Fisher's F with degrees of freedom

$$\frac{2[E(Z_m) - E(Z_r)]^2}{\mathrm{Var}(Z_m) - 2\mathrm{Cov}(Z_m, Z_r) + \mathrm{Var}(Z_r)} \quad \text{and} \quad \frac{2}{l_{r,n}},$$

where $l_{r,n}\xi^2 = \xi^2 [E(L\xi) / [1 - E(L\xi)]]$ is the variance of ξ^*.

One can calculate prediction intervals based on a sample of size about 10 to 25 using Tables 5.3 and 7.2 or extensions of these appearing in Mann (1967a) and Mann, Scheuer, and Fertig (1973). To obtain the variances and covariances of Z_m and Z_r, it is necessary to refer to Table 3.11 in Mann (1968a). For very large sample sizes, about 100 or more, ξ^{**}, in place of ξ^*, can be calculated by using a table of Bain (1972) corresponding to Table 5.11. The covariance of Z_m and Z_r can be approximated by using the expression for their asymptotic covariance, as discussed in Section 5.2.2(b) following Equation (5.18). This yields

$$\mathrm{Cov}(Z_r, Z_m) \cong \frac{(r/n^2)/(1 - r/n)}{\ln(1 - r/n)\ln(1 - m/n)}$$

for an extreme-value sample of size greater than or equal to about 100. The large-sample variances of Z_m and Z_r can also be approximated similarly.

Example. Consider a sample of size 12 of composite structures that have been environmentally degraded and stressed until six failures occur within 40 hours after the onset of stress. With probability .80, what is an upper prediction interval for the twelfth and last failure time when application of the stress is continued, if the failure times are assumed to have a two-parameter Weibull distribution?

The value of ξ^* has been calculated, from values in Table 5.3 and the failure data, to be .12; and one sees from this table that the value of the variance $l_{r,n}$ of ξ^*/ξ is $.14695/.85305 = .17226$. From Table 7.2, one sees that

$$E(Z_{12} - Z_6) = \sum_{i=6}^{11} E(Z_{i+1} - Z_i) = 1.5704.$$

Thus

$$F_m = \frac{\ln(t_{(m)}) - \ln(40)}{.12(1.5704)}$$

has approximately an F distribution with degrees of freedom $2 \times (1.5704)^2/\text{Var}(Z_{12} - Z_6)$ and $2/.17226 \cong 11.6$. Using Table 3.11 in Mann (1968a), one finds for $n = 12$, $\text{Var}(Z_{12} - Z_6) = .18626 - 2(.04106) + .15249 = .25297$.

The 80th percentile of F_m with noninteger parameters can be approximated as for the approximate F variates used in obtaining exponential prediction intervals in Section 5.1.1 or in obtaining approximate confidence bounds on x_P from (5.22). Alternatively, one can simply use tables of the F distribution.

It should be noted, before proceeding to a discussion of warranty periods, that Pyke's asymptotic result applies also to differences of adjacent Gaussian-order statistics. Thus exactly the same procedure as described above can be applied to the lognormal distribution. For obtaining the linear weights for estimating μ and σ and the necessary means, variances, and covariances, tables in Sarhan and Greenberg (1962), discussed in Section 5.4.2, can be used.

(b) **Warranty Periods.** A warranty period is, in effect, a prediction interval for $T_{(1)} = \exp X_{(1)}$ for a future lot of size N. Recall, from the discussion leading up to definition (5.22) of the approximate F variate, F_x, for obtaining lower confidence bounds on x_P, that $cQ = (\eta^{**} - x_P - \xi^{**} B_{r,n}/l_{r,n})/\xi$ is approximately a weighted chi square. Since η^{**} is equal to $X_{(r)} - E(Z_r)\xi^{**}$, cQ involves a weighted sum of $X_{(1)} - x_P$ and $X_{(i)} - X_{(i-1)}$, $i = 2, \ldots, r$. Thus $(X_{(1)} - x_P)/\xi$ must have approximately a weighted chi-square distribution when F_x is approximately an F variate.

Therefore one forms the sum of weighted chi squares

$$\frac{\eta^{**} - x_P - \xi^{**} B_{r,n}/l_{r,n}}{\xi} - \frac{X_{(1)} - x_P}{\xi} = \frac{\eta^{**} - X_{(1)} - \xi^{**} B_{r,n}/l_{r,n}}{\xi},$$

which we expect will have a weighted chi-square distribution for prior sample size and lot size sufficiently large. The example below shows that, even for prior sample size as small as 2, the chi-square approximation applies in certain cases and that an F approximation applies for

$$F_1 = \frac{\eta^{**} - X_{(1)} - \xi^{**} B_{r,n}/l_{r,n}}{\xi^{**}(\gamma + \ln N - B_{r,n}/l_{r,n})}, \qquad (5.25)$$

where γ is Euler's constant, approximately .5772156. The degrees of freedom for F_1 are $2(\gamma + \ln N - B_{r,n}/l_{r,n})^2/(A_{r,n} - B_{r,n}^2/l_{r,n} + \pi^2/6)$, where $\pi^2/6 \cong 1.6449$ is the variance of $X_{(1)}/b$ and N is the size of the lot of interest. Recall also that $A_{r,n}$ and $l_{r,n}$ are the variances of η^{**}/ξ and ξ^{**}/ξ, and $B_{r,n}$ is their covariance. In the following example, this approximation is applied to η^* and ξ^*, the best linear unbiased estimators of η and ξ.

Example. From Table III of the 1968 report upon which the article of Mann and Saunders (1969) is based, one finds that for sample size $n = 2$ a 75% warranty period for the first failure in a future lot of size $N = 20$ is given by $\exp[x_1 - 5.81(x_2 - x_1)]$. The expression is optimum for $n = r = 2$, since it is the unique expression that is based on the only two observations, x_1 and x_2, and corresponds to the specified assurance level of .75. From Equation (5.25) one obtains, as a 75% lower confidence bound for the random variate $X_{(1)}$ corresponding to a lot size of 20, $\eta^* - \xi^* B_{r,n}/l_{r,n} - F_{1,.75}\xi^*(\gamma + \ln 20 - B_{r,n}/l_{r,n})$. The (unique, since $r = 2$) unbiased estimates of η and ξ are $\eta^* = x_1 + .91637(x_2 - x_1)$ and $.72135(x_2 - x_1)$, respectively. The value of $B_{r,n}/l_{r,n} = E(CP)/E(L\xi)$, from Table 5.3, is .09036. The degrees of freedom for F_1 are $2(3.572 - .09036)^2/(.6595 - .0058 + 1.6449) \cong 10.6$ and $2/.71186 \cong 2.81$, where $A_{r,n} = .6595$ is $E(L\eta) + E(CP)B_{r,n}$ and $B_{r,n}$ is $E(CP)/[1 - E(L\xi)]$. The appropriate value for the 75th percentile of F_1 is 2.52. Thus the approximate 75% lower confidence bound for the first failure in a lot of size 20 is $x_1 + [.91637 - .72135 (.09036) - 2.52(.72135)(3.4818)](x_2 - x_1)$, or $x_1 - 5.48(x_2 - x_1)$.

This represents surprisingly good agreement with the exact value, since the asymptotic result of Pyke (1965) has been applied to a sample of size 2. The chi-square approximation for such a small sample size does not provide as good agreement, however, for higher assurance levels. One might expect sufficiently good agreement for assurance levels as high as .95 for sample sizes of 4 or more.

PROBLEMS FOR SECTION 5.2

5.10.*a.* Determine an expression for the Cramér-Rao lower bound for the variances of regular unbiased estimators of extreme-value parameters η and ξ, based on complete samples.

b. Using this result, find the Cramér-Rao lower bound for the mean-squared error of regular invariant estimators of ξ.

c. Calculate the values of the bounds from these expressions for complete samples of size 5, and compare your results with the mean-squared errors of best linear unbiased and best linear invariant estimators, whichever are appropriate.

5.11. For $X_{(1)},...,X_{(n)}$, ordered observable sample variates from an extreme-value distribution with scale parameter ξ with $X_{(k)} \leqslant X_{(k+1)}$, $k=1,...,n$, show, by using Equation (5.16), that $E(X_{(2)} - X_{(1)}) = n \ln[1 + 1/(n-1)]\xi$.

5.12. Show that the mean, median, and mode of a three-parameter Weibull distribution are given by $\mu + \delta \Gamma(1 + 1/\beta)$, $\mu + \delta(\ln 2)^{1/\beta}$, and $\mu + \delta(1 - 1/\beta)^{1/\beta}$, $\beta > 1$, respectively, for μ, δ, and β, the Weibull location, scale, and shape parameters, respectively.

5.13. Show that, for the two-parameter Weibull distribution with density given by Equation (5.6) and $\mu = 0$, the maximum-likelihood equations can be put into the forms given by (5.11) and (5.12).

5.14. For $r = 2$, show that the maximum-likelihood estimators of ξ and η are $\hat{\xi} = c_n(X_{(2)} - X_{(1)})$ and $\hat{\eta} = a_n X_{(2)} + (1 - a_n) X_{(1)}$, respectively, where c_n can be obtained by iteratively solving the equation

$$2\left[1 + (n-1)\exp\left(\frac{1}{c_n}\right)\right] + \frac{1}{c_n}\left[1 - (n-1)\exp\left(\frac{1}{c_n}\right)\right] = 0$$

and where $a_n = c_n \ln[1 + (n-1)\exp(1/c_n)]$. *Hint*: Divide both sides of (5.11) by t_1.

5.15. Show that, if T has a two-parameter Weibull density given by Equation (5.6) with $\mu = 0$, then $X = \ln T$ has the distribution function given by (5.13).

5.16. Consider the reduced random variate $Z_i = (X_{(i)} - \eta)/\xi$, and let the variance of Z_i be σ_{ii} and the covariance of Z_i and Z_j be σ_{ij}. Show that any unbiased estimator of ξ of the form $\sum_{i=1}^{r} C_{i,r,n} X_{(i)}$ must be such that

$$\sum_{i=1}^{r} C_{i,r,n} = 0 \quad \text{and} \quad \sum_{i=1}^{r} C_{i,r,n} E(Z_i) = 1.$$

Show that, to find the best unbiased estimator of ξ of the form $\sum_{i=1}^{r} C_{i,r,n} X_{(i)}$, one must find the C's that, under the constraints specified above, simultaneously minimize $\sum_{i=1}^{r} \sum_{j=1}^{r} C_{i,r,n} C_{j,r,n} \sigma_{ij}$.

5.17. For the definitions given in Problem 5.16, determine the constraints necessary so that the biases of the estimators of η and ξ are independent of η.

5.18. Demonstrate that, for $X_{(1)}$ and $X_{(2)}$, the smallest two of n sample variates having an extreme-value distribution, the density of $\exp(-X_{(2)} + X_{(1)})$ is given by Equation (5.16).

5.19.*a*. Given that, for a reduced extreme-value distribution, the distribution percentile $(x_\lambda - \eta)/\xi$ is equal to $\ln\{\ln[1/(1-\lambda)]\}$, use Mosteller's results to calculate the approximate variances and the covariance of the 100th and 200th smallest observable variates in a sample of size 1000 from such a distribution.

b. Show that, for $n = 100$, $.33[X_{98} - X_{17}]$ has a variance approximately equal to $\xi^2(.608/100)/.66 = .0092\xi^2$.

5.20. Demonstrate that the asymptotic three-order-statistic estimator, $X_{([.40n]+1)} - .473(X_{([.98n]+1)} - X_{([.13n]+1)})$, of $X_{.10}$ is asymptotically unbiased.

5.3 THE GAMMA DISTRIBUTION

The two-parameter gamma distribution is derived in Section 4.3 by considering the time to the kth successive arrival in a Poisson process. If no arrivals can occur before time μ, the three-parameter form of the distribution results. The probability density function of a random variate X having the three-parameter gamma distribution is given by

$$f_X(x) = \begin{cases} \left[\left(\dfrac{1}{\theta}\right)^{\alpha} \dfrac{(x-\mu)^{\alpha-1}}{\Gamma(\alpha)} \right] \exp\left[\dfrac{-(x-\mu)}{\theta} \right], & x \geqslant \mu, \\[2em] 0, & x < \mu; \theta, \alpha > 0; \mu \geqslant 0. \end{cases}$$

$$(5.26)$$

The parameter μ specifies a threshold, θ is a scale parameter, and the parameter α determines the shape of the distribution. For $\mu = 0$, the parameter α is also the square of the reciprocal of the distribution *coefficient of variation*, that is, the ratio of the distribution standard deviation and mean. The distribution mean is $\mu + \alpha\theta$, and its variance is $\alpha\theta^2$. If α is an integer, the distribution is also known as the Erlang distribution. If $\alpha = 1$, the distribution reduces to an exponential. Unless $\alpha = 1$, the distribution function cannot be expressed in closed form as is the case for the Weibull distribution.

5.3.1 Estimation for the Three-Parameter Gamma Distribution

If all three of the gamma parameters are unknown, the only efficient method of parameter estimation is that of maximum likelihood. The iterative procedure of Harter and Moore (1965), described in Section 5.2.1 with reference to the three-parameter Weibull distribution, can be used to obtain the maximum-likelihood estimates from samples censored at the rth smallest observation. The method can be applied exactly as described in Section 5.2.1 with θ substituted for δ and α for β. The logarithm of the gamma likelihood function L for $X_{(r_0+1)} \leqslant \cdots \leqslant X_{(r)}$ and the three maximum-likelihood equations to which the procedure is to be applied are

$$\ln L = \ln\left[f_{X_{(r_0+1)},\ldots,X_{(r)}}(x_{r_0+1},\ldots,x_r) \right]$$

$$= \ln n! - \ln(n-r)! - \ln r_0! - n\ln\Gamma(\alpha) - (r-r_0)\alpha\ln\theta$$

$$+ (\alpha-1) \sum_{i=r_0+1}^{r} \ln(x_i-\mu) - \frac{\displaystyle\sum_{i=r_0+1}^{r} (x_i-\mu)}{\theta}$$

$$+ (n-r)\ln\left[\Gamma(\alpha) - \Gamma\left(\alpha; \frac{x_r - \mu}{\theta}\right)\right] + r_0\ln\left[\Gamma\left(\alpha; \frac{x_{r_0+1} - \mu}{\theta}\right)\right];$$

$$\frac{\partial \ln L}{\partial \theta} = -\frac{(r-r_0)\alpha}{\theta} + \frac{\displaystyle\sum_{i=r_0+1}^{r} (x_i - \mu)}{\theta^2}$$

$$+ \frac{(n-r)(x_r - \mu)^{\alpha}\exp[-(x_r - \mu)/\theta]}{\theta^{\alpha+1}\{\Gamma(\alpha) - \Gamma[\alpha; (x_r - \mu)/\theta]\}}$$

$$- \frac{r(x_{r_0+1} - \mu)^{\alpha}\exp[-(x_{r_0+1} - \mu)/\theta]}{\theta^{\alpha+1}\Gamma[\alpha; (x_{r_0+1} - \mu)/\theta]}, \tag{5.27}$$

$$\frac{\partial \ln L}{\partial \alpha} = -(r-r_0)\ln\theta + \sum_{i=r_0+1}^{r} \ln(x_i - \mu) - \frac{n\Gamma'(\alpha)}{\Gamma(\alpha)}$$

$$+ \frac{(n-r)\{\Gamma'(\alpha) - \Gamma'[\alpha; (x_r - \mu)/\theta]\}}{\Gamma(\alpha) - \Gamma[\alpha; (x_r - \mu)/\theta]}$$

$$+ \frac{r_0\Gamma'[\alpha; (x_{r_0+1} - \mu)/\theta]}{\Gamma[\alpha; (x_{r_0+1} - \mu)/\theta]}, \tag{5.28}$$

and

$$\frac{\partial \ln L}{\partial \mu} = (1-\alpha)\sum_{i=r_0+1}^{r} (x_i - \mu)^{-1} + \frac{r - r_0}{\theta}$$

$$+ \frac{(n-r)(x_r - \mu)^{\alpha-1}\exp[-(x_r - \mu)/\theta]/\theta^{\alpha}}{\Gamma(\alpha) - \Gamma[\alpha; (x_r - \mu)/\theta]}$$

$$- \frac{r_0(x_{r_0+1} - \mu)^{\alpha-1}\exp[(x_{r_0+1} - \mu)/\theta]/\theta^{\alpha}}{\Gamma[\alpha; (x_{r_0+1} - \mu)/\theta]}, \tag{5.29}$$

where the primes indicate differentiation with respect to α, and $\Gamma(\alpha; z) = \int_0^z t^{\alpha-1}\exp(-t)dt$ is the incomplete gamma function.

As indicated in Section 5.2.1, the iterative procedure begins with $r_0 = 0$. Only if $\hat{\alpha}$, the maximum-likelihood estimate of α, is less than or equal to 1, so that $\hat{\mu}$, the maximum-likelihood estimate of μ, is equal to $x_{(1)}$, is r_0 redefined, and the smallest $r_0 > 0$ observations (the smallest observation plus any other observation equal to it) censored. Also, as indicated in Section 5.2.1, any parameters assumed to be known can be eliminated from the estimation procedure with their assumed values substituted for the estimates. A computer program listing for this maximum-likelihood procedure can be obtained from Dr. H. L. Harter, Aerospace Research Laboratories, Wright-Patterson Air Force Base, Ohio.

5.3.2 Estimation for the Two-Parameter Gamma Distribution

If μ is known to be zero and $r = n$ (the sample is not censored), the maximum-likelihood equations reduce to

$$\hat{\theta} = \frac{\sum_{i=1}^{n} x_i}{n\hat{\alpha}} \quad \text{and} \quad g(\hat{\alpha}) = 0,$$

where

$$g(\hat{\alpha}) = \ln\hat{\alpha} - \psi(\hat{\alpha}) - \ln\left(\sum_{i=1}^{n} \frac{x_i}{n}\right) + \frac{1}{n}\sum_{i=1}^{n} \ln(x_i).$$

Here $\psi(z)$ is Euler's psi function, given by

$$\frac{d[\ln\Gamma(z)]}{dz} = \int_0^\infty \frac{e^{-t}/t - e^{-zt}}{1 - e^{-t}} dt = -\gamma + \frac{1}{z} + z\sum_{i=1}^{\infty}[i(i+z)]^{-1},$$

where γ is Euler's constant, equal to .5772157.... For large $\hat{\alpha}$, Linhart (1965) suggests approximating $\ln\hat{\alpha} - \psi(\hat{\alpha})$ in $g(\hat{\alpha})$ by $(2\hat{\alpha} - \frac{1}{3})^{-1}$, so that

$$2\hat{\alpha} \cong \left[\ln\left(\sum_{i=1}^{n} \frac{x_i}{n}\right) - \frac{1}{n}\sum_{i=1}^{n} \ln(x_i)\right]^{-1} + \frac{1}{3}.$$

Choi and Wette (1969) give expressions for the asymptotic variances and the covariance for $\hat{\alpha}$ and $\hat{\lambda} = 1/\hat{\theta}$ in terms of $\psi'(\alpha) = d\psi(\alpha)d\alpha$. Tables given by Choi and Wette can be used to facilitate the general solution of the equation defining $\hat{\alpha}$. These authors suggest the use of either a Newton-Raphson technique or a maximum-likelihood scoring method for determining $\hat{\alpha}$, the solution of $g(\hat{\alpha}) = 0$. The Newton-Raphson technique gives, for the kth approximation to $\hat{\alpha}$,

$$\hat{\alpha}_k = \hat{\alpha}_{k-1} - \frac{g(\hat{\alpha})}{g'(\hat{\alpha})} = \hat{\alpha}_{k-1} - \frac{\ln\hat{\alpha}_{k-1} - \psi(\hat{\alpha}_{k-1}) - M}{1/\hat{\alpha}_{k-1} - \psi'(\hat{\alpha}_{k-1})},$$

where $M = \ln(\sum_{i=1}^{n} x_i / n) - \sum_{i=1}^{n} \ln(x_i)/n$.

Choi and Wette use asymptotic approximations and recurrence relations (see Davis, 1964) to evaluate $\psi(\hat{\alpha})$ and $\psi'(\hat{\alpha})$:

$$\psi(z) = \psi(z+1) - \frac{1}{z} \doteq \ln z - \frac{1 + \left\{1 - \left[\frac{1}{10} - 1/(21z^2)\right]/z^2\right\}/(6z)}{2z}$$

and

$$\psi'(z) = \psi'(z+1) + \frac{1}{z^2} = \frac{1}{z} + \frac{1 + \left\{1 - \left[\frac{1}{5} - 1/(7z^2)\right]/z^2\right\}/(3z)}{2z^2}.$$

Tables of $\psi(z)$ and $\psi'(z)$ are also given by Davis (1964) for integer arguments and for $z = 1.(.005)2$.

Alternatively, the method of moments can be employed for estimating α and θ. Moment estimators, given by

$$\alpha' = \frac{\left(\sum_{i=1}^{n} X_i\right)^2}{\sum_{i=1}^{n} n\left(X_i - \overline{X}\right)^2} \quad \text{and} \quad \theta' = \frac{\sum_{i=1}^{n} X_i}{n\alpha'} = \frac{\sum_{i=1}^{n} \left(X_i - \overline{X}\right)^2}{\sum_{i=1}^{n} X_i},$$

are obtained by equating $\alpha\theta$ and $\alpha\theta^2$ to the sample mean and the variance, respectively. These estimators are known to have a lower asymptotic efficiency than corresponding maximum-likelihood estimators. Monte Carlo results of lilliefors (1971) indicate, however, that, for sample sizes of 20 or less and α larger than about 2, the considerably simpler moment estimators of α and $\lambda = 1/\theta$ tend to have mean-squared errors either slightly smaller or only slightly larger than those of MLE's.

Both types of estimators give biased estimates. Correction factors suggested by Lilliefors (1971) can be used to obtain estimates with approximately 0 bias and with smaller mean-squared error in each case. The approximately unbiased estimators are obtained from $\hat{\alpha}$ by dividing the estimator by $1 + 3/n$ and from α' by dividing α' by $1 + 2/n$ and subtracting $5/n$. Estimators of λ,

$$\frac{n\hat{\alpha}}{(1+3/n)\sum\limits_{i=1}^{n} X_i} \quad \text{and} \quad \frac{n\alpha'/(1+2/n)-5/n}{\sum\limits_{i=1}^{n} X_i},$$

are then approximately unbiased and have smaller mean-squared errors than $\hat{\lambda}=1/\hat{\theta}$ and $\lambda'=1/\theta'$, respectively. The moment estimates can be obtained without the use of a computer and, as noted above, for samples of size 20 or less and α larger than about 2 should have efficiencies approximately equal to those of the maximum-likelihood estimates. For samples as large as 100 and α smaller than about 4, however, the mean-squared errors of moment estimators of α tend to be $1\frac{1}{3}$ to as much as about 5 (for α very small) times as large as those of corresponding MLE's.

We consider a sample of size 3 from a gamma population: 16.2, 20.4, 26.2. The moment estimator of α is $\alpha'=[(62.8)^2/56.0]/3=23.5$. An approximately unbiased estimate of α is, then, $23.5/(1+\frac{2}{3})-\frac{5}{3}=12.4$, where approximately unbiased moment estimators of α have considerably smaller mean-squared error than raw moment estimators. Biased and approximately unbiased moment estimators of λ are, respectively, $23.5/20.9$ $=1.1$ and $12.4/20.9=.60$. Since α is quite large and n is small, moment estimation is as good a procedure as maximum-likelihood estimation. It is also simpler, requiring no iteration.

Because of difficulties involving lack of independence of α for functions of the maximum-likelihood and moment estimators of θ, no general method has been developed for obtaining confidence intervals for this parameter. For confidence sets for α, the statistic

$$S_1 = \frac{\left(\prod\limits_{i=1}^{n} X_i\right)^{1/n}}{\sum\limits_{i=1}^{n} X_i/n}$$

is sufficient (see Problem 5.25) and has a distribution independent of θ. Bartlett (1937) has shown that $\ln S_1$ has approximately the distribution of

$$\left[\frac{1+(1+1/n)/6\alpha}{2n\alpha}\right]\chi^2(n-1), \tag{5.30}$$

and Linhart (1965) has demonstrated that this approximation works well as long as the value of a lower confidence bound, calculated (iteratively) for α by using Equation (5.30), is not less than 1 and is excellent if the bound is 2 or more. Values to aid in the calculation of confidence bounds for α (or for $\alpha^{-1/2}$, the coefficient of variation for the gamma distribution) are given by Linhart (1965).

5.4 THE LOGNORMAL DISTRIBUTION

A random variable subject to a process of change wherein the change in the variable at the nth step of the process is a random proportion of the value of the variable at the $(n-1)$st step is said to obey the so-called law of proportional effect, as discussed in detail in Section 4.6. Consider such a random variable Z, with $Z_n \equiv X$ its value at the nth step of the process, $n \to \infty$. The random variable X has a logarithmic normal (or lognormal) distribution. Suppose that X is such that $\ln(X/\delta)^{1/\sigma}$ has a standard normal distribution (mean 0 and standard deviation 1). Then X has a two-parameter lognormal distribution with median δ and shape parameter σ, and $\ln X$ has a normal distribution with mean $\mu = \ln \delta$ and standard deviation σ. If $X = Y - \tau$ and $\ln[(Y-\tau)/\delta]^{1/\sigma}$ has a standard normal distribution, Y has a three-parameter lognormal distribution with location parameter (sometimes called guarantee-time parameter) τ. The density of $X = Y - \tau$, where Y has a three-parameter lognormal distribution with location parameter τ, will be denoted by $\Lambda(Y - \tau; \mu, \sigma^2)$ with μ the mean of the distribution of $W = \ln X$ and σ^2 its variance. If τ is known to be zero, one may write the density of $X = Y$ as $\Lambda(\mu, \sigma^2)$. Suppose that X is distributed as $\Lambda(Y - \tau; \mu, \sigma^2)$. Then $E(X^k) = E[(Y-\tau)^k] = e^{k\mu + k^2\sigma^2/2}$.

To demonstrate this, one need only consider the moment generating function given in Section 3.3.5 for a normal variate,

$$M_W(\theta) = e^{\mu\theta + \sigma^2\theta^2/2}.$$

Now $E[(Y-\tau)^k] = E(e^{kW})$, where W is distributed as $N(\mu, \sigma^2)$. But $E(e^{kW})$ is just the moment generating function for W, with θ replaced by k. Hence the proof is complete.

Thus $E(Y) = \tau + e^{\mu + \sigma^2/2}$. For $\tau = 0$, or known and set equal to 0 by a translation of the data, the variance of $X = Y$ is equal to $E(X^2) - (EX)^2 = e^{2\mu + 2\sigma^2} - e^{2\mu + \sigma^2} = e^{2\mu + \sigma^2}(e^{\sigma^2} - 1)$. The median of Y for the three-parameter model is $\tau + e^\mu \equiv \tau + \delta$, and the mode is $\tau + e^{\mu - \sigma^2}$. Proof of these two facts is left to the reader.

From the knowledge that $\ln X$ is normally distributed, one can immediately derive the probability density function of $Y = X + \tau$, namely,

$$f_Y(y) = \begin{cases} \dfrac{1}{\sqrt{2\pi}\,\sigma(y-\tau)} \exp\left\{\dfrac{-[\ln(y-\tau)-\mu]^2}{2\sigma^2}\right\}, & y \geqslant \tau, \sigma > 0, \\ 0, & \text{elsewhere.} \end{cases}$$

$$(5.31)$$

5.4.1 Estimation for the Two-Parameter Lognormal Distribution from Complete Samples

If X has a two-parameter logarithmic normal distribution, one can obtain estimates of μ and σ^2 with optimum properties by considering the distribution $W = \ln X$. That is, from Section 3.4.2, the maximum-likelihood estimator $\hat{\mu}$ of μ for complete samples is the sample mean given by

$$\frac{\sum\limits_{i=1}^{n} W_{(i)}}{n} \equiv \frac{1}{n}\sum_{i=1}^{n} \ln(X_{(i)}),$$

where $X_{(i)}$ is the ith ordered observable sample value from $\Lambda(\mu,\sigma^2)$ and n is sample size. The MLE, $\hat{\sigma}^2$, of σ^2 is

$$\frac{\sum\limits_{i=1}^{n} (\ln X_{(i)})^2 - n\left[\sum\limits_{i=1}^{n} \ln(X_{(i)})/n\right]^2}{n}.$$

Since, for complete samples, $\hat{\mu}$ and $\hat{\sigma}^2$ are based on complete sufficient statistics (see Section 3.4), best unbiased estimators of μ and σ^2 can be obtained by determining the expectations of these estimators. Since $E(\hat{\mu}) = \mu$ and $E(\hat{\sigma}^2) = [(n-1)/n]\sigma^2$ (see Problem 5.18), the best unbiased estimators of μ and σ^2 are $\hat{\mu}$ and $s^2 = [n/(n-1)]\hat{\sigma}^2$, respectively. Because μ is a location parameter and σ a scale parameter of the distribution of W, and because it is easily shown that $\text{Cov}(\hat{\mu},\hat{\sigma}) = 0$ (see Section 3.4), $\hat{\mu}$ is also the best invariant estimator of μ. One can demonstrate, using a method of proof similar to that of Mann (1969c) cited in Section 5.1.2(a), that the best invariant estimator of σ^2 is $\tilde{\sigma}^2 = [n/(n+1)]\hat{\sigma}^2$.

Often an estimate of σ^2 is needed simply for the purpose of obtaining a confidence bound on a specified percentile of the distribution. In this case, the constant terms will cancel, and whether one uses a maximum-likelihood, best unbiased, or best invariant estimate is immaterial. If a point estimate is needed, however, s^2, the best unbiased estimator, can be utilized if the estimates are to be averaged. Otherwise, the maximum-likelihood estimator, $\hat{\sigma}^2$, or the best invariant estimator can be used. The MLE gives the estimate that maximizes the likelihood of obtaining the sample from which the estimate is obtained, and the best invariant estimator has smallest mean-squared error among all estimators with mean-squared error independent of μ (or invariant under translation of the data). The three estimators, of course, are asymptotically equivalent, so that, if n is moderately large, it matters very little which of the three is used.

The maximum-likelihood estimator has the convenient property that, for $\hat{\theta}$ the MLE of θ, $g(\hat{\theta})$ is the MLE of $g(\theta)$ (see Section 3.4). Using this fact, one can very easily obtain MLE's of the mean $e^{\mu+\sigma^2/2}$, and the variance, $e^{2\mu+\sigma^2}(e^{\sigma^2}-1)$ of the distribution of X by substituting $\hat{\mu}$ and $\hat{\sigma}^2$ for μ and σ, respectively. The median e^μ and the mode $e^{\mu+\sigma^2}$ are also very easily estimated by maximum-likelihood methods as $e^{\hat{\mu}}$ and $e^{\hat{\mu}+\hat{\sigma}^2}$.

One can also use normal or lognormal probability paper to plot, respectively, $\ln X_{(i)}$ or $X_{(i)}$ versus some estimate of $F_{X_{(i)}}(x_i), i=1,2,\ldots,n$. (Graphical plotting techniques are discussed in more detail in Section 5.2.2.) Chernoff and Lieberman (1956) give tables of estimates of $F_{X_{(i)}}(x_i), i=1,2,\ldots,n$, which, when used as plotting positions, yield graphical approximations to the best linear unbiased and best linear invariant estimates. These authors, incidentally, refer to what is called here the best linear invariant estimate of σ as "the biased linear estimate of σ which has minimum mean square deviation."

Kimball (1960) recommends the use of $(i-\frac{3}{8})/(n+\frac{1}{4})$ as a substitute plotting position [or estimate of $F_{X_{(i)}}(x_i)$] when tables are not available, since this choice yields estimates of σ with small bias as well as small mean-squared error. Using $(i-\frac{1}{2})/n$, however, appears to give graphical estimates of σ with slightly higher bias and slightly smaller mean-squared error. Even for (complete) samples as small as 6, there is very little discrepancy in theoretical mean-squared error in estimating, whether one uses $(i-\frac{3}{8})/(n+\frac{1}{4})$, $(i-1\frac{1}{2})/n$, or the plotting positions yielding the best linear unbiased or best linear invariant estimates.

In plotting on normal probability paper, one uses the relationship $W_P=\mu+Z_P\sigma$, where Z_P is the $100P$th percentile of a normal variate with mean 0 and variance 1, and W_p is the $100P$th percentile of a normal distribution with mean μ and variance σ^2. If $F_{X_{(i)}}(x_i)$ is estimated by $p_{i,n}=(i-\frac{1}{2})/n$, one plots the ith observed value of X or $W=\ln X$ versus $(i-\frac{1}{2})/n$, which is converted by the probability paper to $z_{(i-1/2)/n}$. The line plotted is $z=(1/\sigma)(-\mu)+(1/\sigma)w$, so that an estimate of $(1/\sigma)$ is given by an estimate of the slope of the line, and the p intercept of the plotted line with the principal ordinate is an estimate of $-\mu/\sigma$. If a plot is made and a curve that is concave downward results, an estimate of the threshold parameter τ can be obtained iteratively (by trial and error) by finding a value, $\hat{\tau}$, which, when subtracted from $\exp(w_{(1)})=x_{(1)}$, converts the curve to a line. See Section 5.2.2(b), where an example of a similar type of probability plot is given.

To obtain confidence bounds for parameters of the two-parameter lognormal distribution, one can use the results of Section 3.3.5. It is shown in Section 3.3.5(a) that $\hat{\mu}\equiv\overline{W}$ is normally distributed with mean μ and variance σ^2/n. Hence, if σ is known, a two-sided confidence interval for μ

at confidence level $1 - \alpha$ is given by

$$\phi_{\alpha/2} \leqslant \frac{\overline{W} - \mu}{\sigma/\sqrt{n}} \leqslant \phi_{1-\alpha/2},$$

where ϕ_ζ is the 100ζth percentile of a normal distribution with mean 0 and variance 1, or

$$\overline{W} + \frac{\phi_{\alpha/2}\sigma}{\sqrt{n}} \leqslant \mu \leqslant \overline{W} + \frac{\phi_{1-\alpha/2}\sigma}{\sqrt{n}}$$

(since, because of symmetry considerations, $\phi_{\alpha/2} = -\phi_{1-\alpha/2}$).

If σ is not known, one can use Student's t distribution, discussed in Section 3.3.5(d), to obtain confidence intervals for μ. From Section 3.3.5(b) it can be seen that

$$\frac{n\hat{\sigma}^2}{\sigma^2} = \frac{\sum_{i=1}^{n} (W_i^2) - n\overline{W}^2}{\sigma^2}$$

has a chi-square distribution with $n-1$ degrees of freedom. Thus $(\overline{W} - \mu)$ $/(\sigma/\sqrt{n})/[(n\hat{\sigma}^2/\sigma^2)/(n-1)]^{1/2}$ has Student's t distribution with $n-1$ degrees of freedom, and a two-sided confidence bound for μ is given by

$$\overline{W} + \frac{t_{\alpha/2}\hat{\sigma}}{(n-1)^{1/2}} \leqslant \mu \leqslant \overline{W} + \frac{t_{1-\alpha/2}\hat{\sigma}}{(n-1)^{1/2}},$$

where t_ζ is the 100ζth percentile of Student's t distribution with $n-1$ degrees of freedom, and $t_\zeta = -t_{1-\zeta}$.

Confidence bounds for σ can be obtained by use of the fact that $n\hat{\sigma}^2/\sigma^2$ has a chi-square distribution with $n-1$ degrees of freedom. An upper confidence bound for σ at level $1 - \alpha$ is therefore given by

$$\sigma \leqslant \left[\frac{n\hat{\sigma}^2}{\chi_\alpha^2(n-1)} \right]^{1/2}.$$

The $100P$th percentile of a normal distribution is given by $w_P = \mu + \phi_P\sigma$, where ϕ_P is the $100P$th precentile of a standard normal distribution with mean 0 and variance 1. If $\gamma = 1 - P$ is a specified survival proportion for a lognormal population, a lower bound on reliable life, $x_P = \exp(w_P)$, corresponding to $\gamma = 1 - P$, can be determined from a lower confidence bound on w_P obtained through the use of tables of the *noncentral t* distribution (see Resnikoff and Lieberman, 1957, or Locks, Alexander, and Byars,

1963). If n is not too small (20 or so), one can approximate the distribution of the noncentral t variate,

$$T' = \frac{\overline{W} - \mu - \phi_P \sigma}{\sqrt{\hat{\sigma}^2/(n-1)}}$$

by considering

$$Z = \frac{T'\{1 - [1/4(n-1)]\} - \phi_P}{\{1 + [(T')^2/2(n-1)]\}^{1/2}}.$$

The variate Z is then approximately Gaussian with mean 0 and variance 1.

A lower 90% confidence bound on the 5th percentile of a lognormal distribution from a sample of size 40 is therefore given by $\exp(\overline{W} - t'\sqrt{\hat{\sigma}^2/39})$, where t' is defined implicitly by

$$1.282 = \frac{t'(1 - \tfrac{1}{4}/39) - (-1.645)}{\{1 + [(t')^2/2]/39\}^{1/2}},$$

and -1.645 and 1.282 are the 5th and 90th percentiles, respectively, of a standard normal variate. The quadratic formula can be used to solve for t', the larger solution being used in this case.

5.4.2 Estimation from Censored Samples from the Two-Parameter Lognormal Distribution

Consider a size-n sample subjected to life test until r, $r < n$, lognormally distributed failures occur. One can then use tabulated coefficients or weights for multiplying the logarithms of the first r of n ordered observations to obtain either best linear unbiased or best linear invariant estimates of μ and σ. The weights, published by Sarhan and Greenberg (1962), give the best linear unbiased estimates of the two parameters, but for small n or extensive censoring these estimates can easily be converted to best linear invariant estimates, using the result given in Section 3.6.3. If an estimate of a function $g(\mu, \sigma)$ of μ and σ is needed, it can be obtained as the same function, $g(\tilde{\mu}, \tilde{\sigma})$, of the best linear invariant estimates, $\tilde{\mu}$ and $\tilde{\sigma}$, of the two parameters. In a situation in which $g(\mu, \sigma)$ is a percentage point corresponding to a specified proportion P of the distribution of $W = \ln X$ and estimates of $g(\mu, \sigma)$ are to be averaged over a period of time, one should use the best linear unbiased estimator $W_P^* = \mu^* + z_P \sigma^*$ of $w_P = \mu + z_P \sigma$, where z_P is the $100P$th percentile of the standard normal distribution, and

μ^* and σ^* are the best linear unbiased estimators of μ and σ, respectively.

The methodology described in Section 5.2.3(c) for obtaining confidence bounds for the extreme-value scale parameter ξ from linear estimates of ξ can also be applied to linear estimates of σ. This is possible because, under certain regularity conditions for location-scale parameter distributions, there is a ζ_i that can be used to divide the difference of the $(i+1)$st and ith ordered observation, $i=1,\ldots,r-1$, and to convert these differences to approximate exponential variates with mean and variance approximately 1.

Patnaik's result then guarantees that a linear combination of these variates has approximately a weighted chi-square distribution. As described in Section 5.2.3(c), then, $2(\sigma^*/\sigma)/C$, where $C\sigma^2$ is the variance of σ^*, has approximately a chi-square distribution with $2/C$ degrees of freedom. Use of the Wilson-Hilferty transformation of chi-square to normality gives

$$\frac{\sigma^*}{\left(1-1/9C+\phi_\alpha\sqrt{1/9C}\ \right)^3}$$

as an upper confidence bound for σ at confidence level $1-\alpha$. This approach works also for any unbiased linear estimator of σ, such as that based on the sample range, $X_{(n)}-X_{(1)}$ (see Grubbs, Coon, and Pearson, 1966), as long as the variance of the estimator is known. Another application of the weighted chi-square result for differences of adjacent ordered normal variates is the obtaining of prediction intervals for the mth observation, with $m>r$, as in Section 5.2.4.

If censoring is symmetric, the sample can be *Winsorized*, that is, the missing $(n-r)/2$ observations on the low side can be set equal to the smallest observation, and the $(n-r)/2$ missing observations on the high side can be set equal to the largest observation. Then $\overline{W}$ and s can be calculated just as if the sample were complete. Confidence intervals for μ can be determined just as usual for $r>n-r$ if one multiplies a calculated t value, based on the n (Winsorized) observations, by $(r-1)/(n-1)$ and assumes that the Winsorized t has $r-1$ degrees of freedom (see Dixon and Tukey, 1968). Similarly approximate confidence bounds for σ can be obtained by multiplying the sum of squared deviations from $\overline{W}$ based on all n Winsorized observations by $(n-1)/(r-1)$. This statistic, then, when divided by σ^2, has approximately a chi-square distribution with $r-1$ degrees of freedom.

If the second through seventh observations from a size-8 sample of Gaussian variates are -8.2, -4.3, -1.1, 0.7, 3.7, and 8.7, one Winsorizes by setting the first and eighth observations equal to -8.2 and 8.7, respectively. A two-sided confidence bound for μ at level $1-\alpha$ is given by

$$\overline{w} \pm \frac{8-1}{6-1} t_{1-\alpha/2} \sqrt{\sum_{i=1}^{8} \left[(w_i - \overline{w})^2/8 \right]/(8-1)} \ ,$$

where $t_{1-\alpha/2}$ is the $100(1-\alpha/2)$th percentile of Student's distribution with $6-1=5$ degrees of freedom.

5.4.3 Samples from the Three-Parameter Lognormal Distribution

(a) Maximum-Likelihood Estimation. For the three-parameter lognormal distribution, solutions to the maximum-likelihood equations cannot be obtained explicitly. The likelihood equations for a sample of size n, with the smallest r of the n ordered variates observed as $y_{(1)}, y_{(2)}, \ldots, y_{(r)}$, are as follows:

$$\frac{\partial L}{\partial \mu} = \frac{1}{\sigma} \left[\sum_{i=1}^{r} s_{(i)} + \frac{(n-r)f(s_{(r)})}{1-F(s_{(r)})} \right], \tag{5.32}$$

$$\frac{\partial L}{\partial \sigma} = \frac{1}{\sigma} \left[-r + \sum_{i=1}^{r} s_{(i)}^2 + \frac{(n-r)s_{(r)}f(s_{(r)})}{1-F(s_{(r)})} \right], \tag{5.33}$$

$$\frac{\partial L}{\tau} = \sum_{i=1}^{r} (y_{(i)} - \tau)^{-1} + \frac{1}{\sigma} \left\{ \sum_{i=1}^{r} s_{(i)}/(y_{(i)} - \tau) + \frac{(n-r)f(s_{(r)})}{(y_{(r)} - \tau)[1-F(s_{(r)})]} \right\}, \tag{5.34}$$

where

$$s_{(i)} = \frac{\ln(y_{(i)} - \tau) - \mu}{\sigma},$$

$$F(s_{(i)}) = \int_{-\infty}^{s_{(i)}} f(t)\, dt, \quad \text{and} \quad f(s_{(i)}) = (2\pi)^{-1/2} \exp\left(\frac{-s_{(i)}^2}{2} \right).$$

If $r = n$, Equations (5.32) and (5.33) can be solved as explicit functions of τ for the maximum-likelihood estimates of μ and σ, namely,

$$\hat{\mu} = \frac{\sum_{i=1}^{n} \ln(y_{(i)} - \tau)}{n}$$

and

$$\hat{\sigma} = \frac{\sum_{i=1}^{n} \left[\ln(y_{(i)} - \tau) - \hat{\mu} \right]^2}{n}.$$

If censoring occurs, all three of the original likelihood equations must be used in an iterative procedure for calculating local maximum-likelihood estimates, since the global solution does not give reasonable estimates (see below). Harter and Moore (1966) give the following procedure for computer solution of the likelihood equations. The parameters are estimated one at a time in the cyclic order μ, σ, and τ. Any parameters assumed to be known have their assumed values substituted in the likelihood equations.

Initial estimates are chosen for the unknown parameters. (A probability plot provides a good method for doing this.) Then, at the first step, the rule of false position (iterative linear interpolation) is used to determine the next value of the parameter being estimated that satisfies the value of the appropriate likelihood equation. The new estimate is then substituted in the other likelihood equations (or equation), and the process is continued.

A problem often arises in using this iterative procedure (particularly if the number of observations is small): no estimate, $\tau \leqslant y_1$, can be found that satisfies the third likelihood equation. In such cases, the likelihood function is increasing monotonically and the iterative procedure is proceeding along a path to the global solution of the system of likelihood equations: $\hat{\tau} = y_1$, $\hat{\mu} = -\infty$, and $\hat{\sigma} = +\infty$. When this occurs, a modified procedure can be used to obtain local maximum-likelihood estimates, which are reasonable estimates and appear to possess most of the desirable asymptotic properties usually associated with maximum-likelihood estimates.

The modification involves censoring the smallest observation y_1, as well as all observations equal to y_1, so that it (or they) no longer contributes to the estimation procedure except as an upper bound on $\hat{\tau}$. Since the likelihood function is now bounded, the iterative procedure can be used without risk of its convergence to the useless global maximum-likelihood estimates obtained before censoring.

(b) Approximations to Best Linear Unbiased Estimators. In order to obtain linear estimates of parameters of the three-parameter lognormal distribution, Munro and Wixley (1970) have reparameterized in the following way. They define

$$\xi \equiv e^{\mu}\sigma \quad \text{and} \quad \lambda \equiv \tau + \frac{\xi}{\sigma} \equiv \tau + e^{\mu}.$$

Thus the three parameters are the shape parameter, σ, the scale parameter ξ, and the location parameter λ, which is the distribution median. The density function of $T = Y/\xi$ is then given by

$$f_T(t) = \frac{1}{\sqrt{2\pi}\,(1+\sigma t - \sigma\lambda/\xi)\xi}\exp\left\{-\frac{1}{2\sigma^2}\left[\ln\left(1+\sigma t - \frac{\sigma\lambda}{\xi}\right)\right]^2\right\},$$

$$t > (\lambda-\xi)/\sigma; \xi, \sigma > 0. \tag{5.35}$$

The standard normal deviate, $(W - \mu)/\sigma$, is therefore equal to $\ln(1 + \sigma T - \sigma\lambda/\xi)/\sigma$. With this parameterization, as σ tends to zero with λ and ξ fixed, the distribution of T tends to the normal with mean λ and standard deviation ξ.

This particular parameterization also allows one to obtain, by iterative procedures, approximations to best linear unbiased estimates of the three parameters λ, ξ, and σ (see Section 3.6.2). The iterative methods described by Munro and Wixley (1970) for obtaining these estimates use asymptotic approximations for the expectations, the variances, and the covariances of the reduced ordered observable variates, $Z_{(i)} = (T_{(i)} - \lambda)/\xi$.

If a new computer program is to be written for obtaining estimates of the three lognormal parameters from censored data, the selection of which of the two methods described here is to be used might be left to the computer programmer, who could consider the problems involved in each procedure. If precise estimates are not required, the values obtained by graphical plotting techniques may suffice; certainly they are a great deal simpler to obtain than analytical estimates.

5.5 THE BIRNBAUM-SAUNDERS FATIGUE-LIFE DISTRIBUTION

The derivation of the Birnbaum-Saunders distribution from plausible considerations of the physical behavior of fatigue-crack growth under repeated loading is shown in Section 4.11. This two-parameter distribution of a nonnegative random variate T is defined by

$$F_T(t;\alpha,\beta) = \Phi\left\{ \left(\frac{1}{\alpha}\right)\left[\left(\frac{t}{\beta}\right)^{1/2} - \left(\frac{t}{\beta}\right)^{-1/2} \right] \right\}, \quad t>0; \alpha,\beta>0, \quad (5.36)$$

where $\Phi(x) = \int_{-\infty}^{x} (1/\sqrt{2\pi})\exp(-y^2/2)dy$ is the distribution function of the standard normal variate. The parameter α determines the shape of the distribution, while the median β is a scale parameter. The distribution mean, $\beta(1 + \alpha^2/2)$, and variance, $(\alpha\beta)^2(1 + 5\alpha^2/4)$, are derived in Section 4.11.1. Point estimation of α and β has been investigated at length by Birnbaum and Saunders (1969b).

5.5.1 Estimation of the Parameters α and β

The procedure suggested by Birnbaum and Saunders for general point estimation of the parameters α and β is that of maximum-likelihood. In order to define the maximum-likelihood estimators of α and β, one first defines, for a given set of positive random variables $T_1, \ldots, T_n$, the arithmetic and harmonic means, respectively,

$$S = \frac{1}{n} \sum_{i=1}^{n} T_i, \qquad R = \left(\frac{1}{n} \sum_{i=1}^{n} \frac{1}{T_i} \right)^{-1}$$

and the harmonic mean function,

$$K(x) = \left[\frac{1}{n} \sum_{i=1}^{n} (x + T_i)^{-1} \right]^{-1}.$$

Thus, if $T_1, \ldots, T_n$ represents a sample of independent random variables, each with distribution given by Equation (5.36), the maximum-likelihood estimator $\hat{\beta}$ of β is the unique positive solution of the random equation $g(x) = 0$, where

$$g(x) = x^2 - x[2R + K(x)] + R[S + K(x)].$$

The maximum likelihood estimator $\hat{\alpha}$ of α is then given in terms of $\hat{\beta}$, S, R:

$$\hat{\alpha} = \left[\frac{1}{n} \sum_{i=1}^{n} \xi^2 \left(\frac{T_i}{\hat{\beta}} \right) \right]^{1/2} = \left(\frac{S}{\hat{\beta}} + \frac{\hat{\beta}}{R} - 2 \right)^{1/2},$$

where $\xi(T) = T^{1/2} - T^{-1/2}$.

The simplified estimator $\tilde{\beta}$ of β, which is shown by Birnbaum and Saunders to be consistent and, for small values of α, to be virtually the same as the maximum-likelihood estimator $\hat{\beta}$, is given by

$$\tilde{\beta} = \sqrt{SR}.$$

If the sample size is $n = 2$, the MLE's of α and β are

$$\hat{\alpha} = \left| \xi \left(\sqrt{T_1 / T_2} \right) \right|, \qquad \hat{\beta} = \tilde{\beta} = \sqrt{T_1 T_2}.$$

Note that, for $n = 2$, exact confidence sets and tests of hypotheses concerning β and independent of α are functions of the ratio $\xi(T_2/\beta)/\xi(T_1/\beta)$. For a sample of size 2, there is apparently no function of $\hat{\beta}/\beta$ and $\hat{\alpha}$ to which this ratio is equal. One would infer that functions of $\hat{\alpha}$ and $\hat{\beta}$ (or $\tilde{\beta}$) cannot be used to obtain confidence sets for β. In fact, it has been demonstrated (see Saunders and Mann, 1974) that for $n = 2$ one cannot have confidence sets and tests of hypotheses for either α or β based on functions of $\hat{\alpha}$ and $\hat{\beta}$. That this is true for any n seems a justifiable conjecture.

5.5.2 Confidence Bounds and Tests for β

Consider a set of ordered observations $T_{(1)}, T_{(2)}, \ldots, T_{(n)}$ from the distribution defined by (5.36). A method investigated by Saunders and Mann (1974) for obtaining confidence bounds for β for small complete samples will now be described. Observe that $T_{(i)}/\beta < 1$ implies

$$X_{(i)} = \xi\left(\frac{T_{(i)}}{\beta}\right) = \left(\frac{T_{(i)}}{\beta}\right)^{1/2} - \left(\frac{\beta}{T_{(i)}}\right)^{1/2} < 0,$$

and $T_{(i)}/\beta > 1$ implies $X_{(i)} > 0$. Notice also that $\xi(T_{(i)}/\beta)$ is an order-preserving transformation. Note that, for a sample of size 2, $\mathrm{Prob}[X_{(2)} < 0]$ $= \mathrm{Prob}[X_{(1)} > 0] = (\frac{1}{2})^2$. Hence $\mathrm{Prob}[\beta > T_{(2)}] = \mathrm{Prob}[\beta < T_{(1)}] = (\frac{1}{2})^2$. One can, therefore, determine iteratively, by Monte Carlo procedures, a value $u(\epsilon)$ such that the joint event $X_{(2)}/X_{(1)} > u(\epsilon)$ and $X_{(2)} < 0$ occurs with probability ϵ. Then an upper confidence bound on β at level $1 - \epsilon$ is given by $\max[T_{(2)}, g]$, where g is equal to $[u(\epsilon)T_{(1)}^{1/2} - T_{(2)}^{1/2}]/[u(\epsilon)/T_{(1)}^{1/2} - 1/T_{(2)}^{1/2}]$. Because of the symmetry of the normal distribution, a lower confidence bound on β at level $1 - \epsilon$ is given by

$$\min\left\{ T_{(1)}, g'[u(\epsilon), T_{(1)}, T_{(2)}] \right\}$$

with

$$g' = \frac{u(\epsilon)T_{(2)}^{1/2} - T_{(1)}^{1/2}}{u(\epsilon)/T_{(2)}^{1/2} - 1/T_{(1)}^{1/2}}.$$

Clearly, because $\mathrm{Prob}[X_{(1)} > 0] = (\frac{1}{2})^n$, procedures based on the assumption $X_{(1)} > 0$ or $X_{(n)} < 0$ will encounter difficulties if n is large, unless $1 - \epsilon$ is extremely close to 1.

For samples of size n one can, therefore, determine $l(\epsilon)$ such that the probability that the event

$$nS(\beta) \equiv \sum_{i=1}^{n} X_{(i)} < 0 \quad \text{and} \quad W_1(\beta) \equiv \frac{\left(1/\sqrt{n-1}\,\right)\sum_{i=2}^{n} X_{(i)}}{X_{(1)}} > l(\epsilon)$$

$$(5.37)$$

occurs is ϵ. Then an upper confidence bound for β at level $1 - \epsilon$ is given by

$$\eta_n = \max\left\{ \frac{\sum_{i=1}^{n} T_{(i)}^{1/2}}{\sum_{j=1}^{n} (1/T_{(j)})^{1/2}}, h[l(\epsilon), T_{(1)}, \ldots, T_{(n)}] \right\}, \qquad (5.38)$$

where

$$h[l(\epsilon), T_{(1)}, \ldots, T_{(n)}] = \frac{\sqrt{n-1}\, l(\epsilon) T_{(1)}^{1/2} - \sum_{i=2}^{n} T_{(i)}^{1/2}}{\sqrt{n-1}\, l(\epsilon)/T_{(1)}^{1/2} - \sum_{j=2}^{n} (1/T_{(j)})^{1/2}}.$$

From considerations of symmetry, one obtains a lower confidence bound for β at level $1 - \epsilon$ as

$$\omega_n = \min\left\{ \frac{\sum_{i=1}^{n} T_{(i)}^{1/2}}{\sum_{j=1}^{n} (1/T_{(j)}^{1/2})}, h'[l(\epsilon), T_{(1)}, \ldots, T_{(n)}] \right\}, \qquad (5.39)$$

where

$$h'[l(\epsilon), T_{(1)}, \ldots, T_{(n)}] = \frac{\sqrt{n-1}\, l(\epsilon) T_{(n)}^{1/2} - \sum_{i=1}^{n-1} T_{(i)}^{1/2}}{\sqrt{n-1}\, l(\epsilon)/T_{(n)}^{1/2} - \sum_{j=1}^{n-1} (1/T_{(j)})^{1/2}}.$$

A test of the hypothesis $H_1 : \beta \geqslant \beta_0$ versus $H_2 : \beta < \beta_0$, based on $W_1(\beta_0)$, rejects H_1 at significance level ϵ if

$$\frac{\sum_{i=2}^{n} \xi(T_{(i)}/\beta_0)}{\xi(T_{(1)}/\beta_0)} > l(\epsilon) \qquad \text{and} \qquad nS(\beta_0) \equiv \sum_{i=1}^{n} \xi\left(\frac{T_{(i)}}{\beta_0}\right) < 0;$$

a test of $H_2 : \beta \leqslant \beta_0$ versus $H_1 : \beta > \beta_0$, based on W_n:

$$W_n(\beta_0) = \frac{(1/\sqrt{n-1}) \sum_{i=1}^{n-1} \xi(T_{(i)}/\beta_0)}{\xi(T_{(n)}/\beta_0)},$$

rejects H_2 at significance level ϵ if

$$W_n(\beta_0) > l(\epsilon) \quad \text{and} \quad nS(\beta_0) > 0.$$

A two-sided test at significance level 2ϵ rejects $H_3 : \beta = \beta_0$ versus $H_4 : \beta \neq \beta_0$ if

$$nS(\beta_0) > 0 \quad \text{and} \quad W_n(\beta_0) > l(\epsilon)$$

or

$$nS(\beta_0) < 0 \quad \text{and} \quad W_1(\beta_0) > l(\epsilon).$$

In like manner one can obtain two-sided confidence intervals at confidence level $1 - 2\epsilon$ as

$$\omega_n < \beta < \eta_n.$$

Ordinarily, however, since β is the median of the distribution, one would want only a lower bound on β or would want to test only H_1 versus H_2.

Clearly, there are many other functions of the ordered X_i's that one might consider. It was felt by Saunders and Mann, however, that, for very small sample sizes and relatively small values of ϵ, it is appropriate to base tests and confidence sets on W_1 and W_n, with constraints either on nS or on $X_{(1)}$ and $S_{(n)}$. A Monte Carlo investigation of the accuracy of confidence intervals, based on three different test statistics that are functions of W_1 and W_n, with constraints as specified, was made by these authors. The confidence interval $[\omega_n, \eta_n]$ was found to be the most accurate of the three studied. Values of $l(\epsilon)$ needed to calculate the bounds given in (5.38) and (5.39) and to test various hypotheses concerning β were computed by Saunders and Mann, using a Monte Carlo sample of 10,000, and are shown in Table 5.13. The methods used in the Monte Carlo generation are discussed in Section 7.2.4(c). The authors have shown that the tests corresponding to the critical values $l(\epsilon)$ are unbiased and that the power of the tests increases with sample size for the sample sizes considered.

Table 5.13   Critical values of $l(\epsilon)$

ϵ / n	.005	.01	.025	.05	.10
2	.97	.94	.85	.74	.51
3	1.13	1.04	.85	.67	.44
4	1.15	1.02	.81	.65	.43
5	1.11	1.00	.81	.64	.41

An illustration of the use of Table 5.13 is now given. Let $t_1, \ldots, t_5$ be the ordered failure times of a device in life testing. Under the assumption that the unordered failure times would be a random sample from the two-parameter distribution defined by (5.36), a $100(1-\epsilon)\%$ lower confidence bound on β is given by (5.39), namely,

$$
\min \left[\frac{\displaystyle\sum_{i=1}^{5} t_i^{1/2}}{\displaystyle\sum_{i=1}^{5} t_i^{-1/2}} , \frac{2l(\epsilon)\sqrt{t_5} - \displaystyle\sum_{1}^{4} t_i^{1/2}}{2l(\epsilon) t_5^{-1/2} - \displaystyle\sum_{1}^{4} t_i^{-1/2}} \right].
$$

Suppose that the life times in hours were

$$t_1 = 48{,}310,$$

$$t_2 = 55{,}154, \qquad t_3 = 61{,}273,$$

$$t_4 = 58{,}110, \qquad t_5 = 67{,}769,$$

and that $\epsilon = .1$. One computes a lower confidence bound for β at level .90 as equal to 45,375 hours.

PROBLEMS FOR SECTIONS 5.3–5.5

5.21. From the density function given by (5.26), with $\mu = 0$, determine the equations defining the maximum-likelihood estimators $\hat{\alpha}$ and $\hat{\theta}$.

5.22. Using the fact that the variance of the unbiased estimator of $\sigma^2, s^2 = n\hat{\sigma}^2/(n-1)$, is $2\sigma^4/(n-1)$, show, for the density function given by (5.31) with $\tau = 0$, that $(n-1)s^2/(n+1)$ has minimum mean-squared error among estimators of σ^2 of the form ks^2.

5.23. Determine expressions for the median and the mode of a three-parameter lognormal variate.

5.24. Show that, for a Gaussian distribution with mean μ and variance σ^2, the maximum-likelihood estimators of μ and σ^2 are independent.

5.25. Show that the statistic

$$
S_1 = \frac{\left(\displaystyle\prod_{i=1}^{n} x_i \right)^{1/n}}{\displaystyle\sum_{i=1}^{n} x_i / n}
$$

is a sufficient statistic for obtaining confidence sets for the gamma shape parameter, α, when $\mu = 0$.

5.26. Derive the equations defining the maximum-likelihood estimators of the two parameters of the Birnbaum-Saunders distribution.

5.27. Show that, for a Gaussian distribution with mean μ and variance σ^2, $E(\hat{\mu}) = \mu$ and $E(\hat{\sigma}^2) = [(n-1)/n]\sigma^2$ if $\hat{\mu}$ and $\hat{\sigma}^2$ are the maximum-likelihood estimators of μ and σ^2, respectively.

REFERENCES FOR SECTION 5.1

Bartholomew, D. J. (1963), The Sampling Distribution of an Estimate Arising in Life Testing, *Technometrics*, Vol. 5, pp. 361-374.

Basu, A. P. (1964), Estimates of Reliability for Some Distributions Useful in Life Testing, *Technometrics*, Vol. 6, pp. 215-219.

Bracken, Jerome (1966), Percentage Points of the Beta Distribution for Use in Bayesian Analysis of Bernoulli Processes, *Technometrics*, Vol. 8, pp. 687-694.

Brown, G. G. and H. C. Rutemiller (1971), A Sequential Stopping Rule for Fixed-Sample Acceptance Tests, *Operations Research*, Vol. 19, pp. 970-976.

Cox, D. R. (1953), Some Simple Approximate Tests for Poisson Variates, *Biometrika*, Vol. 40, pp. 354-360.

Epstein, B. and M. Sobel (1953), Life Testing, *Journal of the American Statistical Association*, Vol. 48, pp. 486-502.

Epstein, B. and M. Sobel (1954), Some Theorems Relevant to Life Testing from an Exponential Distribution, *Annals of Mathematical Statistics*, Vol. 25, pp. 373-381.

Grubbs, Frank E. (1971), Fiducial Bounds on Reliability for the Two Parameter Negative Exponential Distribution, *Technometrics*, Vol. 13, pp. 873-876.

Harter, H. L. (1964), *New Tables of the Incomplete Gamma Function Ratio and of Percentage Points of the Chi-Square and Beta Distributions*, U. S. Government Printing Office, Washington, D. C.

Johnson, Leonard G. (1964), *The Statistical Treatment of Fatique Experiments*, Elsevier Publishing Company, New York.

Lawless, J. F. (1971), A Prediction Problem Concerning Samples from the Exponential Distribution, with Application in Life Testing, *Technometrics*, Vol. 13, pp. 725-730.

Lieblein, J. and M. Zelen (1956), Statistical Investigation of the Fatigue Life of Deep Groove Ball Bearings, Research Paper 2719, *Journal of Research, National Bureau of Standards*, Vol. 57, pp. 273-316.

Mann, Nancy R. and Frank E. Grubbs (1973), Simple, Efficient Approximations for Beta Percentiles, Reliability-Parameter Confidence Bounds and Exponential Prediction Intervals (submitted for publication).

Patnaik, P. B. (1949), The Non-Central χ^2 and F Distributions and Their Applications, *Biometrika*, Vol. 36, pp. 202-232.

Pierce, Donald A. (1972), Fiducial, Frequency, and Bayesian Inference on Reliability for the Two-Parameter Negative Exponential Distribution, *Technometrics*, Vol. 15, pp. 249-253.

Pugh, E. L. (1963), The Best Estimate of Reliability in the Exponential Case, *Journal of the Operations Research Society of America*, Vol. 11, pp. 56-61.

Saunders, S. C. (1968), On the Determination of a Safe Life for Distributions Classified by Failure Rate, *Technometrics*, Vol. 10, pp. 361-377.

Wilson, E. B. and M. M. Hilferty (1931), The Distribution of Chi-Square, *Proceedings of the National Academy of Science (U.S.)*, Vol. 17, pp. 684-688.

Zacks, S. and M. Even (1966), The Efficiencies in Small Samples of the Maximum Likelihood and Best Unbiased Estimators of Reliability Functions, *Journal of the American Statistical Association*, Vol. 61, pp. 1033-1051.

Zelen, Marvin and Mary C. Dannemiller (1961), The Robustness of Life Testing Procedures Derived from the Exponential Distribution, *Technometrics*, Vol. 3, pp. 29-49.

REFERENCES FOR SECTION 5.2

Bain, Lee J. (1972), Inferences Based on Censored Sampling from the Weibull or Extreme-Value Distribution, *Technometrics*, Vol. 14, pp. 693-702.

Billman, Barry R., Charles L. Antle, and Lee J. Bain (1971), Statistical Inference from Censored Weibull Samples, *Technometrics*, Vol. 14, pp. 831-840.

Bogdanoff, David A. and Donald A. Pierce (1973), Bayes Fiducial Inference for the Weibull Distribution, *Journal of the American Statistical Association*, Vol. 68, pp. 659-664.

Chan, L. K. and A.B.M. Lutful Kabir (1969), Optimum Quantities for the Linear Estimation of the Parameters of the Extreme-Value Distribution in Complete and Censored Samples, *Naval Research Logistics Quarterly*, Vol. 16, pp. 381-404.

Cohen, A. Clifford, Jr. (1965), Maximum Likelihood Estimation in the Weibull Distribution Based on Complete and on Censored Samples, *Technometrics*, Vol. 7, pp. 579-588.

Dubey, Satya D. (1966), Hyper-efficient Estimator of the Location Parameter of the Weibull Laws, *Naval Research Logistics Quarterly*, Vol. 13, pp. 253-263.

Dubey, Satya D. (1967), Some Percentile Estimators of Weibull Parameters, *Technometrics*, Vol. 9, pp. 119-129.

Englehardt, Max and Lee J. Bain (1973), Some Complete and Censored Sampling Results for the Weibull or Extreme-Value Distribution, *Technometrics*, Vol. 15, pp. 541-549.

Harter, H. Leon and Satya D. Dubey (1967), Theory and Tables for Tests of Hypotheses Concerning the Mean and Variance of a Weibull Population, *Aerospace Research Laboratories Report* ARL 67-0059, Office of Aerospace Research, U. S. Air Force, Wright-Patterson Air Force Base, Ohio.

Harter, H. Leon and Albert H. Moore (1965), Maximum Likelihood Estimation of the Parameters of the Gamma and Weibull Populations from Complete and from Censored Samples, *Technometrics*, Vol. 7, pp. 639-643.

Harter, H. Leon and Albert H. Moore (1968), Maximum Likelihood Estimation, from Doubly Censored Samples, of the Parameters of the First Asymptotic Distribution of Extreme Values, *Journal of the American Statistical Association*, Vol. 63, pp. 889-901.

Johns, M. V., Jr., and G. J. Lieberman (1966), An Exact Asymptotically Efficient Confidence Bound for Reliability in the Case of the Weibull Distribution, *Technometrics*, Vol. 8, pp. 135-175.

Johnson, Leonard G. (1951), The Median Ranks of Sample Values in Their Population with an Application to Certain Fatigue Studies, *Industrial Mathematics*, Vol. 2, pp. 1-6.

Kao, J. H. K. (1959), A Graphical Estimation of Mixed Weibull Parameters in Life Testing of Election Tubes, *Technometrics*, Vol. 1, pp. 389-407.

Kao, J. H. K. (1960), A Summary of Some New Techniques on Failure Analysis, *Proceedings of the Sixth National Symposium on Reliability and Quality Control in Electronics*, pp. 190-201.

Kimball, B. F. (1949), An Approximation to the Sampling Variance of an Estimated Maximum Value of Given Frequency Based on Fit of Doubly Exponential Distribution of Maximum Values, *Annals of Mathematical Statistics*, Vol. 17, pp. 299-309.

Kimball, B. F. (1960), On the Choice of Plotting Positions on Probability Paper, *Journal of the American Statistical Association*, Vol. 55, pp. 546-560.

Lawless, J. F. (1973a), Conditional versus Unconditional Confidence Intervals for the Parameters of the Weibull Distribution, *Journal of the American Statistical Association*, Vol. 68, pp. 665-669.

Lawless, J. F. (1973b), On the Estimation of Safe Life When the Underlying Life Distribution Is Weibull, *Technometrics* (to appear).

Lemon, G. H. (1974), Maximum-Likelihood Estimation for the Three-Parameter Weibull Distribution Based on Censored Samples, *Technometrics* (to appear).

Lieblein, J. (1954), A New Method of Analyzing Extreme-Value Data, *Technical Note* 3053, National Advisory Committee for Aeronautics.

Lieblein, J. and M. Zelen (1956), Statistical Investigation of the Fatigue Life of Deep Groove Ball Bearings, Research Paper 2719, *Journal of Research, National Bureau Standards*, Vol. 57, pp. 273-316.

Mann, Nancy R. (1967a), Results on Location and Scale Parameter Estimation with Application to the Extreme-Value Distribution, *Aerospace Research Laboratories Report* ARL 67-0023, Office of Aerospace Research, U. S. Air Force, Wright-Patterson Air Force Base, Ohio.

Mann, Nancy R. (1967b), Tables for Obtaining the Best Linear Invariant Estimates of Parameters of the Weibull Distribution, *Technometrics*, Vol. 9, pp. 629-645.

Mann, Nancy R. (1968a), Results on Statistical Estimation and Hypothesis Testing with Application to the Weibull and Extreme-Value Distributions, *Aerospace Research Laboratories Report* ARL 68-0068, Office of Aerospace Research, U.S. Air Force, Wright-Patterson Air Force Base, Ohio.

Mann, Nancy R. (1968b), Point and Interval Estimation Procedures for the Two-Parameter Weibull and Extreme-Value Distributions, *Technometrics*, Vol. 10, pp. 231-256.

Mann, Nancy R. (1969a), Exact Three-Order Statistic Confidence Bounds on Reliable Life for a Weibull Model with Progressive Censoring, *Journal of the American Statistical Association*, Vol. 64, pp. 306-315.

Mann, Nancy R. (1969b), Cramér-Rao Efficiencies of Best Linear Invariant Estimators of Parameters of the Extreme-Value Distribution under Type II Censoring from Above, *SIAM Journal of Applied Mathematics*, Vol. 17, pp. 1150-1162.

Mann, Nancy R. (1969c), Optimum Estimators for Linear Functions of Location and Scale Parameters, *Annals of Mathematical Statistics*, Vol. 40, pp. 2149-2155.

Mann, Nancy R. (1970a), Extension of Results Concerning Parameter Estimators, Tolerance Bounds and Warranty Periods for Weibull Models, *Aerospace Research Laboratories Report* ARL 70-0010, Office of Aerospace Research, U.S. Air Force, Wright-Patterson Air Force Base, Ohio.

Mann, Nancy R. (1970b), Estimation of Location and Scale Parameters under Various Models of Censoring and Truncation, *Aerospace Research Laboratories Report* ARL 70-0026, Office of Aerospace Research, U.S. Air Force, Wright-Patterson Air Force Base, Ohio.

Mann, Nancy R. (1970c), Estimators and Exact Confidence Bounds for Weibull Parameters Based on a Few Ordered Observations, *Technometrics*, Vol. 12, pp. 345-361.

Mann, Nancy R. (1970d), Warranty Periods Based on Three Ordered Sample Observations from a Weibull Population, *I.E.E.E. Transactions on Reliability*, Vol. 19, pp. 167-171.

Mann, Nancy R. (1971), Best Linear Invariant Estimation for Weibull Parameters under Progressive Censoring, *Technometrics*, Vol. 13, pp. 521-533.

Mann, Nancy R. and Kenneth W. Fertig (1973), Tables for Obtaining Confidence Bounds and Tolerance Bounds Based on Best Linear Invariant Estimates of Parameters of the Extreme-Value Distribution, *Technometrics*, Vol. 15, pp. 87-101.

Mann, Nancy R. and Kenneth W. Fertig (1974a), Simplified Efficient Point and Interval Estimators for Weibull Parameters, *Technometrics* (to appear).

Mann, Nancy R. and Kenneth W. Fertig (1974b), A Goodness-of Fit Test for the Two-Parameter Weibull Distribution Against Three-Parameter Weibull Alternatives; Confidence Bounds for a Weibull Threshold Parameter (submitted for publication).

Mann, Nancy R., Kenneth W. Fertig, and Ernest M. Scheuer (1971), Confidence and Tolerance Bounds and a New Goodness-of-Fit Test for Two-Parameter Weibull or Extreme-Value Distributions with Tables for Censored Samples of Size 3(1)25, *Aerospace Research Laboratories Report* ARL 71-0077, Office of Aerospace Research, U.S. Air Force, Wright-Patterson Air Force Base, Ohio.

Mann, Nancy R. and Sam C. Saunders (1969), On Evaluation of Warranty Assurance When Life Has a Weibull Distribution, *Biometrika*, Vol. 56, pp. 615-625.

Mann, Nancy R., Ernest M. Scheuer, and Kenneth W. Fertig (1973), A New Goodness-of-Fit Test for the Two-Parameter Weibull or Extreme-Value Distribution with Unknown Parameters, *Communications in Statistics*, Vol. 2, pp. 383-400.

Mosteller, F. (1946), On Some Useful Inefficient Statistics, *Annals of Mathematical Statistics*, Vol. 17, pp. 377-407.

Pyke, R. (1965), Spacings, *Journal of the Royal Statistical Society B*, Vol. 27, pp. 395-449.

Rockette, Howard, Charles Antle, and Lawrence A. Klimko (1973), Maximum Likelihood Estimation with the Weibull Model, *Journal of the American Statistical Association* (to appear).

Thoman, D. R., L. J. Bain, and C. E. Antle (1969), Inferences on the Parameters of the Weibull Distribution, *Technometrics*, Vol. 11, pp. 445-460.

Thoman, D. R., L. J. Bain, and C. E. Antle (1970), Reliability and Tolerance Limits in the Weibull Distribution, *Technometrics*, Vol. 12, pp. 363-371.

Van Montfort, M.A.J. (1970), On Testing That the Distribution of Extremes Is of Type I When Type II Is the Alternative, *Journal of Hydrology*, Vol. 11, pp. 421-427.

White, John S. (1967), The Moments of Log-Weibull Order Statistics, *General Motors Research Publication* GMR-717, General Motor Corporation, Warren, Michigan.

REFERENCES FOR SECTIONS 5.3–5.5

Bartlett, M. S. (1937), Properties of Sufficiency and Statistical Tests, *Proceedings of the Royal Society A*, Vol. 160, pp. 268-282.

Birnbaum, Z. W. and Sam C. Saunders (1969a), A New Family of Life Distributions, *Journal of Applied Probability*, Vol. 6, pp. 319-327.

Birnbaum, Z. W. and Sam C. Saunders (1969b), Estimation for a Family of Life Distributions with Applications to Fatigue, *Journal of Applied Probability*, Vol. 6, pp. 328-347.

Chernoff, Herman and Gerald J. Lieberman (1956), The Use of Generalized Probability Paper for Continuous Distributions, *Annals of Mathematical Statistics*, Vol. 27, pp. 806-818.

Choi, S. C. and R. Wette (1969), Maximum Likelihood Estimation of the Parameters of the Gamma Distribution and Their Bias, *Technometrics*, Vol. 11, pp. 683-690.

Davis, Philip J. (1964), Gamma Function and Related Functions, in *Handbook of Mathemati-*

cal Functions, edited by M. Abramowitz and I. A. Stegun, National Bureau of Standards Washington, D.C.

Dixon, W. J. and John W. Tukey (1968), Approximate Behavior of the Distribution of Winsorized t (Trimming/Winsorization 2), Technometrics, Vol. 10, pp. 83-98.

Grubbs, F. E., Helen J. Coon, and E. S. Pearson (1966), On the Use of Patnaik Type Approximations to the Range in Significance Tests, Biometrika Vol. 53, pp. 248-252.

Harter, H. Leon and Albert H. Moore (1965), Maximum Likelihood Estimation of the Parameters of the Gamma and Weibull Populations from Complete and from Censored Samples, Technometrics, Vol. 7, pp. 639-643.

Harter, H. Leon and Albert H. Moore (1966), Local Maximum Likelihood Estimation of the Parameters of Three-Parameter Lognormal Populations from Complete and Censored Samples, Journal of the American Statistical Association, Vol. 61, pp. 842-851.

Kimball, B. F. (1960), On the Choice of Plotting Positions on Probability Paper, Journal of the American Statistical Association, Vol. 55, pp. 546-560.

Lilliefors, Hubert W. (1971), Reducing the Bias of Estimates of Parameters for the Erlang and Gamma Distributions (submitted for publication).

Linhart, H. (1965), Approximate Confidence Limits for the Coefficient of Variation of Gamma Distributions, Biometrics, Vol. 21, pp. 733-738.

Locks, M. O., M. J. Alexander, and B. J. Byars (1963), New Tables of the Noncentral t Distribution, Aerospace Research Laboratories Report ARL 63-19, Aeronautical Research Laboratories, Wright-Patterson Air Force Base, Ohio.

Munro, A. H. and R.A.J. Wixley (1970), Estimators Based on Order Statistics of Small Samples from a Three-Parameter Lognormal Distribution, Journal of the American Statistical Association, Vol. 65, pp. 212-225.

Resnikoff, George J. and Gerald J. Lieberman (1957), Tables of the Noncentral t Distribution, Stanford University Press, Stanford, California.

Sarhan, Ahmed E. and B. G. Greenberg (1962), Contributions to Order Statistics, John Wiley and Sons, New York.

Saunders, Sam C. and Nancy R. Mann (1974), On Small Sample Confidence Intervals for the Parameters of a New Life Distribution, Naval Research Logistics Quarterly (to appear).

CHAPTER 6

Testing Reliability Hypotheses

In a book that uses methods from one discipline (statistics) to solve problems in another discipline (reliability), certain compromises of nomenclature are sometimes necessary. The title of this chapter is a case in point. In virtually all statistical applications, the material to be covered in this chapter would be known as methods of testing statistical hypotheses. On the other hand, many of the statistical hypothesis test applications in reliability are called reliability demonstration tests. Finally, in view of the relation between confidence interval estimates and statistical hypothesis tests to be discussed in Section 6.5.1, the question of why this chapter is here at all might be asked. We try to explain this in the following discussion. Suppose that a random sample $(X_1, \ldots, X_n)$ has been obtained from a probability distribution, say an exponential distribution, and that it is desired to test hypotheses about the mean of the distribution. By means of the methods presented in Chapter 5 and this chapter, a test statistic is selected, regions of acceptance and rejection are set up, and risks of wrong decisions are *calculated*. However, the key idea in this chapter is that the *risks of wrong decision* are specified *before* the sample $(X_1, \ldots, X_n)$ is obtained, and in this case n, the sample size, is generally to be determined. When the sample size must be determined in advance to satisfy certain criteria (risks), an additional degree of complexity is introduced.

In summary, then, the general problem in hypothesis testing concerns not only what (test) statistic to use and how to select an accept/reject region, but also how much data to gather. When the last point is addressed (in reliability), the hypothesis test is often called a *reliability demonstration test*. If for one reason or another n is already determined, the accept/reject criteria given here can be used.

6.1 STATISTICAL HYPOTHESIS TESTS

Throughout this chapter the concern will be mostly with parametric hypotheses, that is, hypotheses about the parameters in probability distributions occurring in reliability work. The parameter(s) could be the mean of an exponential distribution, the shape parameter of a Weibull distribution, and many others. An hypothesis is called *simple* if it specifies unique values for all the unknown parameters. Otherwise, it is termed *composite*. For example, if

$$H_0 : \theta = \theta_0, \tag{6.1}$$

$$H_0 : \theta > \theta_0, \tag{6.2}$$

and θ is the mean of an exponential distribution, (6.1) is a simple hypothesis and (6.2) is a composite hypothesis. If δ is the Weibull scale parameter and β is the Weibull shape parameter,

$$H_0 : \delta = \delta_0, \quad \beta = \beta_0,$$

is simple, whereas

$$H_0 : \delta = \delta_0, \quad \beta > \beta_0,$$

is composite. A simple hypothesis completely specifies the distribution.

In deciding whether to accept or reject the null hypothesis, i.e. H_0 the procedure is to find a region S_α in the sample description space, $Sx \ldots xS = S^{(n)}$, such that $S_\alpha \subset S^{(n)}$, and, if $(x_1, \ldots, x_n) \in S_\alpha$, then H_0 is *rejected*. This region S_α is called the *critical region*; and if $0 < \alpha < 1$ is chosen so that

$$P[(X_1, \ldots, X_n) \in S_\alpha | H_0] \leqslant \alpha, \tag{6.3}$$

the probability of rejecting H_0 when it is true is less than or equal to α. The number α, called the *significance level* of the test, is generally selected to be small, usually $\alpha < \frac{1}{2}$. Rejection of the null hypothesis (H_0) when it is true is called a *Type I error*. The probability of committing it is often referred to as the *size* of the test. When equality holds in (6.3), the size of the test is α and the test is *exact*. In particular, if H_0 is simple, equality holds in (6.3).

Unfortunately, there are many different regions S_α that will yield a fixed value of α. In order to select one region, S_α, of size α, it is necessary to specify another kind of error. Since we have discussed the error of rejecting H_0 when it is true (a Type I error), it is obvious that the other kind of error is the acceptance of H_0 when it is false. This is called a *Type II error*, and the probability of committing it,

$$P[(X_1, \ldots, X_n) \in \bar{S}_\alpha | H_0 \text{ false}] = \beta, \tag{6.4}$$

is denoted by β. Recall here that the overhead bar, $^-$, is the complementation operator. The probability in (6.4) is generally a function of "how false" H_0 is; that is, if $H_0:\theta=\theta_0$ as in (6.1), for most useful tests β will depend on what value of θ obtains if $\theta\neq\theta_0$. Thus, if we specify an alternative simple hypothesis, called H_1,

$$H_0:\theta=\theta_0 \quad \text{versus} \quad H_1:\theta=\theta_1, \quad \theta_1<\theta_0, \text{ is tested.}$$

For a fixed sample size n and a fixed value of α, the value of β (in 6.4) at $\theta=\theta_1$ is generally determined. The question still remains: having selected α,n, which of the many S_α should be picked? The natural answer is to select S_α so that (6.4) is minimized. The same S_α will also maximize the accuracy of a confidence bound for θ. In testing a simple null versus a simple alternative hypothesis, Neyman and Pearson (a) showed such an S_α always exists if the underlying family of the probability distribution is specified, and (b) showed how to find S_α.

Because a large part of this chapter is concerned with simple versus simple hypotheses, we will give the result here. It is known as the *Neyman-Pearson lemma*. To this end, let $f(x_1,\ldots,x_n|H_0)\equiv L(x_1,\ldots,x_n|H_0)$ be the joint density of the sample observations when H_0 is true. Similarly, define $f(x_1,\ldots,x_n|H_1)$, the joint distribution of the sample when H_1 is true. The best region, that is, the set S_α that minimizes β for fixed α,n, is then the set of points $(x_1,\ldots,x_n)$ satisfying

$$\frac{L(x_1,\ldots,x_n|H_0)}{L(x_1,\ldots,x_n|H_1)} \leqslant C_\alpha \tag{6.5}$$

when H_0 holds. In other words, H_0 is rejected whenever (6.5) holds.

This is called the *likelihood ratio*, and C_α is a constant generally depending on α. If the distribution of the random variable X is continuous, there will be a C_α for every α. If X is discrete, we may have to settle for an α, say α', near the desired α. The remarkable feature of (6.5) is that often the likelihood-ratio test turns out to be determined by a single function of the sample values (e.g., the sample mean $\overline{X}$), and in this case this function is called the *test statistic*. This is an important simplification, since it is not necessary to determine C_α if the sampling distribution of the test statistic is known. In view of this fact, the question might be asked: why not proceed directly to one of the "good" statistics discovered in Chapter 5 (e.g., the maximum-likelihood estimate of the particular parameter)? The answer is that this method (finding a good estimator and using it as a test statistic) generally gives the best S_α, but not always. The likelihood-ratio test, on the other hand, always gives the best region S_α of size α.

To fix some of the ideas, particularly to show that it is not necessary to determine C_α, consider the following simple example.

Example. Suppose that the reliability function $R(t)$, t fixed, is unknown, and it is desired to test

$$H_0: R(t) = p_0 \quad \text{versus} \quad H_1: R(t) = p_1, \quad p_1 < p_0,$$

and that n independent trials of survival/nonsurvival for time t can be observed. Then this is a Bernoulli process, and

$$f(x_1, \dots, x_n; p_0) = p_0^{\Sigma x_i} (1 - p_0)^{n - \Sigma x_i},$$

$$f(x_1, \dots, x_n; p_1) = p_1^{\Sigma x_i} (1 - p_1)^{n - \Sigma x_i},$$

where

$$X_i = \left\{ \begin{array}{ll} 1 & \text{if survival is observed at the } i\text{th trial,} \\ 0 & \text{if nonsurvival is observed at the } i\text{th trial.} \end{array} \right\}$$

The likelihood-ratio test is then

$$\frac{p_0^{\Sigma x_i} (1 - p_0)^{n - \Sigma x_i}}{p_1^{\Sigma x_i} (1 - p_1)^{n - \Sigma x_i}} \leqslant C_\alpha. \tag{6.6}$$

Taking natural logarithms of both sides and making algebraic changes, one finds that this is equivalent to

$$\sum x_i \leqslant \frac{\log C_\alpha - n \log[(1 - p_0)/(1 - p_1)]}{\log[p_0(1 - p_1)/p_1(1 - p_0)]}. \tag{6.7}$$

That is, when $(x_1, \dots, x_n) \in S_\alpha$ and hence (6.6) is satisfied, so is (6.7) satisfied and conversely.

Note the following:

1. Under both hypotheses, Σx_i has a binomial distribution, that is, if $T = \Sigma x_i$,

$$f(T) = \binom{n}{T} p_i^T (1 - p_i)^{n - T}, \quad i = 0, 1.$$

2. Since n, α, p_0, p_1 are all specified numbers, other than C_α, the right-hand side of (6.7) is known.
3. Because of the equivalence of (6.6) and (6.7),

$$P[(X_1, \dots, X_n) \in S_\alpha | H_0] = P(\sum x_i \leqslant T_c | H_0) = \alpha.$$

Since $\sum x_i$ has a binomial distribution,

$$P\left(\sum x_i = T \leqslant T_c | H_0\right) = \sum_{T=0}^{T_c} \binom{n}{T} p_0{}^T (1-p_0)^{n-T} = \alpha.$$

Since T must be an integer, we may not find a T_c such that a test of size α is obtained exactly. In this case T_c may be selected to give a value α' as near α as possible. For example, if $p_0 = .90$, $n = 10$, $\alpha = .05$, the binomial tables (National Bureau of Standards) yield

$$P(\sum x_i = T \leqslant 7) = .0702,$$

$$P(\sum x_i = T \leqslant 6) = .0128,$$

so that T_c could be taken as 7 and hence $\alpha' = .0702$. In this case the test (of size .0702) could be summarized thus:

If $\sum x_i \leqslant 7$, reject H_0; otherwise accept H_0.

A randomization procedure could be found that would yield the desired $\alpha = .05$. In fact, suppose that a random uniform number y is drawn in the interval $[0, 1]$ and agree that if $y \leqslant .352$ $T_c = 6$ will be used and if $y > .352$ we will use $T_c = 7$. Then α is a two-valued random variable with mean

$$(.0128)(.352) + .0702(1 - .352) = .05.$$

However, in reliability practice these randomization procedures are usually not worth the trouble and confusion that arise in application.

4. By hypothesis $p_1 < p_0$; thus

$$P(\sum x_i > T_c | H_0) = 1 - \alpha > P(\sum x_i > T_c | H_1) = \beta.$$

Hence β is such that $1 - \beta > \alpha$ for all $p_1 < p_0$. Tests that have this property are called *unbiased* tests.

5. It was not necessary to determine C_α.

6. The likelihood-ratio test works even if the X_i are not independent. All that is required is that $f(x_1, \ldots, x_n)$ be available. This is a particularly important observation, for example, when population size is finite and the hypergeometric distribution, rather than the binomial distribution, is applicable to the above example.

7. The test statistic is the usual maximum-likelihood estimator of np.

8. With n determined, it was not necessary to specify p_1.

PROBLEMS

6.1. In the example of Section 6.1 take $n=20$, $\alpha=.10$, $p_0=.90$ and find the critical region, that is, the region for rejection of H_0. Also, find β for $p_1=.70$. Suppose that the observed T is 16 survivals; should H_0 be rejected?

6.2. In the example of Section 6.1 take $p_0=.90$, $\alpha=.10$, $p_1=.86$, $\beta=.05$. Find the smallest n such that the true α, say α', is $\leqslant.10$.

6.3. In Problem 6.2 plot β as a function of p_1 for at least five values of p_1.

6.4. A test is said to be *consistent* if

$$\lim_{n\to\infty} P\left[\,(X_1,\ldots,X_n)\in S_\alpha|H_1\right]=1,$$

that is, the probability of rejecting H_0 when it is false approaches 1 as n becomes arbitrarily large. Present an argument why this is true of the test given in the example of Section 6.1.

6.5. For the example of Section 6.1, discuss the behavior of n as

(a) $p_0/p_1\to1(p_0>p_1)\alpha,\beta$ fixed,

(b) α and/or $\beta\to0(p_0,p_1$ fixed).

In the following table we compare some of the statistical notation and reliability notation for hypothesis tests. The power curve or the operating characteristic curve provides a very important description of how the test performs under various values of the alternative hypothesis. In particular, in reliability work, *no test is completely described without the operating characteristic curve.*

Symbol	Statistical Name	Reliability Name
α	Size of Test	Producer's risk
β	$1-\beta$ is called the power of the test	Consumer's risk
Graph of β vs. value of H_1	Graph of $1-\beta$ is called the power curve	Graph of β is called the operating characteristic curve

The reader who is already acquainted with statistical hypothesis tests may wonder why the likelihood-ratio test is not seen more often in elementary books. The answer is simple enough: for a large number of types of probability distributions, the best statistic is already worked out (based on the likelihood ratio) and there is little need to repeat the development. Moreover, for many situations the best statistic turns out to be the maximum-likelihood estimate of the parameter(s) in question.

Before proceeding to tests for the mean of an exponential distribution a final remark on the example given above should be made: the test is in one sense parametric and in another sense nonparametric. It is parametric in that it is a test of the parameter in a binomial distribution. On the other hand,

$$R(t) = \int_t^\infty f(x)\, dx,$$

where X is time to failure. If the underlying failure distribution is known, say to be exponential, then

$$R(t) = e^{-t/\theta},$$

where $\theta =$ mean time to failure, and for fixed t a test equivalent to the one of the example would be

$$H_0 : \theta = \theta_0,$$

$$H_1 : \theta = \theta_1, \tag{6.8}$$

where θ_i is the solution to $p_i = e^{-t/\theta_i}$, $i = 0, 1$.

Thus the likelihood-ratio test of the example could have been used to test (6.8). The point is that, if only survival for time t is observed, the test given in the example is the best test for an $R(t)$ of unknown form. If the actual failure times are available, however, the test for θ [and hence for $R(t)$ if the distribution is known to be exponential] is superior in the sense that the sample size required will be smaller for fixed (α, β). This point is discussed in Section 5.1.1. However, when testing for the mean of an exponential distribution, it is often economical to observe only survival or nonsurvival for fixed time t, rather than actual failure times. Section 6.2 deals with hypothesis tests for the mean of an exponential distribution.

6.2 TESTS FOR THE MEAN OF AN EXPONENTIAL DISTRIBUTION AND UNKNOWN RELIABILITY FUNCTIONS

At the outset we recall that for an exponential distribution the mean time between failures (MTBF), θ, is the reciprocal of the hazard rate, so that all the tests of this section can be used for hazard rate, λ. Also this section can be used to test hypotheses about a quantile, x_p, since $x_p = -\ln(1-p)\theta$. It should be noted too that if the form of $R(t)$ is unknown the methods of Sections 6.2.1–6.2.3 are used by testing $R_0 = 1 - p_0$ versus $R_1 = 1 - p_1$.

6.2.1 A Fixed-Sample-Size Test When Survival-Nonsurvival for Time t is Observed

We first introduce a change, the purpose of which is to avoid confusion. In

Section 6.1 it was seen that the hypothesis test

$$H_0 : \theta = \theta_0, \quad \text{versus} \quad H_1 : \theta = \theta_1$$

could be made equivalent to a test on the parameter in a binomial distribution, since in n observations of survival/nonsurvival for fixed time t the number of survivals is a binomial random variable with parameter $R(t) = e^{-t/\theta}$. In virtually all the literature on binomial tests and in tables, the values of the parameter p are chosen so that $p \leqslant \frac{1}{2}$. Since $R(t) = e^{-t/\theta}$ is generally $> \frac{1}{2}$ for most practical problems, in what follows we will deal with $1 - R(t)$ and the number of *nonsurvivals* in n observations. Thus to test

$$H_0 : \theta = \theta_0 \text{ versus } H_1 : \theta = \theta_1, \quad \theta_0 > \theta_1, \tag{6.9}$$

find p_0, p_1 as follows:

$$p_i = 1 - e^{-t/\theta_i}, \quad i = 0, 1$$

and test the equivalent hypothesis,

$$H_0 : p = p_0 \text{ versus } H_1 : p = p_1, \quad p_1 > p_0. \tag{6.10}$$

Thus, in view of this minor change, define a random variable,

$$X = \begin{cases} 1 & \text{if nonsurvival for time } t \text{ is observed,} \\ 0 & \text{if survival for time } t \text{ is observed.} \end{cases}$$

Then, in n independent observations, $T = \Sigma x_i$ records the total number of nonsurvivals. The likelihood-ratio test, as was seen in Section 6.1, results in $T = \Sigma x_i$ as the test statistic.

Suppose that it is desired to test (6.9) with

$$\alpha = P(\text{rejection of } H_0 | \theta_0) \text{ and } \beta = P(\text{acceptance of } H_0 | \theta_1) \text{ fixed.}$$

It remains to determine n, the sample size (or number of trials), so that the desired pair (α, β) is obtained. Since, under $H_0, T = \Sigma x_i =$ the total number of nonsurvivals is a binomial random variable with $p_0 = 1 - e^{-t/\theta_0}$, we must find an integer c, commonly called the acceptance number, such that

$$\alpha = P(\text{rejection of } H_0 | \theta_0) = P(\text{rejection of } H_0 | p_0)$$

$$= \sum_{T=c+1}^{n} \binom{n}{T} p_0^{T}(1-p_0)^{n-T}$$

and for the β risk

$$\beta = P(\text{acceptance of } H_0 | \theta_1) = P(\text{acceptance of } H_0 | p_1)$$

$$= \sum_{T=0}^{c} \binom{n}{T} p_1^{T}(1-p_1)^{n-T} \tag{6.11}$$

with the stipulation that

$$\text{if } T = \sum x_i \leqslant c, \text{ accept } H_0,$$

$$\text{if } T > c, \text{ reject } H_0.$$

In the notation of Section 6.1, c was denoted by $n - T_c - 1$.

In the language of hypothesis testing the critical region S_α is the set of vectors $(x_1, \ldots, x_n)$ such that $\sum x_i > c$. The notation c has been adopted here because it is the commonplace designation for the maximum number of allowed nonsurvivals. The proper value of c can be found from tables of the binomial distribution or Poisson tables if that approximation is applicable or from normal tables if that approximation is applicable. Note that there will generally be no pair (n, c) such that (α, β) will be obtained exactly because of the discreteness of the binomial random variable. We will often follow the convention of finding the pair (n, c) such that n is the smallest integer for which the actual pair (α, β), say (α', β'), is such that $\alpha' \leqslant \alpha, \beta' \leqslant \beta$. Other conventions are available to suit various tastes.

Example. Suppose that it is desired to test

$$H_0 : \theta = \theta_0 = 1000 \text{ hours,}$$

$$H_1 : \theta = \theta_1 = 250 \text{ hours,}$$

with $\alpha = .05$, $\beta = .10$, and $t = 50$ hours. Then

$$p_0 = 1 - e^{-t/\theta_0} = 1 - e^{-50/1000} = .049,$$

$$p_1 = 1 - e^{-t/\theta_1} = 1 - e^{-50/250} = .181.$$

Thus (6.11) becomes

$$.05 = \sum_{T=c+1}^{n} \binom{n}{T} (.049)^{T} (.951)^{n-T},$$

$$.10 = \sum_{T=0}^{c} \binom{n}{T} (.181)^{T} (.819)^{n-T}.$$

Because n is likely to be large and p_0, p_1 are relatively small, we use the Poisson approximation to the binomial. Then the mean of the Poisson is np_0 and np_1 under the null and alternative hypotheses, respectively. Also note that the ratio of the means $np_0/np_1 = p_0/p_1$, so that we need only look for Poisson means in the ratio

$$\frac{p_0}{p_1} = \frac{.049}{.181} = .27.$$

From the tables of Molina (1942) find that

$$P(T > c = 5 | np_0 = 2.5) = .042 < .05$$

and

$$P(T \leq c = 5 | np_1 = 9.3) = .099 < .10.$$

Solving $np_0 = 2.5$ gives $n = 51$. Solving $np_1 = 9.3$ for n does not give exactly $n = 51$; because of the graduations in the tables we cannot obtain $p_0/p_1 = .27$ exactly. However, it does not matter a great deal, since the Poisson is an approximation in this case anyway. Thus the solution pair is $(n = 51, c = 5)$, and, if $T = \sum x_i \leq c$, H_0 is accepted; otherwise H_0 is rejected.

Sobel and Tischendorf (1959) have discussed this method and tabulated a number of plans for this type of test. They give a large number of choices of t/θ_1 (what we call θ_1, they call θ_0) for $1 - \beta = .75$, .80, .85, .90, and .95 and acceptance numbers $c = 0(1)14$. In their tables the value of θ_0 is unspecified and so is α. There is a table in Sobel and Tischendorf to aid in computing θ_0 when $\alpha = .05$ for $c = 0(1)10$. Bowker and Lieberman (1955) also give some very useful tables. If these aids do not suffice, the user must either select one of the c's given and compute α for the desired θ_0 or forget the tables and proceed directly as indicated in this section, using binomial or Poisson tables.

Finally an example is given of computing a point on the operating characteristic (OC) curve.

Example. Suppose that the probability of acceptance is desired at $\theta = 500$ hours. This corresponds to

$$p = 1 - e^{-t/\theta} = .095 \quad \text{and} \quad np = 51(.095) = 4.85.$$

Using Molina's tables again,

$$P(T \leqslant c = 5 | np = 4.85) = .64.$$

This point, in addition to the pairs $(\theta = \infty, 1.0), (\theta = 1000, .95)(\theta = 250, .10), (\theta = 0, 0)$, could be used to sketch a rough OC curve.

In this test situation, the sample size n is fixed. Actually, this is not quite true, since testing can be discontinued (and the unit rejected) as soon as $c + 1$ nonsurvivals have occurred. However, acceptance is the hoped for situation, and acceptance cannot occur before $n - c$ observations and possibly not until n. Thus, for acceptance, at least $n - c$ units must survive time t, and hence at least $(n - c)t$ but not more than nt total unit hours will be required. Since c is generally small with respect to n, the test time and costs are pretty well identified in advance. However, there are situations in which the cost of test depends more on $T = \Sigma x_i =$ number of nonsurvivals than on n and t. We can then let $T = \Sigma x_i$ (which is a random variable in the test of this section) be fixed, and let n be a random variable, $n \geqslant T$. This is done in the next section.

PROBLEMS

6.6. Compute the points on the OC curve for the text example at $\theta = 800$ and $\theta = 350$.

6.7. Find (n, c) necessary to test $(\theta_0 = 1000, \theta_1 = 500, \alpha = .10, \beta = .10, t = 100$ hours). Compute the OC curve with at least five points. Assume an exponential failure distribution.

6.8. Suppose that the lifetime of each expensive part in a lot of size $N = 30$ has the same exponential distribution and that $\theta_0 = 47.62, \theta_1 = 22.52, \alpha = \beta = .10, t = 5$ hours. Using the hypergeometric distribution (Lieberman and Owen, 1961), find (n, c) (these values of θ_0, θ_1 result in $p_0 = .10, p_1 = .20$).

6.9. The tests of this section are known in the fields of quality control and inspection as single sampling plans by attributes. Two famous tables of these are the Dodge-Romig tables (1944) and the MIL-STD-105D plans (Duncan, 1959) Find out what name Dodge and Romig give to p_1 and what name is used in the MIL-STD-105D plans for p_0.

6.10. When a test based on $(\theta_0, \alpha, \theta_1, \beta)$ has been passed, it is often said that a θ of at least θ_1 has been demonstrated with $1 - \beta$ confidence. What can be said about θ when such a test is failed?

6.11. The median, say $\tilde{x}$, of a continuous probability distribution $f(x)$ is given by

$$\frac{1}{2} = \int_{-\infty}^{\tilde{x}} f(x)\, dx$$

when the median is unique.

    a. Show that for the exponential distribution

$$\tilde{x} = \theta \ln 2,$$

where $\theta =$ the mean.

    b. Show how to construct a test of

$$H_0 : \tilde{x} = \tilde{x}_0, \qquad H_1 : \tilde{x} = \tilde{x}_1,$$

based on survival/nonsurvival data.

    6.12. Discuss the relationship between n and t for the test of this section.

6.2.2 A Fixed-Number-of-Nonsurvivals Test

We indicated in the preceding section that, instead of fixing the number of observations and observing the random variable $T = \Sigma x_i =$ total number of nonsurvivals in n observations, it might be desirable to fix T and let n be a random variable. This could be the situation when the costs depend largely on the number of units destroyed, that is, nonsurvivals.

    Immediately, as before, convert the hypothesis

$$\left.\begin{array}{cc} H_0 : \theta = \theta_0 & H_0 : p = p_0 \\ \\ \text{to} \\ \\ H_1 : \theta = \theta_1, & H_1 : p = p_1 \end{array}\right\} \; p_0 < p_1 ; \alpha, \beta \text{ fixed,}$$

where $p_i = 1 - e^{-t/\theta_i}, i = 0, 1$, is the probability of a nonsurvival in time t. If a random variable

$$X = \begin{cases} 1 & \text{if nonsurvival is observed,} \\ 0 & \text{if survival is observed,} \end{cases}$$

is again constructed, the probability that n_1 observations are required to obtain one nonsurvival is

$$f(n_1; p) = p(1-p)^{n_1 - 1}, \quad n_1 = 1, 2, \ldots . \tag{6.12}$$

This is the geometric distribution. A random sample of size T nonsurvivals has joint probability distribution

$$f(n_1, \ldots, n_T; p) = \prod_{i=1}^{T} p(1-p)^{n_i - 1} = p^T (1-p)^{\Sigma n_i - T}. \tag{6.13}$$

This process is generally called Pascal sampling (and is sometimes termed inverse binomial sampling), as compared with the binomial sampling of the preceding section. If one denotes $\Sigma n_i = N$, the likelihood-ratio test is then as follows: Reject H_0 whenever

$$\frac{p_0^T(1-p_0)^{N-T}}{p_1^T(1-p_1)^{N-T}} \leqslant C_\alpha. \tag{6.14}$$

This is equivalent to

$$N \leqslant \frac{\log C_\alpha + T\log[\,p_1(1-p_0)/p_0(1-p_1)\,]}{\log[\,(1-p_0)/(1-p_1)\,]}. \tag{6.15}$$

Again recall that T (the number of nonsurvivals) is fixed, and hence the random variable N is the waiting time until the Tth nonsurvival and has the negative binomial distribution

$$f(N) = \binom{N-1}{T-1} p^T(1-p)^{N-T}, \quad T \geqslant 1, N = T, T+1, \ldots, . \tag{6.16}$$

We thus seek a pair (T, N_c) such that, if $N > N_c$, H_0 is accepted, that is, the test is passed, and if $N \leqslant N_c$, H_0 is rejected and

$$\alpha = P(N \leqslant N_c | p_0) = \sum_{N=T}^{N_c} \binom{N-1}{T-1} p_0^T(1-p_0)^{N-T},$$

$$\beta = P(N > N_c | p_1) = \sum_{N=N_c+1}^{\infty} \binom{N-1}{T-1} p_1^T(1-p_1)^{N-T}. \tag{6.17}$$

In words, H_0 is rejected if the number of trials to observe T nonsurvivals is too small. Generally, there will not be a pair (T, N_c) that will satisfy (6.17) exactly. In that case, select the pair with smallest T such that the true risks (α', β') give $\alpha' \leqslant \alpha, \beta' \leqslant \beta$. There will be many pairs that give $\alpha' \leqslant \alpha, \beta' \leqslant \beta$, and the selection of the pair with the smallest T is a good choice, since the expected value of N is

$$E_p(N) = \frac{T}{p}$$

and $E_p(N)$ is a minimum at the smallest T for all p.

Tables of the negative binomial distribution are not generally available and for good reason: the cumulative negative binomial distribution is related to the cumulative binomial distribution. Consider the following two events:

1. More than N^* trials are required to obtain exactly T nonsurvivals.
2. Less than T nonsurvivals have occurred in N^* trials.

These two events imply each other, so that

$$P(N > N^*) = \sum_{N=N^*+1}^{\infty} \binom{N-1}{T-1} p^T (1-p)^{N-T} = P(Y < T)$$

$$= \sum_{y=0}^{T-1} \binom{N^*}{y} p^y (1-p)^{N^*-y}.$$

Thus (6.17) becomes

$$\alpha = \sum_{N=T}^{N_c} \binom{N-1}{T-1} p_0^T (1-p_0)^{N-T} = \sum_{y=T}^{N_c} \binom{N_c}{y} p_0^y (1-p_0)^{N_c-y},$$

$$\beta = \sum_{N=N_c+1}^{\infty} \binom{N-1}{T-1} p_1^T (1-p_1)^{N-T} = \sum_{y=0}^{T-1} \binom{N_c}{y} p_1^y (1-P_1)^{N_c-y}.$$

$$(6.18)$$

The right sides of (6.18) may be obtained from binomial tables.

PROBLEMS

6.13. Take $p_0 = .10$, $p_1 = .20$, $\alpha = \beta = .10$, $t = 100$ hours.
a. Find θ_0, θ_1.
b. Find the pair (T, N_c).
c. What are the actual risks α', β'?

6.14. Sketch the operating characteristic curve for Problem 6.13 in terms of p and θ.

6.15. Generally, the negative binomial distribution gives the probability that exactly N observations are required to observe exactly T occurrences of an event of constant probability p. For us, the event is nonsurvival. In any case, show that the mean value of N is

$$E_p(N) = \frac{T}{p}.$$

There are a number of ways to obtain this result; a good one is given in the following.
Hint: Let n_i be the number of observations from the $(i-1)$st to the ith nonsurvival, $i = 1, 2, \ldots, T$. Observe that
(i) $N = \sum n_i$.
(ii) n_i has a geometric distribution for each i.

6.16. Try to develop the negative hypergeometric distribution.

6.2.3 A Sequential Test When Survival/Nonsurvival for Time t Is Observed

Again, the concern is with testing about an exponential mean, that is,

$$H_0 : \theta = \theta_0, \qquad H_1 : \theta = \theta_1, \quad \theta_0 > \theta_1. \tag{6.19}$$

Immediately convert this to

$$H_0 : p = p_0; \qquad H_1 : p = p_1; \quad p_0 < p_1,$$

by using, as before,

$$p_i = 1 - e^{-t/\theta_i}, \quad i = 0, 1.$$

Again assume that the number of nonsurvivals in n observations, $T = \Sigma x_i$, has a binomial distribution.

In the middle 1940's Professor Abraham Wald introduced the idea of testing simple hypotheses in a sequential manner. This idea is discussed in detail in his famous, readable book (Wald, 1947). In sequential testing the idea is that the observations $(X_1, \ldots)$ become available one at a time, and when the nth observation is made, $n = 1, 2, 3, \ldots$, one of three decisions is made.

1. Accept H_0.
2. Reject H_0.
3. Take another sample.

Wald assumed that random samples are taken from a probability distribution (assumed known with respect to family), $f(x; \theta)$. The test statistic that he proposed is

$$\frac{f(x_1, \ldots, x_n; \theta_1)}{f(x_1, \ldots, x_n; \theta_0)} = q_n. \tag{6.20}$$

Since he assumed random sampling in most of his work (although this is not necessary for his method to be applicable), the above statistic can be written as

$$q_n = \frac{\prod\limits_{i=1}^{n} f(x_i; \theta_1)}{\prod\limits_{i=1}^{n} f(x_i; \theta_0)}. \tag{6.21}$$

The numerator and the denominator represent the probability density of the sample $(x_1, \ldots, x_n)$ under the alternative and null hypotheses, respectively. For this reason the test is often called the Wald sequential probability ratio test (SPRT). It is to be noted that θ may be a vector of parameters, provided only that the hypotheses remain simple. One of the remarkable results that Wald obtained is that, under not too restrictive conditions on $f(x; \theta)$,

(i) if H_0 is accepted and the test stopped at the first n for which $q_n \leqslant \beta/(1-\alpha)$,

(ii) if H_0 is rejected and the test stopped at the first n for which $q_n \geqslant (1-\beta)/\alpha$,

or

(iii) if the test is continued as long as

$$\frac{\beta}{1-\alpha} < q_n < \frac{1-\beta}{\alpha}, \quad \alpha + \beta < 1,$$

the desired (α, β) will be obtained to an approximation satisfactory for most applications. This result is true even if the X_i are not independent, provided that the test terminates with probability 1. Since n (the sample size at which a decision to stop is made) is a random variable, a new dimension is added to the analysis. In most applications it is sufficient to look at the mean value of $n, E(n)$; since it is generally a function of θ, there will be, in addition to the usual operating characteristic (OC) curve, an average sample number (ASN) curve. The ASN curve is a graph of the pairs of points $[E_\theta(n), \theta]$. The advantage of sequential testing is very important in reliability work, where the usual test environment involves a tight budget and limited time.

Generally, for $\theta \geqslant \theta_0$ or $\leqslant \theta_1$, $E_\theta(n)$ is less than the corresponding value for the fixed-sample-size test with the same set $(\theta_0, \alpha, \theta_1, \beta)$. This advantage is extremely important because the cost of obtaining observations is often very high in reliability testing. The SPRT is somewhat more difficult to apply, however, as far as administrative problems are concerned. Actually, reliability applications have given "new" impetus to the SPRT, since very often the observations (say, failure times) occur one at a time and then the sequential test is the natural way to proceed.

Wald showed that under general conditions the sequential test would terminate with probability 1. Now no one (in most applications) could be expected to use a sequential test that does not eventually terminate; and although Wald showed that it does, apparently this has not been good enough in practice, that is, virtually all sequential tests in reliability are truncated. By *truncation*, one means that a value of n, say n_t, is preselected so that, if a decision has not been reached at or before the n_tth observation, the test is stopped at the n_tth observation and an accept or reject decision is made. It is worth noting that generally, irrespective of what the accept/

reject rule is at the n_tth observation, when n_t is relatively small (early truncation) the (α,β) risks may be seriously altered: α and β may both increase. A detailed discussion of truncation is beyond the scope of this book, and the reader is referred to the article by Epstein (1954).

We now return to testing (6.19). For the Bernoulli process,

$$f(x_1,\ldots,x_n;p)= \prod_{i=1}^{n} f(x_i;p)=p^{T}(1-p)^{n-T},$$

where, as usual, $T=\sum x_i$ and p is the probability of nonsurvival. Then, according to the Wald procedure, the "continue test" region is given by

$$\frac{\beta}{1-\alpha}<\frac{p_1^{T}(1-p_1)^{n-T}}{p_0^{T}(1-p_0)^{n-T}}<\frac{1-\beta}{\alpha}. \tag{6.22}$$

Taking logarithms of both sides and making some other simplifications leads to the equivalent "continue test" region,

$$\frac{\log[\beta/(1-\alpha)]}{\log(p_1/p_0)-\log[(1-p_1)/(1-p_0)]}$$

$$+n\frac{\log[(1-p_0)/(1-p_1)]}{\log(p_1/p_0)-\log[(1-p_1)/(1-p_0)]}<T$$

$$=\sum x_i<\frac{\log[(1-\beta)/\alpha]}{\log(p_1/p_0)-\log[(1-p_1)/(1-p_0)]}$$

$$+n\frac{\log[(1-p_0)/(1-p_1)]}{\log(p_1/p_0)-\log[(1-p_1)/(1-p_0)]}. \tag{6.23}$$

Note that, if n is regarded as a real number (rather than the positive integer it is), the boundary for the "continue test" region is two straight lines in T and n with common slope

$$s=\frac{\log[(1-p_0)/(1-p_1)]}{\log(p_1/p_0)-\log[(1-p_1)/(1-p_0)]}.$$

The T intercept for the accept line is

$$h_0=\frac{\log[\beta/(1-\alpha)]}{\log(p_1/p_0)-\log[(1-p_1)/(1-p_0)]},$$

and the T intercept for the reject line is

$$h_1 = \frac{\log[(1-\beta)/\alpha]}{\log(p_1/p_0) - \log[(1-p_1)/(1-p_0)]},$$

where h_0, h_1, s are the commonly used notations.

Thus, for each integer n, the number of nonsurvivals that cause acceptance is

$$a_n = h_0 + sn, \tag{6.24}$$

and the number of nonsurvivals that cause rejection at n is

$$r_n = h_1 + sn. \tag{6.25}$$

If a_n of (6.24) is not an integer, it must be replaced with $[a_n] \equiv$ the largest integer $< a_n$. Similarly, if r_n of (6.25) is not an integer, it must be replaced with $[r_n + 1]$.

The OC and ASN curves are difficult, if not impossible, to compute directly, but thanks to Wald's ingenuity good approximations are available. We deal first with the OC curve. Denote the points on this curve by the pair $[p, P_p(A)]$, where $P_p(A)$ is the probability of acceptance of H_0, given p. Wald showed that this pair could be defined parametrically in terms of the real variable h as follows:

$$P_p(A) = \frac{[(1-\beta)/\alpha]^h - 1}{[(1-\beta)/\alpha]^h - [\beta/(1-\alpha)]^h}$$

and

$$p = \frac{1 - [(1-p_1)/(1-p_0)]^h}{(p_1/p_0)^h - [(1-p_1)/(1-p_0)]^h}. \tag{6.26}$$

For any value of h, (6.26) may be solved for the pair $[p, P_p(A)]$. In fact, for $h = 1$, we note that (6.26) gives

$$p = p_0 \quad \text{and} \quad P_{p_0}(A) = 1 - \alpha.$$

Other points are as follows:

$$h = +\infty: \quad p = 0, \quad P_0(A) = 1,$$

$$h = -\infty: \quad p = 1, \quad P_1(A) = 0,$$

$$h = -1: \quad p = p_1, \quad P_{p_1}(A) = \beta.$$

Fortunately, a fifth point is also available; at $h=0$, (6.26) leads to

$$p=s \quad \text{and} \quad P_s(A) = \frac{h_1}{h_1 + |h_0|}.$$

Of course, exponential means are being tested; hence the five p's: $0, p_0, s, p_1, 1$, must be converted to θ's, which are, respectively, $\theta = \infty, \theta_0, \theta_s, \theta_1, \theta = 0$.

Additional points on the OC curve may be found by choosing h judiciously. For example, if $P_\theta(A)$ is desired for a θ between θ_s and θ_1, h should be chosen between $h=0$ and $h=-1$. Similarly, if a value of $P_\theta(A)$ is desired for $\theta_s < \theta < \theta_0$, h should be chosen between $h=0$ and $h=1$.

The ASN curve is a graph of the pair $[p, E_p(n)]$ or, in terms of θ, $[\theta, E_\theta(n)]$. Wald has shown that the expected value of n (which depends on p) is given by

$$E_p(n) = \frac{P_p(A)\log[\beta/(1-\alpha)] + [1 - P_p(A)]\log[(1-\beta)/\alpha]}{p\log(p_1/p_0) + (1-p)\log[(1-p_1)/(1-p_0)]}, \quad (6.27)$$

approximately. The approximation is due to neglecting the excess (or deficit) of $T = \Sigma x_i$ over the boundary when termination occurs. Since

$$s = \frac{\log[(1-p_0)/(1-p_1)]}{\log(p_1/p_0) - \log[(1-p_1)/(1-p_0)]},$$

it is clear that for $p=s$ (6.27) is indeterminate, and in that case the following equation should be used:

$$E_s(n) = -\frac{\{\log[\beta/(1-\alpha)]\}\{\log[(1-\beta)/\alpha]\}}{[\log(p_1/p_0)]\{\log[(1-p_0)/(1-p_1)]\}}.$$

For other values of p (6.27) can be used. In particular, for $p=0$,

$$E_0(n) = \frac{\log[\beta/(1-\alpha)]}{\log[(1-p_1)/(1-p_0)]}$$

and, for $p=1$,

$$E_1(n) = \frac{\log[(1-\beta)/\alpha]}{\log(p_1/p_0)}.$$

Example. For comparative purposes, we choose the example of the preceding section, namely, $\theta_0 = 1000$, $\theta_1 = 250$, $\alpha = .05$, $\beta = .10$, and $t = 50$. Then, solving $p_i = 1 - e^{-t/\theta_i}, i = 0, 1$, we obtain, as before, $p_0 = .049, p_1 = .181$. Using the preceding equations, one obtains

$$h_0 = -1.546, \qquad h_1 = 1.985, \qquad s = .103,$$

so that

$$a_n = -1.546 + .103n, \qquad r_n = 1.985 + .103n.$$

The accompanying table gives the results for $n = 1(1)20$.

n	a_n	r_n
1	—	—
2	—	—
3	—	3
4	—	3
5	—	3
6	—	3
7	—	3
8	—	3
9	—	3
10	—	4
11	—	4
12	—	4
13	—	4
14	—	4
15	—	4
16	0	4
17	0	4
18	0	4
19	0	4
20	0	5

A dash in the accept (a_n) column means that no accept decision can be made at that n. A similar statement holds for the reject column.
Points are already available on the OC curve at $p = 0, p_0, p_1, p = 1$, and they are, of course, $1, 1 - \alpha, \beta, 0$. A fifth point at $p = s = .103$ is

$$P_s(A) = \frac{h_1}{h_1 + |h_0|} = .562,$$

and this corresponds to the solution θ to $.103 = 1 - e^{-50/\theta}$, which is approximately $\theta = 454$. In the fixed-sample-size test of this example it was found in the preceding section that $n = 51$. It will be interesting to compare this with $E_{p_0}(n)$ and $E_{p_1}(n)$. Using (6.27), we find

$$E_{p_0}(n) = 25.57, \qquad E_{p_1}(n) = 20.82;$$

$E_{p_0}(n)$ is larger than $E_{p_1}(n)$ because $\alpha < \beta$.

PROBLEMS

6.17. For the example in the text, compute the point on the OC and ASN curves at $p = .08$. Find the corresponding θ. Sketch the OC and ASN curves in θ.

6.18. Design a sequential test for $\theta_0 = 500, \theta_1 = 100, t = 25, \alpha = .01, \beta = .01$. Sketch the OC and ASN curves (in θ).

6.19. Show that, if $\alpha = \beta$,

$$P_s(A) = \tfrac{1}{2}.$$

6.20. Show that, when $p = s$,

$$\lim_{h \to 0} P_s(A) = \frac{h_1}{h_1 + |h_0|}.$$

6.21. Design a SPRT for Pascal sampling (see Section 6.2.2). In other words, for each nonsurvival $T = 1, 2, 3, \ldots,$ there will be a pair (N_r, N_a) such that, if N_T is the number of observations required to observe T nonsurvivals, the "continue test" region is $N_r < N_T < N_a$.

6.2.4 Fixed-Sample-Size and Fixed-Time Tests Based on Observed Lifetimes

In the preceding sections the only information available was that a unit did or did not survive a fixed time period in order to test a hypothesis about the mean of an exponential distribution. Now assume that n identical (i.e., identically distributed) units are available for test and that actual lifetimes can be observed. Several possibilities then come to mind for terminating testing.

1. Stop testing after r_0 failures occur. This is a fixed-sample-size test; if $r_0 < n$, it is also called a censored test because only r_0 of the n possible lifetimes have been observed.

2. Stop testing after a fixed time t_0 has elapsed. This is called a truncated (in time) test.

In Chapter 5, possibilities 1 and 2 were called Type II and Type I censoring, respectively. In addition, there are the following options:

a. Not replace units that failed during the test. This is called the nonreplacement case.

b. Replace units that failed during the test. This is called the replacement case.

Notation will be changed to conform with common usage. The number of failures is now called r instead of T, as in preceding sections, and later in this section T will be used for total unit test time. Thus there are four important possibilities, and although there is considerable overlap among the resulting methods, the cases will be treated separately. Some points are worth noting. In a fixed-sample-size test (Type II censoring) the number of failures to be observed, r_0, is fixed, and the total test time (i.e., the time required to observe r_0 failures) is a random variable. On the other hand, in the Type I censored test the test time is fixed, and the number of failures occurring is a random variable.

Selection of the type of test is largely an engineering decision based on costs, time available for test, and other considerations. It is worth noting, however, when comparing the variables (actual lifetimes observed) tests with the attributes (unit did or did not survive t hours) tests of the preceding sections, that the variables tests of this section are more efficient. In other words, for a fixed set $(\theta_0, \alpha, \theta_1, \beta)$ the total test time required is larger for the attributes tests than for the variables tests. This is true generally in statistical hypothesis testing as long as the variables-attributes comparison is made in terms of the best test for each. Much of the development that follows is due to Epstein (1953, 1960) and Epstein and Sobel (1955), who contributed a series of papers of fundamental importance.

(a) Type II Censoring—the Nonreplacement Case. In what follows we will, for notational convenience, take the set $(\theta_0, \alpha, \theta_1, \beta)$ to mean that the null hypothesis is $\theta = \theta_0$ with Type I error probability (producer's risk) α and that the alternative hypothesis is $\theta = \theta_1$ with Type II error probability (consumer's risk) β. It is assumed that n units are put on test, all with the same exponential distribution with unknown parameter θ. If testing is stopped after $r \leqslant n$ failures are observed, the failure times occur naturally in the order:

$$X_{1,n} < X_{2,n} < \ldots < X_{r,n}$$

where $X_{i,n}$ is the ith failure time out of n units and is generally called the ith order statistic. If testing is stopped after exactly r failures, no failure times are available on the $n-r$ nonfailing units. In any event, the joint density of the sample $(X_{1,n}, \ldots, X_{r,n})$ is

$$f(x_{1,n}, \ldots, x_{r,n}; \theta) = \frac{n!}{(n-r)!} \left[\prod_{i=1}^{r} \left(\frac{1}{\theta} \right) \exp\left(\frac{-x_{i,n}}{\theta} \right) \right] [\exp(-x_{r,n}/\theta)]^{(n-r)}$$

$$= \frac{n!}{(n-r)!}\left(\frac{1}{\theta}\right)^{r}\exp\left\{-\frac{\left[\sum_{i=1}^{r}x_{i,n}+(n-r)x_{r,n}\right]}{\theta}\right\}. \quad (6.28)$$

Briefly, the $n!/(n-r)!$ term arises because that is the number of permutations of the r out of n failure times; the product term is the joint distribution of a random sample of size r from an exponential distribution; the last term accounts for the fact that $n-r$ units lived $x_{r,n}$ or longer. The likelihood ratio is then

$$\frac{f(x_{1,n},\ldots,x_{r,n};\theta_0)}{f(x_{1,n},\ldots,x_{r,n};\theta_1)}=\left(\frac{\theta_1}{\theta_0}\right)^{r}\exp\left[-T_{r,n}\left(\frac{1}{\theta_0}-\frac{1}{\theta_1}\right)\right],\quad \theta_0>\theta_1, \quad (6.29)$$

where $T_{r,n}=\Sigma_{i=1}^{r}X_{i,n}+(n-r)X_{r,n}=$ the total number of hours lived by all units on test.

Also $T_{r,n}/r$ is the maximum-likelihood estimate, $\hat{\theta}_{r,n}$, of θ, as might have been expected (see Chapter 5). Thus, when

$$\left(\frac{\theta_1}{\theta_0}\right)^{r}\exp\left[-T_{r,n}\left(\frac{1}{\theta_0}-\frac{1}{\theta_1}\right)\right]\leqslant C_\alpha,$$

H_0 is rejected. This is equivalent to

$$T_{r,n}\leqslant\frac{\log\left[C_\alpha(\theta_0/\theta_1)^{r}\right]}{1/\theta_1-1/\theta_0}, \quad (6.30)$$

but it has already been shown that $2T_{r,n}/\theta$ has a chi-square distribution with $2r$ degrees of freedom (see Chapter 5), and also (6.30) is equivalent to

$$\frac{2T_{r,n}}{\theta}\leqslant\frac{2\log\left[C_\alpha(\theta_0/\theta_1)^{r}\right]}{\theta(1/\theta_1-1/\theta_0)}. \quad (6.31)$$

It is clear, then, that we need a pair (r_0,T_c) such that

$$P\left(T_{r_0,n}\leqslant T_c|\theta=\theta_0\right)=\alpha \quad (6.32)$$

and

$$P\left(T_{r_0,n}\leqslant T_c|\theta=\theta_1\right)=1-\beta, \quad (6.33)$$

and H_0 is rejected whenever $T_{r_0,n} \leqslant T_c$.

Since $2T_{r_0,n}/\theta$ is $\chi^2_{2r_0}$, it is easily seen that Equation (6.32) means

$$P\left(\frac{2T_{r_0,n}}{\theta_0} \leqslant \frac{2T_c}{\theta_0}\right) = P\left(\frac{2T_{r_0,n}}{\theta_0} \leqslant \chi^2_{2r_0,\alpha}\right) = \alpha, \qquad (6.34)$$

where $\chi^2_{\cdot,\alpha}$ is the αth quantile of the chi-square distribution. From Equation (6.34) it is obvious that $2T_c/\theta_0 = \chi^2_{2r_0,\alpha}$, so that

$$T_c = \frac{\theta_0 \chi^2_{2r_0,\alpha}}{2}. \qquad (6.35)$$

We need r_0 before we can find T_c numerically. Now, (6.33) means that

$$P\left(\frac{2T_{r_0,n}}{\theta_1} \leqslant \frac{2T_c}{\theta_1}\middle|\theta = \theta_1\right) = P\left(\frac{2T_{r_0,n}}{\theta_1} \leqslant \chi^2_{2r_0,1-\beta}\middle|\theta = \theta_1\right) = 1 - \beta,$$

and obviously $2T_c/\theta_1 = \chi^2_{2r_0,1-\beta}$, which means that, in addition to (6.35),

$$T_c = \frac{\theta_1 \chi^2_{2r_0,1-\beta}}{2}. \qquad (6.36)$$

Equations (6.35) and (6.36) imply that

$$\frac{\chi^2_{2r_0,\alpha}}{\chi^2_{2r_0,1-\beta}} = \frac{\theta_1}{\theta_0}. \qquad (6.37)$$

The ratio, $(\theta_1/\theta_0)^{-1} = \theta_0/\theta_1$ is called the discrimination ratio, and its magnitude affects the size of the required r_0 (see Problem 6.25). We can proceed to χ^2 tables, dividing the entry in the α column by the entry in the $(1-\beta)$ column until their ratio is θ_1/θ_0. However, it will generally be found that the ratio cannot be obtained exactly; moreover,

$$\lim_{r_0 \to \infty}\left(\frac{\chi^2_{2r_0,\alpha}}{\chi^2_{2r_0,1-\beta}}\right) = 1$$

so that there are many r_0's for which $(\chi^2_{2r_0,\alpha}/\chi^2_{2r_0,1-\beta}) \geqslant \theta_1/\theta_0$. We will select the smallest r_0 that gives α exactly [i.e., T_c will be found from Equation (6.35)] and a true $\beta' \leqslant \beta$. The smallest r_0 is a good choice because the expected value of the length of the test, that is, the expected value of $X_{r_0,n}$, is

$$E(X_{r_0,n}) = \theta \sum_{j=1}^{r_0} \frac{1}{n-j+1}, \qquad (6.38)$$

and the smaller r_0, the smaller will be the expected length of test. Table 5.1 can be used to assist in computing (6.38).

Points on the OC curve are easily obtained by computing for the point $\theta = \theta'$, say:

$$P_{\theta'}(A) = 1 - P\left(T_{r_0,n} \leqslant T_c \mid \theta = \theta'\right) = 1 - P\left(\frac{2T_{r_0,n}}{\theta'} \leqslant \frac{2T_c}{\theta'}\right), \qquad (6.39)$$

and noting that, if $\theta = \theta', 2T_{r_0,n}/\theta'$ is $\chi^2_{2r_0}$ distributed and that T_c is given by $\theta_0 \chi^2_{2r_0,\alpha}/2$.

Before proceeding to an example, two remarks are in order. First, it is very important to note that the magnitude of n had nothing whatever to do with the test criterion: the sampling distribution of the test statistic, $T_{r,n}$, depended only on θ and r. Hence the "statistical" sample size is r_0, the number of failure times observed, not n, the number of units on test. The choice of n must then be made on economic grounds—in particular, on the expected length of time to observe r_0 failure times. It has already been pointed out that this is given by

$$E(X_{r_0,n}) = \theta \sum_{j=1}^{r_0} \frac{1}{n-j+1}. \qquad (6.40)$$

The development of Equation (6.40) is given in Problem 6.27. The right side is a decreasing function of increasing n, so that the larger n is chosen [r_0 already fixed by Equation (6.37)] the smaller the expected test time.

The second important remark is that a glance at Equation (6.37) shows that all tests with the same pair (α, β) and same *ratio* θ_1/θ_0 have the same r_0. Thus, when designing test tables, as was done in MIL-STD 781B (1967) and H-108 (1960), it is necessary only to index the tables on $(\alpha, \beta, \theta_1/\theta_0)$ instead of $(\theta_0, \alpha, \theta_1, \beta)$. This saves a great deal of time and space.

Example. Suppose that, as in the example of Section 6.2.1, $\theta_0 = 1000, \alpha = .05, \theta_1 = 250, \beta = .10$. Then $\theta_1/\theta_0 = .25$. Proceeding to the χ^2 table, Table I in the Appendix of this book, one obtains

$$\frac{\chi^2_{11,.05}}{\chi^2_{11,.90}} = \frac{4.575}{17.275} = .265 > .25$$

and

$$\frac{\chi^2_{10,.05}}{\chi^2_{10,.90}} = \frac{3.94}{15.987} = .246 < .25.$$

Thus $2r = 10$ is too small, and since $2r = 11$ results in r not an integer, take $2r = 12$, which means $r_0 = 6$. We can still keep α exactly satisfied merely by using Equation (6.35) to find T_c instead of (6.36). Thus

$$T_c = \frac{\theta_0 \chi^2_{2r_0,\alpha}}{2} = \frac{1000(5.226)}{2} = 2613.0.$$

This selection for T_c guarantees α exactly because

$$P\left(\frac{2T_{r_0,n}}{\theta_0} \leqslant \frac{2T_c}{\theta_0} = \chi^2_{2r_0,\alpha}\right) = \alpha.$$

We now calculate the Type II error probability, β'. From Equation (6.39)

$$\beta' = 1 - P\left[\frac{2T_{r_0,n}}{\theta_1} \leqslant \frac{2T_c}{\theta_1} = \left(\frac{\theta_0}{\theta_1}\right)\chi^2_{2r_0,\alpha} = 20.904\right].$$

Proceeding to the χ^2 table in the Appendix, with $2r_0 = 12$ degrees of freedom, one finds

$$P\left(\frac{2T_{r_0,n}}{\theta_1} \leqslant 20.904\right) \cong .94$$

so that

$$\beta' \cong 1 - .94 = .06 < .10.$$

This calculation also suffices to show how points on the OC curve are obtained.

We now illustrate the effect of the magnitude of n on the expected test time. If $n = r_0 = 6$,

$$E(X_{6,6}) = \theta \sum_{j=1}^{6} \frac{1}{6-j+1} = 2.45\theta,$$

but, if $n = 10, E(X_{6,10}) = .846\theta$. Thus, on the hypothesis that $\theta = \theta_0 = 1000$ hours, the expected test time is reduced from 2450 to 846 hours. This reduction in time may well be worth the additional part and setup costs necessary to go from $n = 6$ to $n = 10$. In Epstein (1960) a table is given to assist in computing $E(X_{r,n})$ for various values of n and r. Also, Table 5.1 gives $E(X_{r,n})/E(X_{r,r})$ for selected r, n.

PROBLEMS

6.22. For the text example, compute the points on the OC curve at $\theta = 373.2$ and 578.5 hours, and graph the OC curve.

6.23 For the text example use Equation (6.36) to determine T_c, and show that the value $\beta = .10$ is obtained exactly. Find the true $\alpha = \alpha'$.

6.24. For $\theta_0 = 100, \theta_1 = 50, \alpha = \beta = .01$,

(a) find T_c and r_0;

(b) sketch the OC curve for at least five points;

(c) compute the expected test time at θ_0, θ_1 for $n = r$.

6.25. Show that, for (α, β) fixed as $\theta_0/\theta_1 \to 1, r$, the required number of failures, approaches infinity. Also show that, for fixed θ_0/θ_1 as α and/or $\beta \to 0$, r approaches infinity.

6.26. Discuss the economics of the example of Section 6.2.1 ($n = 51$) versus the example of this section, and try to give arguments as to when one might be preferred over the other.

6.27. Show that

$$E(X_{r,n}) = \theta \sum_{j=1}^{r} \frac{1}{n-j+1}.$$

Hint: (i) $X_{r,n} = X_{1,n} + \cdots + (X_{j,n} - X_{j-1,n}) + \cdots + (X_{r,n} - X_{r-1,n})$.
(ii) Clearly,

$$E(X_{j,n} - X_{j-1,n}) = \frac{\theta}{n-j+1}, \quad j = 1, 2, \ldots, r; X_{0,n} \equiv 0.$$

(iii) The expectation of a sum is the sum of the expectations.

There have been quite a few calculations and formulae thus far in this chapter. It is easy to lose sight of the basic ideas of hypothesis testing and to do the arithmetic by rote. For this reason, we will review here, in words, the underlying ideas of the hypothesis test and the steps involved. We start with the premise that $(\theta_0, \alpha, \theta_1, \beta)$ have been selected.

1. It is desired to determine whether $\theta = \theta_0$ is a reasonable possibility.

2. A statistic T is selected which is a function of the (random) sample values and *whose distribution depends on the unknown parameter*, θ. The Neyman-Pearson lemma gives us this statistic; in Section 6.2.4 (a) it was $T_{r,n}$.

The two important criteria for selection of the statistic are that its distribution (sometimes called a sampling distribution) is known or tabulated and that the statistic results in a good OC curve.

3. We agree then to the following decision rule: Look at the observed value of the statistic T *under the hypothesis that* $\theta = \theta_0$. This is possible,

since it has been required that the distribution of T depend on θ. If the observed value of T is "too" small (large) to have arisen from the sampling distribution (with $\theta = \theta_0$), it is concluded that $\theta \neq \theta_0$. This is the classical statistical approach: inductive reasoning.

4. In order to quantify what is "too" small above, α is selected, usually small, and some critical value of the statistic, say T_c, such that

$$P(T \leqslant T_c | \theta = \theta_0) = \alpha.$$

5. We now find that what has been done so far, although good, does not determine the sample size uniquely. In particular, if the data already exist and hence the sample size is determined, T_c can be determined. If not, then T_c cannot be determined [see Equation (6.35)]. In other words, generally, the criterion for what values of the test statistic are too small (large) depends on the sample size. Consider the example of tossing a coin and the hypothesis that it is a fair (probability of head $= \frac{1}{2}$) coin. If 10 tosses yield 6 heads, that is, 60% heads, this result does not cast doubt on the hypothesis. On the other hand, if 100,000 tosses are made and 60,000 heads are obtained; that is, 60% heads, surely the coin is not a fair one.

6. To determine sample size uniquely, specify a value θ_1 such that, if $\theta = \theta_1 \neq \theta_0$, the hypothesis $\theta = \theta_0$ is to be *rejected* with high probability $(1 - \beta)$.

7. The sample size and the accept/reject criterion, say T_c, are then determined.

(b) Type II Censoring—the Replacement Case. Now consider the case in which n units are put on test, the units have identical exponential failure distributions, failed units are replaced, and testing is to be stopped after r failures, (r, n) to be determined. It is clear that, since failed items are replaced, the time between the $(i-1)$st and ith failures, say $(X_{i,n} - X_{i-1,n})$, has an exponential distribution with mean θ / n. These times are also independent, so that

$$f(x_{1,n}, x_{2,n} - x_{1,n}, \ldots, x_{r,n} - x_{r-1,n}; \theta)$$

$$= \prod_{i=1}^{r} \left\{ \left(\frac{n}{\theta} \right) \exp\left[\frac{-n(x_{i,n} - x_{i-1,n})}{\theta} \right] \right\},$$

where $X_{0,n} \equiv 0$. Since $X_{r,n} = X_{1,n} + (X_{2,n} - X_{1,n}) + \cdots + (X_{r,n} - X_{r-1,n})$,

$$f(x_{1,n}, x_{2,n} - x_{1,n}, \ldots, x_{r,n} - x_{r-1,n}; \theta) = \left(\frac{n}{\theta} \right)^r \exp\left(\frac{-nx_{r,n}}{\theta} \right).$$

The change of variable,

$$Y_{1,n} = X_{1,n},$$

$$Y_{2,n} = X_{1,n} + (X_{2,n} - X_{1,n}),$$

$$\vdots$$

$$Y_{r,n} = X_{1,n} + (X_{2,n} - X_{1,n}) + \cdots + (X_{r,n} - X_{r-1,n}),$$

means

$$Y_{1,n} = X_{1,n}, \; Y_{2,n} = X_{2,n}, \ldots, Y_{r,n} = X_{r,n},$$

so that the joint density of $(X_{1,n}, \ldots, X_{r,n})$ is

$$f(x_{1,n}, \ldots, x_{r,n}; \theta) = \left(\frac{n}{\theta}\right)^r \exp\left(\frac{-nx_{r,n}}{\theta}\right). \tag{6.41}$$

The best test statistic is easily seen to be

$$T_{r,n} = nX_{r,n}, \tag{6.42}$$

and $T_{r,n}/r$ is the maximum-likelihood estimate of θ.

The probability distribution of $2T_{r,n}/\theta$ is, as in the nonreplacement case, χ^2_{2r}. Since

$$X_{r,n} = X_{1,n} + (X_{2,n} - X_{1,n}) + \cdots + (X_{r,n} - X_{r-1,n}),$$

then

$$E(X_{r,n}) = E(X_{1,n}) + E(X_{2,n} - X_{1,n}) + E(X_{r,n} - X_{r-1,n})$$

$$= \frac{\theta}{n} + \frac{\theta}{n} + \cdots + \frac{\theta}{n}$$

$$= \frac{r\theta}{n}, \tag{6.43}$$

and this is the expected test time.

We again choose $T_c = \theta_0 \chi^2_{2r,\alpha}/2$ as in Equation (6.35), and the smallest r is chosen, as before, so that

$$\frac{\chi^2_{2r,\alpha}}{\chi^2_{2r,1-\beta}} \geqslant \frac{\theta_1}{\theta_0}. \tag{6.44}$$

The smallest $r = r_0$ of course minimizes the expected test time for fixed n.

The larger n, the smaller is the expected test time, and again n is to be selected on economic grounds, since only r_0 is determined by $(\theta_0, \alpha, \theta_1, \beta)$. Although only n units are on test at any one time, a total of $n + r_0 - 1$ units will be required: n original units and $r_0 - 1$ replacement items; the last failed item need not be replaced.

PROBLEMS

6.28. Suppose that $n = 10, \theta_0 = 1200, \alpha = .10, \theta_1 = 120, \beta = .10$.
a. Show that $r_0 = 2$.
b. Find T_c.
c. Suppose that $x_{1,10} = 10.8$ hours, $x_{2,10} = 34.9$ hours. Is the test passed?

6.29. Discuss the economics of a nonreplacement versus a replacement test. Choose an example if you like.

In the preceding development n units were placed on test, each with the same exponential distribution, and it was agreed to stop testing after r_0 failures and to make a decision using two statistics:

$$T_{r_0,n} = \sum_{i=1}^{r_0} X_{i,n} + (n - r_0)X_{r_0,n}(\text{the nonreplacement case}),$$

$$T_{r_0,n} = nX_{r_0,n} = (\text{the replacement case}).$$

The value r_0 and the value T_c such that, if

$$T_{r_0,n} \leqslant T_c, \text{ reject } \theta = \theta_0,$$

were selected to satisfy $(\theta_0, \alpha, \theta_1, \beta)$. The length of such a test is then a random variable. It may be more convenient, for planning purposes or other reasons, to specify the length of test, say t_0, in advance. These types of tests, called Type I censored tests, are considered next.

(c) Type I Censoring—the Nonreplacement Case. In a Type I censored test the test length is specified to be some fixed number t_0. It remains to determine the values of r_0 and n to satisfy $(\theta_0, \alpha, \theta_1, \beta)$. It turns out that we can determine r_0 from (6.37) and the procedure following that equation, that is r_0 is the smallest r for which

$$\frac{\chi^2_{2r,\alpha}}{\chi^2_{2r,1-\beta}} \geqslant \frac{\theta_1}{\theta_0} \tag{6.45}$$

where again it is customary to call $(\theta_1/\theta_0)^{-1}$ the discrimination ratio. Before turning to the determination of n, the OC curve, and the expected test time, we give the accept/reject procedure.

Again, n units are placed on test (failed units not replaced). If the time of the r_0th failure $X_{r_0,n}$ is such that

$$X_{r_0,n} \geqslant t_0, \text{ the test is passed } (\theta = \theta_0 \text{ accepted});$$

$$X_{r_0,n} < t_0, \text{ the test is failed } (\theta = \theta_0 \text{ rejected}).$$

In other words, if $0 \leqslant r < r_0$ failures occur in t_0, the test is passed. If r_0 failures occur in less than t_0, the test is failed. This last remark indicates how to proceed to determine n, the OC curve, and the expected test time. Since failed items are not replaced, the survival or nonsurvival for t_0 of any particular unit is a Bernoulli trial with probability of failure (nonsurvival) $p = 1 - e^{-t_0/\theta}$, and the dependence of p on the unknown θ should be noted. Thus the number of failures r, has the following mass function,

$$f(r;\theta) = \binom{n}{r}(1 - e^{-t_0/\theta})^r (e^{-t_0/\theta})^{n-r}, \quad r = 0, 1, \ldots, r_0 - 1,$$

$$f(r;\theta) = 1 - \sum_{r=0}^{r_0-1} \binom{n}{r}(1 - e^{-t_0/\theta})^r (e^{-t_0/\theta})^{n-r}, \quad r = r_0,$$

and of course

$$f(r;\theta) = 0, \quad r = r_0 + 1, \ldots, n,$$

since testing is always stopped at or before the r_0th failure. Clearly, then, it is required that

$$P(r < r_0|\theta_0) = 1 - \alpha = \sum_{r=0}^{r_0-1} \binom{n}{r}(1 - e^{-t_0/\theta_0})^r (e^{-t_0/\theta_0})^{n-r},$$

$$P(r < r_0|\theta_1) = \beta = \sum_{r=0}^{r_0-1} \binom{n}{r}(1 - e^{-t_0/\theta_1})^r (e^{-t_0/\theta_1})^{n-r}. \quad (6.46)$$

Generally we will not be able to satisfy either or both (α, β) exactly, therefore we will determine the smallest r for which the true risks (α', β') are such that $\alpha' \leqslant \alpha, \beta' \leqslant \beta$. We can determine r_0 from (6.45) and then, using binomial tables, determine n. For that matter, one could ignore (6.45) and determine both n and r_0 from trial and error in the binomial tables. At the very least, this exercise would make us appreciate (6.45).

Clearly, points on the OC curve $[P_{\theta'}(A), \theta']$ are given by

$$P_{\theta'}(A) = \sum_{r=0}^{r_0-1} \binom{n}{r} \left(1 - e^{-t_0/\theta'}\right)^r \left(e^{-t_0/\theta'}\right)^{n-r}. \tag{6.47}$$

The expected test time is given by

$$E_\theta(t) = \sum_{r=1}^{r_0} \theta f(r;\theta) \left(\frac{1}{n} + \frac{1}{n-1} + \cdots + \frac{1}{n-r+1}\right) \tag{6.48}$$

where $f(r;\theta)$ is given immediately above Equation (6.46).

If time and/or binomial tables are not available, n is approximately given by

$$n = \left[\frac{r_0}{1 - e^{-t_0/C_0}}\right], \tag{6.49}$$

where $C_0 = \theta_0 \chi^2_{2r_0, \alpha}/2r_0$, $[u]$ is the greatest integer less than or equal to u, and the approximation is quite good for θ_0/t_0 large, for example, for $\theta_0/t_0 \geqslant 3$. In view of the equivalence of Equation (6.11) of Section 6.2.1 and Equation (6.46) of this section, it is easy to see that the pair (n, r_0) selected by the methods of this section will be the same as the $(n, c = r_0 - 1)$ of Section 6.2.1. In fact, the only reasons for giving the development here are the following:

1 In application the tests are slightly different. The test here has n items on test at the start. In Section 6.2.1 the n trials may occur sequentially.

2 It was more convenient to introduce Equations (6.45) and (6.49) here. These provide a good short cut to the tedious job of trial and error in the binomial tables (or Poisson tables if that approximation is valid).

The notation here is somewhat different from that of Section 6.2.1 to conform with the notation used in the literature; the notation of Section 6.2.1 comes more from quality control than from reliability.

Example. We choose again the example of Section 6.2.1 (i.e., $\theta_0 = 1000$

hours, $\theta_1 = 250$ hours, $\alpha = .05$, $\beta = .10$, and $t_0 = 50$). Then, applying (6.45), one finds $2r_0 = 12$ and $r_0 = 6$. Also note, using Equation (6.49),

$$n = \left[\frac{6}{1 - e^{-50/435.55}} \right]$$

$$= \left[\frac{6}{.11} \right] = [54.54] = 54 \text{ approximately.}$$

Thus $r_0 - 1 = 5 = c$ of the Section 6.2.1 example, and $n = 54$ is about the same as the $n = 51$ of Section 6.2.1.

(d) Type I Censoring—the Replacement Case. We now modify the preceding section by assuming that failed items are replaced. The accept/reject criterion is still:

$$X_{r_0,n} \geqslant t_0, \text{ the test is passed,}$$

$$X_{r_0,n} < t_0, \text{ the test is failed.}$$

Since there are always n items on test, each with identical exponential probability distributions for lifetime, the probability of exactly r failures in time t_0 is given by the following mass function:

$$f(r;\theta) = \frac{e^{-nt_0/\theta}(nt_0/\theta)^r}{r!}, \quad r = 0, 1, \ldots, r_0 - 1,$$

$$f(r;\theta) = 1 - \sum_{r=0}^{r_0-1} \frac{e^{-nt_0/\theta}(nt_0/\theta)^r}{r!}, \quad r = r_0. \tag{6.50}$$

Again, the appropriate value of r_0 may be determined by choosing the smallest r such that

$$\frac{\chi^2_{2r,\alpha}}{\chi^2_{2r,1-\beta}} \geqslant \frac{\theta_1}{\theta_0}. \tag{6.51}$$

From Section 6.2.4(b) on censored tests with replacement, recall that the accept/reject criterion was

$$T_{r_0,n} = nX_{r_0,n} \leqslant T_c = \frac{\theta_0 \chi^2_{2r_0,\alpha}}{2}.$$

Since termination occurs at $X_{r_0, n}$ or t_0, whichever is smaller, it must be that

$$t_0 = \frac{\theta_0 \chi^2_{2r_0, \alpha}}{2n}.$$ (6.52)

Since t_0, θ_0 are preselected and r_0 is determined from Equation (6.51), it is a simple matter to solve (6.52) for n. Recalling that n must be an integer, one chooses

$$n = \left[\theta_0 \chi^2_{2r_0, \alpha} / 2t_0 \right],$$ (6.53)

where again $[u]$ is the largest integer $\leqslant u$. This procedure may result in $\beta' > \beta$ (slightly) but $\alpha' \leqslant \alpha$.

A more direct but tedious method is available to find r_0, n by using Equation (6.50). Since this equation provides an easy method of computing points on the OC curve, we will explore it here. Clearly, to satisfy the producer's and consumer's risk requirements, that is, α and β, one needs

$$P(\text{acceptance}|\theta = \theta_0) = P(r \leqslant r_0 - 1 | \theta = \theta_0) = 1 - \alpha$$

$$= \sum_{r=0}^{r_0 - 1} \frac{e^{-nt_0/\theta_0} (nt_0/\theta_0)^r}{r!},$$

$$P(\text{acceptance}|\theta = \theta_1) = P(r \leqslant r_0 - 1 | \theta = \theta_1) = \beta$$

$$= \sum_{r=0}^{r_0 - 1} \frac{e^{-nt_0/\theta_1} (nt_0/\theta_1)^r}{r!}.$$ (6.54)

By using Molina's tables, it is possible, by trial and error, to find an r_0 and n such that (6.54) holds. Of course the Poisson is a discrete distribution, so that (α, β) cannot be satisfied exactly. However, r_0, n can be found so that the true risks satisfy $\alpha' \leqslant \alpha, \beta' \leqslant \beta$. The search in the Poisson tables can be made easier by noting that the Poisson means,

$$\mu_0 = \frac{nt_0}{\theta_0}, \qquad \mu_1 = \frac{nt_0}{\theta_1},$$

must be in the ratio $\mu_0 / \mu_1 = \theta_1 / \theta_0$, as could have been expected from Equation (6.51).

A point on the OC curve at, say, $\theta = \theta'$ is given by

$$P_{\theta'}(A) = \sum_{r=0}^{r_0 - 1} \frac{e^{-nt_0/\theta'} (nt_0/\theta')^r}{r!}.$$ (6.55)

Finally, the expected test time, that is, the expected time to reach a decision, is

$$E_\theta(t) = \left(\frac{\theta}{n}\right) E_\theta(r). \qquad (6.56)$$

where

$$E_\theta(r) = \sum_{r=0}^{r_0} rf(r;\theta).$$

Example.   Suppose that $\theta_0 = 500, \theta_1 = 250, t_0 = 50, \alpha = \beta = .10$. Then, if $\theta = \theta_0$, the number of failures in time t_0 has the Poisson distribution with mean

$$\mu_0 = \frac{nt_0}{\theta_0}$$

and, if $\theta = \theta_1$, the Poisson distribution of the number of failures in time t_0 has mean

$$\mu_1 = \frac{nt_0}{\theta_1}.$$

Thus, applying the direct method with $\mu_0/\mu_1 = \theta_1/\theta_0 = \frac{1}{2}$, one finds from the Poisson tables, for $r_0 = 15$,

$$P(r \leqslant 14 | \mu_0 = 10.2) = .906; \qquad P(r \leqslant 14 | \mu_1 = 20.4) = .095.$$

Thus $.906 \geqslant 1 - \alpha$ and $.095 \leqslant \beta$. To find n note that

$$\mu_0 = 10.2 = \frac{nt_0}{\theta_0} = \frac{n(50)}{500}.$$

Notice that t_0 and n enter this equation as a product and since θ_0 is known, n and t_0 may be "traded-off" as long as the product nt_0 gives the required μ_0 (in this case 10.2).
This means that $n = 102$. Now, applying the method of Equations (6.51) and (6.53), one finds

$$\frac{\chi^2_{29,.10}}{\chi^2_{29,.90}} = \frac{19.77}{39.09} = .506,$$

$$\frac{\chi^2_{28,.10}}{\chi^2_{28,.90}} = \frac{18.94}{37.92} = .499.$$

Since r_0 must be an integer, we choose $2r_0 = 30$ and $r_0 = 15$. Now, applying Equation (6.53), one obtains

$$n = \left[\frac{\theta_0 \chi_{30,\alpha}}{2t_0} \right] = [102.995].$$

$$= 102.$$

Suppose that the point on the OC curve at $\theta = \theta' = 300$ is desired. Then the Poisson mean is

$$\mu' = \frac{nt_0}{\theta'} = \frac{102(50)}{300} = 17$$

and

$$P(r \leqslant 14 | \mu' = 17) = .281.$$

PROBLEM

6.30. Find n and r_0 for a truncated nonreplacement test with $t_0 = 10, \theta_0 = 100, \theta_1 = 40, \alpha = .05, \beta = .10.$

6.2.5 Sequential Tests by Variables

In this section we discuss sequential tests as in Section 6.2.3, except that actual failure times (variables data) will be assumed to be available. In this situation n units are placed on test, and failed units either replaced or not replaced. The majority of applications of sequential tests consist of testing rather "large" units and n is small, so that failed units are usually replaced (e.g., repaired). It is recommended that, unless n can be chosen rather large (so that there is little difference between the replacement and nonreplacement cases and a decision will be made before we "run out" of units), the replacement case be used. The details of the material presented here rather briefly can be found in Epstein and Sobel (1955).

Before beginning the development, note that in sequential tests the length of test (and also the observed number of failures) is a random variable. Although sequential tests are more difficult administratively than fixed-time and fixed-sample-size tests, the expected time (expected number of failures) to reach a decision is generally smaller than that required in fixed-time (fixed-number-of-failures) tests when the true, but unknown, θ is $\geqslant \theta_0$ or $\leqslant \theta_1$.

The sequential test is monitored continuously in time, say t, and continued in the replacement (repair) case as long as $T = nt = $ total accumulated lifetime satisfies

$$-h_1 + rs < T < h_0 + rs, \tag{6.57}$$

where

$$h_0 = \frac{-\log B}{1/\theta_1 - 1/\theta_0}, \qquad h_1 = \frac{\log A}{1/\theta_1 - 1/\theta_0}, \qquad s = \frac{\log(\theta_0/\theta_1)}{1/\theta_1 - 1/\theta_0},$$

$B = \beta/(1-\alpha), A = (1-\beta)/\alpha, r$ is the number of failures, and $\theta_0 > \theta_1$. Acceptance of $H_0 : \theta = \theta_0$ occurs when, for the first time,

$$T \geqslant h_0 + rs, \tag{6.58}$$

and rejection of $H_0 : \theta = \theta_0$ occurs when, for the first time,

$$T \leqslant -h_1 + rs. \tag{6.59}$$

In the nonreplacement case the results are the same as Equations (6.57)–(6.59) except that

$$T = \sum_{i=1}^{r} X_i + (n-r)(t - X_r),$$

where X_i is the time of the ith failure, and $X_0 \equiv 0$.

To avoid trivialities like $A = B$, it is required that $B < 1 < A$. This means that $\alpha + \beta < 1$. This is no restriction at all, since in the context of classical hypothesis testing theory $\alpha < \frac{1}{2}$ and $\beta < \frac{1}{2}$ are the only *reasonable* selections.

The "continue test" region may be regarded as being bounded by two straight lines with common slope s and intercepts $-h_1$ and h_0, so that the test may be portrayed graphically if desired. Also note that, if we choose to change time units from real time t to time, say t', normalized by θ_0 (always known), that is, $t' = t/\theta_0$, Equation (6.57) becomes $-h_1 + rs < nt'\theta_0 < h_0 + rs$, and dividing through by θ_0 shows that the continue test endpoints for nt' depend only on the ratio θ_0/θ_1 and not on θ_0, θ_1 individually. This fact is used to advantage in tabulating tests, as is done in the MIL-STD-781B and H-108 sequential tests. The time given in the tables just cited may be converted to real time by multiplying through by θ_0. Incidentally, the ratio θ_0/θ_1 is called the discrimination ratio.

As in the attributes case, the operating characteristic (OC) curve is given in parametric form:

$$P_\theta(A) = \frac{A^h - 1}{A^h - B^h}, \qquad \theta = \frac{(\theta_0/\theta_1)^h - 1}{h(1/\theta_1 - 1/\theta_0)}. \tag{6.60}$$

However, five points are immediately available to sketch the OC curve: $\theta = 0, P(A) = 0$; $\theta = \theta_1, P(A) = \beta$; $\theta = s, P(A) = \log A/(\log A - \log B)$; θ

$= \theta_0, P(A) = 1 - \alpha;\ \theta = \infty, P(A) = 1$. These correspond, respectively, to h $= -\infty, -1, 0, 1, \infty$. The remaining points may be obtained by varying h.

The expected time to reach a decision is important, since the time to reach a decision is a random variable. It is shown in Epstein and Sobel (1955) that

$$E(T) = \theta E_\theta(r). \tag{6.61}$$

In the replacement case, then, since $T = nt$,

$$E(t) = \frac{\theta E_\theta(r)}{n}, \tag{6.62}$$

where $E_\theta(r)$ is the expected number of failures to termination, and the dependence on θ is explicitly noted. Furthermore

$$E_\theta(r) \begin{array}{l} \sim \dfrac{h_1 - P_\theta(A)(h_0 + h_1)}{s - \theta}, \quad \theta \neq s \\[2em] \sim \dfrac{h_0 h_1}{s^2}, \quad\quad\quad\quad\quad\ \theta = s. \end{array} \tag{6.63}$$

In the nonreplacement case things are not quite as simple, that is, the probability distribution on r is required. In any event,

$$E_\theta(t) = \sum_{k=1}^{n} \left[P(r = k; \theta) E_\theta(X_{k,n}) \right], \tag{6.64}$$

where

$$E_\theta(X_{k,n}) = \theta \sum_{i=1}^{k} \frac{1}{n - i + 1}.$$

Wald showed (1947) that the probability is 1 that the sequential procedure cannot continue forever. However, in practical applications it is often important to know for planning and budgeting purposes, that the tests can last no longer than some fixed time t_0. The way to achieve this is through *truncating* the test (usually at some time t_0). If the key cost figure is number of failures, the test can be truncated at a fixed number of failures r. It is interesting to note that the first applications of sequential tests involved wartime and immediate postwar quality control problems. At that time there was great and impressive opposition, mostly from the theoretical camp, to truncating sequential tests. The reason is simple: truncation of a sequential test changes the (α, β) risks. In fact, if truncation is made too early, it may *increase* both α and β. However, in the last 20 years truncation has been studied in great detail—in fact, in much greater detail than we can afford here. It can be said that "early" truncation times (e.g., $t_0 \leqslant \theta_0$) are likely to increase both α and β. To minimize this source of trouble a sequential test of hypothesis should not be truncated any earlier than $t_0 = 3\theta_0$. However, if truncation must occur early, the exact α and β can be calculated and the accept/reject rule at truncation adjusted so that desired risks are achieved. For example, test plan VI of MIL-STD-781B has been truncated at $1.25\theta_0$, but the accept/reject rule (at truncation) has been selected so that $\alpha = \beta \cong .10$.

Finally, we mention one other source of error, usually of no practical consequence in reliability. The numbers (A, B) were selected to be $(1 - \beta)/\alpha, \beta/(1 - \alpha)$, respectively. Really the latter are only approximations to (A, B). In fact, what is known about (A, B) is that $B \geqslant \beta/(1 - \alpha)$ and $A \leqslant (1 - \beta)/\alpha$. Thus in a nontruncated sequential test the (α, β) that are specified are not exactly achieved. The practical implications of this fact are usually negligible; for further discussion the reader is referred to Wald (1947).

PROBLEMS

6.31. Find h_0, h_1, s for $\theta_0 = 100, \alpha = .05, \theta_1 = 50, \beta = .10$. Also, find the point on the OC curve corresponding to $h = \frac{1}{2}$. Sketch the $E_\theta(t)$ curve. Assume the replacement case.

6.32. Give approximate expressions for $E_{\theta_1}(t)$, $E_{\theta_0}(t)$ in the replacement case.

6.33. Discuss the behavior of $E_{\theta_0}(t), E_{\theta_1}(t)$ in the replacement case for fixed (α, β) and a decreasing ratio θ_0/θ_1.

6.34. Suppose fixed $(\theta_0, \alpha', \theta_1, \beta')$. Discuss the behavior of $E_{\theta_0}(t)$, $E_{\theta_1}(t)$ for $(\theta_0, \alpha < \alpha', \beta < \beta')$.

6.3 TESTS FOR THE TWO-PARAMETER EXPONENTIAL DISTRIBUTION

6.3.1 Tests without Replacement (Type II Censoring)

As noted at the beginning of Chapter 5, the two-parameter exponential probability density function is given by

$$f(x) = \lambda \exp[-\lambda(x - \mu)], \quad x \geqslant \mu, \lambda \geqslant 0,$$

$$= 0, \quad \text{elsewhere.} \tag{6.65}$$

As before, μ is called the location parameter. The discussion will be brief in this section because the relatively few instances in which $\mu \neq 0$ occur mostly in component parts applications. In turn, hypothesis tests do not find quite the large applications for components that they do for equipment/subsystems/systems. In what follows we deal with λ in terms of

$\theta = \lambda^{-1}$. First, censored sampling is assumed. We dismiss more or less immediately two cases. (1) If μ is known (or is 0), a change of variable $Y = X - \mu$ can be made, as discussed in Chapter 5, and then Y has the one-parameter exponential distribution and the methods of Section 6.2 can be used. (2) If μ is unknown, θ is known, and the testing is censored without replacement, it is known from section 5.1.2 that $2n(X_{(1)} - \mu)/\theta$ is distributed as chi square with 2 degrees of freedom (d.f). Here $X_{(1)}$ (the maximum-likelihood estimator for μ) is the first order statistic, that is, the first failure time, and θ of course is known. A test for μ could be based on the above statistic. However, it is probably easier, denoting $X_{(1)}$ by Y for notational convenience, to observe that the cumulative distribution function of $X_{(1)} = Y$ is

$$F_Y(y) = 1 - \exp\left[\frac{-n(y - \mu)}{\theta} \right] \qquad (6.66)$$

and so to test $H_0 : \mu = \mu_0$ versus $H_1 : \mu = \mu_1, \mu_1 > \mu_0$, with (α, β) selected, we need to find a critical value, say y_c, so that

$$P(Y \geqslant y_c | \mu = \mu_0) = \alpha = \exp\left[\frac{-n(y_c - \mu_0)}{\theta} \right], \qquad (6.67)$$

with acceptance of H_0 whenever $\mu_0 \leqslant y \leqslant y_c$ and rejection otherwise. In addition, we need to satisfy the β risk requirement when H_1 is true, that is, when $\mu = \mu_1$. It is

$$P(Y \geqslant y_c | \mu = \mu_1) = 1 - \beta = \exp\left[\frac{-n(y_c - \mu_1)}{\theta} \right]. \qquad (6.68)$$

These two equations may be solved simultaneously for n and y_c, since $(\alpha, \beta, \mu_0, \mu_1, \theta)$ are known. From Equations (6.67) and (6.68),

$$y_c = \frac{[\ln(1 - \beta)]\mu_0 - (\ln\alpha)\mu_1}{\ln(1 - \beta) - \ln\alpha}, \quad \alpha + \beta < 1, \qquad (6.69)$$

and

$$n = \frac{-(\theta\ln\alpha)}{y_c - \mu_0}. \qquad (6.70)$$

The expected test time, that is, $E(Y)$, is the mean of the distribution of Y

and is clearly $(\theta/n)+\mu$. The OC curve may be computed from Equation (6.66) by taking $y=y_c$ and $\mu=\mu'$, the desired point on the OC curve. We note that, for $\mu \geqslant y_c, P_\mu(A)\equiv 0$. It should also be noted that for $y<\mu_0$ neither hypothesis is tenable, since always $Y \geqslant \mu$. The case $\mu_0>\mu_1$ is interesting and is left for the reader to work out in the problems.

The situation when both μ and θ are unknown is straightforward. In Chapter 5 it was pointed out that the two statistics

$$\frac{2(r-1)\theta^*}{\theta} \quad \text{and} \quad \frac{2n(X_{(1)}-\mu)}{\theta}$$

are distributed independently as chi-square variates with $2(r-1)$ and 2 d.f., respectively. Now, to test hypotheses regarding θ, recall that we have already dealt with the designing of tests about θ in Section 6.2.4. In particular, we discussed the design of tests when the sampling distribution of the statistic was chi square and the statistic was of the form $2(r-1)\theta^*/\theta$. It need only be recalled that, when μ is unknown,

$$\theta^* = \left(\frac{r}{r-1}\right)\hat{\theta} \quad \text{and} \quad \hat{\theta} = \frac{\left[\sum_{i=1}^{r} X_{(i)} + (n-r)X_r - nX_{(1)}\right]}{r}.$$

It could also have been observed simply that $2r\hat{\theta}/\theta$ is χ^2 with $2(r-1)$ d.f.

In discussing tests for μ when θ is unknown, a glance at the statistic (writing Y for $X_{(1)}$) $2n(Y-\mu)/\theta$ shows that we are not so fortunate as before. In the previous situation the distribution of $2(r-1)\theta*/\theta$ did not depend on μ. However, in the statistic $2n(Y-\mu)/\theta$, whose distribution is chi square with 2 d.f., we find not only μ but also θ. Consequently, the distribution of Y will depend on θ. This was already known, of course, from Equation (6.66). But, as in Chapter 5,

$$W=\frac{2n(Y-\mu)/2\theta}{2(r-1)\theta^*/2(r-1)\theta}=\frac{n(Y-\mu)}{\theta^*}$$

has an F distribution with $2,2(r-1)$ d.f., and the distribution of W is independent of θ. To test $(\mu_0,\alpha,\mu_1,\beta)$, $\mu_0<\mu_1$, y_c is needed so that

$$P(Y \geqslant y_c | \mu=\mu_0)=\alpha$$

and

$$P(Y \geqslant y_c | \mu=\mu_1)=1-\beta. \tag{6.71}$$

But it is known that (where the notation for degrees of freedom has been suppressed)

$$P\left(Y \geqslant \frac{\theta^* F_{1-\alpha}}{n} + \mu_0\right) = \alpha$$

and

$$P\left(Y \geqslant \frac{\theta^* F_\beta}{n} + \mu_1\right) = 1 - \beta. \tag{6.72}$$

It is clear, then, that

$$y_c = \frac{\theta^* F_{1-\alpha}}{n} + \mu_0 = \frac{\theta^* F_\beta}{n} + \mu_1$$

and thus

$$\mu_1 - \mu_0 = \frac{\theta^*}{n}(F_{1-\alpha} - F_\beta), \quad \alpha + \beta < 1. \tag{6.73}$$

The problem still remains of selecting r and n. It is obvious that we must estimate θ *before* the test to determine r and n. Specifically, having picked θ, we can use Equations (6.69) and (6.70) to determine n and then use this n, along with the estimated θ, in Equation (6.73) to determine r. In doing this it is important to note that

$$F_\gamma(u,v) = \left[F_{1-\gamma}(v,u)\right]^{-1}.$$

For that matter it might be more convenient to use the result given in Chapter 5 [good only for $2, 2(r-1)$ d.f.] that

$$F_\gamma(2, 2r-1) = (r-1)\left[(1-\gamma)^{1-1/r} - 1\right]. \tag{6.74}$$

In any event we must use the y_c given above, not the y_c of the θ-known case. The situation here is much like the classic t test for the (normal) mean when the standard deviation is unknown: we must know the standard deviation in order to preselect the sample size. Of course, if the data are simply given (i.e., r and n selected), merely select α, determine y_c, and apply the test. The case $\mu_0 > \mu_1$ is left for the problems.

PROBLEMS

6.35. For the two-parameter exponential distribution with $\theta = \lambda^{-1}$ known, test ($\mu_0 = 10, \mu_1 = 100, \alpha = .05, \beta = .10$) with $\theta = 1000$. Find five points in the OC curve, and sketch in the expected test time curve.

6.36. Devise a test, that is, find the critical value y_c and specify the accept/reject rule, for the two-parameter exponential location parameter μ, with θ known and $\mu_0 > \mu_1$.

6.37. For the two-parameter exponential distribution with μ and θ both unknown, develop the test procedure (i.e., the critical value for θ^*, n, and r) to test $(\theta_0, \alpha, \theta_1, \beta)$.

6.38. For the two-parameter exponential, μ and θ unknown, develop a procedure to test $(\mu_0, \alpha, \mu_1, \beta)$, $\mu_0 > \mu_1$. The determination of n, r will only be approximate, of course.

6.4 TESTS FOR COMPARING TWO EXPONENTIAL DISTRIBUTIONS

6.4.1 A Test for the Equality of Location Parameters

Consider the case of two exponential populations with common, unknown scale parameter θ but possibly differing location parameters μ_0, μ_1. Thus it is desired to test

$$H_0 : \mu_0 = \mu_1 \quad \text{versus} \quad H_1 : \mu_0 \neq \mu_1.$$

For the case of θ known, the methods of Section 6.3 can be used.

The method to be described was first presented by Kumar and Patel (1971). Suppose that the first k order statistics based on a random sample of size m from the exponential distribution with location parameter μ_0 are available. Similarly, suppose that a random sample of size n from the exponential distribution with location parameter μ_1 yields the first l order statistics. This corresponds to a censored sampling situation, and it is assumed that $2 \leqslant k \leqslant m, 2 \leqslant l \leqslant n$.

The test statistic proposed in Kumar and Patel (1971) is

$$T = \frac{(k+l-2)|X_{(1)} - Y_{(1)}|}{(k-1)\theta_0^* + (l-1)\theta_1^*}.$$

For a test of size α the null hypothesis is to be rejected if $T > c$, where c has been selected so that $P(T > c | \mu_0 = \mu_1) = \alpha$. In the statistic T

$$\theta_0^* = \frac{\sum_{i=1}^{k-1} (m-i)(X_{(i+1)} - X_{(i)})}{k-1}; \qquad \theta_1^* = \frac{\sum_{j=1}^{l-1} (n-j)(Y_{(j+1)} - Y_{(j)})}{l-1}.$$

The Kumar-Patel paper gives values of c for $\alpha = .01, .05$ and for $m, n = 2(1)10$ with $d = k + l - 2 = 2(1)18$. For nontabled values of α, m, n, and d, the relationship

$$m\left(1 + \frac{nc}{d}\right)^{-d} + n\left(1 + \frac{mc}{d}\right)^{-d} = (m+n)\alpha$$

may be used.

6.4.2 A Test for the Equality of Two Means

In what follows it is assumed that the two populations have zero location parameter or that they have known location parameters, (so that a simple change of variable leads back to the one-parameter exponential distribution). Suppose that it is desired to test

$$H_0 : \theta_0 = \theta_1 \quad \text{versus} \quad H_1 : \theta_0 = a\theta_1, \quad a > 1,$$

and that a censored situation exists. In other words, from the exponential distribution with mean θ_0 we have the ordered samples

$$X_{(1)}, X_{(2)}, \ldots, X_{(k)}, \quad k \leqslant m$$

and from the distribution with mean θ_1 we have

$$Y_{(1)}, Y_{(2)}, \ldots, Y_{(l)}, \quad l \leqslant n.$$

Assume also that the sampling is random and that the X, Y samples were obtained independently. Then, with

$$T_0 = \sum_{i=1}^{k} X_{(i)} + (m - k) X_{(k)} \quad \text{and} \quad T_1 = \sum_{j=1}^{l} Y_{(j)} + (n - l) Y_{(l)},$$

it is clear that $2T_0 / \theta_0$ and $2T_1 / \theta_1$ each have a chi-square distribution with $2k$ and $2l$ d.f., respectively. Consequently, under the null hypothesis, the ratio $T_0 l / T_1 k$ has an F distribution with $2k$ and $2l$ d.f. As usual, and still under the null hypothesis,

$$P \left[\frac{T_0 l}{T_1 k} > F_{1-\alpha}(2k, 2l) \right] = \alpha. \tag{6.75}$$

On the other hand, if the alternative hypothesis is true, that is, if $\theta_0 = a\theta_1$, the ratio $T_0 l / T_1 k a$, has the same F distribution as above. Thus

$$P \left[\frac{T_0 l}{T_1 k a} > \frac{F_{1-\alpha}(2k, 2l)}{a} \right] = 1 - \beta. \tag{6.76}$$

It is clear that we must have (suppressing degrees of freedom) $a F_\beta = F_{1-\alpha}$, the test statistic is $T_0 l / T_1 k$, and the critical value is $F_{1-\alpha}$. In other words, if $T_0 l / T_1 k > F_{1-\alpha}$, reject H_0. Note that k and l may be found without specifying θ_0, θ_1 individually, only a is needed. Points on the OC curve may be calculated as follows: Suppose that the point at $a' \neq a$ is desired (the point at a is already known). Then we need

$$P\left(\frac{T_0 l}{T_1 k} \leqslant F_{1-\alpha} | \theta_0 = a' \theta_1 \right).$$

Clearly, this is the same as

$$P\left(\frac{T_0 l}{T_1 ka'} \leqslant \frac{F_{1-\alpha}}{a'} | \theta_0 = a' \theta_1 \right)$$

and since, when $\theta_0 = a' \theta_1$, $T_0 l / T_1 ka'$ has an F distribution with $(2k, 2l)$ d.f., the above probability may be found in the F table, Table II of the Appendix.

Example. Suppose that $\alpha = \beta = .10$ and $a = 3.4$. Then $F_{.90}(2k, 2l)$ $= aF_{.10}(2k, 2l) = aF_{.90}^{-1}(2l, 2k)$. This means that

$$a = F_{.90}(2k, 2l) F_{.90}(2l, 2k).$$

Trial and error with the F distribution table in the Appendix leads to

$$F_{.90}(12, 40) F_{.90}(40, 12) = (1.71)(1.99) = 3.4$$

so that $k = 6, l = 20$.

For $0 < a < 1$, interchange the roles of θ_0 and θ_1. A sequential method for testing this hypothesis is available in Schafer and Takenaga (1972).

PROBLEMS

6.39. Discuss (OC curve, test statistic, accept/reject criterion) the test for $H_0 : \theta_0 = a\theta_1$ versus $H_1 : \theta_0 = a'\theta_1$, $a, a' \neq 1$.

6.40. Compute two points on the OC curve for the example of Section 6.4.2.

6.5 TESTS FOR THE WEIBULL DISTRIBUTION

In this section hypothesis tests for the Weibull distribution parameters are discussed. Because a test for a Weibull location, or threshold, parameter μ is discussed in Section 5.2.2(d) and 7.1.2, our consideration here is limited to the two-parameter Weibull distribution, with density function given by

$$f_X(x) = \left(\frac{\beta}{\delta} \right) \left(\frac{x}{\delta} \right)^{\beta - 1} \exp\left[-\left(\frac{x}{\delta} \right)^\beta \right], \quad x \geqslant 0; \ \delta, \beta > 0,$$

as in Equation 5.6 with $\mu \equiv 0$. Numerous estimators for the parameters (β, δ) are available in the literature. However, as pointed out in Section 5.2.3, the best estimators on which to base hypothesis tests are the ones with smallest mean-squared error, namely, the maximum-likelihood and best linear invariant estimators. It should also be noted that, for large n, say 100 or more, both the maximum-likelihood estimators (MLE's) and best linear invariant estimators are normally distributed, and tests of hypotheses based on well-known normal theory may be used. For example, denoting the MLE of β by $\hat{\beta}$, one finds that the random variable $\hat{\beta}/\beta$ is asymptotically normal with mean 1 and variance $.608/n$. For the rate at which $\hat{\beta}/\beta$ approaches normality, the reader is referred to Thoman, Bain, and Antle (1969). In regard to hypothesis tests, things are not nearly as neat for the Weibull distribution as for the exponential distribution, that is, closed-form expressions involving the sample size n are not available. The closed-form expressions given in Section 5.2.2(c) for obtaining approximate confidence intervals for β, δ, reliability and reliable life involve sample size n only implicitly.

Finally, we remark that, if β is known, a change of variable $Y = X^{\beta}$ results in Y being exponential with mean δ^{β}, so that the methods presented earlier in this chapter can be used. It is rare, however, that either parameter is known a priori.

6.5.1 Relationship Between Interval Estimates and Hypothesis Tests

Generally speaking, if an interval estimate for an unknown parameter is available, a hypothesis test may be conducted with little or no additional work. As discussed in Section 3.4.3, interval estimates and hypothesis tests, in the classical sense, are related. Suppose that it is desired to test $H_0 : \rho = \rho_0$ with significance level α, where ρ is an unknown parameter, and that a $100(1 - \alpha)\%$ confidence interval estimate of ρ is available. The result, true under fairly general conditions, is that a necessary and sufficient condition for the acceptance of H_0 is that the interval estimate contain ρ_0. Intuitively, it is clear that values of ρ in the interval are reasonable or they would not be there. Thus, if ρ_0 belongs to the interval, H_0 should be accepted. On the other hand, if H_0 is accepted, ρ_0 should belong to any confidence interval with confidence coefficient equal to 1 minus the significance level of the test.

We now give a little more definitive argument. In what follows, it will be assumed that we are dealing with continuous random variables; the discrete case proceeds similarly. To this end, suppose that a random sample $(x_1, \ldots, x_n)$ is available, and it is desired to test $H_0 : \rho = \rho_0$ versus $H_1 : \rho = \rho_1$, $\rho_1 < \rho_0$, with size α. To test H_0 we select, as always, a value ρ_c^*

such that
$$P(\rho^* < \rho_c^* \mid H_0) = \int_{-\infty}^{\rho_c^*} g(\rho^*; \rho_0) \, d\rho^* = \alpha$$

and H_0 is accepted if, and only if, $\rho^* \geqslant \rho_c^*$. Here g is the sampling distribution of the test statistic (estimator), $\rho^*(X_1, \ldots, X_n)$.

Now suppose that an interval estimate, based on ρ^*, is prepared for ρ with confidence coefficient $1 - \alpha$, that is,

$$P(\rho \leqslant \bar{\rho}) = 1 - \alpha.$$

The reader should assure himself that, since H_1 is $\rho = \rho_1, \rho_1 < \rho_0$, only a one-sided upper bound for ρ makes sense. Now $\bar{\rho}$ is obtained as the largest member of the set of ρ such that

$$\int_{-\infty}^{\rho_{\text{OBS}}^*} g(\rho^*; \rho) \, d\rho^* \geqslant \alpha, \tag{6.77}$$

where ρ_{OBS}^* is the ρ^* obtained for this particular $(x_1, \ldots, x_n)$.

To avoid details that only obscure the argument, it is assumed that the integral of Equation (6.77) is a strictly decreasing continuous function of ρ. Then the set of ρ satisfying (6.77) may be represented as

$$S_{\bar{\rho}} \equiv \{ \rho \mid \rho \leqslant \bar{\rho} \}.$$

Now suppose that H_0 is accepted. Then clearly $\rho_{\text{OBS}}^* \geqslant \rho_c^*$. Moreover, if $\rho_{\text{OBS}}^* \geqslant \rho_c^*$,

$$\int_{-\infty}^{\rho_{\text{OBS}}^*} g(\rho^*; \rho_0) \, d\rho^* \geqslant \int_{-\infty}^{\rho_c^*} g(\rho^*; \rho_0) \, d\rho^* = \alpha$$

so that obviously $\rho_0 \in S_{\bar{\rho}}$. On the other hand, suppose that $\rho_0 \in S_{\bar{\rho}}$. Then

$$\int_{-\infty}^{\rho_{\text{OBS}}^*} g(\rho^*; \rho_0) \, d\rho^* \geqslant \alpha = \int_{-\infty}^{\rho_c^*} g(\rho^*; \rho_0) \, d\rho^*,$$

so that clearly $\rho_{\text{OBS}}^* \geqslant \rho_c^*$ and H_0 is accepted.

Thus a test of $\rho = \rho_0$ may be made by observing whether ρ_0 is contained in the interval estimate, and methods of Chapter 5 can be used. The shortcoming of this approach is that a central problem treated in this chapter, the determination of sample size to satisfy specifications on Type I and Type II errors, cannot be addressed.

Before proceeding to some illustrations for the Weibull distribution we give a simple example based on the exponential distribution.

Example. Suppose that a complete random sample of size n is available and $\hat{\theta}$ is observed. Suppose further that it is desired to test $H_0 : \theta = \theta_0$ versus $H_1 : \theta = \theta_1, \theta_1 < \theta_0$, with size $\alpha = .10$. Then the upper confidence bound $\bar{\theta}$ is $\bar{\theta} = 2n\hat{\theta}/\chi^2_{2n,.10}$. The critical value, $\hat{\theta}_c$, is given by

$$\hat{\theta}_c = \frac{\theta_0 \chi^2_{2n,.10}}{2n}.$$

Now, if $\hat{\theta} \geqslant \hat{\theta}_c$, H_0 is accepted, and this means that

$$\hat{\theta} \geqslant \frac{\theta_0 \chi^2_{2n,.10}}{2n} \Rightarrow \theta_0 \leqslant \frac{2n\hat{\theta}}{\chi^2_{2n,.10}} = \bar{\theta}.$$

On the other hand

$$\theta_0 \leqslant \frac{2n\hat{\theta}}{\chi^2_{2n,.10}} \Rightarrow \hat{\theta} \geqslant \frac{\theta_0 \chi^2_{2n,.10}}{2n} = \hat{\theta}_c,$$

so that H_0 is accepted.

The results of Chapter 5 can be used to test hypotheses for the parameters, reliability function, and quantiles of the Weibull distribution. We will give two illustrations of the approach for testing the hypothesis

$$H_0 : \beta = \beta_0 \quad \text{versus} \quad H_1 : \beta = \beta_1, \beta_1 > \beta_0.$$

First, let us suppose that a complete random sample of size n from a Weibull distribution is available. The method to be used to test the above hypothesis is based on the maximum-likelihood estimate, $\hat{\beta}$, of β. Thoman, Bain, and Antle (1969) have shown that for $\hat{\beta}$ and $\hat{\delta}$ the maximum-likelihood estimators of the Weibull parameters β and δ, respectively, $\hat{\beta}/\beta \equiv \xi/\hat{\xi}$ and $\hat{\beta} ln(\hat{\delta}/\delta) \equiv (\hat{\eta} - \eta)/\hat{\xi}$ have distributions that are parameter free. Mann and Fertig (1973), referenced in Chapter 5, have shown that the same is true of $\tilde{\xi}/\xi$, ξ^*/ξ, $(\tilde{\eta} - \eta)/\tilde{\xi}$, and $(\eta^* - \eta)/\xi^*$, where $\tilde{\xi}$ and $\tilde{\eta}$ and ξ^* and η^* are best linear invariant and best linear unbiased estimators of ξ and η. In fact, any linear estimator $\bar{\xi}$ of ξ (linear in the logarithms of the ordered Weibull variates) whose coefficients add to 0 will be such that $\bar{\xi}/\xi$ has a parameter-free distribution. And for $\bar{\xi}/\xi$ parameter-free, any linear estimator $\bar{\eta}$ of η whose coefficients add to 1 will be such that $(\bar{\eta} - t_m)/\bar{\xi}$ and $(\bar{\eta} - t_p)/\bar{\xi}$ have parameter-free distributions. Here t_m is a specified mission time and t_p is the $100p$th Weibull distribution percentile. One can use tables of distribution percentiles published by Thoman, Bain, and Antle (1969 and 1970) for complete samples or those of Billman, Antle, and Bain (1972) for 25% and 50% censoring to test hypotheses concerning

β, δ, $R(t_m)$, or t_p by means of maximum-likelihood estimates. Alternatively, one can use tables appearing in Chapter 5, or extensions thereof, that apply to (equally efficient, or in some cases, nearly equally efficient) linear estimates based on possibly censored samples.

New results described in Section 5.2.2(c) allow one to determine closed-form expressions for the approximate distributions of ξ^*/ξ, $(\eta^* - t_m)/\xi^*$ and $(\eta^* - t_p)/\xi^*$, where ξ^* and η^* are best, or in fact any, unbiased linear estimators of ξ and η. All that are necessary besides failure data to use these results to test hypotheses concerning $\xi \equiv \beta^{-1}$, $\eta \equiv ln\delta \equiv ln(t_{1 - exp(-1)})$, t_p, and $R(t_m)$ are the variances and covariances of ξ^* and η^*. For many cases of interest these are tabulated or can be determined from tabulated values in Chapter 5.

The tables of Thoman, Bain, and Antle (1969) apply to maximum-likelihood estimates of δ and β based on uncensored samples of size $n = 5(1)20(2)80(5)100(10)120$ and the same percentages as given in Tables 5.7–5.10. Thus a wide choice of n and $1 - \alpha$ is available for the complete sample case.

To test the above hypothesis, we need, as usual, to find $\hat{\beta}_c$ such that

$$P\left(\hat{\beta} > \hat{\beta}_c | H_0 \right) = P\left(\frac{\hat{\beta}}{\beta_0} > \frac{\hat{\beta}_c}{\beta_0} \right) = \alpha,$$

and H_0 is rejected whenever $\hat{\beta} > \hat{\beta}_c$.

As has already been pointed out, we need only construct a $(1 - \alpha)100\%$ lower confidence bound for β; if β_0 is contained in it, H_0 is accepted. Suppose that $n = 10$, and $\alpha = .05$ is chosen. Then, from Table 1 of Thoman et al.,

$$P\left(\frac{\hat{\beta}}{\beta} \leqslant 1.807 \right) = .95 = P\left(\beta \geqslant \frac{\hat{\beta}}{1.807} \right).$$

Thus the lower 95% confidence bound is $\hat{\beta}/1.807$, and if $\beta_0 \geqslant \hat{\beta}/1.807 H_0$ is accepted. Of course, this is the same as $\hat{\beta}/\beta_0 \leqslant 1.807$, so obviously $\hat{\beta}_c = 1.807\beta_0$. In general, the lower $100(1 - \alpha)\%$ confidence bound for β is $\hat{\beta}/k_{1 - \alpha}$, and $\hat{\beta}_c$ is $\beta_0 k_{1 - \alpha}$, where k_u is the uth quantile of the distribution of $\hat{\beta}/\beta$. Similar methods can be used to test hypotheses about δ. It turns out that the power of the above test depends only on the ratio β_1/β_0, and Thoman et al. give the OC curves for β_1/β_0 up to 3 for numerous n, so that the sample size required to satisfy (β_0, α, β_1, probability of a Type II error) can very nearly be obtained.

For the second example, assume that a censored sample, $X_{(1)}, X_{(2)}, \ldots,$ $X_{(r)}$ is available, $r \leqslant n$, and that it is desired to test the same hypothesis.

The best linear invariant estimator of β will be used. It also has small mean-squared error.

In Section 5.2.3 the best linear invariant estimators, $\tilde{\eta}$ and $\tilde{\xi}$, of $\eta = \ln \delta$ and $\xi = 1/\beta$ were used to obtain interval estimates of η and ξ and hence interval estimates of δ and β. As before, the interval estimate permits us to test $H_0 : \beta = \beta_0$ versus $H_1 : \beta = \beta_1, \beta_1 > \beta_0$, which is the same as $H_0 : \xi = \xi_0$ versus $H_1 : \xi = \xi_1$, $\xi_1 < \xi_0$. The necessary weights for computing $\tilde{\xi}$ are given in Table 5.3. In Table 5.7 values of w_α are given such that $P(\tilde{\xi}/\xi = W \leqslant w_\alpha) = \alpha$, but since $1/\xi = \beta$ this statement amounts to $P(\beta \tilde{\xi} \leqslant w_\alpha) = \alpha$. Now a lower one-sided $100(1 - \alpha)\%$ interval estimate of β is clearly given by $\underline{\beta} = w_\alpha/\tilde{\xi}$. Therefore, if $\beta_0 \geqslant \underline{\beta}, H_0$ is accepted. One might also use the approximate chi-square-over-degrees-of-freedom variate ξ^{**}/ξ described in Section 5.2.3(c) to test H_0 when sample size is larger than 20 or so.

6.6 TESTS FOR THE TWO-PARAMETER LOGNORMAL DISTRIBUTION

For a two-parameter lognormal variate X, by definition $\ln X$ has the usual normal distribution. Hypothesis testing for this distribution has been exhaustively covered in numerous textbooks, so that, if hypothesis tests are required on μ and/or σ, procedures are readily available. Although tests of hypotheses for the moments of the lognormal distribution are not available, we can test hypotheses about the median, e^μ, quite easily. For example, if a two-sided $100(1 - \alpha)\%$ confidence interval for μ is given by $(\underline{\mu}, \bar{\mu})$, then $e^{\underline{\mu}}, e^{\bar{\mu}}$ is a two-sided $(100(1 - \alpha)\%$ confidence interval for the median e^μ. Since the mean and the variance of the lognormal distribution are $e^{\mu + \sigma^2/2}$ and $e^{2\mu + \sigma^2}(e^{\sigma^2} - 1)$, respectively, the coefficient of variation = standard deviation/mean = $(e^{\sigma^2} - 1)^{1/2}$, and hypothesis tests regarding the coefficient of variation can be conducted by the usual normal distribution tests for σ^2. Interval estimates for this distribution are discussed in Section 5.4.

6.7 TESTS FOR THE BIRNBAUM-SAUNDERS FATIGUE-LIFE DISTRIBUTION

The Birnbaum-Saunders distribution is discussed in Section 5.5. The two parameters are scale and shape-parameters. Hypothesis tests are available at this time only for the scale parameter. Because this distribution is new and relatively complex, in order to maintain a degree of continuity the hypothesis test procedure for the scale parameter is discussed in Section 5.5.2.

PROBLEMS

6.41. Indicate a test of hypothesis procedure for a Weibull shape parameter for $H_0: \beta = \beta_0$ versus $H_1: \beta = \beta_1, \beta_1 < \beta_0$, based on a censored sample.

6.42. Devise a test of hypothesis procedure for the Weibull scale parameter δ, based on $\hat{\delta}$ and a complete sample.

6.43. Suppose that a random sample of size 10 is obtained from a two-parameter lognormal distribution with parameters μ (unknown) and $\sigma = 1$ and that $\hat{\mu} = 2$. Test the hypothesis that the median of the lognormal distribution is 8 versus the hypothesis that is 6 with $\alpha = .10$.

REFERENCES

Billman, Barry R., Charles E. Antle, and Lee J. Bain (1972), Statistical Inference for Censored Weibull Samples, *Technometric* Vol. 14, pp. 831–840.

Bowker, A. H. and G. J. Lieberman (1955), *Handbook of Industrial Statistics*, Prentice-Hall, Englewood Cliffs, New Jersey.

Dodge, H. F. and H. G. Romig, (1944), *Sampling Inspection Tables*, John Wiley and Sons, New York.

Duncan, A. J. (1959), *Quality Control and Industrial Statistics*, Richard D. Irwin, Homewood, Illinois.

Epstein, B. (1954), Truncated Life Tests in the Exponential Case, *Annals of Mathematical Statistics*, Vol. 25, pp. 555-564.

Epstein, B. (1960), Statistical Life Test Acceptance Procedures, *Technometrics*, Vol. 2, No. 4, pp. 435-446.

Epstein, B. and M. Sobel, (1953), Life Testing, *Journal of the American Statistical Association*, Vol. 48, pp. 486-502.

Epstein, B. and M. Sobel, (1955), Sequential Life Tests in the Exponential Case, *Annals of Mathematical Statistics*, Vol. 26, pp. 82-93.

Kumar, S. and H. I. Patel, (1971), A Test for the Comparison of Two Exponential Distributions, *Technometrics*, Vol. 13, No. 1, pp. 183-189.

Lieberman, Gerald J. and Donald B. Owen, (1961), *Tables of the Hypergeometric Probability Distribution*, Stanford University Press, Stanford, California.

Molina, E. C. (1942), *Poisson's Exponential Binomial Limit*, D. Van Nostrand Company, Princeton, New Jersey.

Nadler, J. (1960), Inverse Binomial Sampling Plans When an Exponential Distribution Is Sampled with Censoring, *Annals of Mathematical Statistics*, Vol 31, pp. 1201-1204.

Reliability Tests: The Exponential Distribution (1967), MIL-STD-781B, Department of Defense.

Robertson, William H. (1960), *Tables of the Binomial Distribution Function for Small Values of p*, Sandia Corporation Monograph.

Romig, Harry G. (1953), *50-100 Binomial Tables*, John Wiley and Sons, New York.

Sampling Procedures and Tables for Life and Reliability Testing (1960), *Quality Control and Reliability Handbook* (Interim) II-108, Office of the Assistant Secretary of Defense, U.S. Government Printing Office, Washington, D.C.

Sarhan, A. E. and B. G. Greenberg, Editors (1962), *Contributions to Order Statistics*, John Wiley and Sons, New York.

Schafer, R. E. and R. Takenaga, (1972), A Sequential Probability Ratio Test for Availability, *Technometrics*, Vol. 14, No. 3, pp. 122-135.

Sobel, M. and J. A. Tischendorf, (1959), Acceptance Sampling with New Life Test Objectives, *Proceedings of the Fifth National Symposium on Reliability and Quality Control*, Philadelphia, Pennsylvania, pp. 108-118.

Tables of the Binomial Probability Distribution, *Applied Mathematics Series* 6, National Bureau of Standards, U.S. Government Printing Office, Washington, D.C.

Tables of the Cumulative Binomial Distribution (1955), Harvard University Press, Cambridge, Massachusetts.

Thoman, Darrel R., Lee J. Bain, and Charles E. Antle, (1969), Inferences on the Parameters of the Weibull Distribution, *Technometrics*, Vol. 11, No. 3, pp. 445-460.

Thoman, Darrel R., Lee J. Bain and Charles E. Antle (1970) Maximum Likelihood Estimation, Exact Confidence Intervals for Reliability, and Tolerance Limits in the Weibull Distribution, *Technometrics*, Vol. 12, No. 2, pp. 363–371.

Wald, A. (1947), *Sequential Analysis*, John Wiley and Sons, New York.

CHAPTER 7

Goodness of Fit, Monte Carlo, and Distribution-Free Methods

7.1 TESTING GOODNESS OF FIT

In Chapters 5 and 6, procedures are given for the estimation and testing of parameters of certain specified lifetime distributions. In this section, we discuss tests of the fit of observed data to these various distributions.

One method often used to determine whether or not a set of observations has been selected from a particular population is graphical plotting. Plotting techniques are discussed in some detail for the Weibull distribution in Section 5.2.2 and in rather less detail for the lognormal distribution in Section 5.4.1. The use of graphical plotting procedures to test the goodness of fit of the data to the distribution assumed usually works well if the assumption is completely inappropriate or if the data plot nearly perfectly into a straight line. Since subjective judgment must be used, however, in deciding whether or not the plot seems to be a straight line, it is often difficult in less clear-cut cases to decide whether or not to reject the hypothesized distribution.

In the following, some analytical tests of fit that have been developed for particular distributions are described, as well as a general chi-square test that can be used to test the fit of data to nearly any hypothesized distribution.

Some of the tests of fit for particular distributions are analogs (incorporating parameter estimates) of classical goodness-of-fit tests such as the Kolmogorov-Smirnov and Cramér-von Mises tests (see Darling, 1957). The classical tests apply to testing the assumption that data belong to a completely specified distribution, that is, all parameter values, as well as the distributional form, are assumed to be known. For such an assumption, one table of critical values (see Massey, 1951, and Birnbaum, 1952) can be used with the test statistic no matter what distributional form and parameter values are assumed, as long as the distribution is continuous and the

335

unordered sample variates are identically distributed. Results of David and Johnson (1948) are used to justify the use of analogs of the classical tests wherein estimates are substituted for location and/or scale parameters.

7.1.1 Tests for Exponentiality or Constant Failure Rate

Because the one-parameter exponential distribution has been so widely used in the analysis of life-test data, two dozen or more tests of exponential fit have been suggested and developed. Recently, Monte Carlo techniques have been used to compare the power of some of these tests under various alternative hypotheses.

The classical Kolmogorov-Smirnov two-sided test statistic is

$$D_n = \sup_x |F(x;\boldsymbol{\theta}) - F_n(x)|,$$

where $F(x;\boldsymbol{\theta})$ is the completely specified continuous null or hypothesized distribution (with parameter values assumed to be known), and $F_n(x)$ is defined as the number, divided by sample size n, of observations less than or equal to x. Also, $F(x,\boldsymbol{\theta})$ is the expected value of $F_n(x)$ for any continuous F.

The computing form of the Kolmogorov-Smirnov statistic is

$$D_n = \max_i (\delta_i),$$

with $\delta_i = \max[i/n - F(X_{(i)};\boldsymbol{\theta}), F(X_{(i)};\boldsymbol{\theta}) - (i-1)/n]$, where, if the distribution postulated is exponential with, for example mean 2, $F(X_{(i)};\theta) = 1 - \exp(-X_{(i)}/2), i = 1,\dots,n$.

For the exponential distribution with unknown mean, Lilliefors (1969) gives, for complete samples, critical values for a Kolmogorov-Smirnov type statistic with the distribution mean replaced by the sample mean $\overline{X} = \sum_{i=1}^{n} X_i/n$. The computing form of Lilliefors' statistic, D_n^*, is defined as

$$D_n^* = \max_i (\delta_i^*),$$

where

$$\delta_i^* = \max\left[\frac{i}{n} - F^*(X_{(i)}), F^*(X_{(i)}) - \frac{i-1}{n} \right],$$

and where $F^*(X) = 1 - \exp(-X/\overline{X})$. The hypothesis of exponentiality is rejected if D_n^* is too large.

J. van Soest (1969) compared the Kac, Kiefer, and Wolfowitz (1955) analog of the classical Cramér-von Mises test (which is similar to the

Kolmogorov-Smirnov analog but involves sums of squares of certain deviations rather than the particular maximum deviations given above) to tests of exponentiality based on criteria arising from probability plotting (see Hahn and Shapiro, 1967), on the conditional distribution of total lives (suggested by Epstein, 1960), and on the likelihood ratio. The Cramér-von Mises analog (with $\overline{X}$ used to estimate θ) is most powerful against various gamma alternatives among those considered by van Soest. Finklestein and Schafer (1971) compared the powers of both the Lilliefors (1969) analog of the Kolmogorov-Smirnov test and the van Soest (1969), Kac, Kiefer, and Wolfowitz (1955) analog of the Cramér-von Mises test for certain alternatives used by these authors with a test based on the statistic

$$S_n^* = \sum_{i=1}^{n} |\delta_i^*|.$$

The test of Finklestein and Schafer based on S_n^* is shown to be at least as powerful as the analogs of the classical tests and, in certain cases, significantly more powerful than the Kolmogorov-Smirnov analog. Since the test statistic S_n^* is very little more difficult to calculate than D_n^* and is considerably simpler to calculate than the Cramér-von Mises analog statistic, its use is recommended for complete samples. Table 7.1 lists critical values of S_n^*.

Thus, if one wished to test the hypothesis that a set of 10 observations is a sample from a one-parameter exponential distribution with unknown mean, he would reject the hypothesis at significance level $\alpha = .10$ if $S^* = \sum_{i=1}^{n} |\delta_i^*|$ were greater than 1.55, where δ_i^* is defined above. An example

Table 7.1   Critical values for S_n^*

n \ α	.20	.15	.10	.05	.01	n \ α	.20	.15	.10	.05	.01
2	.86	.89	.92	.96	.99	12	1.49	1.56	1.65	1.82	2.13
3	.97	1.01	1.07	1.13	1.23	13	1.53	1.61	1.71	1.87	2.19
4	1.05	1.09	1.15	1.23	1.37	14	1.58	1.65	1.76	1.92	2.28
5	1.12	1.17	1.23	1.32	1.50	15	1.61	1.70	1.80	1.98	2.34
6	1.19	1.24	1.30	1.40	1.62	16	1.65	1.73	1.84	2.02	2.39
7	1.24	1.30	1.37	1.48	1.72	17	1.68	1.77	1.89	2.08	2.46
8	1.29	1.35	1.44	1.56	1.80	18	1.72	1.81	1.93	2.12	2.52
9	1.35	1.41	1.49	1.62	1.89	19	1.76	1.85	1.97	2.16	2.56
10	1.40	1.47	1.55	1.70	1.98	20	1.79	1.88	2.01	2.21	2.62
11	1.45	1.52	1.61	1.75	2.04	25	1.95	2.05	2.18	2.42	2.88
						30	2.08	2.20	2.35	2.59	3.10

of the use of Table 7.1 for a set of data is given after a discussion of some tests of exponentiality that can be used with censored data.

Fercho and Ringer (1972) consider tests of fit of data, which may be censored, to a one-parameter exponential distribution with unknown mean. They compare two tests suggested by Epstein (1960) and one by Hartley (1950), based on tests for homogeneity of variances, with a test suggested by Gnedenko et al. (1960), based on an F-distribution criterion. Fercho and Ringer show by Monte Carlo analysis that the test of Gnedenko et al. tends to be the most powerful among these against Weibull alternatives specifying various values of shape parameter β, both larger and smaller than 1. The test of Finklestein and Schafer, based on S_n^*, has been shown by the present authors to be substantially more powerful, for complete samples, for the two values of the Weibull shape parameter considered ($\beta = 2, \beta = \frac{1}{2}$) for the lognormal and chi-square alternatives and gamma alternatives considered by Lilliefors (1969) and von Soest (1969), respectively, than the test suggested by Gnedenko et al. However, since the Gnedenko test is the most powerful among those that can be based on censored samples, and since it is quite simple to implement computationally, its test statistic is given here:

$$Q(r_1, r_2) = \frac{\sum_{i=1}^{r_1} (S_i/r_1)}{\sum_{i=r_1+1}^{r} S_i/r_2},$$

where $S_i = (n - i + 1)(X_{(i)} - X_{(i-1)})$, with $X_0 \equiv 0$ and $r_2 = r - r_1$. Since each S_i has an independent exponential distribution, as shown in Section 5.1.1, $\sum_{i=1}^{k} 2S_i/\theta$ has a chi-square distribution with $2k$ degrees of freedom, and $Q(r_1, r_2)$, from Section 3.3, has an F distribution with $2r_1$ and $2r_2$ degrees of freedom. Exponentiality is rejected at significance level α if $Q(r_1, r_2)$ is greater than $F_{1-\alpha/2}(2r_1, 2r_2)$ or less than $F_{1-\alpha/2}^{-1}(2r_2, 2r_1)$, where $F_\zeta(k_1, k_2)$ is the 100ζth percentile of the F distribution with parameters k_1 and k_2. If $Q(r_1, r_2)$ is large, increasing hazard rate is indicated. If it is small, one might conclude that hazard rate is decreasing ($\beta < 1$). This particular form of two-sided test results from the fact that, if Z has an F distribution with k_1 and k_2 degrees of freedom, $1/Z$ has an F distribution with k_2 and k_1 degrees of freedom. Fercho and Ringer used $r_1 = r_2 = r/2$ for r even and $r_1 = [r/2]$ and $r_2 = [r/2] + 1$ for r odd, where $[x]$ is the greatest integer less than x.

Example. Suppose that $F(x)$ is hypothesized to be a one-parameter exponential distribution with unknown mean, and the following complete sample of size 5 was observed and ordered:

$$x_{(1)} = 7.602, \ x_{(2)} = 14.341, \ x_{(3)} = 15.180, \ x_{(4)} = 27.526, \ x_{(5)} = 32.586.$$

Then

$$F^*(x_{(1)}) = .324, \ F^*(x_{(2)}) = .521, \ F^*(x_{(3)}) = .542,$$

$$F^*(x_{(4)}) = .760, \ F^*(x_{(5)}) = .815,$$

$$\delta_1^* = .324, \ \delta_2^* = .321, \ \delta_3^* = .142, \ \delta_4^* = .160, \ \delta_5^* = .185.$$

Therefore

$$D_5^* = .324, \qquad S_5^* = 1.13,$$

and

$$Q(2,3) = \frac{[5(7.602) + 4(6.739)]/2}{[3(.839) + 2(12.346) + 5.060]/3}$$

$$= 3.03.$$

If $\alpha = .10$, critical values of D_5^*, S_5^*, and $Q(2,3)$ are .406 (see Lilliefors, 1969), 1.23 (from Table 7.1), and $F_{.95}(4,6) = 4.53$, respectively. The hypothesis is not rejected by any of these tests. For $\alpha = .20$, only the test based on S_n^* rejects the hypothesis.

7.1.2 Tests of Fit to Two-Parameter Weibull or Extreme-Value Distribution

Both the test of van Montfort (1970), designed for the analysis of certain rainfall problems, and the test of Mann, Scheuer, and Fertig (1973) to verify that observations are from a first asymptotic distribution of smallest (extreme) values are based on an observation of van Montfort (1970), depending on a result of Pyke (1965), concerning the properties of adjacent ordered observations. This observation is that *leaps* from the extreme-value distribution, that is, differences of adjacent ordered observations divided by their expectations, have distributions that are approximately exponential with expectations exactly 1, variances approximately 1, and covariances essentially 0. Mann et al. observe that, therefore, these leaps, when multiplied by 2, have approximately independent chi-square distributions with 2 degrees of freedom each. Thus, define

$$l_i = \frac{X_{(i+1)} - X_{(i)}}{E(Z_{i+1}) - E(Z_i)}, \quad Z_i = \frac{X_{(i)} - \eta}{\xi},$$

where η and ξ are the location and scale parameters, respectively, of the extreme-value distribution, as defined by Equation (5.13). Then

$$W = \frac{\displaystyle\sum_{i=[r/2]+1}^{r-1} l_i / [(r-1)/2]}{\displaystyle\sum_{i=1}^{[r/2]} l_i / [r/2]}$$

(where $[X]$ is the largest integer less than or equal to X) has approximately an F distribution (see Section 3.3) with $2[(r-1)/2]$ and $2[r/2]$ degrees of freedom.

A test of fit based on the statistic W (or, equivalently, on S, to be defined subsequently as a function of W) was derived by Mann, Scheuer, and Fertig as a consequence of the fact that the left tail of an extreme-value density function is *longer* than most distributions of interest, whereas the right tail is *shorter*. Hence, for example, if one were to form a statistic proportional to the ratio of the last and the first differences of ordered observable variates from a sample of size $n, k(X_{(r)} - X_{(r-1)})/(X_{(2)} - X_{(1)})$, with $r = n$, he would expect this statistic to be smaller under the null hypothesis than under applicable alternative hypotheses. It was found that, especially for small significance levels, the power of a test based on this kind of statistic could be increased by using all r observed leaps as in W, rather than just the first and the last as directly above.

In order that Monte Carlo-generated critical test values would fall in the unit interval rather than in the interval $(0, \infty)$, the test statistic was transformed to

$$S = cW/(1 + cW) = \frac{\displaystyle\sum_{i=[r/2]+1}^{r-1} l_i}{\displaystyle\sum_{i=1}^{r-1} l_i},$$

where $c = [(r-1)/2]/[r/2]$. Because of the relationship between the beta distribution and the F distribution (see Section 3.3), S for n sufficiently large or r sufficiently small has approximately a beta distribution with parameters $[(r-1)/2]$ and $[r/2]$. Percentiles of the distribution of S and values of $E(Z_{i+1}) - E(Z_i), i = 1, \ldots, n-1$, are given in Table 7.2 for sample size $n, n = 3(1)16$. Tables for sample sizes $3(1)25$ are given by Mann, Scheuer, and Fertig (1973). Comparison of the percentile values with those in tables of percentiles of the beta distribution reveals that, for $n \geqslant 16$, the tabulated values and the beta values agree to within a unit in the second decimal place (except for the 99th percentile, for which there is sometimes a discrepancy of as much as 2 units in the second decimal place). Agreement between corresponding percentiles is especially good for r considerably less than n and also for per cent points 90 and lower. For n large, one can therefore use the tables of White (1967), which give expected values of order statistics from the reduced extreme-value distribution for sample sizes 1(1)50(5)100, to form the test statistic. For n very large (but less than 100), one can form W using only the differences of the first two and the last two ordered observations. The ratio of the latter to the former, when multiplied by the appropriate ratio of differences of expected values, has an F distribution with 2, 2 degrees of freedom. Table II of the F distribution in the Appendix or Harter's tables (1964) of beta distribution percentiles can be used to obtain critical values for W or S statistics based on large sample sizes.

Monte Carlo power comparisons (applying to complete samples) of the test based on S and analogs of the Kolmorov-Smirnov, the Kuiper (1960), and the standard version, as well as a weighted version, due to Anderson and Darling (1952), of the Cramér-von Mises tests (all using best linear invariant estimates, essentially equivalent to maximum-likelihood estimates of η and ξ), revealed that the S test is most powerful against the alternatives studied. The alternatives considered are that X is normal, or $T = \exp(X)$ is a two-parameter lognormal (failure) distribution, and that T has a three-parameter Weibull distribution with two different specified values of shape parameter β less than or equal to 1 and one specified value of μ/δ, with the location parameter μ not equal to 0. The latter applies to situations in which the transformation to log space is made and then the transformed variates are tested to determine whether a nonzero location parameter has been overlooked. One rejects at significance level α the hypothesis that X is an extreme-value variate if the calculated value of S exceeds the $100(1 - \alpha)$th distribution percentile of S.

It seems likely that a test more powerful than this one against some particular alternative family of distributions, particularly for complete samples, could be derived. The test based on S or W, however, has the advantage that it is applicable to censored samples of size up to 100 without the generation of many pages of new tables, and applicable to any alternative hypothesis specifying tail properties different from those of the extreme-value distribution.

Mann and Fertig (1974) have modified the test statistic S to increase the power of the test under three-parameter Weibull alternatives. The new test statistic P is defined exactly as is S except that $[r/2]$ is replaced by the integer, k. Values of k, which approximately optimize the power of the test under any three-parameter Weibull alternative, are tabulated by Mann and

Table 7.2 Percentiles of the distribution of S and differences of expected values of reduced extreme-value order statistics

n	i	$Ez_{i+1} - Ez_i$	0.75	0.80	0.85	0.90	0.95	0.99
3	1	1.216395	0.75	0.79	0.84	0.90	0.95	0.99
	2	0.863046						
	3							
4	1	1.150727	0.74	0.79	0.85	0.90	0.95	0.99
	2	0.706698	0.50	0.55	0.60	0.67	0.76	0.89
	3	0.679596						
	4							
5	1	1.115718	0.75	0.80	0.85	0.90	0.95	0.99
	2	0.645384	0.50	0.56	0.61	0.68	0.77	0.89
	3	0.532445	0.67	0.71	0.75	0.79	0.86	0.94
	4	0.583273						
	5							
6	1	1.093929						
	2	0.612330						

n		0.75	0.80	0.85	0.90	0.95	0.99
3	0.474330	0.75	0.80	0.85	0.90	0.95	0.99
4	0.442920	0.50	0.55	0.61	0.68	0.76	0.89
5	0.522759	0.67	0.71	0.75	0.80	0.86	0.93
6		0.54	0.57	0.61	0.66	0.73	0.84

7	n		0.75	0.80	0.85	0.90	0.95	0.99
	1	1.079055	0.75	0.80	0.85	0.90	0.95	0.99
	2	0.591587	0.50	0.55	0.61	0.68	0.77	0.89
	3	0.442789	0.67	0.71	0.75	0.80	0.86	0.94
	4	0.387289	0.54	0.58	0.62	0.67	0.74	0.85
	5	0.387714	0.64	0.67	0.70	0.74	0.80	0.88
	6	0.480648						
	7							

8	n		0.75	0.80	0.85	0.90	0.95	0.99
	1	1.068752	0.75	0.80	0.85	0.90	0.95	0.99
	2	0.577339	0.50	0.55	0.61	0.68	0.77	0.90
	3	0.422889	0.67	0.71	0.75	0.80	0.86	0.94
	4	0.356967	0.54	0.58	0.62	0.67	0.74	0.85
	5	0.334089	0.64	0.67	0.70	0.74	0.80	0.89
	6	0.349907	0.55	0.58	0.61	0.65	0.71	0.81
	7	0.449338						
	8							

TABLE 7.4 (Continued)

n	i	$Ez_{i+1}-Ez_i$	0.75	0.80	0.85	0.90	0.95	0.99
9	1	1.060046	0.75	0.80	0.85	0.90	0.95	0.99
	2	0.566942	0.50	0.55	0.61	0.68	0.77	0.89
	3	0.409157	0.67	0.71	0.75	0.80	0.86	0.94
	4	0.337763	0.54	0.58	0.62	0.67	0.75	0.86
	5	0.304777	0.63	0.67	0.70	0.74	0.80	0.89
	6	0.297949	0.55	0.58	0.61	0.66	0.72	0.82
	7	0.322189	0.62	0.64	0.67	0.71	0.76	0.85
	8	0.424958						
	9							
10	1	1.053606	0.75	0.80	0.85	0.90	0.95	0.99
	2	0.559013	0.50	0.55	0.61	0.68	0.77	0.90
	3	0.399100	0.67	0.71	0.75	0.80	0.86	0.94
	4	0.324470	0.54	0.58	0.62	0.68	0.75	0.85
	5	0.286163	0.63	0.67	0.71	0.75	0.81	0.89
	6	0.269493	0.55	0.58	0.62	0.66	0.72	0.81
	7	0.271645	0.62	0.65	0.68	0.71	0.76	0.85
	8	0.300869	0.55	0.58	0.61	0.64	0.69	0.79
	9	0.405316						
	10							
11	1	1.048411	0.75	0.80	0.85	0.90	0.95	0.99
	2	0.552769	0.49	0.55	0.61	0.68	0.77	0.90
	3	0.391410	0.67	0.71	0.75	0.80	0.86	0.94
	4	0.314705	0.54	0.58	0.63	0.68	0.75	0.86
	5	0.273745	0.64	0.67	0.71	0.75	0.81	0.89
	6	0.251386	0.55	0.58	0.62	0.66	0.72	0.82
	7	0.243928						
	8	0.251548						

	9	0.283879	0.62	0.64	0.68	0.71	0.77	0.85
	10	0.389071	0.55	0.58	0.61	0.64	0.70	0.79
	11		0.60	0.63	0.65	0.69	0.74	0.82
12	1	1.044137						
	2	0.547721						
	3	0.385338	0.75	0.79	0.84	0.90	0.95	0.99
	4	0.307221	0.50	0.55	0.61	0.68	0.78	0.89
	5	0.263737	0.67	0.71	0.75	0.80	0.86	0.94
	6	0.238797	0.54	0.58	0.62	0.67	0.74	0.85
	7	0.226264	0.64	0.67	0.70	0.75	0.81	0.89
	8	0.224477	0.55	0.58	0.62	0.66	0.72	0.82
	9	0.235630	0.62	0.64	0.68	0.71	0.77	0.85
	10	0.269966	0.55	0.58	0.61	0.65	0.70	0.79
	11	0.375356	0.60	0.63	0.66	0.69	0.74	0.82
	12		0.55	0.57	0.60	0.63	0.68	0.76
13	1	1.040555						
	2	0.543556						
	3	0.380417	0.75	0.80	0.85	0.90	0.95	0.99
	4	0.301300	0.50	0.55	0.61	0.68	0.77	0.89
	5	0.256437	0.67	0.71	0.75	0.80	0.86	0.94
	6	0.229515	0.54	0.58	0.63	0.68	0.75	0.86
	7	0.213966	0.64	0.67	0.71	0.75	0.81	0.90
	8	0.207205	0.55	0.58	0.62	0.66	0.72	0.82
	9	0.209131	0.62	0.65	0.68	0.72	0.77	0.85
	10	0.222667	0.55	0.58	0.61	0.65	0.70	0.79
	11	0.258323	0.60	0.63	0.66	0.69	0.74	0.82
	12	0.363582	0.55	0.57	0.60	0.64	0.68	0.76
	13		0.59	0.61	0.64	0.67	0.72	0.79

Table 7.2 (continued)

n	i	$Ez_{i+1}-Ez_i$	0.75	0.80	0.85	0.90	0.95	0.99
14	1	1.037513	0.75	0.79	0.85	0.90	0.95	0.99
	2	0.540059	0.49	0.54	0.61	0.68	0.77	0.90
	3	0.376352	0.67	0.71	0.75	0.80	0.86	0.94
	4	0.296496	0.54	0.58	0.62	0.68	0.74	0.86
	5	0.250650	0.64	0.67	0.71	0.75	0.81	0.89
	6	0.222377	0.55	0.58	0.62	0.66	0.73	0.82
	7	0.204885	0.62	0.65	0.68	0.72	0.77	0.85
	8	0.195165	0.55	0.58	0.61	0.65	0.70	0.79
	9	0.192209	0.60	0.63	0.66	0.69	0.74	0.82
	10	0.196679	0.55	0.57	0.60	0.64	0.68	0.77
	11	0.211875	0.59	0.61	0.64	0.67	0.72	0.79
	12	0.248409	0.55	0.57	0.59	0.62	0.67	0.75
	13	0.353334						
	14							
15	1	1.034894	0.75	0.80	0.84	0.90	0.95	0.99
	2	0.537085	0.51	0.56	0.62	0.69	0.78	0.90
	3	0.372934	0.68	0.71	0.76	0.80	0.86	0.94
	4	0.292518	0.54	0.58	0.62	0.67	0.75	0.86
	5	0.245947	0.64	0.67	0.71	0.75	0.81	0.89
	6	0.216712						
	7	0.197893						

Left block (rows 8–15):

k	value	c1	c2	c3	c4	c5	c6
8	0.186266	0.82	0.72	0.66	0.62	0.58	0.55
9	0.180402	0.85	0.77	0.72	0.68	0.65	0.62
10	0.180072	0.79	0.70	0.65	0.61	0.58	0.55
11	0.186347	0.82	0.74	0.69	0.66	0.63	0.61
12	0.202727	0.77	0.68	0.64	0.60	0.57	0.55
13	0.239842	0.79	0.72	0.67	0.64	0.62	0.59
14	0.344309	0.75	0.67	0.63	0.60	0.57	0.55
15		0.77	0.70	0.66	0.63	0.61	0.59

Right block (rows 1–16):

k	value	c1	c2	c3	c4	c5	c6
1	1.032617	0.99	0.95	0.90	0.85	0.80	0.75
2	0.534521	0.89	0.78	0.69	0.62	0.56	0.51
3	0.370021	0.94	0.86	0.80	0.76	0.72	0.68
4	0.289169	0.86	0.75	0.68	0.63	0.58	0.54
5	0.242049	0.89	0.81	0.75	0.71	0.67	0.64
6	0.212103	0.82	0.72	0.66	0.62	0.58	0.55
7	0.192338	0.85	0.77	0.72	0.68	0.65	0.62
8	0.179407	0.79	0.71	0.65	0.61	0.58	0.55
9	0.171667	0.82	0.74	0.69	0.66	0.63	0.60
10	0.168476	0.77	0.69	0.64	0.60	0.58	0.55
11	0.170026	0.80	0.72	0.68	0.64	0.62	0.60
12	0.177619	0.75	0.67	0.63	0.60	0.57	0.55
13	0.194859	0.77	0.70	0.66	0.63	0.61	0.59
14	0.232350	0.73	0.66	0.62	0.59	0.57	0.55
15	0.336283						
16							

16

Fertig for small r. For $r \geqslant 15$, k is equal to the integer closest to $r/3$. The test statistic $P = \sum_{i=k+1}^{r-1} l_i / \sum_{i=1}^{r-1} l_i$, which is discussed in more detail in Section 5.2.2(d), has a distribution which more closely approximates a beta than does S. In fact, beta tables, or the approximation described in Section 5.1.2(a) for distribution percentiles of a negative logarithm of a beta variate, can be used for r as small as 3.

The test of van Montfort (1970) makes use of his conclusions concerning the leaps and is designed specifically to test that untransformed observations (such as largest rainfall amounts) are from a Type I as opposed to a Type II distribution of largest values. Because of symmetry considerations, the test is also applicable for testing that *untransformed* data are from a (Type I) extreme-value distribution of smallest values versus the alternative that they are from a three-parameter Weibull distribution. The test statistic, which is approximately normally distributed, involves the sample correlation coefficient of $n-1$ modified leaps and functions of i and n related to the leaps.

An example of the use of Table 7.2 is now given.

Example.   Listed below are ordered observations $t_{(i)}$ from a sample (ignition times) of size 16, censored at the ninth smallest observation, that is, $n = 16$, $r = 9$. One wishes to test the hypothesis that this sample was drawn from a two-parameter Weibull distribution, with a three-parameter Weibull an appropriate alternative. The significance level is .20. Accordingly, one first forms

$$l_i = \frac{x_{(i+1)} - x_{(i)}}{E(Z_{i+1}) - E(Z_i)}, \quad i = 1, \ldots, 8,$$

where $x_{(i)} = \ln(t_{(i)})$, and the $E(Z_{i+1}) - E(Z_i)$ are read from Table 7.2 and then computes a value for

$$S = \frac{\sum_{5}^{8} l_i}{\sum_{1}^{8} l_i}.$$

For the data:

$$t_{(1)} = 15.5, \quad t_{(4)} = 20.5, \quad t_{(7)} = 26.5,$$
$$t_{(2)} = 16.5, \quad t_{(5)} = 23.1, \quad t_{(8)} = 33.8,$$
$$t_{(3)} = 19.5, \quad t_{(6)} = 23.5, \quad t_{(9)} = 33.9,$$

one calculates $S = .68$, and therefore the hypothesis that these data were drawn from a two-parameter Weibull distribution would be rejected at a 20% significance level. (One rejects the hypothesis if the calculated value of S *exceeds* its tabulated percentile at 1 minus the appropriate level of significance.)

To use P to test this hypothesis, one first finds from Table III of Mann and Fertig (1974) for $r = 9$ and $\alpha = .20$ that $k = 2$. Then, one rejects the hypothesis if $\sum_{i=3}^{8} l_i / \sum_{i=1}^{8} l_i$ exceeds the 85th distribution percentile of a beta variate with parameters $r - k - 1 = 5$ and $k = 3$. The calculation of the value of the statistic P for these data is left as an exercise for the reader.

7.1.3 Tests of Normality

Lilliefors (1967) has developed an analog of the Kolmogorov-Smirnov test applicable to distributions that are assumed to be, under H_0, normal with unknown mean and variance. It involves the estimate of $F_0(x_i; \boldsymbol{\theta})$ incorporating unbiased estimates of the location and scale parameters.

The computing form of Lilliefors' test statistic is

$$\hat{D}_n = \max_{1 \le i \le n} \left(\hat{\delta}_i \right),$$

with

$$\hat{\delta}_i = \max \left[F_0\left(X_{(i)}; \overline{X}, s\right) - \frac{i-1}{n}, \frac{i}{n} - F_0\left(X_{(i)}; \overline{X}, s\right) \right],$$

where s^2 is $\sum_{i=1}^{n}(X_{(i)} - \overline{X}^2)/(n-1)$ and $F_0(X; \overline{X}, s)$ is the cumulative standard normal distribution function $\Phi(Y)$, with $Y = (X - \overline{X})/s$. If the value of $\hat{D}_n$ exceeds the critical value in Table 7.3 corresponding to the test significance level α, one rejects the hypothesis that the observations are from a normal, or Gaussian, distribution.

The values tabulated by Lilliefors are based on a Monte Carlo sample size of only 1000 but have been subjected to some smoothing.

As an example of the use of Lilliefors' analog of the Kolmogorov-Smirnov test of normality, consider eight life-test failure times whose natural logarithms are .26, .53, .88, 1.22, 1.76, 2.44, 3.41, and 4.90. To test that failure times are from a two-parameter lognormal population, one first calculates $\overline{x} = 1.92$ and $s^2 = [47.30 - 8(1.92)^2]/7 = 2.55$. Then, for each observed log failure time $x_{(i)}, i = 1, \ldots, 8$, one determines $y_i = (x_{(i)} - \overline{x})/s : y_1 = -1.03$, $y_2 = -.87$, $y_3 = -.65$, $y_4 = -.44$, $y_5 = -.10$, $y_6 = .33$, $y_7 = .93$, $y_8 = 1.86$. Tables of normal deviates are then used to compute the corresponding values of $F_0(x_{(i)}; \overline{x}, s), i = 1, \ldots, 8 : .152, .192, .258, .330, .460, .629, .824, .968$. The value of D_n is therefore given by $\frac{4}{8} - .330 = .170$—not sufficiently large to reject at confidence level .20 an hypothesis of normality for log failure time.

Table 7.3 **Critical values of $\hat{D}_n$ corresponding to test significance level**

Sample Size n	Significance Level				
	.20	.15	.10	.05	.01
4	.300	.319	.352	.381	.417
5	.285	.299	.315	.337	.405
6	.265	.277	.294˙	.319	.364
7	.247	.258	.276	.300	.348
8	.233	.244	.261	.285	.331
9	.223	.233	.249	.271	.311
10	.215	.224	.239	.258	.294
11	.206	.217	.230	.249	.284
12	.199	.212	.223	.242	.275
13	.190	.202	.214	.234	.268
14	.183	.194	.207	.227	.261
15	.177	.187	.201	.220	.257
16	.173	.182	.195	.213	.250
17	.169	.177	.189	.206	.245
18	.166	.173	.184	.200	.239
19	.163	.169	.179	.195	.235
20	.160	.166	.174	.190	.231
25	.142	.147	.158	.173	.200
30	.131	.136	.144	.161	.187
Over 30	$\dfrac{.736}{\sqrt{N}}$	$\dfrac{.768}{\sqrt{N}}$	$\dfrac{.805}{\sqrt{N}}$	$\dfrac{.886}{\sqrt{N}}$	$\dfrac{1.031}{\sqrt{N}}$

7.1.4 Chi-Square Test for Goodness of Fit

The chi-square test for goodness of fit can be applied to any distribution $F_X(x)$. In using this test, one divides the real x axis into k intervals, the first and last being infinite intervals. The points a_i (with $a_0 = -\infty$) dividing the intervals are chosen so that $F_X(x; \boldsymbol{\theta})$ is continuous at each a_i. In the case of a discrete variable, each a_i is chosen as a point of zero probability.

Then p_i, the probability that X will fall in the ith interval, is given by $F_X(a_i; \boldsymbol{\theta}) - F_X(a_{i-1}; \boldsymbol{\theta})$, where here it is assumed that $\boldsymbol{\theta}$ is known. If the observations are regarded as trials with $a_{i-1} < X < a_i$ regarded as a success and $X < a_{i-1}$ or $X > a_i$ regarded as a failure, p_i is the probability of success. If n is the total number of observations of X, the expected number, e_i, of observations falling in the ith interval between a_{i-1} and a_i is given by

$$e_i = n \cdot p_i.$$

If the observable number of observations lying in the ith interval is 0_i, and the intervals are chosen so that, for all p_i, p_i is greater than 0, then

$$W = \sum_{i=1}^{k} \frac{(0_i - e_i)^2}{e_i}$$

has a distribution that approaches chi-square with $k-1$ degrees of freedom as $n \rightarrow \infty$. In general, n should be large enough so the e_i are greater than 5 for all i. Thus, if each e_i is greater than 5, one rejects, at significance level α, the hypothesis that a sample is from $F_X(x; \boldsymbol{\theta})$ if W, calculated from sample data and the specified distribution function, exceeds the $100(1-\alpha)$ th percentile of a chi-square distribution.

If the parameter values are not specified in testing $F_X(x; \boldsymbol{\theta})$, but are estimated from the data, W has an asymptotic chi-square distribution as above, but the number of degrees of freedom is diminished accordingly.

Suppose that $\boldsymbol{\theta} = \theta_1, \ldots, \theta_s$ and that $\hat{p}_i$ represents $F_X(a_i; \hat{\boldsymbol{\theta}}) - F_X(a_{i-}; \hat{\boldsymbol{\theta}})$, where $\hat{\boldsymbol{\theta}}$ is the maximum-likelihood (or equivalent) estimate of $\boldsymbol{\theta}$, *based on the cell frequencies only*. Then

$$\hat{W} = \sum_{i=1}^{k} \frac{(0_i - n\hat{p}_i)^2}{n\hat{p}_i}$$

has an asymptotic chi-square distribution with $k-s-1$ degrees of freedom.

Chernoff and Lehmann (1954) prove a theorem specifying the asymptotic distribution of W^*, $\hat{W}$, with maximum-likelihood estimates of the parameters *based on all the observations*, as a sum of weighted chi-square variates, where the weights may depend on the s unknown parameters. From results of Patnaik (1949), discussed in Sections 5.1.2 and 5.2.3, it can be seen that W^* is approximated by an asymptotic weighted chi-square distribution that in turn could be approximated for a particular set of data by a central chi-square distribution if the asymptotic mean and variance of W^* were known. Chernoff and Lehmann, in fact, show that between 0 and s chi-square degrees of freedom are gained by using W^* rather than $\hat{W}$. Exactly how many are gained, however, cannot be predicted from the sample. Lilliefors (1967) states that, in making Monte Carlo comparisons of chi-square and Kolmogorov analogs for the normal distribution with unknown mean and standard deviation, he found a test based on W^*, using four intervals and a sample of size 20, to have rejected a true hypothesis of normality with probability .11 for chi-square with $1 = 4 - 2 - 1$ degrees of freedom when the significance level was .05. The actual number of degrees of freedom for this sample appears to be, as predicted by Chernoff and Lehmann, between 1 and 3, where presumably the true mean was 0 and the variance 1.

Suppose that one observes ordered failure times of 1.3, 1.4, 1.5, 1.9, 2.3, 2.4, 2.4, 2.9, 3.1, 3.6, 3.8, 3.9, 4.3, 4.6, 4.8, 4.8, 5.1, 5.2, and 6.8 hours. A chi-square test that these failure times belong to an exponential population is made by, for example, letting $a_1 = 2.1$, $a_2 = 4.7$, and $a_3 = 15.0$. Then $\bar{x}$, based on the midpoints of the three cells, is $[5(1.05) + 10(3.4) + 5(9.85)]/20 = 4.42$, and

$$\hat{p}_1 = 1 - \exp\left(\frac{-2.1}{4.42}\right) - 0 = .38, \qquad \hat{p}_2 = 1 - .34 - (1 - .62) = .28,$$

$$\text{and} \quad \hat{p}_3 = 1 - (.38 + .28) = .34,$$

so that

$$\chi^2(1) = \frac{(7.6 - 5)^2}{7.6} + \frac{(5.6 - 10)^2}{5.6} + \frac{(6.8 - 5)^2}{6.8} = 4.82.$$

If $\alpha = .05$, the hypothesis is rejected, since $\chi^2_{.95}(1) = 3.84$. If θ is estimated by $\bar{x} = 3.40$, based on all the observations, one obtains

$$\hat{p}_1 = 1 - \exp\left(\frac{-2.1}{3.40}\right) = .46, \qquad \hat{p}_2 = 1 - .25 - (1 - .53) = .29,$$

$$\text{and} \quad \hat{p}_3 = 1 - (1 - .25) = .25,$$

so that the chi-square test statistic is equal to $(9.2 - 5)^2/9.2 + (5.8 - 10)^2/5.8 + (5 - 5)^2/5 = 5.44$. Here the number of degrees of freedom is somewhere between 1 and 2. Since $\chi^2_{.95}(2)$ is 5.99, one would be reluctant to reject the hypothesis at significance level .05 on the basis of this result.

PROBLEMS

7.1. Show that, if Z has an F distribution with k_1 and k_2 degrees of freedom, respectively,

$$P\left[\frac{1}{F_{1-\alpha/2}(k_2, k_1)} \leqslant Z \leqslant F_{1-\alpha/2}(k_1, k_2)\right] = 1 - \alpha.$$

7.2. Show that, if Z has an F distribution with $2k_1$ and $2k_2$ degrees of freedom, respectively,

$$S = \frac{(k_1/k_2)Z}{1 + (k_1/k_2)Z}$$

has a beta distribution with parameters k_1 and k_2, respectively.

7.2 MONTE CARLO SIMULATION OF PROBABILITY DISTRIBUTIONS

Because readers of this book may find it necessary on occassion to simulate various probability distributions by Monte Carlo computer techniques, certain aspects of such simulation are discussed in this section. The generation of random numbers or pseudo-random and quasi-random numbers from uniform and various other probability distributions is considered, along with the determination of Monte Carlo sample size requirements and several methods of obtaining percentiles of distributions of estimators, based on the generated variates.

7.2.1 Random Number Generators

Early random number generators were typically based on gambling devices. Because the most famous gambling Mecca was, at least in earlier days, Monte Carlo, the method took on the name of this Riviera town. Among more sophisticated random number generators are those based on successive digits of π (see Hall, 1873) and the "needle of Buffon" (see Mantel, 1953), which is theoretically dropped on a surface so as to yield a digit where the needle falls. These methods are far too slow, however, to be used in connection with modern electronic computers.

Random number generators that have been employed successfully on electronic computers are automated roulettes, such as that of the RAND Corporation (1955), and electronic circuits, which use various types of *noise*. These generators, too, tend to be quite slow and to require frequent testing for stability. Another disadvantage is the nonrepeatability of computations.

Because of the problems involved with true random number generators, the current most common source of uniformly distributed random numbers is a computer algorithm of some sort which generates a pseudo-random sequence. The method of this type most often used today is the *multiplicative congruential* or *power-residue* method, or some variation thereof. In the straightforward multiplicative congruential method, one multiplies the current random integer I by a constant multiplier k and keeps as the new random integer J the difference between the product Ik and a specified modulus m. The fraction $U = J/m$ provides a uniformly distributed pseudo-random variate on the unit interval.

Marsaglia (1968) has pointed out the following serious defect of the multiplicative congruential method. If k-tuples $(u_1, \ldots, u_k)$, $(u_2, \ldots, u_{k+1})$, ... of numbers generated by this method are viewed as points of a unit cube of k dimensions, all the points will be found to lie in a relatively small number of parallel hyperplanes. Such a property clearly makes these sequences of numbers unsatisfactory for certain Monte Carlo problems.

Because of Marsaglia's discovery, the use of mixed-congruential procedures has been widely adopted.

Marsaglia and Bray (1968) suggest the use of a composite generator that mixes three congruential generators; one to fill, for example, 128 storage locations, one to choose a location from the 128, and a third thrown in for good measure. Under the assumption that $N_1, N_2, ..., N_{128}, L, M$, and K have been assigned initial odd random integer values, a FORTRAN H routine that can be used in conjunction with an IBM 360 computer to generate realizations of uniformly distributed random variates $\{U_i\}$ on the interval $0 < U < 1$ is given by

$$L = L * ML$$
$$M = M * MM$$
$$J = 1 + IABS(L)/1677216$$
$$U = .5 + FLOAT(N(J) + L + M) * .23283064E - 9$$
$$K = K * MK$$
$$N(J) = K,$$

where $.23283064E - 9 = 2^{-32}$. Here *ML, MM,* and *MK* can be chosen in the form $8m \pm 3$ to ensure long periods before the process begins to repeat, for example, $ML = 65539, MM = 33554433$, and $MK = 362436069$.

Marsaglia and Bray also give examples of composite FORTRAN generators that work well for IBM 7094 and SRU 1108 computers. Tausworthe (1965) gives a procedure, involving primitive polynomials, of generating numbers by linear recurrence methods that do not have the unfortunate unit-cube property pointed out by Marsaglia (1968) for the simple congruential method. The sequences generated by Tausworthe's procedure actually consist of *quasi-random* numbers, numbers that lay no claim to randomness but are extremely useful for certain classes of Monte Carlo calculations. Other methods of generating quasi-random numbers are reviewed by Halton (1970).

Many statistical tests for randomness have been applied to pseudo-random sequences, for example, the test of Kendall and Smith (1954) for proportion of occurrences of digit values, a test for serial correlation suggested by McLaren and Marsaglia (1965), and the test of Gruenberger and Mark (1951) for the distribution of distances between points in the unit square. It should be pointed out, however, that a sample of numbers, no matter how random, cannot be expected to pass a large number of statistical tests at an arbitrary level of significance.

7.2.2 Generation of Samples from Specified Distributions

Once a uniformly distributed random, pseudo-random, or quasi-random number has been generated, it may be converted to a sample variate from some other particular distribution. Ordinarily, one desires to draw samples of size n from a specified density function $f_X(x)$. The number N of size n samples to be drawn is discussed in Section 7.2.3.

As discussed by Spanier and Gelbard (1969), one constructs a sequence of unordered numbers $x_1, x_2, \ldots,$ which may be thought of as realizations of random variates $X_1, X_2, \ldots, -\infty < X_i < \infty$, such that, for any k,

$$P[a < X_i \leqslant b] = \int_a^b f_X(x)\,dx, \quad -\infty < a < b < \infty, \tag{7.1}$$

and

$$P[a < X_{i_1}, \ldots, X_{i_k} \leqslant b] = \left[\int_a^b f_X(x)\,dx\right]^k, \text{for } i_1, \ldots, i_k \text{ all different.} \tag{7.2}$$

This can be accomplished provided that one has available $u_1, u_2, \ldots,$ which may be thought of a realizations of uniformly distributed random variates $U_1, U_2, \ldots, 0 \leqslant U_i \leqslant 1$, such that, for any k,

$$P[a < U_i \leqslant b] = \int_a^b dx = b - a, \quad 0 < a < b < 1, \tag{7.3}$$

and

$$P[a < U_{i_1}, \ldots, U_{i_k} \leqslant b] = (b - a)^k, \text{for } i_1, \ldots, i_k \text{ all different.} \tag{7.4}$$

Clearly, if Equations (7.3) and (7.4) are not satisfied because the u's generated are realizations of pseudo-random or quasi-random numbers that are not truly random, (7.1) and (7.2) may not be satisfied. In the following, it is assumed that (7.3) and (7.4) can be satisfied for all practical purposes. The reader is referred to the discussion in Section 7.2.1 for methods of accomplishing this.

To convert the uniformly distributed random number to one with another specified density function, $f_X(x)$, one uses the fact that $F_X(x)$, the cumulative distribution function associated with $f_X(x)$, is uniformly distributed over the unit interval. Thus, for U a random variate uniform over $(0, 1)$, and u a realization of U,

$$u = \int_{-\infty}^x f_X(t)\,dt = F_X(x).$$

If X has a cumulative distribution function that can be expressed in closed form, it is simple to convert values $u_1, u_2, \ldots$ to values $x_1, x_2, \ldots$ by direct inversion of the distribution function. For example, if

$$u_i = 1 - \exp\left(\frac{-x_i}{\theta}\right),$$

then

$$x_i = -\theta \ln(1 - u_i), \quad i = 1, 2, \ldots.$$

If X has a two-parameter Weibull distribution,

$$u_i = 1 - \exp\left[-\left(\frac{x_i}{\delta}\right)^\beta \right],$$

so that

$$x_i = \delta \left[-\ln(1 - u_i) \right]^{1/\beta}.$$

Also, if X has a first asymptotic distribution of smallest values,

$$x_i = \ln\left\{ \delta \left[-\ln(1 - u_i) \right]^{1/\beta} \right\} \equiv \eta + \xi \ln\left[\ln\left(\frac{1}{1 - u_i}\right) \right].$$

In practice, incidentally, one can use u_i in place of $1 - u_i$ above, since both are theoretically from a distribution that is uniform over $(0, 1)$.

Many distributions of interest cannot be expressed in closed form and thus cannot be generated in such a direct manner. Moreover, even for distributions that can be expressed in closed form, methods are often available that are more complicated but faster to execute on a computer than direct inversion of the distribution function. In the following sections, efficient random number generation from various distributions treated in this book is discussed.

(a) **Exponential, Weibull, and Extreme-Value Distributions.** The following device suggested by Marsaglia (1961) allows one to generate an exponentially distributed random variate in considerably less time than is required to calculate the negative of the logarithm of a uniformly distributed variate. In using this method, one selects the minimum of a random number k of random variates uniformly distributed on $(0, 1)$ and then adds a random integer m. Let the random integers k and m take on values according to the following schedule:

Value of k	Probability	Total	Value of m	Probability	Total
1	.58	.58	0	.63	.63
2	.29	.87	1	.23	.86
3	.10	.97	2	.09	.95
4	.02	.99	3	.03	.98
.			.		
.			.		
.			.		

If $U_1, U_2, \ldots$, is a sequence of independent uniform random variables on $(0,1)$, the random variable

$$T = m + \min(U_1, \ldots, U_k)$$

has the exponential distribution with scale parameter 1. The variate T also has a Weibull distribution with $\mu = 0$, $\delta = 1$, and $\beta = 1$ (a reduced parameter-free Weibull distribution). One can, of course, construct a variate X from the reduced extreme-value distribution (with location parameter η equal to 0 and scale parameter ξ equal to 1) by forming the logarithm of T.

(b) Normal, Lognormal, and Birnbaum-Saunders Distributions. Many procedures have been suggested for generating normal or Gaussian variates from uniformly distributed random variates. The reader is referred to Hull and Dobell (1962) for a survey of such methods. The procedure given here is half again as long as the very fastest methods (see Marsaglia, McLaren, and Bray, 1964), but is much easier to program. Furthermore, it requires very little storage space. The method, due to Marsaglia and Bray (1964), consists of generating a normally distributed random variate X from uniformly distributed random variates $U_1, U_2, \ldots$, on $(0,1)$ by putting $X = 2(U_1 + U_2 + U_3 - 1.5)$ 86.38% of the time and $X = 1.5(U_1 + U_2 - 1)$ 11.07% of the time. For the remaining 2.55% of the time a more complicated procedure is used to guarantee that the resulting mixture is correct.

The 2.55% of the time during which the simple, fast procedures are not used is broken into two parts. With probability .0228002039, X is generated by what is known as a rejection technique, that is, one first defines the density $g_3(x)$ as

$$g_3(x) = \begin{cases} ae^{-(1/2)x^2} - b(3 - x^2) - c(1.5 - |x|), & |x| < 1, \\ ae^{-(1/2)x^2} - d(3 - |x|)^2 - c(1.5 - |x|), & 1 < |x| < 1.5, \\ ae^{-(1/2)x^2} - d(3 - |x|)^2, & 1.5 < |x| < 3, \\ 0, & |x| > 3, \end{cases}$$

where $a = 17.49731196$, $b = 4.73570326$, $c = 2.15787544$, and $d = 2.36785163$. Then one generates realizations y and z of the pair of random numbers Y and Z, respectively, where $Z = 6U_4 - 3$, $Y = .358U_5$, and U_4 and U_5 are uniform on $(0, 1)$. If $Y < g_3(z)$, one puts $X = Z$. Otherwise, one chooses a new pair (Y, Z) and repeats the process. Approximately 47% of the time y will be less than $g_3(z)$.

The method of generating the tail of the distribution is also a rejection procedure, used only $.26997961\%$ of the time, and is as follows. One forms pairs (y, z), realizations of (Y, Z) such that

$$z = v_1 \left[\frac{9 - 2\ln(v_1^2 + v_2^2)}{v_1^2 + v_2^2} \right]^{1/2},$$

$$y = v_2 \left[\frac{9 - 2\ln(v_1^2 + v_2^2)}{v_1^2 + v_2^2} \right]^{1/2},$$

where v_1 and v_2 are realizations of random variates uniform on $[-1, 1]$, conditioned by $v_1^2 + v_2^2 < 1$. Then one puts $x = z$ if $|z| \geq 3$ or $x = y$ if $|y| \geq 3$. If $|y|$ and $|z|$ are both less than 3, a new pair is generated. If both $|z| \geq 3$ and $|y| \geq 3$, then either $x = z$ or $x = y$.

The resulting X generated by the specified mixture of the four methods has a normal distribution with mean 0 and standard deviation 1. To transform this to a normal variate W with mean μ_0 and standard deviation σ_0, one puts $W = \mu_0 + \sigma_0 X$. To transform X to a reduced lognormal variate V, one puts $V = \exp(X)$. Finally, if T is a variate from a reduced Birnbaum-Saunders distribution [given by Equation (5.36)], with $X = T^{1/2} - T^{-1/2}$, then $T^{1/2} = X/2 + (1 + X^2/4)^{1/2} > 0$.

If one is unconcerned with the speed of the process of generation, approximate Gaussian variates can be generated with the aid of the central limit theorem. In particular,

$$X = \sum_{i=1}^{m} \pm U_i$$

becomes normally distributed with mean 0 and variance $m/3$, as $m \to \infty$. A value of 12 is often used for m.

A simple rejection technique that generates realizations $\{x\}$ of half-normal variates is given by Kahn (1954). In this procedure, realizations (u_1, u_2) of two uniform random variates (U_1, U_2) are generated. Then $y = -\ln u_1$ and $z = -\ln u_2$. If $(y - 1)^2/2 \leq z$, then $x = y$; otherwise a new set

(u_1, u_2) is generated and the process is repeated. This technique has an efficiency (1-proportion of rejections) of .76.

(c) Gamma, Chi-Square, F, and Beta Distributions. Gamma-distributed random variates with integer shape parameters can be formed as sums of exponentially distributed random variates, that is, if

$$Y = \sum_{i=1}^{m} X_i,$$

where each X is from a distribution with density $f_X(x) = \lambda \exp(-\lambda x)$, the probability density function of Y is given by

$$f_Y(y) = \frac{\lambda^m y^{m-1}}{\Gamma(m)} \exp(-\lambda y).$$

Since an exponential variate, when multiplied by 2, is a chi-square variate with 2 degrees of freedom, chi-square variables with even degrees of freedom can be generated in the same way.

Random variates from an F distribution with even degrees of freedom can then of course be generated as ratios of appropriate independent chi-squares over their respective (even) degrees of freedom.

Beta variates can be obtained from chi-square variates, using

$$V(a,b) = \frac{\chi^2(2a)}{\chi^2(2a) + \chi^2(2b)},$$

where $V(a,b)$ is beta with parameters a and b, and $\chi^2(2a)$ and $\chi^2(2b)$ are independent chi squares with $2a$ and $2b$ degrees of freedom, respectively.

(d) Poisson Distribution. To generate a Poisson variate X with

$$P(X=k) = \frac{\lambda^k \exp(-\lambda)}{k!}, \quad \lambda > 0, k = 0, 1, \ldots,$$

a rejection technique can be used. Let $j=0$ and $y_j = u_0$, with U_j uniform on $(0,1)$, $j=0,1,\ldots$. If $y_j \leqslant \exp(-\lambda)$, one lets $k=j$. Otherwise, one increments j by 1 and lets $y_j = u_j y_{j-1}$. Then one returns to the test on y_j, and continues. The average number of $\{u_j\}$ used in each selection of k is $\lambda + 1$.

7.2.3 Monte Carlo Sample-Size Requirements

An important consideration in the generation of samples from specified probability distributions is the number N of samples to be generated in order to achieve some required precision in the final results. A theorem that can be helpful in determining the size of N is the following, due to Mosteller (1946). If an ordered sample $Y_{(1)}, \ldots, Y_{(N)}$ is drawn from $g_Y(y)$, where $g_Y(y)$ is continuous and does not vanish in the neighborhood of y_λ, and where

$$\lim_{N \to \infty} \frac{i}{N} = \lambda,$$

then $Y_{(i)}$, $i = 1, \ldots, k$, is asymptotically normally distributed with mean y_λ, defined by

$$\int_{-\infty}^{y_\lambda} g_Y(t) \, dt = \lambda,$$

and variance

$$\sigma_\lambda^2 = \frac{\lambda(1 - \lambda)}{N \left[g_Y(y_\lambda) \right]^2}.$$

Since λ represents a distribution proportion, one need only estimate $g(y_\lambda)$ in order to estimate the variance, as a function of N, of his generated estimate $Y_{(i)}$ of y_λ, the 100λth distribution percentile. For example, the statistic S, described in Section 7.1.2 for use in a goodness-of-fit test for the extreme-value or two-parameter Weibull distribution, was known to have approximately a beta distribution with integer parameters $\nu_1 = (r-1)/2$ and $\nu_2 = (r-1)/2$ or $\nu_1 = (r-2)/2$ and $\nu_2 = r/2$. For the largest value of λ of interest (.99), the combination of parameters yielding the smallest value of the beta density function corresponding to the 99th beta percentile, $v_{.99}$, was determined to be $\nu_1 = 1$ and $\nu_2 = 2$, that is, $r = 4$. (For $r = 3$, S has a uniform distribution.) Hence a value of 20,000 was determined as the Monte Carlo sample size. This guaranteed with 95% confidence (through the use of asymptotic normal theory) that the calculated value of $s_{.99} \cong v_{.99}$ would be correct to within a unit in the second significant digit, including roundoff error, for the smallest sample size or most severe censoring ($r = 3$ excluded) of the original samples of size n.

In general, of course, the difficulty lies in estimating $g_Y(y_\lambda)$. A few general rules of thumb are offered here, however, for determining N without actually estimating σ_λ^2. These apply to unimodal distributions or distributions with no mode and to values of λ such that $.01 < \lambda < .99$. For estimating percentiles of distributions of statistics based on a small sample size n or a small number r of ordered observations (10 or less), the Monte Carlo sample size N required to guarantee meaningful accuracy in lieu of information about σ_λ^2 is usually about 20,000. To guarantee the same accuracy for r or $r = n$ between 10 and 50, N should be around 10,000. One can, of course, obtain some notion of the precision of a Monte Carlo-generated distribution percentile by checking on the repeatability of his results for a given N. In Section 7.24 following, some techniques are described for minimizing required computer time in the generation of Monte Carlo samples of specified size.

7.2.4 Methods of Determining Distribution Percentiles

If, in generating an approximation to the distribution $F_S(s)$ of a statistic S, the number N of realizations of S generated is small, values of S can be ordered in obtaining Monte Carlo estimates of its distribution percentiles. The usual procedure, then, is to determine $S_{(i)}$ such that $i/(N+1)=\lambda$ in estimating s_λ, since $i/(N+1)$ is the expected value of $F_{S_{(i)}}(s)$.

If, however, a serious attempt is being made to simulate the distribution of the statistic S, the Monte Carlo sample size, N, will be of the order of several thousand. Then it is usually necessary, because of computer storage limitations, to devise other methods of determining percentiles of the distribution of S. One means of doing this is to keep count of the number of statistics S in the compartments of a histogram. The gradations can be uniformly distributed over the range of S (if it is finite), or finer gradations can be assigned to the areas of either greatest frequency or greatest interest.

(a) **Use of Histograms and Sampling over a Semi-infinite Range.** In simulating the distribution of $\tilde{\xi} > 0$ (see Section 5.2.3) over the positive half of the real line, the investigators generated and counted values in the compartments of a histogram on the unit interval through the use of a linear transformation performed on each statistic that sent prior approximations to the 1st and 99th percentiles to .1 and .9, respectively. The prior approximations were made by generating 1000 samples of $\tilde{\xi}$ and saving the 10th smallest and 10th largest values. The .1 and .9 points of the unit interval were chosen as the image points of the asymptotic approximations to the 1st and 99th percentiles in order that any reasonable errors in these approximations would not cause loss of information in the tails of the distribution. In determining the percentiles of $\tilde{\xi}$, a Monte Carlo sample size of 20,000 was used, and the number of histogram cells was 300.

(b) **Use of Generated Tables.** An alternative method of determining distribution percentiles can be used when the calculation of the statistic of interest is complicated and expensive in terms of computer time. In the Monte Carlo simulation described by Mann (1970) of the posterior distribution of series-system reliability, given subsystem data and prior distributions for subsystems, the jth value for subsystem reliability, $R_j, 0 \leqslant R_j \leqslant 1$, corresponding to a generated uniformly distributed variate must be obtained iteratively from the expression for its posterior density. Since this is very time consuming, efficiency can be achieved by calculating and tabulating 101 values of R_j corresponding to equally spaced values of $F(R_j)$ over the interval $(0, 1)$ and then sampling from the table by generating uniformly distributed random variates. Cubic interpolation should be used in this particular instance in determining the appropriate value of R_j corresponding to a particular value of any uniformly distributed random number u.

For values of u close to 0 or 1 (how close is determined by the data) a different table is sampled. In generating values for this second table, values of $F(R_j)$ are calculated directly from values of R_j so that no iteration is necessary, but the values of $F(R_j)$ are not equally spaced (making interpolation more difficult). The second table, which contains values of $F(R_j)$ much closer together than the one used for nonextreme values of R_j, is necessary because of the steepness of the curve relating R_j and $F(R_j)$ for values of R_j close to 0 or 1.

(c) **"Brunkizing" or Correcting Reversals.** In determining the percentiles listed in Table 5.13 of the distribution of the statistic used for obtaining confidence bounds for the scale parameter of the Birnbaum-Saunders distribution, Saunders and Mann (1971) used a procedure sometimes called "Brunkizing" (see Brunk, 1955). This method can be applied to independent samples that form a sequence of independent observations of a decreasing function.

Saunders and Mann generated the distribution of Z, where

$$D_Z(z) = P[Y<0, W_1>z]$$

with Y and W_1 appropriate random variates. They estimated $D_Z(z)$ by

$$\tilde{D}_Z(z) = \frac{1}{N} \sum_{j=1}^{N} \{ Y^{(j)}<0, W_1^{(j)}>z \},$$

where $\{A\}$ is the indicator of the event A, equal to 1 if true and 0 otherwise, and the superscript j indicates that the statistics Y and W_1 belong to the jth sample generated. In such a procedure, it is possible that $\tilde{D}_Z(z_1)=\epsilon_1$ and $\tilde{D}_Z(z_2)=\epsilon_2$, with $z_1<z_2$ and $\epsilon_1<\epsilon_2$, although $D_Z(z)$ is known to be monotonically decreasing in z. In such a case, both ϵ_1 and ϵ_2 are replaced by $(\epsilon_1+\epsilon_2)/2$. This process of correcting *reversals*, actually a maximum-likelihood procedure, is applied until a nondecreasing sequence is obtained.

(d) **Smoothing and Use of Prior Information.** A method of saving computer time that has been used successfully by, for example, Lilliefors in the generation of the critical values exhibited in Table 7.3 is *smoothing*. Lilliefors used a Monte Carlo sample size of only 1000 to generate the necessary critical values. The raw values were *smoothed*, however, by being fitted with a curve. If smoothing is to be used in obtaining distribution percentiles, one should use larger Monte Carlo sample sizes to spot-check for several widely spaced sample sizes.

Another method of saving computer time is to make use of prior information. For example, for any of the Kolmozorov-Smirnov analogs applying to distributions with unknown parameters, it is clear that critical values will be smaller than those applying to the classical test for completely specified distributions. Prior information obtained from initial investigations based on moderate Monte Carlo sample sizes can also be used. Such information allows one the option of ordering only the observations that are larger (or smaller) than some specified value. Since ordering is a very time-consuming computer operation, a great deal of expense can often be avoided in this way.

7.3 DISTRIBUTION-FREE METHODS

As discussed in Section 4.1, it is often difficult to single out a specific model to characterize the failure behavior of a system or a device. Consequently, a less conventional approach wherein the failure behavior is characterized merely by a property of the hazard rate is found to be quite useful. Such an approach has not only alleviated the task of specifying failure models but also initiated the development of a very comprehensive theory of reliability (Barlow and Proschan, 1965). Some basic properties of the hazard-rate function and their application to reliability problems are presented in this section. The material here is based on Barlow and Proschan (1965) and references given therein.

In Section 4.1, the failure-rate function (or the hazard rate, as it was referred to therein) was defined and briefly discussed. The hazard rate is a function of time and can be either monotone (increasing or decreasing) or nonmonotone. The monotone increasing hazard-rate function, which will be referred to as IHR, includes the class of failure models whose hazard rates increase with time. Such hazard rates are mainly due to wear, and typical failure models that give rise to these are the gamma and the Weibull disbributions with a shape parameter greater than 1. Similarly, DHR will describe the monotone decreasing hazard-rate function; such hazard rates are brought about mainly by a debugging or a work-hardening phenomenon. Typical failure models that give rise to a DHR function are the gamma and the Weibull distributions with a shape parameter less than 1. The exponential distribution, which has a constant hazard rate, is by convention included in either the IHR or the DHR class.

Hazard rates that are not monotone are rare in occurrence. However as pointed out in Section 4.6, if it is reasonable to assume that the failure model is logarithmic normal, then a nonmonotonic hazard rate is apt to occur. A situation wherein a nonmonotone failure rate can be observed is in the study of lifetimes of high-speed steel drills. It has been empirically verified by Singpurwalla and Keubler (1966) that the time to failure of such drills follows a logarithmic normal distribution with a hazard rate that first increases and then decreases. A physical interpretation for the decreasing hazard rate is that, after some time of operation, the drill could resharpen itself and thereby prolong its life, implying a decrease in the hazard rate.

7.3.1 Properties of Monotone Failure Rates

In Section 4.1, a probabilistic interpretation of the hazard (failure) rate was given. Following this interpretation, one can easily verify that the hazard rate can also be defined as

$$\frac{F(x+\Delta x)-F(x)}{\overline{F}(x)},$$

where $\overline{F}(x) = 1 - F(x)$.

A continuous distribution function F is defined to be IHR (DHR) if and only if

$$\frac{F(x+\Delta x)-F(x)}{\overline{F}(x)}$$

is increasing (decreasing) in x for all $\Delta x > 0$, $x \geqslant 0$, and $F(x) < 1$.

Equivalently, it can be shown that, if F has a density f, then F is IHR (DHR) if and only if $h(x)$ is increasing (decreasing) in x. The proof of this is left as an exercise for the reader.

Several alternative criteria for ascertaining whether F is IHR (DHR) exist; two of these are as follows:

1. If F is IHR (DHR), then $\log [1 - F(x)]$ is concave (convex) for all $x \geqslant 0$, such that $F(x) < 1$.
2. If F is IHR (DHR), then $[1 - F(x)]^{1/x}$ is decreasing (increasing) in x.

Proof. 1. From Equation (4.2) it follows that

$$1 - F(x) = \exp\left[-\int_0^x h(s)\,ds \right] = \exp[-R(x)],$$

where $R(x) = \int_0^x h(s)\,ds$.

$$\therefore \qquad \frac{F(x+\Delta x) - F(x)}{\bar{F}(x)} = 1 - \exp[-R(x+\Delta x) - R(x)].$$

If F is IHR (DHR), then $R(x+\Delta x) - R(x)$ is increasing (decreasing) in x for all $\Delta x > 0$.

From the definition of concavity (convexity), it follows that, if F is IHR (DHR), then $R(x)$ is convex (concave). Thus, if F is IHR (DHR), $\log \bar{F}(x)$ is concave (convex) for all $x \geq 0$, so that $F(x) < 1$.

2. From the definition of concavity (convexity) and result 1 above, it follows that, if F is IHR (DHR),

$$\frac{\log \bar{F}(x) - \log \bar{F}(0)}{x - 0}$$

is decreasing (increasing) in x. This implies that $[\bar{F}(x)]^{1/x}$ is decreasing (increasing) in x.

7.3.2 Comparison with the Exponential Distribution

Since the exponential distribution has a constant hazard rate, it can serve as a bound for the IHR and the DHR distributions.

In particular, the exponential distribution provides natural bounds on the survival probability of IHR and DHR distributions. These bounds have some elegant consequences and have stimulated a number of interesting developments in the nonparametric approach to reliability. References concerning the importance of these bounds were made in Sections 4.2 and 5.1.

A useful lower bound on the survival probability, where F is IHR with a mean μ_1, is given as

$$\bar{F}(t) \geq \begin{cases} e^{-t/\mu_1}, & t < \mu_1, \\ 0, & \text{otherwise}. \end{cases}$$

Proof. A proof of this bound follows from *Jensen's inequality* for concave functions. Therefore it is first necessary to establish this inequality.

If $G(y)$ is a concave function,

$$E[G(y)] \leq G[E(y)].$$

Expanding $G(y)$ in a Taylor's series about y_0 gives

$$G(y) = G(y_0) + G^1(y_0)(y - y_0) + G^2(y_0)\frac{(y - y_0)^2}{2!} + \cdots,$$

where $G^i(y_0)$ denotes the ith derivative of $G(y)$. Since $G(y)$ is a concave function, $G^2(y_0) \leqslant 0$, and hence

$$G(y) \leqslant G(y_0) + (y - y_0)G^1(y_0).$$

Setting $y_0 = E(y)$, one obtains

$$G(y) < G[E(y)] + [y - E(y)]G^1[E(y)].$$

When expectations are taken on both sides, the inequality follows.

Since F is IHR with mean μ_1, $\log \bar{F}(t)$ is concave in t, and from Jensen's inequality it follows that

$$E\left[\log \bar{F}(t)\right] \leqslant \log \bar{F}(\mu_1).$$

If F is continuous, $\bar{F}(t)$ is uniform on $(0, 1)$, and integration by parts establishes that

$$E\left[\log \bar{F}(t)\right] = \int_0^1 \log\left[\bar{F}(t)\right] dt = -1$$

or that

$$\bar{F}(\mu_1) \geqslant e^{-1}.$$

Also, since F is IHR,

$$\left[\bar{F}(t)\right]^{1/t} > \left[\bar{F}(\mu_1)\right]^{1/\mu_1}, \quad t < \mu_1,$$

or

$$\bar{F}(t) \geqslant e^{-t/\mu_1}.$$

For $t \geqslant \mu_1$, $\bar{F}(t) \leqslant [\bar{F}(\mu_1)]^{t/\mu_1}$, and hence the only meaningful bound on $\bar{F}(t)$ for $t \geqslant \mu_1$ is 0. Since e^{-t/μ_1} is the survival probability for an exponential distribution with mean μ_1, the result is established.

It is easy to verify that, for F IHR, $\bar{F}(t)$ crosses $e^{-\alpha t}$ at most once, unless they coincide, and that the crossing is necessarily from above. A plot of $\log \bar{F}(t)$ for F IHR versus αt is shown in Figure 7.1.

The convexity of $\log \bar{F}(t)$ establishes that there can be at most one intersection, necessarily from above, as is shown.

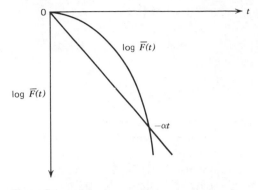

Figure 7.1

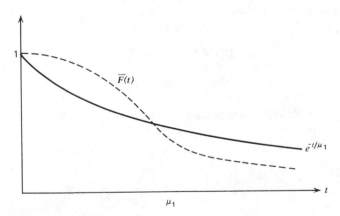

Figure 7.2

It can also be established that for F IHR, $\bar{F}(t)$ crosses e^{-t/μ_1} exactly once, and necessarily from above. This follows from the fact that μ_1, the mean, is the area under the reliability function, and since μ_1 is the same for both the IHR survival curve and the exponential survival curve the two curves must cross, as shown in Figure 7.2.

7.3.3 Applications of the Exponential Bound

Some interesting and useful applications of the lower bound developed in the preceding section are presented here.

For example, suppose that a system consists of n independent components in series with IHR distributions F_i with corresponding means μ_i, $i = 1, 2, \ldots, n$. Clearly, then, the reliability of the system to time t can be represented as follows:

$$R(t) = \prod_{i=1}^{n} \bar{F}_i(t).$$

Since

$$F_i(t) \leqslant 1 - e^{-t/\mu_i}, \quad t < \mu_i, \quad 1 - R(t) = \prod_{i=1}^{n} F_i(t) \leqslant \prod_{i=1}^{n} (1 - e^{-t/\mu_i}), \quad t < \min_i (\mu_i).$$

Bounds on the reliability function of a series or a parallel system can be obtained from minimal assumptions regarding the failure distribution of each component.

Inequalities on the moments of IHR distributions and bounds on the mean life of a series or a parallel system with IHR components have been obtained by Barlow and Proschan (1965, p. 32). In the interest of brevity, these are stated here without proof.

1. If F is IHR with rth moment μ_r, then

$$\mu_r \leqslant \Gamma(r+1)\mu_1^r, \quad r > 1,$$

$$\geqslant \Gamma(r+1)\mu_1^r, \quad 0 \leqslant r \leqslant 1.$$

2. If F_i is IHR with mean μ_i and $G_i(x) = e^{-x/\mu_i}, i = 1, 2, \ldots, n$, then
(a) the mean time to failure of a series system

$$\int_0^{\infty} \prod_{i=1}^{n} \bar{F}_i(x)\,dx \geqslant \left(\sum_{i=1}^{n} \frac{1}{\mu_i}\right)^{-1},$$

and
(b) the mean time to failure of a parallel system

$$\int_0^{\infty} \left[1 - \prod_{i=1}^{n} F_i(x)\right] dx < \int_0^{\infty} \left[1 - \prod_{i=1}^{n} G_i(x)\right] dx.$$

If $\mu_i = \mu_1$, the right-hand side of the above inequality becomes $\mu_1(\sum_{k=1}^{n} 1/k)$. Similar bounds for DHR distributions are obtained by reversing all the inequalities given above.

It is hoped that the foregoing examples illustrate the usefulness of the exponential distribution in reliability studies. Clearly, results obtained by using an exponential distribution can give meaningful bounds for certain quantities of interest when the underlying distribution has a monotone failure rate. Applications of the exponential bound in problems of replacement, spare parts provisioning, and redundancy optimization, as well as in the study of coherent structures, are illustrated by Barlow and Proschan (1965). Methods of obtaining confidence bounds on system reliability for exponential-failure-time models are discussed in Section 10.4.

7.4 INTERVAL ESTIMATES FOR DISTRIBUTIONS, DISTRIBUTION QUANTILES, AND RELIABILITY

7.4.1 Confidence Bands for Continuous Distribution Functions

In some situations it is desirable to prepare a confidence band for a distribution function, $F(x)$, in its entirety. In view of the discussion in Section 6.5.1 on the relationship between confidence intervals and hypothesis tests, it is not surprising that the hypothesis test statistic can be used to compute the desired confidence band.

If it is assumed only that the unknown $F(x)$ is continuous, the Kolmogorov-Smirnov statistic,

$$D_n = \sup_x |F_0(x) - F_n(x)|,$$

can be used to test $F(x) = F_0(x)$ against two-sided alternatives, and if $D_n > D_n(\alpha)$ the null hypothesis is rejected with test size α. Here $F_n(x)$ = (number of $x_i \leqslant x$)/n, that is, $F_n(x)$ is the empirical distribution function. Thus

$$P[F(x) - D_n(\alpha) \leqslant F_n(x) \leqslant F(x) + D_n(\alpha)] = 1 - \alpha,$$

and this statement may evidently be inverted to obtain the confidence statement,

$$P[F_n(x) - D_n(\alpha) \leqslant F(x) \leqslant F_n(x) + D_n(\alpha)] = 1 - \alpha. \tag{7.5}$$

Moreover, the sample size necessary to obtain a required closeness to the true distribution function may be obtained. Since the width of the above interval is $2D_n(\alpha)$ and the closeness needed may be related to $D_n(\alpha)$, it is required only to find the $D_n(\alpha)$ of necessary size; n is determined from tables of the critical values of the Kolmogorov-Smirnov statistic, as given, for example, in Massey (1951). A more detailed discussion of this procedure is available in Kendall and Stuart (1967).

In many cases it turns out that more is known (or assumed) about $F(x)$ than continuity. In particular, the family may be known but the parameter(s) unknown. In discussing this case, it can be said at the outset that in general, if the family of $F(x)$ is known and a confidence band is available based on the known family, it will be more efficient to use this than the

Kolmogorov-Smirnov statistic employed above. In Section 7.1 a goodness-of-fit test for the exponential distribution due to Lilliefors (1969) was presented. The statistic was

$$D_n^* = \sup_x |F^*(x) - F_n(x)|.$$

Again, this may be inverted to obtain the confidence statement,

$$P[F_n(x) - D_n^*(\alpha) \leqslant F(x) \leqslant F_n(x) + D_n^*(\alpha)] = 1 - \alpha, \qquad (7.6)$$

for the entire distribution function, and since $D_n^*(\alpha) < D_n(\alpha)$ the band will be narrower than in the previous case. The sample size required for a certain closeness of the confidence band can be determined in exactly the same way as the Kolmogorov-Smirnov statistic was used.

It should be noted that in both of the above cases the statistics were "maximums," and that this was critical to obtain bounds for the entire distribution function. For example, the test due to Srinivasan (1970) can be used because it is based on a maximum distance statistic, whereas (for the Weibull distribution) the statistic S presented in Section 7.1 cannot be used.

At this writing a maximum distance statistic has not been developed for testing a Weibull fit, although this is easy to do since the distribution of

$$D_n^* = \sup_x |F^*(x) - F_n(x)|,$$

where $F^*(x) = 1 - \exp[-(x/\hat{\delta})^{\hat{\beta}}], \hat{\delta}, \hat{\beta}$ the maximum-likelihood estimators of δ, β, is parameter free. However, a maximum distance statistic may never be developed for two reasons.

1. The S test of fit presented in Section 7.1 is quite powerful for testing the Weibull fit for most alternatives of interest.

2. A procedure due to Srinivasan (1972) gives better confidence bands than the methods previously discussed.

There is no reason the same statistic must be used for an interval estimate as was used for a hypothesis test. Now, note that the empirical distribution function, $F_n(x)$, was employed in the previous discussion in the test statistics, and this is reasonable since the use of $F_n(x)$ provides a consistent test (i.e., the probability of rejecting a false null hypothesis approaches 1 as n becomes large). However, confidence bands are a different matter. Srinivasan (1972) has suggested the use of the statistic

$$L = \max_x |F(x) - \hat{F}(x)|, \qquad (7.7)$$

where $F(x)$ is the null distribution, and $\hat{F}(x)$ is the null distribution with estimates inserted (usually maximum-likelihood estimates) for the unknown parameters. He has pointed out that for many distributions (e.g., exponential, Weibull, normal) the statistic L is parameter free, so that, as for the Lilliefors statistic D_n^*, one set of tables suffices for each family of distributions. For example, for the exponential distribution the parameter can be taken as equal to 1, so that

$$F(x) = 1 - \exp(-x) \quad \text{and} \quad \hat{F}(x) = 1 - \exp\left(\frac{-x}{\bar{x}}\right),$$

where $\bar{x}$ is the ordinary sample mean. The distribution of the statistic L is obtained relatively easily by Monte Carlo methods, so that

$$P\left[\hat{F}(x) - L_\alpha \leqslant F(x) \leqslant \hat{F}(x) + L_\alpha\right] = 1 - \alpha. \tag{7.8}$$

Note that, in general, $\hat{F}(x)$ is not a consistent estimator when the family assumed is incorrect, but this is of no importance for the confidence band. Preliminary results by Srinivasan indicate that this band is narrower, on the average, than the band obtained by using, say, the Lilliefors statistic. Finally, the sample size required for a certain closeness may be obtained by using the critical values as before.

Of course, all the foregoing bands "suffer" from the same shortcoming: they are wider than necessary. Since a fixed distance is added and subtracted from either $\hat{F}(x)$ or $F_n(x)$, certain ambiguities are obtained. For example, it is known that $F(\infty) = 1$ and $F(x) < 1$ for $x < \infty$, but since the distance statistic is always positive the upper band will have the value 1 on a set of x such that $x < \infty$. These bands may be "whittled" down, depending on the particular family of distributions assumed, by eliminating impossible cases.

PROBLEMS

7.3. Suppose that it is required to have a confidence band of width .50 with confidence .90. What n is required, using the Kolmogorov-Smirnov statistic?

7.4. Solve Problem 7.3 assuming that the underlying distribution is exponential and using the Lilliefors statistic.

7.4.2 Distribution-Free Interval Estimates of Distribution Quantiles

For a continuous random variable X, the pth quantile, x_p, is a number such that $F(x_p) = p, 0 < p < 1$. If $F(x)$ is known, finding x_p is a question of solving the above equation—usually an easy matter. In reliability work, however, $F(x)$ is often unknown, even with respect to family, and x_p must be estimated. Often p is taken small, say .05; and, since $1 - F(x_p) = 1 - p, x_p$ is called the reliable life. Now suppose that a lower $100(1 - \alpha)\%$ confidence bound on x_p, say $\underline{x}_p$, is sought, that is

$$P(x_p \geqslant \underline{x}_p) = 1 - \alpha = P[F(x_p) \geqslant F(\underline{x}_p)]. \tag{7.9}$$

Now suppose that a random sample of size n is available from the unknown distribution function $F(x)$. Then clearly the sample can be ordered so that the order statistics $X_{(1)}, \ldots, X_{(n)}$ are obtained. The probability distribution of $X_{(i)}, 1 \leqslant i \leqslant n$, is

$$f_{X_{(i)}}(y) = \frac{n!}{(i-1)!(n-i)!} f(y) [F(y)]^{i-1} [1 - F(y)]^{n-i}.$$

The change of variable $U_{(i)} = F(X_{(i)})$ leads to

$$f_{U_{(i)}}(u) = \frac{n!}{(i-1)!(n-i)!} u^{i-1} (1-u)^{n-i}. \tag{7.10}$$

This is a beta probability distribution with parameters $(i, n-i+1)$. In particular, for the first order statistic (i.e., $i = 1$),

$$f_{U_{(1)}}(u) = n(1-u)^{n-1} \quad \text{and} \quad F_{U_{(1)}}(u) = 1 - (1-u)^n.$$

If one now takes $\underline{x}_p = x_{(1)} = $ the first (observed) order statistic, $F(\underline{x}_p) = F(x_{(1)})$ has the cumulative distribution $1 - (1-u)^n$, so that the right term in (7.9) requires

$$P[U_{(1)} = F(X_{(1)}) \leqslant F(x_p) = p] = 1 - \alpha;$$

therefore

$$1 - \alpha = 1 - (1-p)^n \quad \text{and hence } \alpha = (1-p)^n. \tag{7.11}$$

If p and α are specified, n can be found. It should be recalled that the lower confidence bound for x_p is $x_{(1)}$.

Example. Suppose that a lower $100(1 - \alpha)\%$ confidence interval is desired for reliable life x_p with $p = .10$ and $\alpha = .05$. Then n must satisfy $.05 = (.90)^n$, and the smallest n that does this is $n = 29$. In fact, the solution to this equation is not usually an integer, so that if $x_{(1)}$ is the first order statistic observed, based on a sample size $n = 29$,

$$P(x_{.10} \geqslant x_{(1)}) \geqslant .95.$$

Upper and two-sided confidence bounds can be obtained in a similar way. Other order statistics, for example, $X_{(2)}$, could be used, but the computations become more complex, requiring tables of the beta distribution or of the binomial distribution. If n is determined when one approaches the problem and p is selected, one will have to take whatever α he can get. Also, it should be noted that Equation (7.11) is nonsensical for $1 - p < \alpha$ and hence for $p > 1 - \alpha$. If the latter situation is required, a larger order statistic should be used, along with the fact that, for any order statistic $i, F(X_i)$ has a beta probability distribution. In any case, confidence bounds on x_p can be obtained that do not depend on the unknown $F(x)$. This subject is treated in more detail in Wilks (1962).

PROBLEM

7.5. Prepare a one-sided upper confidence bound for $x_{.90}$, using $X_{(n)}$.

7.4.3 Interval Estimates for Reliability

Again presume that the distribution function of lifetime $X, F_X(x)$, is unknown. Suppose that it is desired to obtain an interval estimate of $R(t_0) = 1 - F_X(t_0)$ for fixed t_0, that is, to obtain an interval estimate of the probability that a single lifetime will equal or exceed t_0, with $R(t_0)$ unknown. Since, generally, a lower bound $\underline{R}$ is desired, that is,

$$P[R(t_0) \geqslant \underline{R}] \geqslant 1 - \alpha, \qquad (7.12)$$

we will proceed with this example, the upper and two-sided cases being similar. Eventually it will be necessary to use tables of the binomial distribution, and because of the symmetry in p and $(1-p)$ the binomial distribution is tabulated only for $p \leqslant \frac{1}{2}$. On the other hand, $R(t_0) > \frac{1}{2}$ in most cases of interest. Hence an upper one-sided confidence bound for $1 - R(t_0)$,

$$P[1 - R(t_0) \leqslant \bar{p}] = P[R(t_0) \geqslant 1 - \bar{p}] \geqslant 1 - \alpha, \qquad (7.13)$$

will be obtained, so that $\underline{R} = 1 - \bar{p}$.

Suppose now that n independent trials are observed with probability of failure $1 - R(t_0)$. The observed number of failures, X, occurring in n trials has the binomial distribution

$$P(X - x) = \binom{n}{x} p^x (1 - p)^{n - x},$$

where $p = 1 - R(t_0)$.

Since X is a discrete random variable, an interval with probability exactly $1 - \alpha$ of containing p cannot be found without additional work. This will be discussed shortly, at the end of the section. In any event, and as usual, the smallest value of p such that

$$P(X > x; p) = \sum_{i=x+1}^{n} \binom{n}{i} p^i (1-p)^{n-i} \geqslant 1 - \alpha \qquad (7.14)$$

is selected and is designated as $\bar{p}$; then $P(p \leqslant \bar{p}) \geqslant 1 - \alpha$. The lower confidence bound for $R(t_0)$ is $\underline{R} = 1 - \bar{p}$.

Clopper and Pearson (1934) have given extensive tables of $\underline{p}, \bar{p}$ for various values of n and $\alpha = .05, .025$. The sum in (7.14) is, of course, the cumulative binomial distribution. There exists an important relation between the cumulative binomial distribution and the cumulative beta distribution, which for various reasons (not the least of which is computational convenience) arises quite frequently. Although this relation is given by Equation (8.20), it is repeated here for convenience. It will be recalled that the lower-case x in (7.14) is the observed number of failures. It turns out that the sum in (7.14) satisfies

$$\sum_{i=x+1}^{n} \binom{n}{i} p^i (1-p)^{n-i} = \frac{\Gamma(n+1)}{\Gamma(x+1)\Gamma(n-x)} \int_0^p u^x (1-u)^{n-x-1} \, du.$$

$$(7.15)$$

The right-hand side of (7.15) is the cumulative beta distribution, mentioned in Chapter 3, with parameters $x + 1, n - x$. Thus, in finding the smallest p that satisfies the inequality in (7.14), tables of the cumulative beta distribution (see, e.g., Harter, 1964) can be very useful.

It should be noted that the upper (lower) one-sided interval is always bounded on the lower (upper) side by 0 (1).

Example 1. Suppose that $n = 20$, $\alpha = .10$, the observed number of failures is $x = 1$, and $\underline{R}$ is desired, so that

$$P[R(t_0) \geqslant \underline{R}] \geqslant .90.$$

Using (7.14) and proceeding to binomial tables, one finds

$$P(X > x = 1; p = .181) = .900 + .$$

Thus $P(p \leqslant .181) \geqslant .90$, and since $\underline{R} = 1 - \bar{p} = .819$,

$$P[R(t_0) \geqslant .819] \geqslant .90.$$

Example 2. Often $R(t_0)$ can be expected to be quite large, so that $1 - R(t_0)$ is small. If n is large also, the Poisson approximation may be used. Suppose again that $n = 20$, $\alpha = .10$, and $x = 1$. From Molina's tables one finds

$$P(X > x = 1; \mu) = \sum_{x=2}^{\infty} \frac{e^{-\mu}\mu^x}{x!} = .9008 \quad \text{for } \mu = 3.9.$$

Since, when n is large and p is small, the quantity $np \cong \mu$, $\bar{p} = 3.9/20 = .195$. The approximation is not bad at all. The lower confidence bound for $R(t_0)$ can be calculated as before. The reader interested in confidence bounds for the negative binomial parameter should now be able to obtain these for himself.

PROBLEM

7.6. Show that for Example 1 of this section an upper 90% confidence bound on $R(t_0)$ is $\bar{R} = .995$, and hence a two-sided 80% interval is $(.819, .995)$.

We have already noted that the interval estimates prepared for $R(t_0)$ (more generally, for the parameter p in a binomial distribution) do not have exactly $1 - \alpha$ probability of containing p because of the discreteness of the random variable X. However, a method is available to obtain exactly $1 - \alpha$; this is due to Stevens (1950) and is discussed by Kendall and Stuart (1967). The method involves, after observing x failures in n trials, drawing a random number uniformly distributed on $(0, 1)$. A new random variate is then $Y = X + U$, where U is the random uniform number and the distribution of U is parameter free. Thus the distribution of Y depends only on the parameter p in the binomial distribution (and the fixed n), and the range of Y is $0, \ldots, n+1$. The distribution of Y is *continuous* and is

$$P(Y \geqslant y_0 = x_0 + u_0) = P(X > x_0) + P(X = x_0)P(U \geqslant u_0)$$

$$= \sum_{i=x_0+1}^{n} \binom{n}{i} p^i (1-p)^{n-i} + \binom{n}{x_0} p^{x_0} (1-p)^{n-x_0} (1 - u_0)$$

$$= u_0 \sum_{i=x_0+1}^{n} \binom{n}{i} p^i (1-p)^{n-i} + (1 - u_0) \sum_{i=x_0}^{n} \binom{n}{i} p^i (1-p)^{n-i}.$$

376 GOODNESS OF FIT, MONTE CARLO, DISTRIBUTION-FREE METHODS

Exact confidence bounds for p can be set by using this continuous distribution. These intervals, called *randomized intervals*, are not appealing to some, who feel that the introduction of a random uniform number having nothing to do with the observed results is artificial. Nonetheless, the procedure provides interval estimates of p with probability of containing p exactly equal to the assigned confidence level. Tables of two-sided intervals are given for $\alpha = .01, .05$ and $n = 2(1)24(2)50$ by Blyth and Hutchinson (1960).

REFERENCES FOR SECTION 7.1

Anderson, T. W. and D. A. Darling (1952), Asymptotic Theory of Certain Goodness of Fit Criteria Based on Stochastic Processes, *Annals of Mathematical Statistics*, Vol. 23, pp. 193-212.

Birnbaum, Z. W. (1952), Numerical Tabulation of the Distribution of Kolmogorov's Statistic for Finite Sample Size, *Journal of the American Statistical Association*, Vol. 47, pp. 425-441.

Chernoff, Herman and E. L. Lehmann (1954), The Use of Maximum Likelihood Estimates in χ^2 Tests for Goodness of Fit, *Annals of Mathematical Statistics*, Vol. 25, pp. 579-586.

Darling, D. A. (1957), The Kolmogorov-Smirnov, Cramér-von Mises Tests, *Annals of Mathematical Statistics*, Vol. 28, pp. 823-838.

David, F. N. and N. L. Johnson (1948), The Probability Integral Transform When Parameters Are Estimated from the Sample, *Biometrika*, Vol. 35, pp. 823-838.

Epstein, B. (1960), Tests for the Validity of the Assumption That the Underlying Distribution of Life Is Exponential, *Technometrics*, Vol. 2, pp. 83-101.

Fercho, W. W. and L. J. Ringer (1972), Small Sample Power of Some Tests of the Constant Failure Rate, *Technometrics*, Vol. 14, pp. 713-724.

Finklestein, J. M. and R. E. Schafer (1971), Improved Goodness of Fit Tests, *Biometrika*, Vol. 58, pp. 641-645.

Gnedenko, B., et al. (1969), *Mathematical Methods of Reliability Theory*, Academic Press, New York.

Hahn, G. J. and S. S. Shapiro (1967), *Statistical Models in Engineering*, John Wiley and Sons, New York.

Harter, H. Leon (1964), *New Tables of the Incomplete Gamma Function Ratio and of Percentage Points of the Chi-square and Beta Distributions*, U.S. Government Printing Office, Washington, D.C.

Hartley, H. O. (1950), The Maximum F Ratio as a Short-cut Test for Homogeneity of Variance, *Biometrika*, Vol. 37, pp. 308-312.

Kac, M., J. Kiefer, and J. Wolfowitz (1955), On Tests of Normality and Other Tests of Fit Based on Distance Methods, *Annals of Mathematical Statistics*, Vol. 26, pp. 189-211.

Kuiper, N. H. (1960), Tests Concerning Random Points on a Circle, *Proceedings of the Koninklijke Nederlandske Akademie van Wetenschappen* A, Vol. 63, pp. 38-47.

Lilliefors, Hubert W. (1967), On the Kolmogorov-Smirnov Test for Normality with Mean and Variance Unknown, *Journal of the American Statistical Association*, Vol. 62, pp. 399-402.

Lilliefors, Hubert W. (1969), On the Kolmogorov-Smirnov Test for the Exponential Distribution with Mean Unknown, *Journal of the American Statistical Association*, Vol. 64, pp. 387-389.

Mann, Nancy R. and Kenneth W. Fertig (1974), A Goodness of Fit Test of the Two-Parameter Weibull Distribution Against Three-Parameter Weibull Alternatives; Confidence Bounds for a Weibull Threshold Parameter (submitted for publication).

Mann, Nancy R., Ernest M. Scheuer, and Kenneth W. Fertig (1973), A New Goodness-of-Fit Test for the Two-Parameter Weibull or Extreme-Value Distribution with Unknown Parameters, *Communications in Statistics*, Vol. 2, pp. 383-400.

Massey, F. J. (1951), The Kolmogorov-Smirnov Test for Goodness of Fit, *Journal of the American Statistical Association*, Vol. 46, pp. 68-78.

Patnaik, P. B. (1949), The Non-Central χ^2 and F Distributions and Their Applications, *Biometrika*, Vol. 36, pp. 202-232.

Pyke, R. (1965), Spacings, *Journal of the Royal Statistical Society B*, Vol. 27, pp. 395-449.

Srinivasan, R. (1970), An Approach to Testing the Goodness of Fit of Incompletely Specified Distributions, *Biometrika*, Vol. 57, pp. 605-610.

Van Montfort, M. A. J. (1970), On Testing That the Distribution of Extremes Is of Type I When Type II Is the Alternative, *Journal of Hydrology*, Vol. 11, pp. 421-427.

Van Soest, J. (1969), Some Goodness of Fit Tests for Exponential Distributions, *Statistica Neerlandica*, Vol. 23, pp. 41-51.

White, John S. (1967), The Moments of Log-Weibull Order Statistics, *General Motors Research Publication* GMR-717, General Motors Research Laboratories, Warren, Michigan.

REFERENCES FOR SECTION 7.2

Brunk, H. D. (1955), Maximum Likelihood Estimates of Monotone Parameters, *Annals of Mathematical Statistics*, Vol. 29, pp. 437-454.

Gruenberger, F. and A. M. Mark (1951), The d^2 Test, *Mathematical Tables and Computations*, Vol. 5, pp. 109-110.

Hall, A. (1873), On an Experimental Determination of π, *Messeng. Math.*, 2, 113-114.

Halton, John H. (1970), A Retrospective and Prospective Survey of the Monte Carlo Method, *SIAM Review*, Vol. 12, pp. 1-64.

Hull, T. E. and A. R. Dobell (1962), Random Number Generators, *SIAM Review*, Vol. 4, pp. 230-254.

Kahn, Herman (1954), Applications of Monte Carlo, AECU-3259, U.S. Atomic Energy Commission Technical Information Service Extension, Oak Ridge, Tennessee.

Kendall, M. G. and B. Babington Smith (1954), Tables of Random Sampling Numbers, *Tracts for Computers*, No. 24, Cambridge University Press, Cambridge, England.

Mann, Nancy R. (1970), Computer-Aided Selection of Prior Distributions for Generating Monte Carlo Confidence Bounds on System Reliability, *Naval Research Logistics Quarterly*, Vol. 17, pp. 41-53.

Mantel, Nathan (1953), An Extension of the Buffon Needle Problem, *Annals of Mathematical Statistics*, Vol. 24, pp. 674-677.

Marsaglia, G. (1961), Generating Exponential Random Variables, *Annals of Mathematical Statistics*, Vol. 32, pp. 899-900.

Marsaglia, G. (1968), Random Numbers Fall Mainly in the Planes, *Proceedings of the National Academy of Science*, Vol. 60, pp. 25-28.

Marsaglia, G. and T. A. Bray (1964), A Convenient Method for Generating Normal Variables, *SIAM Review*, Vol. 6, pp. 260-264.

Marsaglia, G. and T. A. Bray (1968), One-Line Random Number Generators and Their Use in Combinations, *Communications Assoc. Comput. Mach.*, Vol. 11, pp. 757-759.

Marsaglia, G., M. D. McLaren, and T. A. Bray (1964), A Fast Procedure for Generating Normal Random Variables, *Communications, Assoc. Comput. Mach.*, Vol. 7, pp. 4-10.

McLaren, M. D. and G. Marsaglia (1965), Uniform Random Number Generators, *J. Assoc. Comput. Mach.*, Vol. 12, pp. 83-89.

Mosteller, Frederick (1946), On Some Useful "Inefficient" Statistics, *Annals of Mathematical Statistics*, Vol. 17, pp. 377-407.

RAND Corporation (1955), *A Million Random Digits With 100,000 Normal Deviates*, Free Press, Glencoe, Illinois.

Saunders, S. C. and N. R. Mann (1971), On Invariant Confidence Intervals for the Parameters of a New-Life Distribution, *Naval Research Logistics Quarterly*,

Spanier, Jerome and Ely M. Gelbard (1969), *Monte Carlo Principles and Neutron Transport Problems*, Addison-Wesley Publishing Company, Reading, Massachusetts.

Tausworthe, Robert C. (1965), Random Numbers Generated by Linear Recurrence Modulo Two, *Mathematics of Computation*, Vol. 19, pp. 201-209.

REFERENCES FOR SECTION 7.3

Barlow, R. E. and F. Proschan (1965), *Mathematical Theory of Reliability*, John Wiley and Sons, New York.

Singpurwalla, N. D. and A. A. Kuebler (1966), A Quantitative Evaluation of Drill Life, *American Society of Mechanical Engineers Publication* 66-WA/PROD-11.

REFERENCES FOR SECTION 7.4

Blyth, C. R. and D. W. Hutchinson (1960), Table of Neyman-Shortest Unbiased Confidence Intervals for the Binomial Parameter, *Biometrika*, Vol. 47, p. 381.

Clopper, C. J. and E. S. Pearson (1934), The Use of Confidence or Fiducial Limits Illustrated in the Case of the Binomial, *Biometrika*, Vol. 25, p. 404.

Harter, H. Leon (1964), *New Tables of the Incomplete Gamma-Function Ratio and of Percentage Points of the Chi-Square and Beta Distributions*, U.S. Government Printing Office, Washington, D.C.

Kendall, M. G. and Alan Stuart (1967), *The Advanced Theory of Statistics*, Vol. 2, Hafner Publishing Company, New York.

Massey, F. J. (1951), The Kolmogorov-Smirnov Test for Goodness of Fit, *Journal of the American Statistical Association*, Vol. 46, p. 68.

Srinivasan, R and Kanofsky, Paul (1972), An Approach to the Construction of Parametric Confidence Bands on Cumulative Distribution Functions, *Biometrika*, Vol. 59, 3.

Stevens, W. L. (1950), Fiducial Limits of the Parameter of a Discontinuous Distribution, *Biometrika*, Vol. 37, p. 117.

Wilks, S. S. (1962), *Mathematical Statistics*, John Wiley and Sons, New York.

CHAPTER 8

Bayes Methods in Reliability

Chapters 5 and 6 presented "classical" methods of estimation and hypothesis testing. The present chapter gives a Bayes approach to these problems in the context of reliability applications. The Chapter begins with a comparison of the two approaches.

8.1 COMPARISON OF BAYES AND CLASSICAL METHODS

8.1.1 Differences in Criteria

A large part of the statistical problem in reliability involves the estimation of parameters in failure models. Each of the methods of obtaining point estimates previously given has certain statistical properties that make it desirable, at least from a theoretical point of view. Not surprisingly, point estimates are often made (particularly in reliability) because decisions are to be based on them. The consequences of the decisions based on the estimates often involve money or, more generally, some form of utility. Hence the decision maker is more interested in the practical consequence of his estimate than in its theoretical properties. In particular, he may be interested in making estimates that minimize expected loss (cost). To illustrate the implications of this criterion, consider the following somewhat simplified examples.

Example 1. Suppose that in the reliability function

$$P(X > t) = R(t)$$

t is fixed, and $R(t)$ for a particular unit can take on only two values: $R(t) = \theta_1 = .90$ or $R(t) = \theta_2 = .99$. The problem, then, is to estimate for this unit which value the parameter takes: $\theta_1 = .90$ or $\theta_2 = .99$.
Two courses of action can be identified:

$$A_1 \equiv \text{estimate } R(t) \text{ as } \theta_1,$$

$$A_2 \equiv \text{estimate } R(t) \text{ as } \theta_2.$$

The accompanying table gives arbitrarily selected losses, $L(\theta_i, A_j)$ for each combination (θ_i, A_j), $i = 1, 2; j = 1, 2$.

	Action	
	A_1	A_2
$R(t) = \theta_1$	0	5
$R(t) = \theta_2$	3	0

The units of the losses are of no interest in this discussion, and the zeros on the main diagonal reflect the fact that no losses are incurred when the correct decision is made. The $5 > 3$ reflects the fact that it is evidently more costly to take action A_2 incorrectly than to take action A_1 incorrectly. With precisely this amount of information, the action that minimizes expected losses cannot be chosen, since there is no way to compute the expected loss. However, notice that, if action A_1 is taken, the loss can be held to no more than 3. Action A_1 is called the *minimax solution* because

$$\underset{j}{\text{Min}} \, \underset{i}{\text{max}} \, [L(\theta_i, A_j)] = L(\theta_2, A_1) = 3.$$

Example 2. Since $R(t)$ can take on two values, it is a random variable. Now suppose that the probability distribution of $R(t)$ is known; that is, suppose that $f(\theta_1) = \frac{1}{3}; f(\theta_2) = \frac{2}{3}$. Now, if action A_1 is selected, the expected loss

$$E[L(\Theta, A_1)] = L(\theta_1, A_1)f(\theta_1) + L(\theta_2, A_1)f(\theta_2)$$

$$= 0 \cdot \tfrac{1}{3} + 3 \cdot \tfrac{2}{3}$$

$$= \tfrac{6}{3},$$

and for action A_2 the expected loss is

$$E[L(\Theta, A_2)] = L(\theta_1, A_2)f(\theta_1) + L(\theta_2, A_2)f(\theta_2)$$

$$= 5 \cdot \tfrac{1}{3} + 0 \cdot \tfrac{2}{3}$$

$$= \tfrac{5}{3}.$$

Thus action A_2 is the one that minimizes the expected loss.

One important consideration is that, in order to choose the appropriate action, the probability distribution of $R(t)$ was required. It should be noted that when a probability distribution is assigned to a parameter (i.e., the parameter is considered to be a random variable) it is customary to call that probability distribution the *prior* (or the *a priori*) *probability distribution*. The modifier prior refers to the fact that this distribution is the one that exists before any sampling of $f(\cdot|\theta)$. Indeed, the next modification of the example will be to acquire some data.

Example 3. Suppose that the particular unit under consideration is now tested for t units of time, and it is observed whether or not the unit has survived. This is a binomial sampling situation with a sample of size 1 and a parameter θ_1 or θ_2, depending on which state prevails. Thus, if $Z=1$ is recorded for survival and $Z=0$ is recorded for nonsurvival,

$$f(Z=1|\theta_1)=.90, \quad f(Z=0|\theta_1)=.10,$$

$$f(Z=1|\theta_2)=.99, \quad f(Z=0|\theta_2)=.01.$$

The joint probability $f(Z=1,\theta_1)$ is given by

$$f(Z=1,\theta_1)=f(Z=1|\theta_1)f(\theta_1)=.90(\tfrac{1}{3})=.30.$$

The accompanying table gives the calculated values of the four joint probabilities:

$$f(z,\theta_i)$$

	θ_1	θ_2
$Z=1$	.3000	.6600
$Z=0$	.0333	.0067

The marginal distribution of Z is easily obtained by summing across the rows, that is,

$$f(Z=1)=.30+.66=.96,$$

$$f(Z=0)=.04.$$

The problem is, having observed $Z=1$, to decide between A_1 and A_2 or, having observed $Z=0$, to decide between A_1 and A_2.

First, suppose that $Z=1$ is observed and action A_1 is taken; then the expected loss is:

$$E[L(\Theta,A_1)]=L(\theta_1,A_1)f(\theta_1|Z=1)+L(\theta_2,A_1)f(\theta_2|Z=1).$$

The probabilities $f(\theta_1|Z=1)$ and $f(\theta_2|Z=1)$ are called the *posterior probabilities*, since they represent the probability distribution on Θ, given the observed data, and are computed using Bayes's theorem:

$$f(\theta_1|Z=1) = \frac{f(\theta_1, Z=1)}{f(Z=1)} = \frac{.30}{.96} = .3125.$$

Similarly, $f(\theta_2|Z=1) = .6875$, so that

$$E[L(\Theta, A_1)] = 0(.3125) + 3(.6875) = 2.0625.$$

Similar calculations show that, when $Z=1$ has been observed and action A_2 is chosen,

$$E[L(\Theta, A_2)] = 5(.3125) + 0(.6875)$$

$$= 1.5625.$$

Thus, when $Z=1$ has been observed, action A_2 is preferred.
Similar computations show that, if $Z=0$ is observed and action A_1 is chosen,

$$E[L(\Theta, A_1)] = 0(.8333) + 3(.1667)$$

$$= .50,$$

and, when action A_2 is chosen,

$$E[L(\Theta, A_2)] = 5(.8333) + 0(.1667)$$

$$= 4.1667.$$

Thus, when $Z=1$ is observed, we would choose action A_2, and when $Z=0$ is observed, action A_1. If $Z=1$ is observed, the expected loss is 1.5625; if $Z=0$ is observed, the expected loss is .50. Since Z is a random variable with a probability mass $f(Z=1) = .96$; $f(Z=0) = .04$; the unconditional expected loss is:

$$E[L(\Theta)] = 1.5625(.96) + .50(.04)$$

$$= 1.52.$$

It is to be noted that in the case where only the prior distribution was available (see Example 2), the expected loss was $5/3 = 1.67$. With the additional information provided by the data (one trial), the expected loss was reduced to 1.52. Whether this reduction was worth the cost of obtaining the data is an important question that is not addressed here.

The methods used to solve Examples 2 and 3 are called Bayes methods, and the solutions are termed *Bayes solutions*. Statistical methods applied to reliability problems that involve parameter estimation in which the parameter is treated as a random variable with a prior probability distribution are generally called *Bayes methods*. To use the Bayes methods outlined here, we must be able to do three things.

1. Assess the losses or costs associated with various combinations of courses of action and parameter values.
2. Consider the parameter as a random variable.
3. Assign the prior probability distribution on the parameter space.

All of these things are done in obtaining the confidence intervals on reliability in Sections 5.1.2 and 10.4.1. In these cases, however, the prior densities implicitly assigned on the parameter space are *improper*, that is, they cannot be normalized to integrate to 1. Thus, these procedures are often termed *extended Bayes*, or simply *decision theoretic*. Although many statisticians argue that it is not meaningful to consider a parameter as a random variable, and that it certainly is a nontrivial problem to assign a prior distribution on the parameter space, the assignment of losses or costs seems to be the largest problem encountered in applying Bayes methods to reliability problems. Often the particular unit under study is merely part of a much larger system, and the costs due to certain actions are extremely hard to assess.

8.1.2 Differences in Philosophy

Parameters in probability distributions rarely, if ever, are known exactly, that is, with probability 1. The natural way to proceed, then, is to make a statement to the effect that the parameter lies in some interval, called an interval estimate (see Chapter 5). The classical and Bayes approaches to interval estimates are widely different and indicate unmistakably the differences in philosophy between the two methods.

Suppose that a random sample $(X_1, X_2, \ldots, X_n)$ of size n is available from a single-parameter probability distribution, say $f(x; \theta)$. Suppose, further, that the data are adequately described by a statistic S, where $S = g(X_1, X_2, \ldots, X_n)$, and the probability distribution of S depends on θ. This probability distribution, say $f_S(s; \theta)$, is called the sampling distribution of S. By way of example, if the random sample is obtained from the exponential distribution and S is chosen to be $\hat{\theta}$ (the maximum-likelihood estimate of θ), $f(\hat{\theta}; \theta)$ is a gamma distribution. In any event, the classical approach to constructing the interval is to use $f(s; \theta)$ to determine two numbers, depending on θ, such that

$$P[s_*(\theta) \leqslant S \leqslant s^*(\theta)] = 1 - \gamma. \tag{8.1}$$

Equation (8.1) is a valid probability statement but has one drawback: we do not care about probability statements made on sample statistics; we desire a probability statement on the parameter, in this case θ. However, since s_* and s^* are functions of θ, these functions can be inverted and solved for θ to obtain two new functions, say $h_*(s)$ and $h^*(s)$, such that

$$P[h_*(s) \leqslant \theta \leqslant h^*(s)] = 1 - \gamma. \tag{8.2}$$

It is precisely this inversion step that creates a problem of interpretation. If (8.2) is interpreted as a probability statement, it is implied that θ is a random variable, and this is contrary to the hypothesis that θ is fixed. As a matter of fact, the correct interpretation of (8.2) is that $100 \times (1 - \gamma)\%$ of the intervals constructed in this manner will contain θ. Thus the word confidence is associated with the interval (h_*, h^*), and (8.2) as it reads is, strictly speaking, incorrect. This is particularly unfortunate in regard to the problems of estimation encountered in reliability, since it is often desired to combine confidence intervals in various ways. Since confidence is not probability, the methods of combining confidence are not well defined. The reader might further illuminate the problem for himself by verifying the following fact: the inversion of (8.1) to obtain (h_*, h^*) of (8.2) is an application of inductive reasoning, that is, h^* is the largest that θ could have been without the observed s having been too (probability $\leqslant \gamma/2$) small, and h_* is the smallest that θ could have been without the observed s having been too (probability $\leqslant \gamma/2$) large.

The Bayes approach to interval estimates is much more direct, that is, deductive. Since in Bayes methods the parameter Θ is assumed to have a prior probability density, say $g(\theta)$, by an application of Bayes's theorem (see Section 2.9) the posterior density,

$$g(\theta|s) = \frac{f(s|\theta)g(\theta)}{f(s)}, \tag{8.3}$$

is well defined at all points where $f(s) \neq 0$. Then, when s has been observed, a probability statement is easily obtained by solving the two equations

$$\int_{-\infty}^{\theta_*} g(\theta|s)\,d\theta = \frac{\gamma}{2}$$

and

$$\int_{\theta^*}^{\infty} g(\theta|s)\,d\theta = \frac{\gamma}{2} \tag{8.4}$$

for θ_* and θ^*, so that $P(\theta_* \leqslant \Theta \leqslant \theta^*) = 1 - \gamma$. The interval (θ_*, θ^*) is a Bayes interval estimate. Equation (8.4) is a valid probability statement and can be combined with other statements by use of the rules of probability.

8.1.3 Summary

In summary, the non-Bayesian (classical) approach considers an unknown parameter as fixed. This means that classical interval estimation and hypothesis testing must lean on inductive reasoning through $f(s;\theta)$. In point estimation, the classical approach must depend on estimates, the criteria for which often are not based on the practical consequences of the estimate. On the other hand, Bayes procedures assume a prior distribution on the parameter space, that is, consider the parameter as a random variable, and hence the posterior distribution is available. This creates the possibility of a whole new class of criteria for estimation, namely, minimization of expected loss, probability intervals, and others.

However, in reliability problems, in view of the difficulty in assessing utility or costs in complex situations, it appears that the largest role for Bayes methods is in providing a means of combining previous data (expressed as the prior distribution) with observed data [expressed in $f(s|\theta)$] to obtain better estimates of parameters by using the posterior density, $f(\theta|s)$. It is these methods that will be emphasized in the rest of this chapter. Four probability densities will be of interest.

$g(\theta)$: the prior density on Θ. This is sometimes called the *mixing* or the *compounding density*.

$f(x_1,\ldots,x_n|\theta)$: the joint density of a sample of size n from $f(x|\theta)$; $f(x|\theta)$ is called the *sampling distribution* of the random variable X.

$f(x_1,\ldots,x_n) = \displaystyle\int_{-\infty}^{\infty} f(x_1,\ldots,x_n,\theta)d\theta$, the *joint marginal density* of the sample observations.

$g(\theta|x_1,\ldots,x_n) = \dfrac{f(x_1,\ldots,x_n,\theta)}{f(x_1,\ldots,x_n)}$, the *posterior density* of Θ.

Here the statistic S is

$$S = (X_1,\ldots,X_n). \tag{8.5}$$

It turns out that the statistics $\{S\}$ that are important and useful are the sufficient statistics (see Chapter 3), and any sufficient statistic for Θ used as the conditioning random variable in the posterior distribution of Θ results in the same posterior distribution as that obtained when (8.5) is used. Thus in what follows we will always use $(X_1,\ldots,X_n)$ as the conditioning (sufficient) statistic. The general problem in reliability to which Bayes methods are applicable is reviewed next.

A parameter Θ (say, the mean time to failure) is regarded as a random variable. The sample observations $(X_1, \ldots, X_n)$ arise as follows:

1. One random observation is obtained from the prior distribution on Θ, with Θ generally *unobservable*.
2. A random sample is drawn from $f(x|\theta)$.

It is desired to estimate from the observations $(X_1, \ldots, X_n)$, each drawn from $f(x|\theta)$, the value of θ.

One of the most useful ideas in statistics is that of obtaining a sample from a probability distribution. Generally, a sample cannot be obtained from the posterior distribution. This *does not* mean that the posterior distribution is not a bona fide probability distribution in the relative frequency sense of Chapter 2. However, if the prior distribution is assigned "subjectively" the posterior distribution, while retaining its mathematical properties, loses the relative frequency interpretation.

8.2 LOSS FUNCTIONS AND BAYES ESTIMATES

As previously indicated, Bayes methods in Reliability involve a known prior distribution, and data $(X_1, \ldots, X_n)$ to estimate the particular value of Θ drawn. It was also indicated that, since the prior distribution is known, the posterior distribution is available, as well as estimates based on minimizing the expected loss. The problem is that generally the loss function is difficult, if not impossible, to write down. However, there is a loss function that is rather popular and gives simple results.

Suppose that $\tilde{\theta}$ is an estimate of Θ and that the loss function is

$$L(\tilde{\theta}, \theta) = (\tilde{\theta} - \theta)^2. \tag{8.6}$$

This function states that the loss is equal to the square of the distance of $\tilde{\theta}$ from θ. The Bayes approach is to select the estimate of Θ that minimizes the expected loss with respect to the posterior distribution. The estimate that accomplishes this is the posterior mean, that is,

$$\tilde{\theta} = E(\Theta|x_1, \ldots, x_n) = \int_{-\infty}^{\infty} \theta g(\theta|x_1, \ldots, x_n) \, d\theta. \tag{8.7}$$

The above loss function is often called the quadratic loss function, and the posterior mean is termed the Bayes estimate.

If the loss function is of the form

$$L(\tilde{\theta}, \theta) = |\tilde{\theta} - \theta|,$$

the estimate of Θ that minimizes the expected loss is the median of the posterior distribution. In general, any estimate that minimizes the expected value of a loss function with respect to the posterior distribution is called a *Bayes estimate.*

Beyond these two simple cases, things become difficult in regard to loss functions for two reasons.

1. The loss function is different for each particular situation and is usually quite complex and unknown.

2. For worthwhile loss functions (i.e., functions that truly represent losses) the Bayes estimate is usually difficult to find.

The expected loss is generally a random variable since it depends on the conditioning statistic S. The unconditional expectation (with respect to S) of the expected loss is called the *Bayes risk* and is minimized by the Bayes estimate.

8.3 BAYES POINT AND INTERVAL ESTIMATES FOR RELIABILITY

8.3.1 The Bernoulli Process: Binomial Sampling to Estimate Unit Reliability

The binomial distribution was discussed in Chapter 3 as arising from n independent observations on an event that has a constant probability of occurrence, say θ, from trial to trial. The process that generates these independent observations, called the Bernoulli process, has a probability distribution assigned on the sample description space $(0, 1)$:

$$f(x|\theta) = \theta^x (1-\theta)^{1-x}, \quad 0 < \theta < 1, \quad x = 0, 1. \tag{8.8}$$

Now suppose that, for a particular unit, survival to a fixed time t is the measure of reliability of interest and that n observations on the event "survival for time t" can be made. These observations can be made by renewing the unit after each trial or by obtaining identical units. Further suppose that the unit in question, although having a fixed and unknown $R(t) = \Theta$, was drawn at random from a set of units and that the varying θ's are described by a prior distribution, $g(\theta)$. The usual estimate of Θ is (number of successes)$/n$.

(a) The Uniform Prior Distribution. A common starting point for a model for $g(\theta)$ is the uniform distribution. This has little of practical value to recommend it, but it illustrates well the ideas under discussion and is interesting to compare to the classical case. Thus

$$g(\theta) = 1, \quad 0 < \theta < 1,$$

$$= 0, \quad \text{elsewhere.} \tag{8.9}$$

Then the joint density of (X, Θ) is given by

$$f(x, \theta) = \theta^x (1 - \theta)^{1-x}. \tag{8.10}$$

The ranges of (X, Θ) are omitted, since they are well defined by Equations (8.8) and (8.9).

Since the sampling is at random from $f(x|\theta)$,

$$f(x_1, \ldots, x_n | \theta) = \prod_{i=1}^{n} f(x_i | \theta) = \prod_{i=1}^{n} \theta^{x_i} (1 - \theta)^{1-x_i}$$

$$= \theta^T (1 - \theta)^{n-T}, \tag{8.11}$$

where $T = \Sigma x_i$. Note that $T = \Sigma x_i$ gives the total number of times that the event "survival for time t" has occurred in n trials.

The joint density of $(X_1, \ldots, X_n, \Theta)$ is

$$f(x_1, \ldots, x_n, \theta) = \theta^T (1 - \theta)^{n-T}$$

$$= \theta^{(T+1)-1} (1 - \theta)^{(n-T+1)-1},$$

so that

$$f(x_1, \ldots, x_n) = \int_0^1 \theta^{(T+1)-1} (1 - \theta)^{(n-T+1)-1} \, d\theta.$$

Now

$$\int_0^1 \theta^{(T+1)-1} (1 - \theta)^{(n-T+1)-1} \, d\theta = \frac{\Gamma(T+1)\Gamma(n-T+1)}{\Gamma(n+2)},$$

where $\Gamma(u)$ is the gamma function of u. When u is an integer, as is the case here, $\Gamma(u) = (u-1)!$. Thus

$$f(x_1, \ldots, x_n) = \frac{\Gamma(T+1)\Gamma(n-T+1)}{\Gamma(n+2)}, T \leqslant n.$$

Then the posterior density of Θ is

$$g(\theta|x_1,\ldots,x_n) = \frac{\Gamma(n+2)}{\Gamma(T+1)\Gamma(n-T+1)}\theta^{(T+1)-1}(1-\theta)^{(n-T+1)-1}.$$

$$(8.12)$$

The density in (8.12) is the beta density discussed in Chapter 3. In the literature it is very often seen with the notation $p = T+1, q = n - T + 1$.

In any case, the posterior mean is the mean of the beta distribution,

$$E(\Theta|x_1,\ldots,x_n) = \int_0^1 \theta g(\theta|x_1,\ldots,x_n)\,d\theta$$

$$= \frac{T+1}{n+2}.$$

$$(8.13)$$

It is to be noted again that $T = \Sigma x_i$ and that, as expected, the posterior mean is a function not only of n but also of T. This shows how the observed data, T, enter into estimating Θ.

It is instructive to compare this estimator of Θ with T/n, the maximum-likelihood estimator (MLE) (no prior distribution assumed). For T near 0 or near n, the two are close. For small n and moderate T, the two may differ somewhat.

Equation (8.13) can be interpreted as a probability. Laplace did this when, having observed n successive sunrises (thus $T = n$), he gave the probability of the next sunrise as

$$\frac{T+1}{n+2} = \frac{n+1}{n+2}.$$

This result, called Laplace's rule of succession, shows that the uniform prior distribution is better for illustrating ideas than for practical use.

Example 1. Suppose that $n = 5$ trials have been conducted to observe the survival of a unit, and that $T = 4$ survivals have been observed. The Bayes estimate of Θ when the prior distribution is uniform is

$$E(\Theta|x_1,\ldots,x_n) = \frac{T+1}{n+2} = \frac{5}{7}.$$

The classical MLE is

$$\frac{T}{n} = \frac{4}{5}.$$

Determining a probability interval for Θ is a problem involving the use of tables of the cumulative beta distribution. The cumulative beta distribution was tabulated by Pearson (1934).

Suppose that a probability interval of the form $P(\theta_* \leqslant \Theta \leqslant \theta^*) = 1 - \gamma$ is desired, with γ assigned and (θ_*, θ^*) to be determined. We will assume that the γ is to be split equally in each tail. This assumption is made to illustrate the method and causes no loss of generality. By using Pearson's tables, the two equations

$$\int_0^{\theta_*} g(\theta | x_1, \ldots, x_n) \, d\theta = \frac{\gamma}{2} \quad \text{and} \quad \int_{\theta^*}^1 g(\theta | x_1, \ldots, x_n) \, d\theta = \frac{\gamma}{2}$$

can be solved for (θ_*, θ^*).

Example 2. Suppose, as previously, that $T = 4$ and $n = 5$, and that $\gamma = .10$ has been chosen. Using Pearson's tables (p. 46) with $T + 1 = p = 5$ and $n - T + 1 = q = 2$, we find

$$\int_0^{.42} g(\theta | x_1, \ldots, x_n) \, d\theta = .05$$

and

$$\int_{.94}^1 g(\theta | x_1, \ldots, x_n) \, d\theta = .05,$$

so that

$$P(.42 \leqslant \Theta \leqslant .94) = .90.$$

The classical method yields $(.35, .99)$ for a 90% confidence interval.

Suppose that (θ_*, θ^*) have been previously selected and that $P(\theta_* \leqslant \Theta \leqslant \theta^*)$ is to be determined. Then, by looking up

$$a = \int_0^{\theta_*} g(\theta | x_1, \ldots, x_n) \, d\theta \quad \text{and} \quad b = \int_{\theta_*}^1 g(\theta | x_1, \ldots, x_n) \, d\theta$$

in Pearson's tables, we obtain the desired probability as $1 - a - b$.
The problem of the one-sided interval follows analogously.

When θ_*, θ^* are preselected and the probability determined, the results must be interpreted carefully because the probability is then a random variable with expectation (with respect to the conditioning statistic) which is the á priori probability that $\theta_* \leqslant \Theta \leqslant \theta^*$. In particular, the interval

(θ_*, θ^*) is *not* an interval estimate in the sense previously defined.

(b) The Truncated Uniform Prior Distribution. The practical utility of the uniform prior distribution can be improved by restricting the range of Θ, that is, suppose Θ is known to be such that $\theta_0 < \Theta < 1$. The distribution of θ is then

$$f(\theta; \theta_0) = \frac{1}{1 - \theta_0}, \quad \theta_0 > 0, \theta_0 < \theta < 1,$$

$$= 0, \quad \text{elsewhere.} \tag{8.14}$$

The reader may wish to verify that is the same density that would be obtained if we started with a uniform prior distribution on Θ and truncated it, that is, computed $f(\theta | \theta \geq \theta_0)$. Actually, this is the way it should be done, since, generally, truncating a distribution changes its family. By the formal operations previously given, it is easy to show that the posterior density is

$$g(\theta | x_1, \ldots, x_n) = \frac{\dfrac{\Gamma(n+2)}{\Gamma(T+1)\Gamma(n-T+1)} \theta^{(T+1)-1}(1-\theta)^{(n-T+1)-1}}{\displaystyle\int_{\theta_0}^{1} \dfrac{\Gamma(n+2)}{\Gamma(T+1)\Gamma(n-T+1)} \theta^{(T+1)-1}(1-\theta)^{(n-T+1)-1} d\theta}.$$

$$\tag{8.15}$$

The constant factor has not been canceled because the denominator represents $1 - F(\theta_0)$, where $F(\theta)$ is the cumulative beta distribution.

The posterior mean is easily seen to be

$$E(\Theta | x_1, \ldots, x_n) = \frac{(T+1)}{(n+2)} \frac{1 - I_{\theta_0}(T+2, n-T+1)}{1 - I_{\theta_0}(T+1, n-T+1)}. \tag{8.16}$$

Here the notation I_{θ_0} is traditional for the cumulative beta distribution evaluated at θ_0.

It is comforting to note that the Bayes estimate given by (8.16) can never estimate Θ to be less than θ_0, since

$$E(\Theta | x_1, \ldots, x_n) = \int_{\theta_0}^{1} \theta g(\theta | x_1, \ldots, x_n) \, d\theta > \int_{\theta_0}^{1} \theta_0 g(\theta | x_1, \ldots, x_n) \, d\theta = \theta_0.$$

(c) The Beta Prior Distribution. The most versatile prior distribution for a Bernoulli process is the beta distribution. In fact, the uniform prior distribution is a special beta distribution with $p = q = 1$. It turns out that, if $g(\theta)$ is a beta density, the posterior density of Θ is also a beta density. In this situation it is said that the beta distribution is *closed* under the Bernoulli process or that the beta prior distribution is conjugate. In general, a prior distribution that results in a posterior distribution of the same family is called a *conjugate prior distribution*.

We continue to assume a Bernoulli process for generating the independent observations $(X_1, \ldots, X_n)$, that is,

$$f(x_1, \ldots, x_n | \theta) = \theta^T (1 - \theta)^{n - T}, \quad T = \Sigma x_i.$$

Now, if the prior distribution is beta,

$$g(\theta; l, m) = \frac{\Gamma(l + m)}{\Gamma(l)\Gamma(m)} \theta^{l - 1} (1 - \theta)^{m - 1}, \quad l, m > 0, \ 0 < \theta < 1. \quad (8.17)$$

Here we have departed from the tradition of Greek letters for parameters; in fact, the semicolon on the left-hand side denotes the fact that l, m are parameters that are not considered as random variables. The marginal density is

$$f(x_1, \ldots, x_n) = \frac{\Gamma(l + m)\Gamma(l + T)\Gamma(m + n - T)}{\Gamma(l)\Gamma(m)\Gamma(l + m + n)}.$$

Thus the posterior density is

$$g(\theta | x_1, \ldots, x_n) = \frac{\Gamma(l + m + n)}{\Gamma(l + T)\Gamma(m + n - T)} \theta^{l + T - 1} (1 - \theta)^{m + n - T - 1}. \quad (8.18)$$

This probability density is again a beta density. It is also easily shown that the posterior mean is

$$E(\Theta | x_1, \ldots, x_n) = \frac{l + T}{l + m + n}. \quad (8.19)$$

It is to be noted that when $n = 0$ Equation (8.19) is the mean of the prior distribution, $l/(l + m)$, as expected. Also, as n becomes arbitrarily large, the Bayes estimate (i.e., the posterior mean) approaches T/n, the classical maximum-likelihood (and consistent) estimator. This is sometimes paraphrased by saying that the influence of the prior distribution, in this case through (l, m), becomes less and less as the sample size grows large. Note also that if $l = 0, m = 1$, the prior density of Θ is improper (cannot be normalized to integrate to 1) and is proportional to θ^{-1}. This is the prior density function associated with the optimum nonrandomized classical lower confidence bound for θ discussed in Section 7.4.3 and defined by Equation 7.15.

Before proceeding to examples, some computational aspects of the beta distribution will be discussed. For this discussion we use standard notation for the cumulative beta distribution:

$$I_\theta(p,q) = \int_0^\theta \frac{\Gamma(p+q)}{\Gamma(p)\Gamma(q)} s^{p-1}(1-s)^{q-1}\,ds.$$

To save space, Pearson's tables give the above integral only for $p \geqslant q$. For $p < q$ the relation

$$I_\theta(p,q) = 1 - I_{1-\theta}(q,p)$$

$$= 1 - \int_0^{1-\theta} \frac{\Gamma(p+q)}{\Gamma(p)\Gamma(q)} s^{q-1}(1-s)^{p-1}\,ds$$

must be used.

The beta distribution is not as well tabulated as the binomial distribution. For this reason the following two relations are often useful:

$$\int_0^\theta \frac{\Gamma(p+q)}{\Gamma(p)\Gamma(q)} s^{p-1}(1-s)^{q-1}\,ds = \sum_{z=p}^{p+q-1} \binom{p+q-1}{z} \theta^z (1-\theta)^{p+q-1-z},$$

$$\theta \leqslant \tfrac{1}{2},$$

$$\int_\theta^1 \frac{\Gamma(p+q)}{\Gamma(p)\Gamma(q)} s^{p-1}(1-s)^{q-1}\,ds = \sum_{z=q}^{p+q-1} \binom{p+q-1}{z} (1-\theta)^z \theta^{p+q-1-z},$$

$$\theta \geqslant \tfrac{1}{2}. \tag{8.20}$$

These are easily proved by integration by parts when p and q are integers.

Another computational aid (used several times in Chapter 5) is available when the sampling process is Bernoulli and the prior distribution is beta. When the posterior distribution is beta with parameters $l+T$ and $m+n-T$, then the posterior distribution of the random variable,

$$U = \left(\frac{m+n-T}{l+T}\right)\left(\frac{\Theta}{1-\Theta}\right),$$

is an F distribution with, $2(l+T)$, $2(m+n-T)$ degrees of freedom. We will illustrate the use of this result by calculating a lower $1-\gamma$ probability interval for Θ:

$$P(U \geqslant F_\gamma) = 1 - \gamma = P\left[\left(\frac{m+n-T}{l+T}\right)\left(\frac{\Theta}{1-\Theta}\right) \geqslant F_\gamma\right]$$

$$= P\left[\left(\frac{1-\Theta}{\Theta}\right) \leqslant \left(\frac{m+n-T}{l+T}\right)\left(\frac{1}{F_\gamma}\right)\right]$$

$$= P\left[\frac{1}{\Theta} \leqslant \left(\frac{m+n-T}{l+T}\right)\left(\frac{1}{F_\gamma}\right) + 1\right]$$

$$= P\left\{\Theta \geqslant \left[\left(\frac{m+n-T}{l+T}\right)\left(\frac{1}{F_\gamma}\right) + 1\right]^{-1}\right\}.$$

Thus the lower $1-\gamma$ probability limit, say θ_*, is given by

$$\theta_* = \left[\left(\frac{m+n-T}{l+T}\right)\left(\frac{1}{F_\gamma}\right) + 1\right]^{-1}. \tag{8.21}$$

A procedure for approximating percentiles of the distribution of the negative logarithm of a beta variate is described in Section 5.1.2(a).

Example 1. Suppose that a unit has $R(t) = \theta$, and this θ was drawn from a beta distribution with $l=3, m=2$. Suppose further that $n=5$ observations on survival for time t lead to 4 survivals, that is, $T = \Sigma x_i = 4$. The Bayes estimate of Θ is given by

$$\frac{l+T}{l+m+n} = \frac{7}{10}.$$

Example 2. Suppose that it is desired to calculate the (posterior) probability, $P(\Theta \geqslant .61)$, that is, the probability Θ equals or exceeds $\theta_* = .61$ is desired. This is given by

$$\int_{.61}^{1} \frac{\Gamma(l+m+n)}{\Gamma(l+T)\Gamma(m+n-T)} \theta^{l+T-1}(1-\theta)^{m+n-T-1} \, d\theta.$$

Then $2(m+n-T)=14$ and $2(l+T)=6$. Using Equation (8.21), one obtains

$$.61 = \left[\left(\frac{3}{7} \right) \frac{1}{F_\gamma} + 1 \right]^{-1},$$

which leads to $F_\gamma = .67$, and hence $1 - \gamma = .75$.

8.3.2 The Bernoulli Process—Pascal Sampling to Estimate Unit Reliability

In the Bernoulli process, the number of samples was held fixed at n and this was called binomial sampling. In obtaining samples it is often advantageous for planning and costing purposes to hold n fixed. However, it may turn out that costs depend more on $T = \sum x_i$ = the number of survivals than on n. In this situation it is advantageous to fix T in advance and to consider n as a random variable. This is known as Pascal sampling, and the procedure is as follows. The idea is that

$$X = \begin{cases} 1 & \text{if a unit survives until time } t, \\ 0 & \text{if a unit does not survive until time } t. \end{cases}$$

Then the probability that n units are required to obtain one survival is

$$f(n|\theta) = \theta(1-\theta)^{n-1}, \quad 0 < \theta < 1, \quad n = 1, 2, \dots. \tag{8.22}$$

This distribution is called the geometric distribution (see Chapter 3). It is obtained by requiring that the first $n-1$ units do not survive to time t [an event of probability $(1-\theta)^{n-1}$] and that the nth unit survives to time t (an event of probability θ). Now, if T survivals are required, a random sample of size T, namely, $n_1, n_2, \dots, n_T$, is obtained from Equation (8.22) so that

$$f(n_1, \dots, n_T | \theta) = \prod_{i=1}^{T} f(n_i | \theta)$$

$$= \theta^T (1-\theta)^{\sum n_i - T}. \tag{8.23}$$

It is to be remembered that the random variable on the right side is $\sum n_i$, not T. We relegate the uniform prior distribution to the problems, and proceed directly to the most useful prior distribution.

The Beta Prior Distribution. Suppose that

$$g(\theta; l, m) = \frac{\Gamma(l+m)}{\Gamma(l)\Gamma(m)} \theta^{l-1} (1-\theta)^{m-1}, \quad l, m > 0, 0 < \theta < 1,$$

$$= 0, \quad \text{elsewhere.} \tag{8.24}$$

Then the joint density of Θ and $(n_1, \ldots, n_T)$ is

$$f(n_1, \ldots, n_T, \theta) = f(n_1, \ldots, n_T | \theta) g(\theta)$$

$$= \frac{\Gamma(l+m)}{\Gamma(l)\Gamma(m)} \theta^{l+T-1} (1-\theta)^{m+\Sigma n_i - T - 1}. \qquad (8.25)$$

Thus the marginal density of the sample is

$$f(n_1, \ldots, n_T) = \frac{\Gamma(l+m)}{\Gamma(l)\Gamma(m)} \frac{\Gamma(l+T)\Gamma(m+\Sigma n_i - T)}{\Gamma(l+m+\Sigma n_i)}. \qquad (8.26)$$

Finally, then,

$$g(\theta | n_1, \ldots, n_T) = \frac{f(n_1, \ldots, n_T, \theta)}{f(n_1, \ldots, n_T)}$$

$$= \frac{\Gamma(l+m+\Sigma n_i)}{\Gamma(l+T)\Gamma(m+\Sigma n_i - T)} \theta^{l+T-1} (1-\theta)^{m+\Sigma n_i - T - 1}. \qquad (8.27)$$

This is again a beta density with posterior mean

$$E(\Theta | n_1, \ldots, n_T) = \frac{l+T}{l+m+\Sigma n_i}. \qquad (8.28)$$

The probability intervals proceed exactly as before (beta posterior distribution) except that it must be remembered that now T is fixed and Σn_i is observed.

8.3.3 Remarks on the Estimation of Reliability

In Sections 8.3.1 and 8.3.2, reliability was estimated using attributes data, that is, failure times were ignored except insofar as they were needed to decide whether or not survival to time t was achieved. This type of sampling is generally inefficient in that we can learn more by utilizing the failure times themselves (for a fixed sample size). It turns out that estimates of reliability $= R(t) = \Theta$ can be made using the observed failure times. Not much will be done in this chapter, however, regarding this situation. The reason is given in what follows.

Let, as always, $R(t) = \Theta$ and assume a uniform prior distribution of Θ, that is,

$$g(\theta) = 1, \quad 0 < \theta < 1,$$

$$= 0, \quad \text{elsewhere.} \qquad (8.29)$$

Schafer (1970) shows that the uniform prior distribution for the reliability function $R(t)$ must be used with great caution. In particular he points out that if $g(\theta)$ is uniform for $\Theta = R(t_1)$ there will be another distribution for $\Theta = R(t_2)$ if $t_1 \neq t_2$. In the following it will be shown that because of the functional relationship $R(t) = \exp(-\Lambda t)$, choosing a prior distribution for Λ is equivalent to choosing a prior for $R(t)$.

Very often the reliability function, $\Theta = R(t)$, is obtained by assuming an exponential time-to-failure distribution, so that

$$R(t) = e^{-\Lambda t}. \tag{8.30}$$

This means that

$$\frac{-\ln \Theta}{t} = \Lambda \qquad \text{and} \qquad \left| \frac{d\Theta}{d\Lambda} \right| = t e^{-\Lambda t},$$

so that the prior density of Λ is easily obtained by substituting in Equation (8.29):

$$g(\lambda) = t e^{-\lambda t}, \quad t \geqslant 0, \lambda > 0,$$

$$= 0, \quad \text{elsewhere.} \tag{8.31}$$

This density is a gamma prior density with shape parameter 1 and scale parameter t. A random sample of size n (failure times) then has the conditional density

$$f(x_1, \ldots, x_n | \lambda) = \prod_{i=1}^{n} \lambda e^{-\lambda x_i} = \lambda^n e^{-\lambda \Sigma x_i}. \tag{8.32}$$

It is then easy to show that

$$g(\lambda | x_1, \ldots, x_n) = \frac{(\Sigma x_i + t)^{n+1}}{\Gamma(n+1)} \lambda^n e^{-(\Sigma x_i + t)\lambda}. \tag{8.33}$$

Hence, for a uniform prior distribution on $\Theta = R(t)$ and exponential sampling, the posterior distribution of Λ is gamma with scale parameter $\Sigma x_i + t$ and shape parameter $n+1$. What has been illustrated is that there always exists a functional relationship between the reliability, $\Theta = R(t)$, and the parameter(s) in the conditional distribution of lifetimes. For the exponential distribution this relation is

$$\Theta = R(t) = e^{-\Lambda t}$$

so that a simple change of variable in the posterior density of Λ [given by Equation (8.33) as an example] leads to the posterior density of R, that is, to $g(\theta|x_1,\ldots,x_n)$. Also, when $g(\lambda|x_1,\ldots,x_n)$ is available, a (Bayes) probability interval is easily constructed on Θ as follows.

Suppose that

$$P(\lambda_* \leqslant \Lambda \leqslant \lambda^*) = 1 - \gamma, \quad (\lambda_*, \lambda^*) \text{ known}.$$

Then, since $\Theta = e^{-\Lambda t}$, the event $\lambda_* \leqslant \Lambda \leqslant \lambda^*$ implies the event $e^{-\lambda^* t} \leqslant e^{-\Lambda t} \leqslant e^{-\lambda_* t}$ and conversely, so that

$$P(\theta_* \leqslant \Theta \leqslant \theta^*) = 1 - \gamma, \qquad \theta_* = e^{-\lambda^* t}, \qquad \theta^* = e^{-\lambda_* t}.$$

For probability intervals for reliability, then, the posterior distribution of Λ is sufficient knowledge when reliability is functionally related to Λ, and this functional relation is available in closed form. Therefore, in the rest of this chapter only prior distributions on Λ (and $1/\Lambda$) will be considered. The shortcoming is that the parameters in the posterior distribution of $\Theta = R(t)$ may be difficult to interpret physically.

It should be noted the uniform prior distribution on $\Theta = R(t)$ leads to a nonuniform prior distribution on Λ. This means that, if we consider each possible θ equally likely, we implicitly assume (when $\Theta = e^{-\Lambda t}$) that the prior distribution of Λ is not uniform. This also illustrates the fact that sometimes seemingly innocuous assumptions (e.g., a uniform prior distribution on Θ) have important implications (e.g., the prior distribution on Λ is not uniform).

Example. The posterior mean of $R(t)$ can be obtained by taking $E(e^{-\Lambda t})$ with respect to the posterior distribution of Λ. Using (8.30) and (8.39) it is $[(\alpha + T)/(\alpha + T + t)]^{n+\beta}$.

Finally, in some of the literature, reliability is defined to be the per cent, fraction, or number of nondefective items occurring in finite lots. This quantity indeed is a random variable, and Bayesian methods are applicable. However, because there are limitations on the size of this book, and because the treatment of attributes sampling plans belongs to the field of sampling inspection and quality control rather than reliability, these methods will not be covered here. For Bayes methods in attributes sampling from finite lots the reader is referred to the papers by Hald (1960), Wallenuis (1969), and Schafer (1967).

8.4 BAYES POINT AND INTERVAL ESTIMATES FOR FAILURE RATE

8.4.1 The Poisson Process: Gamma Sampling to Estimate the (Constant) Failure Rate

A Poisson process has already been identified (see Chapter 3) as a process that generates random times between failure, X, with the exponential probability density,

$$f(x|\lambda) = \lambda e^{-\lambda x}, \quad 0 < x < \infty, \lambda > 0,$$

$$= 0, \quad \text{elsewhere.} \tag{8.34}$$

A random sample $(X_1, \ldots, X_n)$ from this distribution results in a sufficient statistic, $T = \sum x_i$, which has a gamma distribution with scale parameter Λ and shape parameter n. This explains why the sampling process is called gamma sampling. Nonetheless we continue here to use the statistic $(X_1, \ldots, X_n)$ instead of $T = \sum x_i$ as the conditioning statistic. Actually, the matter is of little importance, since the posterior distribution of Λ is the same in either case. We limit our discussion to a gamma prior distribution on Λ, which happens to be the natural conjugate. In what follows we discuss the complete sample case. If the sample is censored at $r < n$ failures or at time t_0 the conditioning (sufficient) statistic is, respectively.

$$\sum_{i=1}^{r} X_i + (n-r)X_r \quad \text{or} \quad \sum_{i=1}^{r} X_i + (n-r)t_0.$$

If the failure times occur in order, i.e., $X_1 \leqslant X_2 \leqslant \ldots \leqslant X_n$, the posterior distribution is unaffected.

The Gamma Prior Distribution. Suppose that

$$g(\lambda; \alpha, \beta) = \frac{\alpha^\beta}{\Gamma(\beta)} \lambda^{\beta-1} e^{-\alpha\lambda}, \quad \alpha, \beta, \lambda > 0,$$

$$= 0, \quad \text{elsewhere.} \tag{8.35}$$

For $\alpha = \beta = 0$, the prior density for Λ is proportional to λ^{-1} the (improper) prior density yielding the uniformly most accurate upper confidence bound for λ, discussed in Sections 5.1.1 and 10.4.1. Then, since random sampling is assumed,

$$f(x_1, \ldots, x_n|\lambda) = \prod_{i=1}^{n} \lambda e^{-\lambda x_i}$$

$$= \lambda^n e^{-\lambda T}, \tag{8.36}$$

where $T = \sum x_i$, so that

$$f(x_1, \ldots, x_n, \lambda) = \frac{\alpha^\beta}{\Gamma(\beta)} \lambda^{\beta + n - 1} e^{-\lambda(\alpha + T)}, \qquad (8.37)$$

and then

$$f(x_1, \ldots, x_n) = \frac{\alpha^\beta}{\Gamma(\beta)} \frac{\Gamma(\beta + n)}{(\alpha + T)^{\beta + n}}. \qquad (8.38)$$

Thus

$$g(\lambda | x_1, \ldots, x_n) = \frac{(\alpha + T)^{\beta + n}}{\Gamma(\beta + n)} \lambda^{\beta + n - 1} e^{-\lambda(\alpha + T)}. \qquad (8.39)$$

The posterior distribution of Λ is thus gamma, with scale parameter $\alpha + T$ and shape parameter $\beta + n$. As previously mentioned, if the sufficient statistic $T = \sum x_i$ is used as the conditioning statistic,

$$f(T|\lambda) = \frac{\lambda^n}{\Gamma(n)} T^{n-1} e^{-\lambda T},$$

$$f(T, \lambda) = \frac{\alpha^\beta}{\Gamma(\beta)\Gamma(n)} \lambda^{\beta + n - 1} T^{n-1} e^{-\lambda(\alpha + T)}.$$

Thus

$$f(T) = \frac{\alpha^\beta}{\Gamma(\beta)\Gamma(n)} \frac{T^{n-1} \Gamma(\beta + n)}{(\alpha + T)^{\beta + n}},$$

so that

$$g(\lambda | T) = \frac{(\alpha + T)^{\beta + n}}{\Gamma(\beta + n)} \lambda^{\beta + n - 1} e^{-\lambda(\alpha + T)}$$

which is identical to (8.39).

The mean of a gamma distribution is the shape parameter divided by the scale parameter (see Chapter 3), so that the posterior mean is

$$E(\Lambda | x_1, \ldots, x_n) = \frac{\beta + n}{\alpha + T}. \qquad (8.40)$$

For large n the posterior mean is approximately n/T, which is the classical maximum-likelihood estimate of Λ.

Example 1. Suppose that $\alpha = 100$, $\beta = 2$, $n = 8$, and $T = \sum x_i = 200$. Then the Bayes estimate of Λ is

$$\frac{\beta + n}{\alpha + T} = .0333.$$

The classical maximum-likelihood estimate is

$$\hat{\lambda} = \frac{n}{T} = \frac{8}{200} = .0400.$$

Example 2.   Suppose again that $\alpha = 100$, $\beta = 2$, $n = 8$, and $T = 200$ and that it is desired to find λ^* such that

$$P(\Lambda \leqslant \lambda^*) = .90.$$

This is given by the solution to

$$\int_0^{\lambda^*} \frac{(\alpha + T)^{\beta + n}}{\Gamma(\beta + n)} \lambda^{\beta + n - 1} e^{-\lambda(\alpha + T)} \, d\lambda = .90.$$

With the change of variable $Y = 2\Lambda(\alpha + T)$, it is clear that

$$g(y|x_1,\ldots,x_n) = \frac{1}{\Gamma(\beta + n) 2^{(\beta + n)}} y^{\beta + n - 1} e^{-y/2}.$$

This is the well-tabulated chi-square distribution with $2(\beta + n)$ degrees of freedom. Hence

$$P(\Lambda \leqslant \lambda^*) = P[2\Lambda(\alpha + T) \leqslant 2\lambda^*(\alpha + T)] = P(Y \leqslant \chi^2_{2(\beta + n), .90});$$

also

$$2(\beta + n) = 20 \quad \text{and} \quad \chi^2_{20, .90} = 28.412$$

so that $2\lambda^*(\alpha + T) = 28.412$ implies that $\lambda^* = .0474$.
It is interesting to obtain the classical upper .90 confidence bound. To do this, we recall that $2n\lambda/\hat{\lambda}$ is distributed as chi square with $2n$ degrees of freedom. Then $\hat{\lambda} = .0400$ and $2n = 16$, so that the upper confidence bound is $\lambda^* = .0588$.

8.4.2 The Poisson Process: Poisson Sampling to Estimate the (Constant) Failure Rate λ

In the analysis of the preceding section the number of times to failure observed was held fixed at n, and $T = \Sigma x_i$ was a random variable. This has the disadvantage that it is not known how much time will be required to record the n failure times. For this reason the test time, T, will now be fixed and n, the number of failures, allowed to vary. In this situation the distribution of the number of failures occurring in fixed (operating) time T

is given by the Poisson distribution:

$$f(n|\lambda; T) = \frac{e^{-\lambda T}(\lambda T)^n}{n!}.$$ (8.41)

The Gamma Prior Distribution. Suppose again that

$$g(\lambda; \alpha, \beta) = \frac{\alpha^\beta}{\Gamma(\beta)} \lambda^{\beta-1} e^{-\alpha\lambda}.$$ (8.42)

Then

$$f(n, \lambda) = \frac{\alpha^\beta}{\Gamma(\beta)} \frac{T^n}{n!} \lambda^{\beta+n-1} e^{-\lambda(\alpha+T)},$$

so that

$$f(n) = \frac{\alpha^\beta}{\Gamma(\beta) n!} \frac{T^n \Gamma(\beta+n)}{(\alpha+T)^{\beta+n}}$$

and

$$g(\lambda|n) = \frac{(\alpha+T)^{\beta+n}}{\Gamma(\beta+n)} \lambda^{\beta+n-1} e^{-\lambda(\alpha+T)}.$$ (8.43)

The posterior mean is again

$$E(\Lambda|n) = \frac{\beta+n}{\alpha+T},$$ (8.44)

and probability limits are obtained in precisely the same fashion as in Section 8.4.1.

8.4.3 The Poisson Process: Estimating the Mean Time to Failure $(1/\lambda)$

Let the mean time to failure in the exponential distribution be denoted by $\Theta = 1/\Lambda$. The natural (conjugate) prior density for both the Poisson and the gamma sampling cases is the inverted gamma prior density:

$$g(\theta; \alpha, \beta) = \frac{\alpha^\beta}{\Gamma(\beta)} \left(\frac{1}{\theta}\right)^{\beta+1} e^{-\alpha/\theta}, \quad \alpha, \beta, \theta > 0,$$

$$= 0, \quad \text{elsewhere.}$$ (8.45)

This prior density is easily obtained by making the change of variable $\Theta = 1/\Lambda$ in the gamma prior density of the preceding two sections. As a matter of fact the posterior density of Θ may be obtained directly from the previous (gamma) posterior densities by the change of variable $\Theta = 1/\Lambda$. Thus

$$g(\theta|x_1,\ldots,x_n) = \frac{(\alpha+T)^{\beta+n}}{\Gamma(\beta+n)}\left(\frac{1}{\theta}\right)^{\beta+n+1} e^{-(\alpha+T)/\theta}. \qquad (8.46)$$

There are, however, some loose ends to be tied together. It turns out that the posterior mean is

$$E(\Theta|x_1,\ldots,x_n) = \frac{\alpha+T}{\beta+n-1}, \quad \beta+n>1,$$

$$= \infty, \qquad \text{otherwise.} \qquad (8.47)$$

In general, the Kth moment for an inverted gamma distribution is given by

$$E(\Theta^K) = \frac{\alpha^K}{\displaystyle\prod_{i=1}^{K}(\beta-i)}, \qquad (8.48)$$

where $\alpha =$ scale parameter and $\beta =$ shape parameter, whenever $K \leqslant [\beta]$ = the largest integer smaller than β. When $K > [\beta]$, the Kth moment does not exist. Thus in the prior density of (8.45), if $0 < \beta \leqslant 1$, none of the moments exists. Nonetheless the distribution is well behaved in other respects, and as far as the methods of this chapter are concerned there are no serious practical consequences of this defect. For example, since always $n \geqslant 1$ and $\beta > 0$, the posterior mean is well defined. Again, in regard to the posterior density of (8.46), Equation (8.48) leads to

$$E(\Theta^K|x_1,\ldots,x_n) = \frac{(\alpha+T)^K}{\displaystyle\prod_{i=1}^{K}(\beta+n-i)}, \qquad (8.49)$$

and the first $[\beta+n]$ moments always exist.

8.4.4 Computational Notes for Gamma and Inverted Gamma Distributions

It has already been pointed out that the gamma and inverted gamma distributions are related by a change of variable $U = 1/V$, that is, if V is gamma, U is said to be inverted gamma. Hence we confine our remarks to the gamma posterior distribution.

Suppose that Λ is the random variable in the gamma posterior distribution. Then a change of variable

$$Y = 2\Lambda(\alpha+T)$$

leads to an incomplete gamma function. These integrals (arising when preparing probability intervals) can be obtained on a computer or from tables of the incomplete gamma function. However, if, in the prior distribution, β is of the form $I/2$ or I, I a positive integer, $2(\beta + n)$ is an integer and hence $Y = 2\Lambda(\alpha + T)$ has a chi-square distribution with $2(\beta + n)$ degrees of freedom.

It can be shown by integration by parts that

$$P(n \leqslant N|\lambda) = \sum_{n=0}^{N} \frac{e^{-\lambda T}(\lambda T)^{n}}{n!}$$

$$= \int_{\lambda T}^{\infty} \frac{u^{N} e^{-u}}{N!} \, du$$

$$= P(\chi^2_{2(N+1)} > 2\lambda T|\lambda).$$

Thus, if χ^2 tables are not available, Poisson tables may be used, and conversely.

It has been pointed out that the posterior mean is the estimate which minimizes the expected value of quadratic loss. It should also be noted that the posterior mean is a random variable because it depends on the (random) conditioning statistic S. Hence, generally, the posterior mean has an expected value and variance with respect to S. These latter two quantities should not be confused with posterior mean and posterior variance.

8.5 BAYESIAN RELIABILITY DEMONSTRATION TESTING

The connection between hypothesis testing and confidence intervals was pointed out in Chapter 6, where the relationship between hypothesis tests and reliability demonstration was also discussed. The same type of situation is present in Bayesian reliability demonstration testing. In other words, we may select, on the basis of methods given previously in this chapter, a value of γ and find $\theta*$ such that

$$P(\Theta \geqslant \theta*|x_1, \ldots, x_n) = \int_{\theta*}^{\infty} g(\theta|x_1, \ldots, x_n) \, d\theta = 1 - \gamma.$$

The basic dispute with this kind of analysis is that, if $1 - \gamma$ is selected we must accept what we get for $\theta*$. The reason for this is that the sample size is usually selected on grounds independent of $(\theta*, 1 - \gamma)$. In the preceding sections we have assumed n (the sample size) given and have presented some interesting methods for obtaining probability intervals, but $\theta*$ and $1 - \gamma$ were not both open to choice. In the following sequential method this problem is addressed:

Having specified $(\theta* = \theta_1, 1 - \gamma = K_0)$, how much data should be gathered?

The notation θ_1 is used for consistency with the notation of reliability testing, which calls the minimum acceptable mean time to failure θ_1. The notation K_0 is used for consistency with the notation of Schafer and Singpurwalla (1970). The method is entirely general in that any parameter with prior distribution $g(\theta)$ can be treated. Here we address the mean time to failure, Θ, in the exponential distribution. It turns out there may be no $n = 1, 2 \ldots$ for a preselected pair $(\theta_1 = \theta*, K_0 = 1 - \gamma)$ such that

$$P(\Theta \geqslant \theta_1 | x_1, \ldots, x_n) \geqslant K_0.$$

Thus we need an additional criterion to help us decide when to abandon the idea $(\Theta \geqslant \theta_1)$ and accept the idea $(\Theta < \theta_1)$. To this end two numbers (K_1, K_0) are selected such that $0 < K_1 < K_0 < 1$. Usually K_0 is chosen at .90, .95, or .99, and K_1 at .10, .05, or .01. At the nth step (i.e., the nth observation, x_n), testing is stopped and the hypothesis $(\Theta \geqslant \theta_1)$ is accepted if and only if

$$P_n \equiv P(\Theta \geqslant \theta_1 | x_1, \ldots, x_n) \geqslant K_0.$$

If, at the nth step, $P_n \leqslant K_1$, testing is stopped and the hypothesis $(\Theta \geqslant \theta_1)$ is rejected. Finally, if $K_1 < P_n < K_0$, testing is continued.

It is easy to see what is being done: a Bayes lower probability limit is specified at θ_1 with probability at least K_0. If we cannot achieve this, we settle for an upper limit of θ_1 with probability at least $1 - K_1$. It is not necessary that $K_1 + K_0 = 1$. If before testing

$$P_0 = \int_{\theta_1}^{\infty} g(\theta) \, d\theta \geqslant K_0, \quad \text{accept } (\Theta \geqslant \theta_1)$$

without testing, and if

$$P_0 = \int_{\theta_1}^{\infty} g(\theta) \, d\theta \leqslant K_1, \quad \text{reject } (\Theta \geqslant \theta_1)$$

without testing. This test terminates with probability 1 under general conditions. Computation of the distribution of sample size to termination and of the probability of accepting $(\Theta \geqslant \theta_1)$, $P(A)$, is generally very difficult, and approximations have been developed. In the following we denote the random variable steps to termination by n_t.

1. The probability of accepting $(\Theta \geqslant \theta_1)$, say $P(A)$, is given by

$$P(A) = 1 \quad \text{if} \quad P(\Theta \geqslant \theta_1) = P_0 \geqslant K_0,$$

$$P(A) \cong \left(\frac{P_0 - K_1}{K_0 - K_1} \right) \quad \text{if} \quad K_1 < P_0 < K_0, \qquad (8.50)$$

$$P(A) = 0 \quad \text{if} \quad P_0 \leqslant K_1,$$

where

$$P_0 = \int_{\theta_1}^{\infty} g(\theta) \, d\theta.$$

The approximation $P(A) \cong (P_0 - K_1)/(K_0 - K_1)$ is usually very good when K_1 is less than .10, K_0 is larger than .90, and P_0 is not too near either K_1 or K_0. In other words, the approximation is good when the expected test time is not small. The approximation is easily developed by using the result $E(P_n) = P_0$ and assuming $P_n = K_0$ at acceptance or $P_n = K_1$ at rejection, that is, by neglecting the amount by which P_n exceeds K_0 or is less than K_1 at termination. Then

$$E(P_n) = P_0 \cong P(A) K_0 + [1 - P(A)] K_1. \qquad (8.51)$$

2. It may be of interest for the user of such a procedure to have an answer to the question: What is the probability of accepting $(\Theta \geqslant \theta_1)$ when indeed $(\Theta \geqslant \theta_1)$ is the case? It is given by

$$P(A | \Theta \geqslant \theta_1) = 1, \quad P_0 \geqslant K_0,$$

$$P(A | \Theta \geqslant \theta_1) \cong \left(\frac{K_0}{P_0} \right) \left(\frac{P_0 - K_1}{K_0 - K_1} \right), \quad K_1 < P_0 < K_0, \qquad (8.52)$$

$$P(A | \Theta \geqslant \theta_1) = 0, \quad P_0 \leqslant K_1.$$

Schafer (1971) has shown that always $P(A | \Theta \geqslant \theta_1) \geqslant (P_0 - K_1)/P_0(1 - K_1)$.

3. A lower bound on the probability that $n_t \leqslant n$ for any fixed n is available. It is

$$P(n_t \leqslant n) \geqslant P(P_n \geqslant K_0) + P(P_n \leqslant K_1). \qquad (8.53)$$

The computation of this bound is easier than it may at first appear. For any particular n, a set $S_{0n} = \{(x_1, \ldots, x_n)\}$ may be identified such that, if $(x_1, \ldots, x_n) \in S_{0n}$, then $P_n \geqslant K_0$ and conversely. A similar set, S_{1n}, may be found that is necessary and sufficient for $P_n \leqslant K_1$. The probability assigned to each of these sets (S_{1n} or S_{0n}) may be computed from the marginal distribution $f(x_1, \ldots, x_n)$ by

$$\int_{S_{in}} dF(x_1, \ldots, x_n), \quad i = 0, 1. \tag{8.54}$$

Example. Suppose that $g(\theta)$ is the inverted gamma probability density function

$$g(\theta) = \frac{\alpha^\beta}{\Gamma(\beta)} \left(\frac{1}{\theta} \right)^{\beta+1} e^{-\alpha/\theta}, \quad \theta \geqslant 0;\ \alpha, \beta > 0,$$

$$= 0, \quad \text{elsewhere,} \tag{8.55}$$

and that

$$f(x|\theta) = \left(\frac{1}{\theta} \right) e^{-x/\theta}, \quad \theta \geqslant 0, x \geqslant 0,$$

$$= 0, \quad \text{elsewhere.} \tag{8.56}$$

Then it is easy to show that

$$f(x_1, \ldots, x_n) = \frac{\Gamma(\beta+n)}{\Gamma(\beta)} \left(\frac{\alpha}{\alpha+T} \right)^\beta \frac{1}{(\alpha+T)^n}, \quad T = \sum x_i, \tag{8.57}$$

and that

$$g(\theta|x_1, \ldots, x_n) = \frac{(\alpha+T)^{\beta+n}}{\Gamma(\beta+n)} \left(\frac{1}{\theta} \right)^{\beta+n+1} e^{-(\alpha+T)/\theta} \tag{8.58}$$

is again an inverted gamma density. Thus the inverted gamma distribution is closed under sampling with respect to the exponential conditional p.d.f. In any event

$$P(\Theta \geqslant \theta_1 | x_1, \ldots, x_n) = \int_{\theta_1}^{\infty} g(\theta|x_1, \ldots, x_n)\, d\theta$$

leads, after the change of variable $Z = (\alpha+T)/\Theta$, to

$$P_n = \int_0^{(\alpha+T)/\theta_1} \frac{1}{\Gamma(\beta+n)} z^{\beta+n-1} e^{-z}\, dz. \tag{8.59}$$

The right side is an incomplete gamma function multiplied by $1/\Gamma(\beta + n)$. It is not necessary to compute P_n at each step because the two equations

$$K_0 = \int_0^{v_0} \frac{1}{\Gamma(\beta+n)} z^{\beta+n-1} e^{-z} dz; \qquad K_1 = \int_0^{v_1} \frac{1}{\Gamma(\beta+n)} z^{\beta+n-1} e^{-z} dz$$

may be solved for each n for v_1, v_0 and T_1, T_0, found such that

$$v_0 = \frac{\alpha + T_0}{\theta_1} \qquad \text{and} \qquad v_1 = \frac{\alpha + T_1}{\theta_1}.$$

Because P_n of (8.59) is monotonically increasing in T,

$$P_n \geqslant K_0 \Leftrightarrow T \geqslant T_0 \tag{8.60}$$

and

$$P_n \leqslant K_1 \Leftrightarrow T \leqslant T_1.$$

Here T_0 and T_1 are descriptions of the aforementioned sets, S_{0n} and S_{1n}. It should be noted that there will be a different pair (T_1, T_0) for each n. For a numerical example, suppose that $\beta = 3$, $\alpha = 100$, $\theta_1 = 60$, $K_1 = .10$, and $K_0 = .90$. In this case, it is easy to find the pairs (T_1, T_0) since $2Z$ is distributed as chi square with $2(\beta + n)$ degrees of freedom. The accompanying table gives the partial results. It should be recalled that $T = \Sigma x_i$ is the test statistic.

	Accept Criteria	
	T_1	T_0
n	Accept $\Theta < \theta_1$ if $T \leqslant T_1$	Accept $\Theta \geqslant \theta_1$ if $T > T_0$
1	4.7	302.0
2	46.1	380.0
3	89.0	455.0
4	133.7	533.0
5	179.3	605.0
6	227.0	680.0
7	272.0	752.0
.	.	.
.	.	.
.	.	.

As an additional illustration, we now compute a lower bound on $P(n_t \leqslant 7)$. By Equation (8.60) and the table, we need only compute for $n = 7$

$$P(T \geqslant T_0 = 752.0) + P(T \leqslant T_1 = 272.0).$$

This is given by

$$Q = 1 - \int_{T_1}^{T_0} f(T) \, dT.$$

The easiest way to obtain $f(T)$ is as follows. Since $f(x|\theta)$ is exponential, $f(T = \Sigma x_i | \theta)$ is gamma with parameters $1/\theta =$ scale and $n =$ shape. Thus

$$f(T|\theta) = \frac{1}{\theta^n \Gamma(n)} T^{n-1} e^{-T/\theta}.$$

The joint density of (T, Θ) is then

$$\frac{\alpha^\beta}{\Gamma(\beta)\Gamma(n)} \left(\frac{1}{\theta}\right)^{\beta+n+1} T^{n-1} e^{-(T+\alpha)/\theta},$$

so that

$$f(T) = \frac{\alpha^\beta \Gamma(\beta+n)}{\Gamma(n)\Gamma(\beta)} \frac{T^{n-1}}{(\alpha+T)^{\beta+n}}.$$

Then

$$Q = 1 - \int_{T_1}^{T_0} \alpha^\beta \frac{\Gamma(\beta+n)}{\Gamma(n)\Gamma(\beta)} \frac{T^{n-1}}{(\alpha+T)^{\beta+n}} \, dT.$$

The change of variable, $Y = \alpha/(\alpha + T)$, means that

$$Q = 1 - \int_{y_0}^{y_1} \frac{\Gamma(\beta+n)}{\Gamma(\beta)\Gamma(n)} y^{\beta-1} (1-y)^{n-1} \, dy,$$

where Y is a beta random variable with parameters β and n, and

$$y_0 = \frac{\alpha}{\alpha + T_0}, \qquad y_1 = \frac{\alpha}{\alpha + T_1},$$

$$= \frac{100}{852}, \qquad\qquad = \frac{100}{372},$$

$$= .1173; \qquad\qquad = .2688.$$

The Pearson tables yield

$$Q \cong 1 - .37 = .63,$$

so that

$$P(n_t \leqslant 7) \geqslant .63.$$

Here a remark is in order. The situations in this chapter have been limited to cases in which the (random) parameter is the only unknown parameter in a probability distribution. In general, there may be a random vector of parameters. For example, if the sampling distribution is the two-parameter Weibull, then (unless one of the parameters is regarded as fixed) the prior distribution will be bivariate. Not much has been done in this area, but the complications are of a computational nature only, since the methods of this chapter still apply.

8.6 FITTING PRIOR DISTRIBUTIONS

In virtually all the methods of this chapter the prior density, $g(\theta)$, has been assumed to be known. Procedures for discovering the prior distribution are of considerable interest, even though methods of analysis are available which assume that the prior distribution exists but is unknown. These methods, called empirical Bayes methods, are described, for example, in Robbins and in Rutherford and Krutchkoff (1967). They are not treated in this book, however, because of their apparent limited applicability to reliability sampling situations.

In using Bayesian methods the prior distribution receives a good deal of attention, and rightly so. However, in any analysis involving actual data (X) the conditional distribution of $(X|\theta)$ must also be known. This conditional p.d.f., sometimes called a sampling distribution, receives less attention, however, because usually more is known about it than about the prior distribution. For example, if several machines are turning out a large number of bolts that can be classified only as good or bad, for fixed fraction defective, p, the sampling distribution or conditional distribution is hypergeometric or binomial, depending on the finiteness of the outputs. Thus often the physical process dictates the sampling distribution. In other cases, experience has dictated what to expect for a conditional distribution. That is the situation here. We have assumed and will continue to assume that the process generating failures is a Poisson process and hence the sampling distribution of time to failure is exponential. If it should turn out that the conditional distribution is Weibull, the same general methods would apply except that the prior distribution would be multivariate. The major change then would be that the algebra would become much more intricate.

In this section the methods of fitting prior distributions are discussed for the random variable Θ = meantime to failure (MTTF) and Poisson sampling (see Section 8.4); the general methods are applicable also   other types of sampling with appropriate changes. Also, it is assumed that the family (e.g., inverted gamma) of the prior distribution has been specified. If the analyst is unwilling to specify a family before trying a fit, the available methods are extremely complex and difficult to apply. For more information on these methods the reader is referred to the papers by Tucker (1963), Rolph (1968), and Rutherford and Krutchkoff (1967).

In fitting a prior distribution to Θ = MTTF, there is a source of error generally not present in the classical case of fitting distributions, where a random sample $(X_1,\ldots,X_n)$ is obtained from a probability distribution and a fit made. In fitting $g(\theta)$, θ is usually unknown and hence a random sample is not even available from $g(\theta)$. To obtain a random sample from $g(\theta)$ it would be necessary to know the θ of an equipment exactly, and this is impossible. Hence we must fit $g(\theta)$ witho t sampling from it. More particularly, what we have is a set of independent pairs:

$$(X_1,\Theta_1),(X_2,\Theta_2),\ldots,(X_n,\Theta_n),$$

where (we have assumed Poisson sampling) X_i is the number of failures occurring in fixed (for all i) time T and θ_i is the unobservable MTTF of the ith equipment. It turns out that under mild restrictions there is a unique correspondence between the prior density $g(\theta)$ and the marginal density $f(x)$. For example, all the distributions treated in this chapter have this property. Thus, having specified a given family for the prior distribution, if the empirical frequency distribution of the number of failures $(X_1,X_2,\ldots,X_n)$ (each on a different equipment) is plotted, it should "look like" the marginal distribution $f(x)$. In other words, as $n\to\infty$, the empirical distribution of the X's approaches the true marginal distribution, which is uniquely determined by the family of the prior distribution. The four steps in fitting the prior distribution are then relatively simple.

1. Specify the family of the prior distribution.
2. Derive the (unique) marginal distribution of X (here, the number of failures in fixed time T).
3. Fit the empirical data $(X_1,X_2,\ldots,X_n)$ to the appropriate marginal distribution by a chi-square test.
4. Estimate the parameters in the prior distribution by using $f(x)$.
To continue the example, suppose that

$$f(x|\theta) = \frac{e^{-T/\theta}(T/\theta)^x}{x!},$$

that is, the distribution of the number of failures is Poisson and the prior density is inverted gamma:

$$g(\theta;\alpha,\beta) = \frac{\alpha^{\beta}}{\Gamma(\beta)}\,(1/\theta)^{\beta+1}e^{-\alpha/\theta}.$$

Then the marginal mass of X is negative binomial, that is,

$$f(x) = \frac{\Gamma(\beta+x)}{\Gamma(\beta)x!}\left(\frac{T}{T+\alpha}\right)^{x}\left(\frac{\alpha}{T+\alpha}\right)^{\beta}. \qquad (8.61)$$

For this distribution

$$E(X) = \frac{\beta T}{\alpha} \quad \text{and} \quad \sigma_X^{\ 2} = \frac{\beta T(T+\alpha)}{\alpha^2}. \qquad (8.62)$$

Estimates of (α,β), say $(\tilde{\alpha},\tilde{\beta})$, can be obtained by the method of moments, using (8.62), and a chi-square test run on the distribution (8.61). If the fit is satisfactory, the prior distribution is assumed to be inverted gamma with the parameters $(\tilde{\alpha},\tilde{\beta})$.

Things become more complicated in the following cases:

1. The fixed time T is different for each i. Thus the number of failures on each equipment is based on a time period different from that for any other equipment.

2. The random variable is time to failure, and the number of failures observed on each equipment is different from that for any other equipment.

These are essentially mixed models, and the marginal distribution is a mixed distribution. Although the methods outlined still apply, they are more complex arithmetically. Also, the question of how large n and T (or number of failures per equipment when lifetime is the random variable observed) must be for good fits is extremely important. All of these points are covered in detail in Schafer and Feduccia (1972). Finally, in the authoritative paper by Bhattacharya (1967) many additional prior distributions are considered. In that paper the quantity called the variance of the posterior mean is the variance of the posterior distribution.

Excellent additional reading for this chapter may be found in Lindley (1965) and Raiffa and Schlaifer (1961). Finally, we have ignored herein the important case of the normal distribution (conditional and prior). In this case the posterior distribution is again normal; an excellent treatment of this situation is given in Guttman, Wilks, and Hunter (1971).

Much of the material presented in this chapter is summarized in Table 8.1.

PROBLEMS

8.1. A unit has an unknown $R(t) = \Theta$. Suppose that $n = 10$ and $T = \Sigma x_i = 8$, and that Θ has a beta prior distribution with $l = 3, m = 4$.

a. Compute the Bayes estimate of Θ.

b. Find (θ_*, θ^*) so that

$$P(\theta_* \leqslant \Theta \leqslant \theta^*) = .95;$$

divide .05 equally in each tail.

c. Find $P(\Theta \geqslant .70)$.

8.2. In Problem 8.1 compute the classical .95 two-sided confidence interval.

8.3. Assume a Bernoulli process with n trials. Then the sufficient statistic, $T = \Sigma x_i$, has a binomial mass function:

$$f(T|\theta) = \binom{n}{T} \theta^T (1-\theta)^{n-T}.$$

Show that a beta prior distribution on Θ leads to the same posterior density as is given in Equation (8.18).

8.4. Compute the posterior variance of the beta distribution of Equation (8.18). Inspect its properties as n becomes very large.

8.5. Work Problem 8.1b, using Equation (8.21).

8.6. Show for the uniform prior density

$$g(\theta) = 1, \quad 0 < \theta < 1,$$

$$= 0, \quad \text{elsewhere,}$$

that $g(\theta|n_1, \ldots, n_T)$ is a beta density.

8.7. For a beta prior distribution the posterior mean is given for Pascal sampling by Equation (8.28). Show that, as the sample size T goes to infinity, the posterior mean approaches the classical estimate, $T/\Sigma n_i$.

8.8. Show that the probability that $\Sigma n_i = N$ trials are required to achieve T survivals (fixed) is

$$f(\Sigma n_i = N|\theta) = \binom{N-1}{T-1} \theta^T (1-\theta)^{N-T},$$

and thus, if a beta prior distribution obtains, $g(\theta|\Sigma n_i = N)$ is exactly the same as the $g(\theta|n_1, \ldots, n_T)$ of Equation (8.27).

Hint: Note that a necessary and sufficient way to achieve T survivals in exactly $\Sigma n_i = N$ trials is to obtain $T - 1$ in $\Sigma n_i - 1 = N - 1$ (which is binomial) and the last trial results in success. Recall also that

$$\binom{N-1}{T-1} = \frac{\Gamma(N)}{\Gamma(T)\Gamma(N-T+1)}.$$

8.9. Derive the posterior distribution of $\Theta = R(t)$ from Equation (8.33), and also derive it

Table 8.1 Distribution results useful in Bayesian reliability estimation

Random Parameter	Prior Density, $g(\theta)$	Observed Random Variable	Joint Density of Observable Random Variables
$\Theta = R(t) = $ reliability	1, $0 < \theta < 1$ 0, elsewhere. (Uniform)	Survival $(x=1)$ or nonsurvival $(x=0)$ for t	$\theta^T (1-\theta)^{n-T}, T \leqslant n$ Bernoulli process
$\Theta = R(t) = $ reliability	$\dfrac{\Gamma(l+m)}{\Gamma(l)\Gamma(m)}\theta^{l-1}(1-\theta)^{m-1}$ $l,m > 0, 0 < \theta < 1$ 0, elsewhere. (Beta)	" "	" " "
$\Theta = R(t) = $ reliability	" "	Number of trials, n, necessary to observe 1 survival	$f(n_1,\dots,n_T|\theta) = \theta^T(1-\theta)^{\Sigma n_i - T}$ Pascal process.
$\Theta(=\Lambda) = $ constant failure rate	$\dfrac{\alpha^\beta}{\Gamma(\beta)}\theta^{\beta-1}e^{-\alpha\theta}$, $\alpha, \beta, \theta > 0$ 0, elsewhere. (Gamma)	Time to failure, x	$\theta^n e^{-\theta T}$ Poisson process
$\Theta(=\Lambda) = $ constant failure rate	" "	Number of failures, n, in fixed time T	$f(n|\theta) = \dfrac{e^{-\theta T}(\theta T)^n}{n!}$ Poisson process
$\Theta = 1/\Lambda = $ MTTF	$\dfrac{\alpha^\beta}{\Gamma(\beta)}\theta^{-(\beta+1)}e^{-\alpha/\theta}$, $\alpha, \beta, \theta > 0$ 0, elsewhere. (Inverted gamma)	Time to failure, x	$\theta^{-n}e^{-T/\theta}$ Poisson process
$\Theta = 1/\Lambda = $ MTTF	" "	Number of failures, n, in fixed time T	$f(n|\theta) = \dfrac{e^{-T/\theta}(T/\theta)^n}{n!}$ Poisson process

Note: $T = \sum_{i=1}^n x_i$ except in fifth and seventh rows.

414

Table 8.1 Distribution results useful in Bayesian reliability estimation

Joint Marginal Density of Observable Random Variables	Posterior Density $g(\theta\|x_1,\ldots,x_n)$	Posterior Mean	Posterior Variance
$\dfrac{\Gamma(T+1)\Gamma(n-T+1)}{\Gamma(n+2)}$	$\dfrac{\Gamma(n+2)}{\Gamma(T+1)\Gamma(n-T+1)}\theta^T(1-\theta)^{n-T}$ Beta	$\dfrac{T+1}{n+2}$	$\dfrac{(T+1)(n-T+1)}{(n+2)^2(n+3)}$
$\dfrac{\Gamma(l+m)\Gamma(l+T)\Gamma(m+n-T)}{\Gamma(l)\Gamma(m)\Gamma(l+m+n)}$	$\dfrac{\Gamma(l+m+n)}{\Gamma(l+T)\Gamma(m+n-T)}\theta^{l+T-1}(1-\theta)^{m+n-T-1}$ Beta	$\dfrac{l+T}{l+m+n}$	$\dfrac{(l+T)(m+n-T)}{(l+m+n)^2(l+m+n+1)}$
$f(n_1,\ldots,n_T)=\dfrac{\Gamma(l+m)}{\Gamma(l)\Gamma(m)}\times$ $\dfrac{\Gamma(l+T)\Gamma(m+\sum n_i-T)}{\Gamma(l+m+\sum n_i)}$	$g(\theta\|n_1,\ldots,n_T)=\dfrac{\Gamma(l+m+\sum n_i)}{\Gamma(l+T)\Gamma(m+\sum n_i-T)}$ $\times\theta^{l+T-1}(1-\theta)^{m+\sum n_i-T-1}$ (Beta)	$\dfrac{l+T}{l+m+\sum n_i}$	$\dfrac{(l+T)(m+\sum n_i-T)}{(l+m+\sum n_i)^2(l+m+\sum n_i+1)}$
$\dfrac{\alpha^\beta\Gamma(\beta+n)}{\Gamma(\beta)(\alpha+T)^{\beta+n}}$	$\dfrac{(\alpha+T)^{\beta+n}}{\Gamma(\beta+n)}\theta^{\beta+n-1}e^{-\theta(\alpha+T)}$ (Gamma)	$\dfrac{\beta+n}{\alpha+T}$	$\dfrac{\beta+n}{(\alpha+T)^2}$
$f(n)=\dfrac{\alpha^\beta\Gamma(\beta+n)}{\Gamma(\beta)n!}\dfrac{T^n}{(\alpha+T)^{\beta+n}}$	$g(\theta\|n)=\dfrac{(\alpha+T)^{\beta+n}}{\Gamma(\beta+n)}\theta^{\beta+n-1}e^{-\theta(\alpha+T)}$ (Gamma)	" "	" "
$\dfrac{\alpha^\beta\Gamma(\beta+n)}{\Gamma(\beta)(\alpha+T)^{\beta+n}}$	$\dfrac{(\alpha+T)^{\beta+n}}{\Gamma(\beta+n)}\theta^{-(\beta+n+1)}e^{-(\alpha+T)/\theta}$ (Inverted gamma)	$\dfrac{\alpha+T}{\beta+n-1}$ $\beta+n>1$	$\dfrac{(\alpha+T)^2}{(\beta+n-1)^2(\beta+n-2)}$ $\beta+n>2$
$f(n)=\dfrac{\alpha^\beta\Gamma(\beta+n)}{\Gamma(\beta)n!}\dfrac{T^n}{(\alpha+T)^{\beta+n}}$	" " "	" "	" " "

directly by making a change of variable in

$$f(x|\lambda) = \lambda e^{-\lambda x}.$$

8.10. Find the variance of the posterior gamma distribution with shape parameter $\beta + n$ and scale parameter $\alpha + T$.

8.11. Suppose that $\alpha = 100$, $\beta = 5$, $n = 10$, and $T = \Sigma x_i = 300$.
a. Find the Bayes and the classical estimates of Λ.
b. Find a two-sided .90 probability interval (λ_*, λ^*) for Λ.
c. Find a two-sided .90 confidence interval for Λ.

8.12. The marginal density of $T = \Sigma x_i$ is

$$f(T) = \frac{\alpha^{\beta} \Gamma(\beta+n)}{\Gamma(\beta)\Gamma(n)} \frac{T^{n-1}}{(\alpha+T)^{\beta+n}}, \quad 0 < T < \infty.$$

Show that

$$E(T) = \frac{n\alpha}{\beta - 1}.$$

Hint: In $E(T)$ make the change of variable

$$Y = \frac{T}{\alpha + T}$$

and recognize the familiar beta density.

8.13. Work Problem 8.12 making use of the fact that the expectation of a sum is the sum of the expectations.

8.14. Suppose that

$$g(\theta) = \frac{1}{(\theta_0 - \theta_1)}, \quad \theta_1 < \theta < \theta_0,$$

$$= 0, \quad \text{elsewhere.}$$

For the exponential conditional distribution and a sample of size n, find the following:
(a) the marginal density, $f(x_1, \ldots, x_n)$;
(b) the posterior density;
(c) the posterior mean.

8.15. Given the inverted gamma density,

$$g(\theta; \alpha, \beta) = \frac{\alpha^{\beta}}{\Gamma(\beta)} \left(\frac{1}{\theta}\right)^{\beta+1} e^{-\alpha/\theta},$$

show that the mode is given by $\alpha / (\beta + 1)$.

8.16. In Problem 8.15 take $\beta = 1/2$ and $\alpha = 3/2$.
(a) Give the mode.
(b) Give the median.

8.17. Suppose that $\beta = 5/2$, $\alpha = 10$, $n = 5$, and $\Sigma x_i = 8$. Find a lower .80 probability interval for $\Theta = 1/\Lambda$. Assume a posterior inverted gamma distribution.

8.18. For the example in Section 8.5 for sequential testing,
(a) show that $K_1 < P_0 < K_0$;
(b) compute $P(A)$;
(c) compute a lower bound for $P(n_t \leqslant 10)$.

8.19. Derive the approximation

$$P(A|\Theta \geqslant \theta_1) \cong \left(\frac{K_0}{P_0} \right) \left(\frac{P_0 - K_1}{K_0 - K_1} \right), \quad K_1 < P_0 < K_0.$$

Hint: $P(C|B)P(B) = P(B|C)P(C)$.

8.20. Develop the sequential accept/reject criteria for n up to 5 for a gamma prior distribution on failure rate Λ with parameters $\alpha = 1000, \beta = 1, \lambda_1 = .001, K_1 = .10, K_0 = .95$. Be sure to verify that $K_1 < P_0 < K_0$. Also, compute $P(A), P(A|\Theta \geqslant \theta_1)$.

8.21. Show that the lower bound on $P(n_t \leqslant n)$, that is, $P(P_n \geqslant K_0) + P(P_n \leqslant K_1)$, is a nondecreasing function of n. *Hint*: $P(n_t \leqslant n) \leqslant P(n_t \leqslant n+1)$.

8.22. For gamma sampling and a prior gamma distribution for the (constant) failure rate, develop the posterior distribution for the case in which only $r < n$ failure times are available. i.e., the censored case.

8.23. In the example of Section 8.3.3 show that as n gets large $[(\alpha + T)/(\alpha + T + t)]^{n+\beta}$ approaches the MLE of $R(t)$.

REFERENCES

Bhattacharya, S. K. (1967), Bayesian Approach to Life Testing and Reliability Estimation, *Journal of the American Statistical Association*, Vol. 62, pp. 48–62.

Guttman, Irwin, S. S. Wilks, and J. Stuart Hunter (1971), *Introductory Engineering Statistics*, John Wiley and Sons, New York.

Hald, Anders (1960), The Compound Hypergeometric Distribution and a System of Single Sampling Inspection Plans Based on Prior Distributions and Costs, *Technometrics*, Vol. 2, pp. 275–339.

Lindley, D. V. (1965), *Introduction to Probability and Statistics from a Bayesian Viewpoint*, Vols. I and II, Cambridge University Press, Cambridge, England.

Pearson, K. (1934), Tables of the Incomplete Beta Function, *Biometrika*, London.

Raiffa, Howard and Robert Schlaifer (1961), *Applied Statistical Decision Theory*, Harvard University Press, Cambridge, Massachusetts.

Robbins, H., An Empirical Bayes Approach to Statistics, *Proceedings of the Third Berkeley Symposium on Mathematical Statistics and Probability*, University of California Press, Berkeley, California.

Rolph, John E. (1968), Bayesian Estimation of Mixing Distributions, *Annals of Mathematical Statistics*, Vol. 39 pp. 1289–1302.

Rutherford, J. R. and R. G. Krutchkoff (1967), The Empirical Bayes Approach: Estimating the Prior Distribution, *Biometrika*, Vol. 54, pp. 326–328.

Schafer, R. E. (1967), Bayes Single Sampling Plans by Attributes Based on Posterior Risks, *Naval Research Logistics Quarterly*, Vol. 14, pp. 81–88.

Schafer, R. E. (1969), Bayesian Reliability Demonstration: Phase I-Data for the A Priori Distribution, RADC TR 69–389.

Schafer, R. E. (1970), A Note on the Uniform Prior Distribution for Reliability, *I.E.E.E. Transactions on Reliability*, Vol. R-19, pp. 76–77.

Schafer, R. E. (1971), On Bounds for the Frequency of Misleading Bayes Inferences (Abstract), *Annals of Mathematical Statistics*, Vol. 42, p. 1793.

Schafer, R. E. and Anthony J. Feduccia (1972), Prior Distributions Fitted to Observed Reliability Data, *I.E.E.E. Transactions on Reliability*, Vol. R-21, No. 3, pp. 148–154.

Schafer, R. E. and N. D. Singpurwalla (1970), A Sequential Bayes Procedure for Reliability Demonstration, *Naval Research Logistics Quarterly*, Vol. 17, pp. 55–67.

Tucker, H. G. (1963), An Estimate of the Compounding Distribution of a Compound Poisson Distribution, *Theory of Probability and Its Applications*, Vol. 8 pp. 195–199.

Wallenius, K. T. (1969), Sequential Reliability in Finite Lots, *Technometrics*, Vol. 11, pp. 61–74.

Wetherill, G. B. (1961), Bayesian Sequential Analysis, *Biometrika*, Vol. 48 pp. 281–292.

CHAPTER 9

Accelerated Life Testing and Related Topics

The discussion in Chapters 5, 6, and 8 was based on the results of life tests conducted on equipment or devices. It was tacitly assumed there that these life tests were conducted under environmental conditions that closely resembled those under which the device in question was supposed to operate. However, for reasons that are pointed out later, it is often difficult to conduct life tests under the normal operating conditions and hence one resorts to what are known as *accelerated life tests*. This chapter addresses itself to such life tests and attempts to present the significant results that have appeared to date.

9.1 ACCELERATED LIFE TESTS AND THE PHYSICS OF FAILURE

The more reliable a device is, the more difficult it is to measure its reliability. This is so because many years of testing under actual operating conditions would be required to obtain numerical measures of its reliability. Even if such testing were feasible, the rate of technical advance is so great that parts would be obsolete by the time their reliability had been measured. In addition, many of the components used in practice are subjected to environments that are difficult to simulate in the laboratory. One approach to solving this predicament is to use accelerated life tests, in which parts are operated at higher stress levels than are required for normal use.

The difficulty of accelerated testing lies in observing the performance of a part at a high stress level and then predicting the performance of the part at the normal stress level. Situations of this nature call for the exploitation of whatever knowledge one has regarding the variation of failure behavior with environment, so that life tests conducted under accelerated environments can be used to make inferences about the behavior of a device in use conditions environment. The dependence of the mean time to failure on the environment is adequate if one is interested only in the average behavior of the device, but from the standpoint of reliability considerations certain percentiles of the distribution are often of concern. For this reason, for any inferences to be meaningful it is necessary to consider the relationship between all the parameters of the failure distribution and the environment.

Despite the simplicity of this idea, there are problems associated with the design and analysis of accelerated life tests. First, it is often difficult to obtain or to be assured of the relationships between the parameters of the failure distribution and the environmental stresses. Second, even if such relationships are known or can be reasonably well hypothesized, obtaining estimates of the parameters of the relationships from life-test data, which are usually limited, is difficult. Finally, many of the relationships commonly used are valid only for a certain range of the stress, and for stress values beyond these limits new relationships have to be assumed, posing additional problems of estimation.

This section is devoted to making inferences from accelerated life tests when certain relationships between the parameters of a failure distribution and the environmental conditions can be reasonably hypothesized. These relationships or models, as they will be interchangeably designated in this chapter, are derived from an understanding of the physics of failure of the device under discussion. The use of these models has often been stressed in the literature, especially by those concerned with the mechanisms that eventually induce failure. In the absence of knowledge regarding the suitability of such models, a nonparametric approach exists, and this is briefly discussed in Section 9.4.

More formally, let $f(t; \theta)$ be the probability density function (p.d.f.) of the time-to-failure random variable of a device in an environment defined by a vector of generalized stresses $\mathbf{S}$; θ is a vector of parameters. Two assumptions are necessary.

1. The severity of the stress levels (characterized by $\mathbf{S}$) does not change the type of lifetime distribution, say $f(t; \theta)$, but the stress levels have an influence on the values of the parameters (Pieruschka, 1961).

2. The relationship between $\mathbf{S}$ and θ, say $\theta = g(\mathbf{S}; a, b, \dots)$, is known, except for one or more of the parameters, $a, b, c, \dots$; it is also necessary to assume that the relationship is valid for a certain range of the elements of $\mathbf{S}$.

The objective here is to obtain estimates of the parameters $a, b, c, \dots$ based on life tests conducted at large values of $\mathbf{S}$ (within the range of the validity of the relationship), and then to use these estimates to make an inference about θ in use environment S_u.

It is clear from the above discussion that the success of the procedure depends on an accurate choice of model. The models that are discussed here are based on concepts of statistical mechanics such as activation energies and reaction rate theories. The Arrhenius and the Eyring type models seem to be generally the most favored ones for electronic components such as semiconductors, whereas the power rule model is preferred for paper capacitors. Although the above models will be specifically dealt with here, the procedures suggested are general enough to be adapted for other models that a user may want to consider.

Since the exponential distribution is widely used in practice, most of the discussion in this chapter is limited to this distribution.

Suppose, then, that under the *constant application* of a single stress V_i, that is, $S:(V)$, a device has an exponential failure distribution given by

$$f(t;\lambda_i) = \lambda_i e^{-\lambda_i t}, \quad \lambda_i > 0, t \geqslant 0,$$

$$= 0, \quad \text{otherwise.}$$

As pointed out in Chapter 5, λ_i is the constant hazard rate under a stress V_i; and, if $\theta_i = 1/\lambda_i$, then θ_i is the mean time to failure under the stress V_i. Four relationships between λ_i and V_i have been suggested in the literature.

1. The Power Rule Model. This model, which can be derived via considerations of kinetic theory and activation energy, has found application for accelerated life tests of paper-impregnated dielectric capacitors (Levenbach, 1957). Here

$$\theta_i = \frac{C}{V_i^P}, \quad C > 0,$$

is the model, and this implies that the mean time of failure, θ, decreases as the pth power of the applied voltage V. It is desired to obtain point and interval estimates of C and P, and then to use these estimates to make inferences about θ at use voltage V_u.

2. The Arrhenius Reaction Rate Model. This model expresses the time rate of degradation of some device parameter as a function of the operating temperature. For the general applicability of this model to accelerated life testing, the reader is referred to Thomas (1964) and Thomas and Gorton (1963). Its applicability to semiconductor material has been discussed by Pershing (1964). Here

$$\lambda_i = \exp\left(A - \frac{B}{V_i}\right)$$

is the model, where V_i is the temperature stress, and A and B are unknown parameters that must be estimated. It is desired to make an inference about λ at use temperature, V_u.

3. The Eyring Model for a Single Stress. This model, derived from principles of quantum mechanics, expresses the time rate of degradation of some device parameter as a function of the operating temperature (*McGraw-Hill Encyclopedia of Science and Technology*, Vol. 7, p. 357). Here

$$\lambda_i = V_i \exp\left(A - \frac{B}{V_i}\right)$$

is the model.

4. The Generalized Eyring Model. This model is applicable if the device under consideration is subjected to the constant application of two types of stresses, a thermal and a nonthermal stress. Its applicability to the accelerated life testing of thin films was discussed by Goldberg (1964). Here

$$\lambda_i = A T_i \left[\exp\left(\frac{-B}{KT_i}\right) \right] \left[\exp\left(CV_i + \frac{DV_i}{KT_i}\right) \right]$$

is the model, where T_i is the thermal and V_i the nonthermal stress. Also A, B, C, and D are unknown parameters that must be estimated, and K is the universal Boltzmann's constant, whose value is 1.38×10^{-16} erg/degree Kelvin.

It is to be emphasized that the above models may be valid only for a certain range of the values of the stress and may not apply outside these limits. This is so because a change in the basic failure mechanism may occur, calling for the use of another model. Procedures to analyze the results from an accelerated life test wherein a change in the basic failure mechanism occurs are not discussed in this chapter but can be found in a paper by Singpurwalla (1974).

If the device under consideration has an exponential failure distribution with a guarantee-time parameter γ_i (see Section 4.2.2), such that $f(t; \lambda_i, \gamma_i) = \lambda_i \exp[-\lambda_i(t - \gamma_i)], \gamma_i \geqslant 0, t \geqslant \gamma_i$, a plausible model for γ_i can be $\gamma_i = \alpha - \beta V_i$, for V_i within a certain range. This implies that the guarantee time decreases with the stress. Of course, several other models for γ_i are plausible, but the one given here is chosen for its simplicity. Any one of the above four models, whichever is applicable, can be chosen for λ_i, and the objective is to make inferences about both λ (or θ) and γ at use stress V_u or V_u and T_u, as the case may be.

9.2 CONDUCTING THE ACCELERATED LIFE TEST—SOME BASIC RESULTS

Suppose that it has been decided to conduct life tests at k accelerated values of the stress, $V_j, j = 1, 2, \ldots, k$, on a device, which under a stress V_j has an exponential failure distribution with a scale parameter $\lambda_j = 1/\theta_j$ (and a guarantee-time parameter γ_j, if appropriate). The k values of the stress should fall within the range of the validity of a model for accelerated life testing, and must be sufficiently high enough to induce failures. A value V_i is next chosen, at random, from the k values $V_j, j = 1, 2, \ldots, k$, and n_i items are put on a life test, which is conducted under constant application of the stress V_i. The test is terminated after r_i failures have occurred, and the respective times to failure, $t_{1i}, t_{2i}, \ldots, t_{r_i i}$, are recorded. This procedure is repeated for all the k values of V_j, each time choosing a V_i at random from the remaining values of V_j. Such a procedure (also known as a *randomization procedure*) ensures an independence of the k life tests thus conducted. The k life tests yield, as data, the set $\{ V_i, n_i, r_i, \hat{\theta}_i \}, i = 1, 2, \ldots, k$, where $\hat{\theta}_i$ is an estimator of θ_i.

The unique minimum variance unbiased efficient estimator of $\theta_i, \hat{\theta}_i$, is given as (Epstein and Sobel, 1954; see also Section 5.1)

$$\hat{\theta}_i = \frac{\sum_{j=1}^{r_i} t_{ji} + (n_i - r_i) t_{r_i i}}{r_i}.$$

The p.d.f. of $\hat{\theta}_i$ is a gamma density function with a shape parameter r_i, so that

$$g(\hat{\theta}_i) = \frac{\exp(-r_i \hat{\theta}_i / \theta_i)}{\Gamma(r_i)} \left(\frac{r_i}{\theta_i} \right)^{r_i} (\hat{\theta}_i)^{r_i - 1}, \quad \begin{array}{l} 0 \leqslant \hat{\theta}_i < \infty, \\ r_i \geqslant 1 \end{array}.$$

Clearly

$$E(\hat{\theta}_i) = \theta_i \quad \text{and} \quad \text{Var}(\hat{\theta}_i) = \frac{\theta_i^2}{r_i}.$$

It has been shown by Epstein and Sobel (1954) that, if the exponential distribution has a location parameter $\gamma_i > 0$,

(i) $\hat{\theta}_i = [-(n_i - 1)t_{1i} + t_{2i} + \cdots + (n_i - r_i + 1)t_{r_i i}]/(r_i - 1)$ is the maximum likelihood estimator of θ_i;

(ii) $\hat{\gamma}_i = t_{1i} - \hat{\theta}_i / n_i$ is the minimum variance unbiased estimator of γ_i;

(iii) t_{1i} and $\hat{\theta}_i$ are independent; and

(iv)   the p.d.f. of $\hat{\theta}_i$ is a gamma density function with a shape parameter of $r_i - 1$, such that

$$g(\hat{\theta}_i) = \frac{\exp\left[-(r_i-1)\hat{\theta}_i/\theta_i\right]}{\Gamma(r_i-1)} \left(\frac{r_i-1}{\theta_i}\right)^{r_i-1} (\hat{\theta}_i)^{r_i-2}, \quad \begin{array}{l} 0 \leqslant \theta_i < \infty, \\ r_i \geqslant 2 \end{array}$$

Clearly

$$E(\hat{\theta}_i) = \theta_i \quad \text{and} \quad \text{Var}(\hat{\theta}_i) = \frac{\theta_i^2}{r_i-1}.$$

In this case the k life tests yield, as data, the set $\{V_i, n_i, r_i, \hat{\theta}_i, \hat{\gamma}_i\}$, $i = 1, 2, \ldots, k$.

Since many of the models presented before are given in terms of the hazard rate, λ_i, it is necessary to discuss the estimation of λ_i. Let $\hat{\lambda}_i = 1/\hat{\theta}_i$ be an estimator of λ_i. Then, since $g(\hat{\theta}_i)$ is known, and since $\hat{\lambda}_i$ is monotone in $\hat{\theta}_i$, by a simple change-of-variables technique (see Section 3.3) it is easy to see that the p.d.f. of $\hat{\lambda}_i$ is an inverted gamma density function, given by

$$g(\hat{\lambda}_i) = \frac{\exp(-R_i/\theta_i\hat{\lambda}_i)}{\Gamma(R_i)} \left(\frac{R_i}{\theta_i}\right)^{R_i} \left(\frac{1}{\hat{\lambda}_i}\right)^{R_i+1}, \quad \begin{array}{l} \hat{\lambda}_i \geqslant 0, \\ R_i \geqslant 1, \end{array}$$

where $R_i = r_i$ or $R_i = r_i - 1$ depending on whether or not γ_i is known.

It follows from the above that

$$E(\hat{\lambda}_i) = \frac{\lambda_i R_i}{R_i - 1}, \quad \text{for } R_i > 1,$$

and

$$\text{Var}(\hat{\lambda}_i) = \lambda_i^2 \frac{R_i^2}{(R_i-1)^2(R_i-2)}, \quad \text{for } R_i > 2.$$

Thus $\hat{\lambda}_i$ is biased; however, if R_i is large, $\hat{\lambda}_i$ tends to be unbiased, and also $\text{Var}(\hat{\lambda}_i) \cong \lambda_i^2/R_i$. Large values of R_i imply that r_i should also be large, a condition often difficult to attain for low values of the stress V_i.

The data obtained from an accelerated life test is used to make inferences about the failure behavior of the device under use conditions stress. This is illustrated in the sections that follow.

9.3 ESTIMATION FROM ACCELERATED LIFE TESTS—PARAMETRIC METHODS

When specific assumptions about the failure distribution and the relationships between the parameters of this distribution and the environment can be made, the techniques of estimation for accelerated life tests will be termed *parametric techniques*. When such assumptions cannot be specifically made, the techniques will be termed *nonparametric*; these are briefly discussed in Section 9.4.

In this section the exponential failure distribution is assumed, and estimation procedures appropriate when the power rule, the Arrhenius, or the Eyring model is applicable are discussed.

9.3.1 The Power Rule Model

In Section 9.1, the power rule model was given as

$$\theta_i = \frac{C}{V_i^P}, \qquad C > 0,$$

for all values of V_i within a certain range.

Suppose that an accelerated life test, conducted according to the procedure specified in Section 9.2, yields, as data, the set $\{V_i, n_i, r_i, \hat{\theta}_i\}, i = 1, 2, \ldots,$ k. In order to obtain estimators of C and P that are asymptotically independent, the power rule model must be amended slightly, without changing its basic character, to

$$\theta_i = \frac{C}{\left(V_i / \dot{V}\right)^P},$$

for all values of V_i within the specified range.

$$\dot{V} = \prod_{i=1}^{k} (V_i)^{R_i / \sum_{i=1}^{k} R_i}$$

is *the weighted geometric mean* of the V_i's.

Since the p.d.f. of $\hat{\theta}_i, g(\hat{\theta}_i)$, is known (Section 9.2), and since the randomization scheme discussed in Section 9.2 ensures independence of the $\hat{\theta}_i, i = 1, 2, \ldots, k$, the likelihood function of C and P, $L(C, P | \hat{\theta})$, can be written (see Singpurwalla, 1971a) as

$$\prod_{i=1}^{k} \Gamma^{-1}(R_i) \left[\frac{R_i}{C} \left(\frac{V_i}{\dot{V}} \right)^{P} \right]^{R_i} (\hat{\theta}_i)^{R_i - 1} \exp\left[-\frac{R_i \hat{\theta}_i}{C} \left(\frac{V_i}{\dot{V}} \right)^{P} \right],$$

where $\hat{\boldsymbol{\theta}}$ is the vector $[\hat{\theta}_1, \hat{\theta}_2, \ldots, \hat{\theta}_k]$. The maximum-likelihood estimators (MLE's) of P and C, $\hat{P}$ and $\hat{C}$, respectively, are given by solution of two equations:

$$\hat{C} = \frac{\sum_{i=1}^{k} R_i \hat{\theta}_i (V_i / \dot{V})^{\hat{P}}}{\sum_{i=1}^{k} R_i}$$

and

$$\sum_{i=1}^{k} R_i \hat{\theta}_i \left(\frac{V_i}{\dot{V}} \right)^{\hat{P}} \log \frac{V_i}{\dot{V}} = 0. \qquad (9.1)$$

Since Equations (9.1) are nonlinear, their solution can be numerically obtained.

It is easy to verify that, since

$$-E\left[\frac{\partial^2}{\partial P^2} \log L(C, P | \hat{\boldsymbol{\theta}}) \right] = \sum_{i=1}^{k} R_i \left(\log \frac{V_i}{\dot{V}} \right)^2,$$

$$-E\left[\frac{\partial^2}{\partial C^2} \log L(C, P | \hat{\boldsymbol{\theta}}) \right] = \frac{\sum_{i=1}^{k} R_i}{C^2},$$

$$-E\left[\frac{\partial^2 \log L(C, P | \hat{\boldsymbol{\theta}})}{\partial P \partial C} \right] = 0, \qquad (9.2)$$

the asymptotic variances and the asymptotic covariances of $\hat{P}$ and $\hat{C}$ are

$$\text{Var}(\hat{P}) = \sigma_P^2 = \left[\sum_{i=1}^{k} R_i \left(\log \frac{V_i}{\dot{V}} \right)^2 \right]^{-1},$$

$$\text{Var}(\hat{C}) = \sigma_C^2 = C^2 \left(\sum_{i=1}^{k} R_i \right)^{-1},$$

and

$$\text{Cov}(\hat{P}, \hat{C}) = 0. \tag{9.3}$$

Since the MLE's are asymptotically unbiased and are approximately distributed as bivariate normal variables (see Section 3.4), $\hat{C}$ and $\hat{P}$ are asymptotically independent, because their asymptotic covariance is zero. Had the likelihood function $L(C, P|\hat{\boldsymbol{\theta}})$ been formed using the model as $\theta_i = C/V_i^P$, then $\text{Cov}(\hat{C}, \hat{P}) \neq 0$, and the estimators $\hat{C}$ and $\hat{P}$ would not be asymptotically independent. (The reader should be aware that although independence of random variables implies that they have zero covariance, the converse is true only if the random variables are normally distributed.) The property of asymptotic independence is advantageously exploited later on.

The Relative-Likelihood Function. The standard large-sample theory (as discussed in Section 3.4) insures that, under appropriate regularity conditions, the MLE's are approximately normally distributed. Sprott and Kalbfleisch (1968) have reported that such large-sample approximations can be misleading for inferences and should be checked against the actual likelihood functions. They recommend examining the shape of the *relative-likelihood function* arising from the actual data. If the shape of the relative-likelihood function is skew, the likelihood is considered to be nonnormal and application of the large-sample approximation is inappropriate.

The relative-likelihood function of an unknown parameter $\theta, R(\theta)$, is the ratio of the likelihood function of $\theta, L(\theta)$, to the likelihood function evaluated at $\hat{\theta}$, where $\hat{\theta}$ is the MLE of θ. The relative-likelihood function for more than one parameter is analogously defined. In the case of likelihoods involving more than one parameter, in order to make inferences about a single parameter, say θ_1, Sprott and Kalbfleisch recommend examining what is known as the *maximum relative-likelihood function* of $\theta_1, R_m(\theta_1)$, where

$$R_m(\theta_1) = \max_{\theta_2} [R(\theta_1, \theta_2)].$$

If the shape of the maximum relative-likelihood function is skew, application of the large-sample approximation for θ_1 is not appropriate.

Define a function $J = L(\hat{C}, \hat{P}|\hat{\boldsymbol{\theta}})$, where J is the function $L(C, P|\hat{\boldsymbol{\theta}})$ with C and P replaced by $\hat{C}$ and $\hat{P}$, respectively. Then the relative-likelihood function for C and $P, R(C, P)$ is

$$R(C, P) = \frac{L(C, P|\hat{\boldsymbol{\theta}})}{L(\hat{C}, \hat{P}|\hat{\boldsymbol{\theta}})},$$

and the maximum relative-likelihood functions of C and P are

$$R_m(C) = \max_P \left[R(P,C) \right] \quad \text{and} \quad R_m(P) = \max_C \left[R(P,C) \right],$$

respectively. For a fixed value of P, say P^*, $R_m(P^*)$ can be evaluated as

$$R_m(P^*) = \frac{1}{J} \prod_i^k \Gamma^{-1}(R_i) \left[\frac{R_i}{\widehat{C(P^*)}} \left(\frac{V_i}{\dot{V}} \right)^{P^*} \right]^{R_i} \left(\hat{\theta}_i \right)^{R_i-1} \exp \left[- \frac{R_i \hat{\theta}_i}{\widehat{C(P^*)}} \left(\frac{V_i}{\dot{V}} \right)^{P^*} \right]$$

where

$$\widehat{C(P^*)} = \frac{\displaystyle\sum_{i=1}^{k} R_i \hat{\theta}_i \left(V_i / \dot{V} \right)^{P^*}}{\displaystyle\sum_{i=1}^{k} R_i}.$$

A plot of $R_m(P^*)$ versus P^* would represent the maximum relative likelihood of P arising from the actual data. If this plot is symmetric and bell shaped, the large-sample approximation for the normality of $\hat{P}$ is valid. Plots of $R_m(P^*)$ versus P^* for different values of k are shown in Figure 9.1b. These plots are based on some data that were generated on a computer (see the examples at the end of this section). It is to be noted that, even though $R_m(P^*)$ is skew for $k=5$, it begins to attain symmetry for k as small as 10.

Analogously to $R_m(P^*)$, $R_m(C^*) = \max_P [R(C^*,P)]$ can be obtained as

$$R_m(C^*) = \frac{1}{J} \prod_{i=1}^k \Gamma^{-1}(R_i) \left[\frac{R_i}{C^*} \left(\frac{V_i}{\dot{V}} \right)^{\widehat{P(C^*)}} \right]^{R_i} \left(\hat{\theta}_i \right)^{R_i-1} \exp \left[- \frac{R_i \hat{\theta}_i}{C^*} \left(\frac{V_i}{\dot{V}} \right)^{\widehat{P(C^*)}} \right]$$

where

$$\sum_{i=1}^k R_i \hat{\theta}_i \left(\log \frac{V_i}{\dot{V}} \right) \left(\frac{V_i}{\dot{V}} \right)^{\widehat{P(C^*)}} = 0$$

gives $\widehat{P(C^*)}$.

Details about the development of the equations for $R_m(P^*)$ and $R_m(C^*)$ can be obtained from Singpurwalla (1971b). Plots of $R_m(C^*)$ versus C^* for different values of k are shown in Figure 9.1a for data generated on a

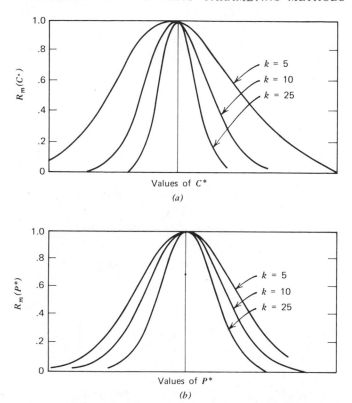

Figure 9.1 Plots of the maximum relative likelihood (*a*) of *C* and (*b*) of *P*.

computer. The symmetrical nature of the plots in Figure 9.1 for k as low as 10 allows one to conclude that the large-sample approximation was reasonable for $\hat{C}$ and $\hat{P}$ for the particular set of data that was used.

In an actual problem the user must plot $R_m(P^*)$ and $R_m(C^*)$ to test for their symmetry. If the result is satisfactory, he can proceed to make an inference about θ_u, the mean time to failure at use conditions stress V_u, along the lines indicated below.

Let $\bar{\theta}_u = \hat{C}(V_u/\dot{V})^{-\hat{P}}$ be the MLE of θ_u. Since $\hat{C}$ and $\hat{P}$ are asymptotically unbiased and asymptotically independent, it follows that if sample sizes are large,

$$E(\bar{\theta}_u) = CE\left[\left(\frac{V_u}{\dot{V}} \right)^{-\hat{P}} \right].$$

If the large-sample approximation for $\hat{P}$ and $\hat{C}$ is valid, then, following

Section 3.4,

$$\hat{C} \overset{.}{\sim} N(C, \sigma_C^2) \quad \text{and} \quad \hat{P} \overset{.}{\sim} N(E, \sigma_P^2),$$

where $\overset{.}{\sim} N(\mu, \sigma^2)$ denotes "approximately normally distributed with a mean μ and a variance σ^2."

If $J_u = (V_u / \dot{V})$, $Y_u = J_u^{\hat{P}}$, and $K_u = \log J_u$, then, for a given value of V_u, it is clear that

$$Y_u \overset{.}{\sim} \Lambda\left[K_u P, (K_u \sigma_P)^2 \right],$$

where $\overset{.}{\sim} \Lambda(\mu, \sigma^2)$ denotes "approximately lognormally distributed with parameters μ and σ^2."

It can be verified (see Problem 9.4) that asymptotically

$$E(Y_u^{-1}) = \exp\left[\tfrac{1}{2}\left((K_u \sigma_P)^2 - K_u P \right) \right],$$

or that

$$E(\bar{\theta}_u) \approx C\left(\frac{V_u}{\dot{V}} \right)^{-P} \exp\left[\tfrac{1}{2}\sigma_P^2 \left(\log \frac{V_u}{\dot{V}} \right)^2 \right].$$

Thus $\bar{\theta}_u$ is a biased estimator of θ_u, but since the bias is in terms of σ_C^2 and σ_P^2, quantities that are known approximately when sample sizes are large, an unbiased estimator of θ_u is

$$\tilde{\theta}_u = \hat{C}\left(\frac{V_u}{\dot{V}} \right)^{-\hat{P}} \exp\left[-\tfrac{1}{2}\sigma_P^2 \left(\log \frac{V_u}{\dot{V}} \right)^2 \right].$$

The variance of $\tilde{\theta}_u$ is

$$\mathrm{Var}(\tilde{\theta}_u) = \exp\left[-\left(\sigma_P \log \frac{V_u}{\dot{V}} \right)^2 \right] \mathrm{Var}\left[\hat{C}\left(\frac{V_u}{\dot{V}} \right)^{-\hat{P}} \right].$$

Since

$$\mathrm{Var}\left[\hat{C}\left(\frac{V_u}{\dot{V}} \right)^{-\hat{P}} \right] = E(\hat{C}^2)E(V_u^{-2}) - \left\{ E\left[\hat{C}\left(\frac{V_u}{\dot{V}} \right)^{-\hat{P}} \right] \right\}^2,$$

and since

$$E(Y_u^{-2}) = \exp\left[2(K_u\sigma_P)^2 - 2K_uP\right]$$

(see Problem 9.4), it follows that

$$\text{Var}\left(\tilde{\theta}_u\right) = \left(\frac{V_u}{\dot{V}}\right)^{-2P}\left\{(\sigma_C^2 + C^2)\exp\left[\sigma_P^2\left(\log\frac{V_u}{\dot{V}}\right)^2\right] - C^2\right\}.$$

In practice C and P are unknown, and hence an estimate of $\text{Var}(\tilde{\theta}_u)$ in terms of σ_C^2 and σ_P^2 can be obtained by using $\hat{C}$ and $\hat{P}$ in place of C and P, respectively.

To obtain approximate confidence bounds for θ_u, it is necessary to ascertain that the maximum relative-likelihood function of $\theta_u, R_m(\theta_u)$, is symmetric and bell shaped. This will establish the fact that the MLE of $\theta_u, \bar{\theta}_u$, is approximately normally distributed. Then $R_m(\theta_u)$ is obtained by setting $C = \theta_u(V_u/\dot{V})^P$ in $L(C,P|\hat{\boldsymbol{\theta}})$ and maximizing $L(C,P|\hat{\boldsymbol{\theta}})$ with respect to P. It follows that

$$R_m(\theta_u) = \frac{1}{J}\prod_1^k \Gamma^{-1}(R_i)\left[\frac{R_i}{\theta_u}\left(\frac{V_i}{V_u}\right)^{\hat{P}(\theta_u)}\right]^{R_i}(\hat{\theta}_i)^{R_i-1}\exp\left[-\frac{R_i\hat{\theta}_i}{\theta_u}\left(\frac{V_i}{V_u}\right)^{\hat{P}(\theta_u)}\right],$$

where $\hat{P}(\theta_u)$ is the solution of

$$\sum_1^k R_i\log\frac{V_i}{V_u} = \frac{1}{\theta_u}\sum_1^k R_i\hat{\theta}_i\left(\frac{V_i}{V_u}\right)^{\hat{P}(\theta_u)}\log\frac{V_i}{V_u}.$$

It is left as an exercise for the reader to show that the asymptotic variance of $\bar{\theta}_u$ is given by

$$\frac{\bar{\theta}_u^2\sum_i^k R_i\log^2(V_i/V_u)}{\sum_1^k R_i\sum_1^k R_i\log^2(V_i/V_u) - \left[\sum_1^k R_i\log(V_i/V_u)\right]^2}.$$

Large-sample confidence bounds on θ_u can now be obtained.

Example. The techniques of this section will be illustrated by means of the following example, the data for which were generated on a computer. First, $C\dot{V}^P = 1000$ and $P = 3$ were arbitrarily chosen, and $k = 5$ values of θ_i were generated, using the model $\theta_i = C/(V_i/\dot{V})^P, i = 1, 2, \ldots$, 5. Corresponding to each value of V_i, n_i and r_i were chosen as shown in Table 9.1. The $\hat{\theta}_i$ were obtained by generating random failure times from an exponential distribution wiht a mean of θ_i, and then applying the methods of Section 9.2. Point estimates of P and C, $\hat{P}$ and $\hat{C}$, respectively, were obtained by solving Equations (9.1) numerically. These estimates were found to be 3.09 and .038, respectively. The variances of $\hat{P}$ and $\hat{C}$ were obtained using Equations (9.3); these turned out to be .0224 and .00001, respectively. The MLE of $\theta_i, \bar{\theta}_i$, was computed using $\theta_i = \hat{C}/(V_i/\dot{V})^P, i = 1, 2, \ldots, 5$, and the corresponding unbiased estimator of $\theta_i, \tilde{\theta}_i$, was obtained using $\tilde{\theta}_i = \hat{C}(V_i/\dot{V})^{-\hat{P}} \exp[-\frac{1}{2}\sigma_p^2 \log(V_i/\dot{V})^2]$. This procedure was repeated for $k = 10$ and $k = 25$, but for brevity only the $k = 5$ values are displayed in Table 9.1. The plots of $R_m(P^*)$ and $R_m(C^*)$ made to ascertain the goodness of the large-sample approximations are shown in Figure 9.1. Predicted values of θ_u at $u \neq i$, based on the $k = 25$ case, are .2.99, 0.17, and .0043 for $V_u = 7$, 39, and 69, respectively.

This example demonstrates the usability of the procedure outlined here. It is clear, however, that the computer is an essential tool in applying such a procedure. The power rule model demonstrates the fact that the problem of inference from accelerated life tests, even when the relationship between the stress and the parameter is simple, becomes quite involved. To obtain any results, one has to resort to the asymptotic theory, whose usefulness in typical life-testing situations with limited data is restricted.

An effort has been made here to illustrate how one can work with results that are asymptotically true. In cases like this the practitioner has two alternatives.

Table 9.1   Monte Carlo results of an accelerated life test

V_i	n_i	r_i	θ_i	$\hat{\theta}_i$	$\bar{\theta}_i$	$\tilde{\theta}_i$
10	30	15	1.000	1.308	1.055	1.041
20	30	15	.125	.078	.124	.123
30	30	20	.037	.030	.035	.035
40	30	25	.016	.017	.014	.014
50	30	25	.008	.008	.007	.007

1. Use a procedure which is quite involved and yields results that are asymptotically true, such as the one given here.
2. Use an ad hoc empirical procedure, as is often done in practice, inferences from which cannot be backed up by statistical reasoning.

9.3.2 The Arrhenius Reaction Rate Model

In Section 9.1 this model was briefly discussed and was given as $\lambda_i = \exp[A - (B/V_i)]$, for V_i within a specified range. Using the set of data $\{V_i, n_i, r_i, \hat{\theta}_i\}, i = 1, 2, \ldots, k$, it is desired to predict a value of λ, say λ_u, under some use conditions stress V_u.

To obtain estimators of the parameters A and B that are asymptotically independent, the model is amended slightly (without changing its basic character) to

$$\lambda_i = \exp\left[A - B\left(V_i^{-1} - \overline{V}\right)\right],$$

for all values of V_i within the specified range.

$$\overline{V} = \frac{\sum_{1}^{k} R_i / V_i}{\sum_{1}^{k} R_i}$$

is the weighted mean of the V_i^{-1}'s.

Since the p.d.f. of $\hat{\lambda}_i, g(\hat{\lambda}_i)$, is known (Section 9.2), and since the randomization scheme discussed in Section 9.2 ensures the independence of the $\hat{\lambda}_i$'s, the likelihood function of A and $B, L(A, B | \hat{\boldsymbol{\lambda}})$, can be written as

$$\prod_{i=1}^{k} \Gamma^{-1}(R_i) \left\{ R_i \exp\left[A - B\left(V_i^{-1} - \overline{V}\right)\right] \right\}^{R_i} (\hat{\lambda}_i)^{-R_i - 1}$$

$$\exp\left\{ -\frac{R_i}{\hat{\lambda}_i} \exp\left[A - B\left(V_i^{-1} - \overline{V}\right)\right] \right\},$$

where $\hat{\boldsymbol{\lambda}}$ is the vector $[\hat{\lambda}_1, \hat{\lambda}_2, \ldots, \hat{\lambda}_k]$.

The MLE's of A and B, $\hat{A}$ and $\hat{B}$, are given by solution of two equations:

$$\sum_{1}^{k} R_i - \sum_{1}^{k} \left(\frac{R_i}{\hat{\lambda}_i}\right) \exp\left[A - B\left(V_i^{-1} - \overline{V}\right)\right] = 0,$$

$$\sum_{1}^{k} \left(\frac{R_i}{\hat{\lambda}_i}\right) \left(V_i^{-1} - \overline{V}\right) \exp\left[A - B\left(V_i^{-1} - \overline{V}\right)\right] = 0. \qquad (9.4)$$

Since Equations (9.4) are nonlinear in A and B, their solution can be numerically obtained (see Singpurwalla, 1973). The large-sample properties of the maximum-likelihood estimators, together with the fact that $\hat{\lambda}_i$ is approximately an unbiased estimator of λ_i for large R_i, give the asymptotic variances and the asymptotic covariance of $\hat{A}$ and $\hat{B}$. These can be obtained as

$$\sigma_a^2 = \text{Var}(\hat{A}) = \left(\sum_1^k R_i \right)^{-1}, \qquad \sigma_b^2 = \text{Var}(\hat{B}) = \left[\sum_i^k R_i \left(V_i^{-1} - \overline{V} \right)^2 \right]^{-1}$$

and

$$\text{Cov}(\hat{A}, \hat{B}) = 0. \tag{9.5}$$

Since the MLE's are asymptotically unbiased and are asymptotically distributed as bivariate normal variables, $\hat{A}$ and $\hat{B}$ are asymptotically independent. Had the likelihood function $L(A, B | \tilde{\boldsymbol{\lambda}})$ been formed using the Arrhenius model as $\lambda_i = \exp(A - B/V_i)$, then $\text{Cov}(\hat{A}, \hat{B}) \neq 0$, and the estimators $\hat{A}$ and $\hat{B}$ would not be asymptotically independent. The property of asymptotic independence is exploited in the subsequent text.

To ascertain the goodness of the large-sample approximation for the normality of the estimators $\hat{A}$ and $\hat{B}$, a procedure similar to that discussed in Section 9.3.1 is followed. Define a function $J = L(\hat{A}, \hat{B} | \tilde{\boldsymbol{\lambda}})$, where J is the likelihood function $L(A, B | \tilde{\boldsymbol{\lambda}})$, with A and B replaced by $\hat{A}$ and $\hat{B}$, respectively. Then the relative-likelihood function of A and $B, R(A, B)$, is defined as $R(A, B) = L(A, B | \tilde{\boldsymbol{\lambda}})/J$, and the maximum relative-likelihood functions of A and B as

$$R_m(A) = \max_B [R(A, B)] \qquad \text{and} \qquad R_m(B) = \max_A [R(A, B)],$$

respectively. For a fixed value of A, say A^*, $R_m(A^*)$ can be evaluated by finding the value of B, say $\widehat{B(A^*)}$, that will maximize $L(A^*, B | \tilde{\boldsymbol{\lambda}})$. Clearly, $\widehat{B(A^*)}$ can be obtained as the solution of Equation (9.4) for B, with A replaced by A^*. Thus $R_m(A^*) = (1/J) L(A^*, \widehat{B(A^*)} | \hat{\boldsymbol{\lambda}})$ can be evaluated for different values of A^*. A plot of $R_m(A^*)$ versus A^* would represent the maximum relative likelihood of A arising from the actual data. If the shape of $R_m(A^*)$ is symmetric and bell shaped, the large-sample approximation for the normality of $\hat{A}$ is valid and one can proceed further, as discussed below. If, on the other hand, the shape of $R_m(A^*)$ is skew, the likelihood is considered to be nonnormal and the application of the large-sample approximation is invalid. Since $R_m(A^*)$ is a function of k, n_i, V, and r_i, in an actual problem the user must check to see whether his data ensure symmetry of $R_m(A^*)$.

Similarly, for a fixed value of B, say B^*, $R_m(B^*)$ can be found by determining the value of A, say $\widehat{A(B^*)}$, that will maximize $L(A, B^*|\hat{\lambda})$.

Plots of $R_m(A^*)$ and $R_m(B^*)$ for different values of k, for some data that were generated on a computer, are shown in Figure 9.2. The symmetrical nature of these plots allows one to conclude that the large-sample approximation is reasonable for $\hat{A}$ and $\hat{B}$ for values of k as low as 5, for the particular set of simulated data.

Should the large-sample approximation be valid, then, following the convention established in Section 9.3.1, one can write that, for large values of k,

$$\hat{A} \dot{\sim} N(A, \sigma_a^2) \quad \text{and} \quad \hat{B} \dot{\sim} N(B, \sigma_b^2).$$

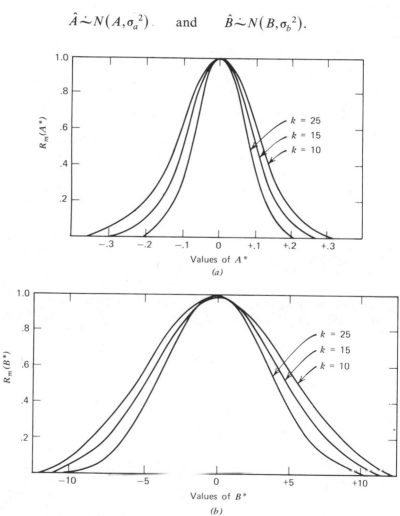

Figure 9.2 Plot of the maximum relative likelihood (a) of A and (b) of B.

If $J_u = V_u^{-1} - \overline{V}$, then, because of the asymptotic independence of $\hat{A}$ and $\hat{B}$,

$$\hat{\Lambda}_u \dot{\sim} \Lambda(A - BJ_u, \sigma_a^2 + J_u^2 \sigma_b^2),$$

where $\hat{\Lambda}_u = \exp[\hat{A} - \hat{B}(V_u^{-1} - \overline{V})]$ is the MLE of λ_u, the hazard rate under use conditions stress V_u.

Since $E(\hat{\Lambda}_u) = \exp(A - BJ_u) + \frac{1}{2}(\sigma_a 2 + J_u^2 \sigma_b^2)$, it follows that $\hat{\Lambda}_u$ is a biased estimator of λ_u with a bias which exists in terms of quantities σ_a^2 and σ_b^2 that are known approximately when sample sizes are large.

An unbiased estimator of λ_u is given by

$$\hat{\Delta}_u = \hat{\Lambda}_u \exp\left[-\tfrac{1}{2}(\sigma_a^2 + J_u^2 \sigma_b^2)\right],$$

and from Theorem 2.1, p. 11 of Aitchison and Brown (1957), it follows that

$$\hat{\Delta}_u \dot{\sim} \Lambda\left[A - BJ_u - \tfrac{1}{2}(\sigma_a^2 + J_u^2 \sigma_b^2), \sigma_a^2 + J_u^2 \sigma_b^2\right],$$

and hence approximate confidence limits for λ_u can be obtained in the usual manner, using tabulated values of the standardized normal distribution.

An example illustrating the use of these techniques is given in Section 9.3.4.

9.3.3 The Eyring Model

Since the Eyring model for a single thermal stress is similar to the Arrhenius model, the development of the pertinent results for this model is left as an exercise to the reader (see Problem 9.6).

The generalized Eyring model with four parameters, A, B, C, D, and the universal Boltzmann's constant, K, is discussed here. For this model

$$\lambda_i = \left[AT_i \exp\left(\frac{-B}{KT_i}\right)\right]\left[\exp\left(CV_i + \frac{DV_i}{KT_i}\right)\right],$$

for all V_i and T_i within a certain range.

Using the set of data $\{V_i, T_i, n_i, r_i, \hat{\theta}_i\}, i = 1, 2, \ldots, k$, one desires to predict a value of λ, say λ_u, under some use conditions stresses V_u and T_u. It is preferable to obtain the least-squares estimators of the parameters of the linear model, since the maximum-likelihood estimators require the solution of four simultaneouus nonlinear equations—a task that could be cumbersome.

For convenience a multiplicative error term is considered, and the model of observations is written as

$$\hat{\lambda}_i = \lambda_i \epsilon_i, \quad i = 1, 2, \ldots, k,$$

or[†]

$$\log \hat{\lambda}_i = \log A + \log T_i - \frac{B}{KT_i} + CV_i + \frac{DV_i}{KT_i} + \log \epsilon_i, \quad i = 1, 2, \ldots, k.$$

If the following are defined:

$$\alpha' = [\log A, C, D, -B],$$

$$\epsilon' = [\log \epsilon_1, \log \epsilon_2, \ldots, \log \epsilon_k],$$

$$\mathbf{Z}' = \left[\log(\hat{\lambda}_1 / T_1), \log(\hat{\lambda}_2 / T_2), \ldots, \log(\hat{\lambda}_k / T_k) \right],$$

and

$$\mathbf{X} = \begin{pmatrix} 1 & V_1 & V_1/KT_1 & (KT_1)^{-1} \\ 1 & V_2 & V_2/KT_2 & (KT_2)^{-1} \\ \cdot & & & \cdot \\ \cdot & & & \cdot \\ \cdot & & & \cdot \\ 1 & V_k & V_k/KT_k & (KT_k)^{-1} \end{pmatrix},$$

then the model of observations in matrix form becomes

$$\mathbf{Z} = X\alpha + \epsilon.$$

A best linear unbiased estimator of the parameter vector α can be obtained if the conditions necessary to invoke the Guass-Markoff theorem (see Section 3.6.1) are satisfied. These conditions are as follows:
   (i) $E(\epsilon) = \mathbf{0}$, and
   (ii) $E(\epsilon\epsilon') = \sigma^2 \mathbf{I}$,
where ϵ' is the transpose of ϵ, σ^2 is a constant, and $\mathbf{I}$ is the identity matrix. Furthermore, if the distribution of the vector $\mathbf{Z}$ is multivariate normal, then confidence bounds for elements of the parameter vector α can be obtained quite easily (Graybill, 1961, p. 122).

[†]All logarithms are to the base e.

Condition (i) above can be essentially established if it can be shown that $E(\log\hat{\lambda}_i) = \log\lambda_i$. The randomization scheme discussed in Section 9.2 ensures that the $\hat{\lambda}_i$ are uncorrelated. Thus condition (ii) is established if it can be shown that $\mathrm{Var}(\log\hat{\lambda}_i)$ is a constant, say σ^2, for all i. The normality of $\mathbf{Z}$ can be verified if it can be shown that $\log\hat{\lambda}_i$ is distributed normally for all i.

Unfortunately, an analytical investigation of the conditions stated above was not possible because of the intractability of the distribution of $\log\hat{\lambda}_i$. Alternatively, a Monte Carlo type of investigation was undertaken, the details of which appear in Singpurwalla and Goldschen (1974). The results of the Monte Carlo investigation reveal that the conditions necessary to invoke the Gauss-Markoff theorem and the desirable property that $\log\hat{\lambda}_i$ is normally distributed hold if, for each accelerated life test, the number of failures is the same, that is, $r_i = r$, for all i.

Let $a_i, i = 1, 2, 3, 4$, be the least-squares estimators of $\log A$, C, D, and $-B$, respectively, and let $\mathbf{a}:[a_1, a_2, a_3, a_4]$. Then the fact that the $\log\hat{\lambda}_i$ are normally distributed and the Gauss-Markoff theorem (Graybill, 1961, pp. 112-113) together imply the following for $r_i = r$, for all i.

1. The minimum variance, estimator among unbiased linear estimators of the vector $\boldsymbol{\alpha}$ is given by $\mathbf{a}$, where

$$\mathbf{a} = (\mathbf{X}'\mathbf{X})^{-1}\mathbf{X}'\mathbf{Z} \quad \text{and} \quad \mathbf{a} \sim N\left[\boldsymbol{\alpha}, \sigma^2(\mathbf{X}'\mathbf{X})^{-1}\right]$$

where σ^2 is a constant.[†] Because of the normality assumption; the elements of $\mathbf{a}$ are also the MLE's of the elements of $\boldsymbol{\alpha}$.

2. $\hat{\lambda}_u = T_u \exp[a_1 + a_2 V_u + (a_3 V_u / KT_u) + (a_4 / KT_u)]$ is an asymptotically unbiased estimator of λ_u under use conditions stresses T_u, V_u.

3. A $100(1-\alpha)\%$ confidence interval for λ_u (often called a *prediction interval*, but not to be confused with those discussed in Chapter 5) is given by

$$P\left[\hat{\lambda}_u \exp(t_{\alpha/2}J_u) \leqslant \lambda_u \leqslant \hat{\lambda}_u \exp(-t_{\alpha/2}J_u)\right] = 1 - \alpha,$$

where

$$J_u = \left\{ S^2\left[1 + \mathbf{X}'_u(\mathbf{X}'\mathbf{X})^{-1}\mathbf{X}_u\right]\right\}^{1/2}, \quad \mathbf{X}_u = \left(1, V_u, \frac{V_u}{KT_u}, \frac{1}{KT_u}\right),$$

$$S^2 = \frac{\mathbf{Z}'\left[\mathbf{I} - \mathbf{X}(\mathbf{X}'\mathbf{X})^{-1}\mathbf{X}'\right]\mathbf{Z}}{k-4},$$

[†] $X \sim N(\mu, \sigma^2)$ denotes the fact that X is distributed normally with a mean μ and a variance σ^2.

and $t_{\alpha/2}$ is the 100 $(\alpha/2)$th percentage point of Student's t distribution with $k-4$ degrees of freedom.

For an example illustrating the use of these techniques the reader is referred to Singpurwalla and Goldschen (1974).

9.3.4 Guarantee Time Decreasing with Stress

For the models discussed in the preceding sections it was assumed that the location parameter of the exponential distribution, γ_i, either was zero or was independent of the stress V_i. In this section it is assumed that $\gamma_i > 0$, and that $\gamma_i = \alpha - \beta V_i$, for V_i within a certain range. Furthermore, it is also assumed that $\lambda_i = \exp[A - B(V_i^{-1} - \bar{V})]$, for V_i within the same range as above. It will be recalled that the above reparameterization for λ_i is the Arrhenius reaction rate model discussed in Section 9.3.2. From the set of data $\{V_i, n_i, r_i, \hat{\theta}_i, \hat{\gamma}_i\}$ (see Section 9.2), $i = 1, 2, \ldots, k$, it is desired to predict λ_u and γ_u, the values of λ and γ under use conditions stress V_u. It is also desired to make inferences about the mean life of the device under use conditions stress μ_u, where $\mu_u = \gamma_u + \lambda_u^{-1}$.

It is preferable to obtain the least-squares estimates of the parameters of the linear model, since maximum-likelihood estimation calls for a solution of two nonlinear equations, which could be cumbersome. The MLE's of the parameters of the Arrhenius model were obtained in Section 9.3.2 because, in order to determine their least-squares estimators and apply the Gauss-Markoff theorem to invoke their desirable properties, it is necessary to know the covariance matrix of the $\hat{\lambda}_i, i = 1, 2, \ldots, k$ (see Graybill, 1961, p. 115). Since the covariance matrix in question involves the unknown parameters A and B, obtaining the least-squares estimators of A and B would not be a straightforward procedure.

To obtain the least-squares estimators of α and β, the following model of observations is considered:

$$t_{1i} = \gamma_i + \epsilon_i, \quad i = 1, 2, \ldots, k,$$

where t_{1i} is the first order statistic from a shifted exponential density function, and ϵ_i is an additive error term. The variance of ϵ_i being θ_i^2 / n_i (heteroscedasticity is implied in the model), a weighting scheme has to be used. A weighted model of observations that leads to independent estimators of α and β is

$$w_i t_{1i} = w_i \left[\alpha - \beta \left(V_i - \dot{V} \right) \right] + w_i \epsilon_i, \quad i = 1, 2, \ldots, k,$$

where the introduction of $\dot{V} = \Sigma w_i V_i / \Sigma w_i$ does not change the basic character of the linear reparameterization for γ_i. The introduction of $\dot{V}$ is a common technique mentioned in Section 3.6.1 and is done here to obtain independent estimators of α and β. The w_i are suitable weights that stabilize the variance in the model.

For the above model, $E(w_i \epsilon_i) = 1/n \cong 0$, if, for all $i, w_i = \lambda_i, n_i = n$, and n is large. Assume for the moment that all of these conditions are satisfied. Also, since $\mathrm{Var}(\lambda_i \epsilon_i) = 1/n^2$, the conditions necessary to apply the generalized Gauss-Markoff theorem in order to obtain the best linear unbiased estimates (BLUE's) of α and β, $\hat{\alpha}$ and $\hat{\beta}$, respectively, are satisfied (see Section 3.6.1). It follows, then, that

$$\hat{\alpha} = \frac{\sum_{i=1}^{k} w_i t_{1i}}{\sum_{i=1}^{k} w_i} \quad \text{and} \quad \hat{\beta} = -\frac{\left[\sum_{i=1}^{k} w_i (V_i - \dot{V}) t_{1i}\right]}{\sum_{i=1}^{k} w_i (V_i - \dot{V})^2}$$

obtained by minimizing $R = \sum_i^k w_i (t_{1i} - \gamma_i)^2$, are the BLUE's of α and β, respectively.

For large values of $r_i, \hat{\lambda}_i$ is an unbiased estimator of λ_i, and hence $\hat{\alpha}$ and $\hat{\beta}$ can be obtained by using $\hat{\lambda}_i$ in place of λ_i.

It is easy to verify that

$$\sigma_1^2 = \mathrm{Var}(\hat{\alpha}) = \frac{k}{\left(n \sum_{i=1}^{k} \lambda_i\right)^2}$$

and

$$\sigma_2^2 = \mathrm{Var}(\hat{\beta}) = \frac{\sum_{i=1}^{k} (V_i - \dot{V})^2 / n^2}{\left[\sum_{i=1}^{k} \lambda_i (V_i - \dot{V})^2\right]^2}. \tag{9.6}$$

Since t_{1i}, the first order statistic from an exponential distribution, also has an exponential distribution (see Section 3.6), it can be shown (see Problem 9.7) that the p.d.f. of $\hat{\alpha}$ is a gamma density function, given by

$$g(\hat{\alpha}) = \exp\left[-\frac{n}{J}(\hat{\alpha} - \alpha)\right] \left(\frac{n}{J}\right)^k (\hat{\alpha} - \alpha)^{k-1} \Gamma^{-1}(k), \quad \text{for} \, \hat{\alpha} \geqslant \alpha, \tag{9.7}$$

where $J = (\Sigma \lambda_i)^{-1}$. This, of course, is true only asymptotically if $w_i = \hat{\lambda}_i, i = 1, \ldots, k$.

The gamma distribution with a shape parameter k can be obtained by convolving k independent, identically distributed exponential variates (see Chapter 4). In view of this fact and by virtue of the central limit theorem, it follows that, for k large,

$$(\hat{\alpha} - \alpha) \overset{\cdot}{\sim} N(0, \sigma_1^{2}).$$

Thus the asymptotic distribution of $\hat{\alpha}$ is approximately a normal one.

The exact distribution of $\hat{\beta}$ cannot be analytically obtained, since it involves convolving k $\lambda_i t_{1i}$'s, and the $\lambda_i t_{1i}$'s, although independent and identically distributed, have different coefficients for each i. The weighted chi-square approximation used in Sections 5.1 and 5.2 applies here, but $v = \sigma_2^{2}$ involves the unknown λ_i's. However, the asymptotic distribution of $\hat{\beta}$ can be obtained by virtue of the central limit theorem if Lyapunov's condition for convergence (Section 3.7.4) is satisfied. Let

$$\hat{\beta}_k = \beta = -\frac{\sum_{1}^{k} w_i (V_i - \dot{V}) t_{1i}}{\sum_{1}^{k} w_i (V_i - \dot{V})^{2}} = \sum_{1}^{k} C_i t_{1i} = \sum_{1}^{k} \hat{t}_i,$$

where

$$C_i = \frac{w_i (V_i - \dot{V})}{-\sum_{1}^{k} w_i (V_i - \dot{V})^{2}} \qquad \text{and} \qquad \hat{t}_i = C_i t_{1i}.$$

The p.d.f. of $\hat{t}_i, g(\hat{t}_i)$, is

$$g(\hat{t}_i) = \left(\frac{n}{\theta_i C_i} \right) \exp\left[-\frac{n}{\theta_i C_i} (\hat{t}_i - C_i \gamma_i) \right], \quad \hat{t}_i \geqslant C_i \gamma_i, \theta_i > 0. \qquad (9.8)$$

Now $\hat{\beta}_k$ is the sum of k independent, nonidentically distributed random varibles $\hat{t}_i$, each with a finite mean β and a finite third central moment (see Problem 9.9). If Lyapunov's condition for convergence is satisifed, then, as $k \to \infty$,

$$(\hat{\beta} - \beta) \overset{\cdot}{\sim} N(0, \sigma_2^{2}).$$

For the sequence $\{\hat{t}_i\}$, it has been shown by Singpurwalla (1968) that, if θ_i decreases in V_i, Lyapunov's condition is satisfied, that is,

$$\underset{k \to \infty}{\text{Lim}} \frac{1}{\sigma^{3}(\hat{\beta}_k)} \sum_{1}^{k} \mu(3; i) = 0,$$

where

$$\sigma^3\left(\hat{\beta}_k\right)=\left[\mathrm{Var}\left(\hat{\beta}\right)\right]^{3/2} \quad \text{and} \quad \mu(3;i)=E\left[|\hat{t}_i-\hat{t}_i|^3\right].$$

Thus the asymptotic distribution of $\hat{\beta}$ is approximately normal.

The independence of $\hat{\alpha}$ and $\hat{\beta}$ can be asserted by examining their variance-covariance matrix, $(\mathbf{V}'\mathbf{WV})^{-1}\mathbf{V}'\mathbf{W}\Sigma\mathbf{W}V(\mathbf{V}'\mathbf{WV})^{-1}$, where $\mathbf{V}$ is the design matrix, Σ a diagonal matrix with diagonal terms $(nw_i)^2, i=1,2,\ldots,k$, and $\mathbf{W}=\Sigma^{-1}$ (see Goodman, 1971).

The variance-covariance matrix obtained as shown above establishes that $\mathrm{Cov}(\hat{\alpha},\hat{\beta})=0$, and this, together with the fact that $\hat{\alpha}$ and $\hat{\beta}$ are asymptotically normally distributed, establishes the asymptotic independence of $\hat{\alpha}$ and $\hat{\beta}$.

In light of the above results, it follows that $\hat{\gamma}_u=\hat{\alpha}-\hat{\beta}(V_u-\dot{V})$, an unbiased estimator of γ_u (the location parameter under use conditions stress V_u), is asymptotically normally distributed with a mean γ_u and a variance given by

$$\mathrm{Var}(\hat{\gamma}_u)=\mathrm{Var}(\hat{\alpha})+\left(V_u-\dot{V}\right)^2\mathrm{Var}\left(\hat{\beta}\right).$$

Thus, large-sample confidence bounds on γ_u can be obtained.

In conclusion, it should be noted that the above results are valid when k, r_i, and n_i are large. It was also assumed here that $n_i=n$, a condition that simplified the algebra.

(a) Joint Prediction Region for γ_u and λ_u. In Section 9.3.2, it was pointed out that

$$\log\hat{\Delta}_u\dot{\sim}N\left(\log\lambda_u-\frac{\sigma_a^2+J_u^2\sigma_b^2}{2},\sigma_a^2+J_u^2\sigma_b^2\right),$$

where

$$\hat{\Delta}_u=\exp\left[\hat{A}-\hat{B}\left(V_u^{-1}-\overline{V}\right)\right]\exp\left[-\tfrac{1}{2}\left(\sigma_a^2+J_u^2\sigma_b^2\right)\right]$$

was an unbiased estimator of λ_u.

Since $\hat{\gamma}_u=\hat{\alpha}-\hat{\beta}(V_u-\dot{V})$, it is clear that $\hat{\gamma}_u$ is a function of the first order statistic, $t_{1i}, i=1,2,\ldots,k$, whereas $\hat{\Delta}_u$ is a function of the estimators $\hat{\theta}_i$. Because of the independence of the t_{1i}'s and the $\hat{\theta}_i$'s (see Section 9.2), it follows from a theorem given by Parzen (1960, p. 295) that $\hat{\gamma}_u$ and $\log\hat{\Delta}_u$ are independent.

Thus the vector $[\hat{\gamma},\log\hat{\Delta}_u]$ is asymptotically distributed approximately as a bivariate normal with a mean vector

$$\left[\gamma_u, \log\lambda_u - \frac{\sigma_a^2 + J_u^2\sigma_b^2}{2} \right]$$

and a variance-covariance matrix given by

$$\begin{pmatrix} \mathrm{Var}(\hat{\gamma}_u) & 0 \\ 0 & \sigma_a^2 + J_u^2\sigma_b^2 \end{pmatrix}.$$

If an estimate of $\mathrm{Var}(\hat{\gamma}_u)$, say $\widehat{\mathrm{Var}(\hat{\gamma}_u)}$, is used, as will often be the case, confidence regions for the mean vector will have to be obtained using Hotelling's-T^2 statistic (Anderson, 1958, p. 101):

$$T^2 = \frac{k(\hat{\gamma}_u - \alpha_u)^2}{\widehat{\mathrm{Var}(\hat{\gamma}_u)}} + \frac{k}{\sigma_a^2 + J_u^2\sigma_b^2}\left(\log\hat{\Delta}_u - \log\lambda_u + \frac{\sigma_a^2 + J_u^2\sigma_b^2}{2} \right)^2.$$

The distribution of $[T^2/(k-1)][(k-2)/2]$ is a Snedocor F distribution (see Chapter 3) with 2 and $k-2$ degrees of freedom (Anderson, 1958, p. 107).

Let $F(1-\gamma)$ denote the $100(1-\gamma)\%$ point of the F distribution with $2, k-2$ degrees of freedom. Also let

$$T_0^2(\gamma) = \frac{2(k-1)}{k-2}F(1-\gamma).$$

Then a $100(1-\gamma)\%$ prediction region is prescribed by the ellipse

$$\frac{k(\gamma - \hat{\gamma}_u)^2}{T_0^2(\gamma)\widehat{\mathrm{Var}(\hat{\gamma}_u)}} + \frac{k\left\{ \log\lambda_u - [\log\Delta_u + (\sigma_a^2 + J_u^2\sigma_b^2)/2]^2 \right\}^2}{T_0^2(\gamma)(\sigma_a^2 + J_u^2\sigma_b^2)} = 1.$$

Hence, if one has a long series of samples, each containing k observations, and if one calculates a series of ellipses with varying centers given by $\hat{\gamma}_u$, $\log\hat{\Delta}_u + (\sigma_a^2 + J_u^2\sigma_b^2/2)$, $100(1-\gamma)\%$ of such ellipses will contain the true values γ_u and $\log\lambda_u$.

(b) Inference about Mean Life at Use Conditions. The results that have been developed thus far will now be used to obtain estimators for the mean time to failure under use conditions stress μ_u. It is easy to verify that

$$\mu_u = \gamma_u + \lambda_u^{-1}.$$

Let $\hat{\mu}_u = \hat{\gamma}_u + (1/\hat{\Delta}_u)$ be an estimator of μ_u. Since $\hat{\Delta}_u$ is approximately lognormally distributed, it follows that

$$E(\hat{\mu}_u) = \gamma_u + \left[\exp(\sigma_a^2 + J_u^2 \sigma_b^2)\right]/\lambda_u,$$

since

$$E\left(\frac{1}{\hat{\Delta}_u}\right) = \exp\left[\frac{\sigma_a^2 + J_u^2 \sigma_b^2}{2} - \left(A - BJ_u - \frac{\sigma_a^2 + J_u^2 \sigma_b^2}{2}\right)\right],$$

where $J_u = V_u^{-1} - \overline{V}$, and σ_a^2 and σ_b^2 are given by Equation (9.5).
An unbiased estimator of μ_u is therefore given by

$$\tilde{\mu}_u = \hat{\gamma}_u + \frac{1}{\hat{\Delta}_u}\exp\left[-(\sigma_a^2 + J_u^2 \sigma_b^2)\right],$$

with

$$\mathrm{Var}(\tilde{\mu}_u) = S_1^2 + \frac{1}{\lambda_u^2}\left[\exp(\sigma_a^2 + J_u^2 \sigma_b^2) - 1\right],$$

where $S_1^2 = \sigma_1^2 + (V_u - \dot{V})^2 \sigma_2^2$, and σ_1^2 and σ_2^2 are given by Equations (9.6).
The distribution of $\tilde{\mu}_u$ cannot be analytically obtained, since it involves the convolution of an asymptotically normally distributed variable, $\hat{\gamma}_u$, with the reciprocal of an asymptotically lognormally distributed variable, $\hat{\Delta}_u$. In view of this fact, it is difficult to obtain confidence bounds for μ_u, and the following alternative approach is suggested.

(c) Plausibility Interval for μ. Sprott and Kalbfleisch (1968) have discussed how the relative-likelihood function can be used as a measure of plausibility for an unknown parameter. For example, all values of an unknown parameter that have at least a 10% relative likelihood can be considered as fairly plausible. A similar approach has been taken by Hudson (1968) and by Singpurwalla (1971b), and it seems to be adequate, at least from a practical point of view.

For convenience, let $\tau = (\sigma_a^2 + J_u^2 \sigma_b^2)/2, 2\tau = S_2^2$. Since $\log\lambda_u = A - BJ_u$, one can write

$$\hat{\Delta}_u \dot\sim (\log\lambda_u - \tau, S_2^2) \qquad \text{and} \qquad \hat{\gamma}_u \dot\sim N(\gamma_u, S_1^2).$$

The independence of $\hat{\Delta}_u$ and $\hat{\gamma}_u$ can be established from the independence of t_{1i} and $\hat{\theta}_1$. Thus the likelihood function of γ_u and

$\lambda_u, L(\gamma_u, \lambda_u | \hat{\gamma}_u, \hat{\Delta}_u)$, can be written as

$$\left(2\pi S_1^2 S_2^2 \hat{\Delta}_u^2\right)^{-1/2} \exp\left\{-\frac{1}{2}\left[\left(\frac{\hat{\gamma}_u - \gamma_u}{S_1}\right)^2 + \left(\frac{\log\hat{\Delta}_u - \log\lambda_u + \tau}{S_2}\right)^2\right]\right\}.$$

This function is a maximum when $\gamma_u = \hat{\gamma}_u$ and $\log\lambda_u = \log\hat{\Delta}_u + \tau$. Therefore, following Sprott and Kalbfleisch, one can obtain the relative-likelihood function, assuming a joint normal distribution of $\hat{\gamma}_u$ and $\log\hat{\Delta}_u$, as

$$R_N(\gamma_u, \lambda_u) = \frac{L\left(\gamma_u, \lambda_u | \hat{\gamma}_u, \hat{\Delta}_u\right)}{L\left(\hat{\gamma}_u, \log\hat{\Delta}_u | \hat{\gamma}_u, \hat{\Delta}_u\right)}$$

$$= \exp\left[-\frac{1}{2}\left(\frac{\hat{\gamma}_u - \gamma_u}{S_1}\right)^2 + \left(\frac{\log\hat{\Delta}_u - \log\lambda_u + \tau}{S_2}\right)^2\right].$$

The maximum relative-likelihood function of λ_u can be written as

$$R_M(\lambda_u) = \max_{\gamma_u}\left[R_N(\gamma_u, \lambda_u)\right] = \exp\left[-\frac{1}{2}\left(\frac{\log\hat{\Delta}_u - \log\lambda_u + \tau}{S_2}\right)^2\right].$$

Since $\mu_u = \gamma_u + (1/\lambda_u)$, $-\log\lambda_u = \log(\mu_u - \gamma_u)$,

$$R_M(\mu_u) = \exp\left\{-\frac{1}{2}\left[\frac{\log\hat{\Delta}_u + \log(\mu_u - \hat{\gamma}_u) + \tau}{S_2}\right]^2\right\}$$

is the maximum relative-likelihood function of μ_u.

Thus $R_M(\mu_u)$ can be plotted for various values of μ_u, and a plausibility interval could consist of the values of μ_u that have, say, at least 10% relative likelihood, given an interval estimate of μ_u. These ideas are best illustrated by the following example.

A Simulated Example. Since data from accelerated life tests are not readily available (often for proprietary reasons), and since frequently the experimenter is not sure whether a certain reparameterization is appropriate for a particular situation, the methods of this section are illustrated by means of simulated data.

First, $A = 5$, $B = 6$, $\alpha = 35$, and $\beta = 1$ were arbitrarily chosen, and five values of λ_i and γ_i were generated, using the Arrhenius model and the linear model, respectively. The estimates $\hat{\lambda}_i$ and $\hat{\gamma}_i$ were obtained by generating random failure times from an exponential distribution with a scale parameter λ_i and location parameter γ_i, and then using the methods of Section 9.2. Point estimators $\hat{A}$ and $\hat{B}$ were obtained by means of the Newton-Raphson method and were found to be 4.728 and 4.168, respectively. Point estimators $\hat{\alpha}$ and $\hat{\beta}$ were obtained using the methods of Section 9.3.4 and were found to be 12.796 and 1.000, respectively. It is to be remarked here that the estimate $\hat{\alpha}$ reflects the effect of amending the linear model by including the $\dot{V}$ term in it. Next, an unbiased estimator of $\lambda_i, \hat{\Delta}_i$, was obtained, using the formula given in Section 9.3.2. This procedure was repeated for $k = 10$, 15, 20, and 25, but for brevity only the case of $k = 5$ is presented in Table 9.2.

To determine the goodness of the large-sample approximation of $\hat{A}$ and $\hat{B}$, the actual maximum relative-likelihood functions of A and B, $R_M(A^*)$ and $R_M(B^*)$, respectively (see Section 9.3.2), were plotted for $k = 10$, 15, and 25. These are shown in Figure 9.2, from which it is clear that their spread narrows with k. The symmetry of these graphs attests to the goodness of the normality approximation of $\hat{A}$ and $\hat{B}$, respectively. It should be remarked at this point that, for a particular application, $k = 10$ may represent too many stress levels at which to conduct accelerated life

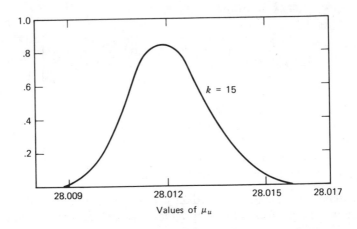

Figure 9.3 Plot of the maximum relative likelihood of μ_u.

Table 9.2

V_i	n_i	r_i	λ_i	γ_i	$\hat{\gamma}_i$	$\bar{\lambda}_i$	Δ_i
10	15	9	81.45	25.0	24.99	92.13	94.07
15	15	9	99.48	20.0	19.99	107.87	108.81
20	15	9	109.94	15.0	15.00	129.47	116.65
25	15	9	116.74	10.0	10.00	138.50	121.49
30	15	9	121.51	5.0	5.00	106.71	124.78

tests. The argument that allows one to use the methodology presented here is the symmetry of the $R_M(A^*)$ and the $R_M(B^*)$, respectively. Since both $R_M(A^*)$ and $R_M(B^*)$ are functions of k, r_i, and V_i, it is possible that for small k, say $k = 5$, one may be able to manipulate the r_i and the V_i's to enforce symmetry of the plots. Hence for an actual problem the user must check to see whether his k, R_i, and V_i take suitable values to ensure symmetry.

The estimators $\hat{A}$, $\hat{B}$, $\hat{\alpha}$, and $\hat{\beta}$ can be used to predict μ_u at a new value $V_u, u \neq i$. For $k = 15$ and $V_u = 7$, an unbiased estimator of $\mu_u = 28.015$ is given as $\tilde{\mu}_u = 28.01$. Note that for $k = 15$, $\Sigma_{i=1}^k r_i = 135$.

A plausibility interval for μ_u can be obtained from a plot of the maximum relative-likelihood function of μ_u, $R_M(\mu_u)$. Such a plot for $V_u = 7$ and $k = 15$ is shown in Figure 9.3. A plausibility interval, obtained by drawing a horizontal line across the 5% relative-likelihood function for μ_u, is given as 28.015 and 28.009. These are the values of μ_u that have a 10% relative likelihood.

Table 9.2 summarizes the results of this example which apply to the case $k = 5$.

9.4 ESTIMATION FROM ACCELERATED LIFE
TESTS—NONPARAMETRIC METHODS

When specific assumptions about the failure distribution, and the relationships between the parameters of this distribution and the environment cannot be made, it is necessary to resort to nonparametric techniques. The nonparametric technique presented here is due to Barlow and Scheuer (1971), and a detailed discussion of it requires familiarity with some concepts and results in nonparametric reliability theory. Since the treatment in this book is primarily parametric in nature, the results of Barlow and Scheuer will merely be highlighted, with little or no discussion of their development.

9.4.1 The Model

Let the device have an unknown failure distribution $G(\cdot)$ in the unaccelerated mode, and let $\mathbf{Y} = (Y_1, Y_2, \ldots, Y_m)$ be m observations on $G(\cdot)$, based on a life test. Suppose that this device has a failure distribution $F(\cdot)$ in the accelerated mode, also unknown, and let $\mathbf{X} = (X_1, X_2, \ldots, X_n)$ be n observations on $F(\cdot)$. Define a *time transformation function* $\alpha(t)$ such that

$$F(t) = G[\alpha(t)],$$

so that, if G^{-1} exists,

$$\alpha(t) = G^{-1}[F(t)].$$

Since acceleration reduces time to failure, it is natural to expect that $\alpha(t) \geq t$. A distribution F is said to be *Increasing Hazard Rate Average IHRA* (Birnbaum, Esary, and Marshall, 1966) if

$$-\frac{\log[1 - F(t)]}{t}$$

is nondecreasing in t.

Thus, given $\mathbf{Y}$ and $\mathbf{X}$, the objective here is to obtain estimates $\hat{G}_m$ and $\hat{F}_n$ such that

(i) $\hat{G}_m$ and $\hat{F}_n$ are IHRA;

(ii) $\hat{G}_m(x) \leq \hat{F}_n(x)$, for all x; and

(iii) $\hat{G}_m$ and $\hat{F}_n$ are closest to G_m and F_n, respectively, in a least-squares sense; here G_m and F_n are the calculated empirical distributions discussed in Section 7.4.1. Also, $\hat{G}_m$ is the estimate of the life distribution in the unaccelerated mode, using data from both the accelerated and the unaccelerated environments.

9.4.2 Least-Squares Estimates for IHRA Distributions

It is desired to obtain an estimate of a failure distribution F that is IHRA, using n observations made upon it, where the n observations represent the lifetimes of the n units. Suppose that k failures are actually observed, and let the ordered times to failure be denoted as $X_0 \equiv 0, X_1 \leq X_2 \ldots < X_k$. Denote by n_j the number of items under observation just before X_j. Then, assuming that there is exactly one failure at each X_i, that is, there are no ties,

$$n_{j+1} = n_j - 1.$$

Define the *product limit* estimate as

$$\bar{F}_n(t) = \begin{cases} 1, & 0 \leqslant t < X_1, \\ \displaystyle\prod_{j=1}^{i} \frac{n_j - 1}{n_j}, & X_i \leqslant t < X_{i+1}, i = 1, \ldots, E-1, \\ 0, & t \geqslant X_E, \end{cases}$$

where

$$E = \begin{cases} k & \text{if } k < n, \\ n & \text{if } k = n, \end{cases}$$

and $\bar{F}_n(t)$ is undefined if $k < n$, for $t \geqslant X_E$.

Since the estimate of F has to have an IHRA property, let

$$\lambda_n(X_i) = \frac{-\log \bar{F}_n(t)}{X_i}, \quad \text{for } i = 1, 2, \ldots, E,$$

and define

$$\hat{\lambda}_n(X_i) = \max_{s \leqslant i} \min_{t \geqslant i} \frac{\displaystyle\sum_{j=s}^{t} \lambda_n(X_j) F_n\{X_j\}}{\displaystyle\sum_{j=s}^{t} F_n\{X_j\}},$$

where

$$F_n\{X_j\} = \bar{F}(X_j-) - \bar{F}_n(X_j+).$$

It can be verified that $\hat{\lambda}_n(X_i)$ will always be nondecreasing in i, and that $\hat{\lambda}_n(X_i)$ is closest to $\lambda_n(X_i)$ in a least-squares sense (Barlow and Scheuer, 1971).

Define

$$\hat{\lambda}_n(t) = \begin{cases} 0, & t < X_1, \\ \hat{\lambda}_n(X_i), & X_i \leqslant t < X_{i+1}, i = 1, 2, \ldots, E-1, \cdot \\ +\infty, & t \geqslant X_E \end{cases}$$

Using $\hat{\lambda}_n(t)$, obtain an estimate of $\bar{F}_n(t)$ as

$$\hat{\bar{F}}_n(t) = \begin{cases} 1, & t < X_1, \\ \exp\left[-\hat{\lambda}_n(t)\right], & X_i \leqslant t < X_{i+1}, t = 1, 2, \ldots, E-1, \cdot \\ 0, & t \geqslant X_E \end{cases}$$

(See Equation 4.2.)

At this point the development of the procedure discussed here is interrupted by a numerical example, inserted purely to illustrate and further clarify the foregoing results. This example is not carried to completion because this can be done only if the necessary computer program is available.

Example. Suppose that $n=5$ items are observed, and that $k=4$ failures occur at times $X_1 < X_2 < X_3 < X_4$; here $E=4$. The product limit estimate is

$$\bar{F}_n(t) = \begin{cases} 1, & 0 \leqslant t < X_1, \\ \dfrac{n_1-1}{n_1} = \dfrac{4}{5}, & X_1 \leqslant t < X_2, \\ \left(\dfrac{n_1-1}{n_1}\right)\left(\dfrac{n_2-1}{n_2}\right) = \left(\dfrac{4}{5}\right)\left(\dfrac{3}{4}\right) = \dfrac{3}{5}, & X_2 \leqslant t < X_3, \\ \left(\dfrac{4}{5}\right)\left(\dfrac{3}{4}\right)\left(\dfrac{2}{3}\right) = \dfrac{2}{5} & X_3 \leqslant t < X_4, \\ \text{undefined}, & X_4 \leqslant t. \end{cases}$$

Therefore

$$\lambda_n(X_1) = \frac{-\log(.8)}{X_1}, \quad \lambda_n(X_2) = \frac{-\log(.6)}{X_2}, \quad \text{and} \quad \lambda_n(X_3) = \frac{-\log(.4)}{X_3},$$

$\lambda_n(X_4)$ being undefined.
Now

$$\hat{\lambda}_n(X_1) = \max_{s \leqslant 1} \min_{t \geqslant 1} \frac{\displaystyle\sum_{j=s}^{t} \lambda_n(X_j) F_n\{X_j\}}{\displaystyle\sum_{j=s}^{t} F_n\{X_j\}}.$$

Since $s \leqslant 1$ implies that $s=0$, $s=1$, and $t \geqslant 1$ implies that $t=1$, $t=2$, $t=3$, one has for $s=0$,

$$\frac{\displaystyle\sum_{j=0}^{1} \lambda_n(X_j) F_n\{X_j\}}{\displaystyle\sum_{j=0}^{1} F_n\{X_j\}}, \qquad \frac{\displaystyle\sum_{j=0}^{2} \lambda_n(X_j) F_n\{X_j\}}{\displaystyle\sum_{j=0}^{2} F_n\{X_j\}}, \qquad \frac{\displaystyle\sum_{j=0}^{3} \lambda_n(X_j) F_n\{X_j\}}{\displaystyle\sum_{j=0}^{3} F_n\{X_j\}},$$

where, for example, $F_n\{X_2\}$ is

$$\bar{F}_n(X_2-)-\bar{F}_n(X_2+)=\tfrac{4}{5}-\tfrac{3}{5}.$$

The next step is to find the minimum of the above three terms, and let this be denoted by T_0. For $s=1$, one has

$$\frac{\lambda_n(X_1)F_n\{X_1\}}{F_n\{X_1\}}, \quad \frac{\sum_{j=1}^{2}\lambda_n(X_1)F_n\{X_j\}}{\sum_{j=1}^{2}F_n\{X_j\}}, \quad \frac{\sum_{j=1}^{3}\lambda_n(X_j)F_n\{X_j\}}{\sum_{j=1}^{3}\lambda_n(X_j)F_n\{X_j\}}.$$

Let T_1 be the minimum of the above three terms. Then

$$\hat{\lambda}_n(X_1)=\max(T_0,T_1).$$

Similarly $\hat{\lambda}_n(X_2)$ and $\hat{\lambda}_n(X_3)$ can be found.
The procedure that was interrupted by this example is continued in the following section.

9.4.3 Procedure for Obtaining Stochastically Ordered Estimates

Suppose that $X_1 \leqslant X_2 \leqslant \ldots \leqslant X_k, k \leqslant n$, are observed failures from F, and $Y_1 \leqslant Y_2 \leqslant \ldots \leqslant Y_r, r \leqslant m$ are observed failures from G, where it is assumed that

(i) F and G are IHRA;
(ii) $\bar{F}(t) \leqslant \bar{G}(t)$ for all $t \geqslant 0$.

It is desired to estimate simultaneously F and G, using the technique given in the preceding section.

Let $\bar{F}_n(t)$ and $\bar{G}_n(t)$ be the product limit estimates of $\bar{F}$ and $\bar{G}$, respectively, and let $\lambda_n(X_i)$ and $\gamma_m(Y_j)$ be the estimates of λ_i and γ_j, respectively. It is desired to obtain the estimates of λ_i and γ_j such that

$$\sum_{i=1}^{k}[\lambda_i-\lambda_n(X_i)]^2F_n\{X_i\}+\sum_{j=1}^{r}[\gamma_j-\gamma_m(Y_j)]^2G_m\{Y_j\}$$

is a minimum, subject to the constraints

$$0\leqslant\lambda_1\leqslant\lambda_2\leqslant\ldots\leqslant\lambda_k, \qquad 0\leqslant\gamma_1\leqslant\gamma_2\leqslant\ldots\leqslant\gamma_r,$$

$$\hat{\bar{G}}_n(t)\geqslant\hat{\bar{F}}_n(t),$$

and also to a constraint for stochastic ordering that is discussed next.

Let $Y_{11} \leqslant Y_{12} \leqslant \ldots \leqslant Y_{1j} < X_1$ denote the set of Y's that are less than X_1. Similarly, let $X_{i-1} \leqslant Y_{i1}, Y_{i2} \ldots Y_{ij_i} < X_i$ denote the Y's in the interval (X_{i-1}, X_i). Then, to ensure stochastic ordering, the problem reduces to performing the minimization subject to the aforementioned constraints and the constraint

$$\gamma_{ij_i} \leqslant \lambda_i, i = 1, 2, \ldots, k.$$

The above minimization scheme constitutes a quadratic programming problem and can be solved by using special codes. The estimates $\lambda_n(X_i)$ and $\gamma_m(Y_j)$ thus obtained can then be used to obtain $\hat{\overline{F}}_n(t)$ and $\hat{\overline{G}}_m(t)$, respectively. This can be done by merely inserting $\lambda_n(X_i)$ and $\gamma_m(Y_j)$, respectively, into the expression for $\overline{F}_n(t)$ given at the end of Section 9.4.2.

Improvements in the estimation procedure result if it can be assumed that $\alpha(t)/t$ is increasing in t, and if $\alpha(t) \geqslant t$.

9.4.4 Remarks

The nonparametric approach discussed here has some advantages in the sense that it does not call for a large number of data (as were required for the parametric approaches presented here), and that it is based on some very general assumptions regarding the failure distribution and the time transformation function. An obvious disadvantage of the method is that it requires some data at the use conditions mode, and it is not always possible to obtain these. Thus, in effect, an absence of data in the unaccelerated mode has to be compensated for by very specific and strong assumptions concerning the model.

From a practical point of view, it should be noted that both the parametric and the nonparametric approach call for procedures that are quite involved.

9.5 ACCELERATED LIFE TESTS WITH CONTINUOUSLY INCREASING STRESS

In Section 9.1 the discussion was based on the fact that life tests were conducted at k accelerated values of the environmental stress. It was stated there that a device with an exponential failure distribution was subjected to the constant application of an accelerated stress (or accelerated stresses). In this section, it is assumed that the stress is an increasing function of time. The material presented here is due to the work of Allen (1959) and should be considered as a special case of inference from accelerated life tests.

The following definitions formalize the notions of accelerated life tests with increasing stress.

Let $P(\cdot)$ and $Q(\cdot)$ be two cumulative distribution functions such that

$$P(0) = Q(0) = 0.$$

Here $Q(\cdot)$ is called an *acceleration* of $P(\cdot)$ if $P(t) \leqslant Q(t)$, for all t.

For any two distributions functions, there exists a function f such that

$$Q(t) = P[f(t)] \quad \text{for all } t.$$

If $Q(t)$ is an acceleration of $P(t)$, then f is called an *accelerating function*; and, if $f(t)$ is strictly increasing in t, the test is termed an *increasing stress test*. Note that in Section 9.4 $f(t)$ was called the time transformation function.

In accelerated life testing one is generally interested in functions $f(t)$ that are monotone (increasing or decreasing) in t. If f is known completely, one can easily make statements about $P(t)$, using the set of observations from $Q(t)$, provided that one can make statements about $P(t)$ from observations on $P(t)$. In this section, it is assumed that the form of f is known but that its parameters are not known and must be estimated from the data.

The concept of hazard rate was introduced in Section 4.1. If $h(t)$ is the hazard rate of $P(t)$, and if $h^*(t)$ is the hazard rate of $Q(t)$, a sufficient condition for $Q(t)$ to be an acceleration of $P(t)$ with an accelerating function $f(t)$ is that

$$h^*(t) \geqslant h(t) \quad \text{for all } t.$$

The proof of this condition is left as an exercise for the reader (see Problem 9.10).

9.5.1 Some Distributions under Specified Accelerating Functions

Let $P(t)$ be an exponential distribution with a hazard rate $\lambda > 0$. Then

$$P(t) = 1 - \exp(-\lambda t) \quad \text{and} \quad h(t) = \lambda \quad \text{for } t \geqslant 0.$$

a. If $h^*(t)$ were given as

$$h^*(t) = \alpha\lambda, \quad \alpha > 1,$$

it follows from Equation (4.2) that

$$Q(t) = 1 - \exp\left[-\int_0^t h^*(\tau)\,d\tau\right] = 1 - \exp(-\alpha\lambda t).$$

Since $P(t) \leqslant Q(t)$ for all t, it follows that $Q(\cdot)$ is an acceleration of $P(\cdot)$ with an accelerating function

$$f(t) = \alpha t.$$

Thus, if a device having an exponential failure distribution $P(t)$ is subjected to an accelerated life test with an *accelerating function* that linearly increases in time, the resulting life distribution $Q(t)$ is also exponential, as shown above.

  b. If $h^*(t)$ were given as

$$h^*(t) = \lambda t^n,$$

then

$$Q(t) = 1 - \exp\left[-\int_0^t h^*(\tau)\,d\tau\right] = 1 - \exp\left(-\frac{\lambda}{n+1}t^{n+1}\right).$$

If one puts $n+1 = \alpha$ and $\lambda/(n+1) = \eta$, then $Q(t) = 1 - \exp(-\eta t^\alpha)$, and this implies that $Q(t)$ is the distribution function of a Weibull distribution with a shape parameter α and a scale parameter η.

For a given value of λ, it is to be noted that $Q(t)$ is greater than $P(t)$ whenever $t^n > n+1$, implying that $Q(t)$ is a *restricted acceleration* of $P(t)$, with an accelerating function given by

$$f(t) = \frac{t^{n+1}}{n+1}.$$

Thus, if a device having an exponential failure distribution with a scale parameter λ is subjected to an accelerated life test with an *accelerating function* given by t^α/α, the resulting life distribution is a Weibull distribution with a shape parameter α and a scale parameter λ/α.

9.5.2 The Power Rule Model under Increased Stress Tests

The power rule model discussed in Section 9.3 under the assumption of a constant stress is now studied for the case in which the stress increases in time.

Let V_1 be the use conditions stress, and V_2 be an accelerated stress. Then, if λ_1 and λ_2 are the respective hazard rates corresponding to these stresses, it follows from the power rule model that

$$\lambda_i^{-1} = CV_i^{-P}, \qquad i = 1, 2,$$

or that

$$\lambda_2 = \lambda_1 \left(\frac{V_2}{V_1} \right)^P,$$

for a certain range of V_i.

a. Now let it be assumed that V_2 linearly increases in time, that is,

$$V_2 = at, \quad t \geqslant 0,$$

and a is a known constant. Then

$$\lambda_2 = \lambda_1 \left(\frac{a}{V_1} \right)^P t^P.$$

If $\alpha = P + 1$, and $\eta = \lambda_1 a^P / [V_1^P (P+1)]$,

$$\lambda_2 = \eta \alpha t^{\alpha - 1},$$

and this is the hazard rate of the Weibull distribution with a scale parameter η and a shape parameter α. If $\hat{\eta}$ and $\hat{\alpha}$ are point estimators of η and α, respectively (see Chapter 5), then

$$\hat{P} = \hat{\alpha} - 1$$

is a point estimator of P, and

$$\hat{\lambda}_1 = \frac{V_1^{\hat{P}} (\hat{P} + 1)}{a^{\hat{P}}}$$

is a point estimator of λ_1, the hazard rate under use conditions stress. The properties of the estimators $\hat{P}$ and $\hat{\lambda}_1$ will depend on the properties of the estimators $\hat{\eta}$ and $\hat{\alpha}$, respectively.

b. If it is assumed that V_2 is a given power of time, such that

$$V_2 = at^b, \quad t \geqslant 0,$$

and a and b are known,

$$\lambda_2 = \lambda_1 \left(\frac{a}{V_1} \right)^P t^{Pb}.$$

If $\alpha = Pb + 1$, and $\eta = \lambda_1 a^P / [V_1^P (Pb + 1)]$,

$$\lambda_2 = \eta \alpha t^{\alpha - 1},$$

which is again the hazard rate of a Weibull distribution with a scale parameter η and a shape parameter α. The estimation of λ_1 follows from the estimation of α and η.

9.6 DESIGN AND ANALYSIS OF ACCELERATED LIFE-TEST EXPERIMENTS

9.6.1 Introduction

The design of accelerated life-test experiments involves the choice of stress levels for experimentation, as well as a determination of the number of tests to be conducted at each stress level. An analysis of results from accelerated life tests could, among other things, lead to determination of the conditions under which inference from such tests is feasible. Section 9.6.2 is based on the work of Zelen (1959) and deals with the establishment of desirable conditions for inference from accelerated life tests. For the design of optimum accelerated life-test experiments, the reader is referred to Mann (1972).

Since this section involves the use of certain concepts from experimental design models, the reader is advised to review these in any of the standard textbooks on the subject, such as Graybill (1961) or Kempthorne (1952).

9.6.2 Factorial Designs in Accelerated Life Testing

Consider a device that is subjected to the simultaneous application of two stresses (constant in time), say V and T. Suppose that the failure distribution of this device is exponential for all combinations of the values of V and T. For simplicity of exposition, the attention here is focused on two types of stresses and the exponential failure times; the extension to more than two stresses and to other failure distributions, however, is straightforward. Let the number of different values that V and T can take be v and τ, respectively.

In experimental design work it is common practice to refer to the stresses or conditions as *factors*, the different values of these factors as *levels*, and a particular combination of these levels as a *treatment combination*. Thus, if i and j refer to the ith and jth levels of V and T, respectively, then $i = 1, 2, \ldots, v$; $j = 1, 2, \ldots, \tau$; and (ij) denotes a treatment combination of V_i and T_j. In particular, if V denotes a voltage stress and T a temperature stress, the total number of different treatment combinations for this experiment will be $v\tau$. Suppose that, for each treatment combination, n

devices are put on a life test, which is terminated after exactly r failures occur. If the values of $V_i, i = 1, 2, \ldots, v$, and $T_j, j = 1, 2, \ldots, \tau$, denote the different conditions of acceleration, what follows is an attempt to determine the conditions under which extrapolation to use condition values will be meaningful.

Let θ_{ij} denote the mean time to failure of the device when the voltage stress is at V_i and the temperature stress at T_j. Clearly, under the assumption of exponentiality,

$$f(t; \theta_{ij}) = \theta_{ij}^{-1} \exp(-t \theta_{ij}^{-1}), \quad t \geqslant 0,$$

$$= 0, \qquad \text{otherwise,}$$

is the failure density under V_i and T_j.

To be able to analyze the effects of the factors V and T on θ_{ij}, suppose that θ_{ij} is written as

$$\theta_{ij} = m v_i \tau_j w_{ij},$$

where the quantities v_i, τ_j, and w_{ij} are positive constants subject to the restrictions

$$\prod_i^v v_i = \prod_j^\tau \tau_j = \prod_i^v w_{ij} = \prod_j^\tau w_{ij} = 1.$$

In experimental design work, the v_i and τ_i are known as the *main effects* and w_{ij} is termed the *interaction*. It follows that

$$\log \theta_{ij} = \log m + \log v_i + \log \tau_i + \log w_{ij},$$

where m is a constant common to all treatment combinations; v_i represents the contribution from level i of the voltage, and τ_j the contribution from level j of the temperature; and w_{ij} is the contribution due to the simultaneous application of V_i and T_j.

An analysis of failure data collected from such a design would involve an estimation of the parameters m, v_i, τ_i, and w_{ij} for all i and j, and a test of the hypothesis $w_{ij} = 1$, which implies that $\theta_{ij} = m v_i \tau_j$. When $\theta_{ij} = m v_i \tau_j$, it is said that the effects of the two factors V and T are *completely multiplicative*, that is, the effects of the temperature can be evaluated independently of the effects of the voltage, and vice versa. A desirable condition for accelerated environmental testing is that the model be completely multiplicative. This is illustrated by the following situation.

Suppose that the use conditions environment of the device can be represented by some low value of the voltage stress, say V_1. It is desired to obtain an estimate of the mean life θ_1 in the use conditions environment. Since conducting a life test under the influence of V_1 alone is time consuming, an accelerated life test is conducted, using both a voltage stress and a temperature stress. Let the levels of the voltage stress and the temperature stress be $V_i, i = 1,2,\ldots,v$, and $T_j, j = 1,2,\ldots,\tau$, respectively. The stress levels T_j are so chosen that every combination of V_1 and $T_j, j = 1,2,\ldots,\tau$, is capable of causing an acceleration in the failure mechanism.

On the basis of the results of the life-test, estimates of m, v_i, τ_j, and w_{ij} are obtained for all values of i and j. If a test of the hypothesis $w_{ij} = 1$ is satisfied for all i and j, the model is completely multiplicative and one can write

$$\log \hat{\theta}_{ij} = \log \hat{m} + \log \hat{v}_i + \log \hat{\tau}_j, \quad i = 1,2,\ldots,v; j = 1,2,\ldots,\tau.$$

A caret ($\hat{}$) is used to indicate that the above quantities are estimated from the data.

It follows from this equation that an estimate of the mean life θ_1 under use conditions is given by

$$\log \hat{\theta}_1 = \log \hat{m} + \log \hat{v}_1,$$

since a temperature stress is not involved under use conditions.

(a) Estimation of Parameters. From the results of Section 9.2 it is clear that $\hat{\theta}_{ij}$, an estimator of θ_{ij}, is given by

$$\hat{\theta}_{ij} = \frac{\left[\sum_{k=1}^{r} t_{ijk} + (n-r)t_{ijr} \right]}{r},$$

or that

$$r\hat{\theta}_{ij} = V_{ij},$$

where V_{ij} is the term in the square brackets.

The p.d.f. of V_{ij} can be written as

$$g(V_{ij}) = \frac{\exp(-V_{ij}/\theta_{ij})}{\Gamma(r)} \frac{(V_{ij})^{r-1}}{\theta_{ij}^r}.$$

Since there are $v\tau$ θ_{ij}'s, the joint likelihood of the V_{ij}'s assuming independence is given by

$$P(V) = [\Gamma(r)]^{-v\tau} \frac{\prod_{i,j} V_{ij}^{r-1}}{\prod_{i,j} \theta_{ij}^r} \exp\left(-\sum_{i,j} \frac{V_{ij}}{\theta_{ij}} \right).$$

Since $\theta_{ij} = m v_i \tau_j w_{ij}$, and

$$\prod_i v_i = \prod_j \tau_j = \prod_i w_{ij} = \prod_j w_{ij} = 1,$$

the joint likelihood reduces to

$$P(V) = [\Gamma(r)]^{-v\tau} \frac{\prod_{i,j} V_{ij}^{r-1}}{m^{v\tau r}} \exp\left(-\sum_{i,j} \frac{V_{ij}}{m v_i \tau_j w_{ij}}\right).$$

The maximum-likelihood estimators of m, v_i, τ_j, and w_{ij}, subject to the constraints on them, can be found using the Lagrange multiplier technique (Kaplan, 1959, p. 129). This technique involves setting equal to 0 the partial derivatives with respect to m, v_i, τ_j, and w_{ij} of

$$Q = L - \lambda_1 \sum_i \log v_i - \lambda_2 \sum_j \log \tau_j - \sum_i \lambda_{3i} \sum_j \log w_{ij}$$

$$- \sum_j \lambda_{4j} \sum_i \log w_{ij},$$

where the λ's are the Lagrange multipliers, and L is the logarithm of $P(V)$; that is,

$$L = v\tau \log \Gamma(r) + (r-1) \sum_{i,j} \log V_{ij} - v\tau r \log m - \sum_{i,j} \frac{V_{ij}}{m v_i \tau_j w_{ij}}.$$

It can be shown that the constrained MLE's of m, v_i, τ_j, and w_{ij} are

$$\hat{m} = \frac{1}{r}\left(\prod_{i,j} V_{ij}\right)^{1/v\tau},$$

$$\hat{v}_i = \frac{1}{r\hat{m}}\left(\prod_j V_{ij}\right)^{1/\tau},$$

$$\hat{\tau} = \frac{1}{r\hat{m}}\left(\prod_i V_{ij}\right)^{1/v},$$

and

$$\hat{w}_{ij} = V_{ij} \left(r \hat{m} \hat{v}_i \hat{\tau}_j \right)^{-1}. \qquad (9.9)$$

(b) The Likelihood-Ratio Test. It was stated before that a desirable condition for accelerated testing is that $w_{ij} = 1$ for all i and j. Thus the hypothesis to be tested is that $w_{ij} = 1$, for all i and j, and this can be tested by forming the likelihood ratio. This is done as follows.

First, the MLE of m, say $\tilde{m}$, when the hypothesis under question is true is found. This involves finding an $\tilde{m}$ that maximizes

$$Q' = L - \lambda_1 \sum_i \log v_i - \lambda_2 \sum_j \log \tau_j,$$

since other terms vanish. Here L is given by

$$L = - v\tau \log \Gamma(r) + (r-1) \sum_{i,j} \log V_{ij} - v\tau r \log m - \sum_{i,j} \frac{V_{ij}}{v_i \tau_j m}.$$

It can be shown that

$$\tilde{m} = \frac{1}{v\tau r} \sum_{i,j} \frac{V_{ij} \left(\prod_{i,j} V_{ij} \right)^{2/v}}{\left(\prod_i V_{ij} \right)^{1/v} \left(\prod_j V_{ij} \right)^{1/\tau}}. \qquad (9.10)$$

Following Mood and Graybill (1963, p. 298), one obtains the generalized likelihood ratio as the quotient

$$\lambda = \frac{L(\tilde{m})}{L(\hat{m})},$$

where $L(x)$ is the logarithm of the likelihood function $P(V)$ with x substituted for m.

It can be shown (Zelen, 1960, p. 512) that

$$\lambda = \left(\frac{\hat{m}}{\tilde{m}} \right)^{v\tau r}.$$

Under the hypothesis $w_{ij} = 1$, for all i and j, the quantity $-2 \log \lambda$ is asymptotically distributed as a chi-square variable with $(v-1)(\tau-1)$ degrees of freedom. This is so because of the asymptotic properties of the generalized likelihood ratio (Mood and Graybill, 1963, p. 301). Appropriate manipulations can be used to show that asymptotically

$$2v\tau r\log\left(\frac{1}{v\tau}\sum_{i,j}\hat{w}_{ij}\right)$$

is distributed as a chi-square variable with $(v-1)(\tau-1)$ degrees of freedom. Thus large values of $-2\log\lambda$ indicate a significant departure from the hypothesis $w_{ij}=1$, for all i and j.

The above test was valid for a large number of observations. Often, in practice, it is not possible to obtain a large number of observations, and an approximation to the above test has been proposed by Zelen (1959) for such cases. This approximation involves the test statistic M, where

$$M=\frac{-2\log\lambda}{\Delta},$$

and

$$\Delta=1+\frac{v\tau+4v+4\tau-11}{6v\tau r}.$$

The quantity M is approximately distributed as a chi-square variable with $(v-1)(\tau-1)$ degrees of freedom. Again, large values of M indicate a significant departure from the hypothesis $w_{ij}=1$, for all i and j.

PROBLEMS

9.1. Verify Equations (9.1) and (9.2).

9.2. Show that the asymptotic variances and the asymptotic covariance of $\hat{C}$ and $\hat{P}$ are given by Equations (9.3).

9.3. Show that, if the power rule model were taken as

$$\theta=C(V)^{-P},$$

the maximum-likelihood estimators of C and P would not be asymptotically independent.

9.4. If Y_i has a lognormal distribution with parameters μ and σ^2, find the expected value and the variance of Y_i^{-1}.

9.5. Verify Equations (9.5), and show that, if the Arrhenius reaction rate model were taken as

$$\lambda=\exp(A-BV_i^{-1}),$$

the MLE's of A and B would not be asymptotically independent.

9.6. Develop results akin to those of Section 9.3.2 for the Eyring model,

$$\lambda_i=V_i\exp(A-BV_i^{-1}).$$

9.7. Using the fact that the p.d.f. of t_{1i} (Section 9.3.4) is an exponential density function given by

$$g(t_{1i}) = \frac{n}{\theta_i} \exp\left[-\frac{n}{\theta_i} (t_{1i} - \gamma_i) \right], \quad t_{1i} \geqslant \gamma_i; \gamma_i, \theta_i > 0,$$

verify Equation 9.7.

9.8. Show that $\hat{\beta}$, given in Section 9.3.4, is an unbiased estimator of β, and verify the expression for the variance of $\hat{\beta}$ given in Equation (9.6).

9.9. Show that the random variable $\hat{t}_i$, whose p.d.f. is given by Equation (9.8), has a finite third central moment, that is, show that $E[|\hat{t}_i - t_i|] < \infty$, where $t_i = E(\hat{t}_i)$.

9.10. If $h(t)$ is the hazard rate of $P(t)$, and if $h^*(t)$ is the hazard rate of $Q(t)$, show that a sufficient condition for $Q(t)$ to be an acceleration of $P(t)$, with an accelerating function $f(t)$ for all t, is that

$$h^*(t) \geqslant h(t) \quad \text{for all } t.$$

9.11. Show that the constrained MLE's of m, v_i, τ_j, and w_{ij}, defined in Section 9.6.1, are given by Equations (9.8). Also, verify Equation 9.9.

9.12. Certain capacitors are known to have an exponential distribution under all conceivable environments. It is desired to predict the mean life of such a device at a use conditions temperature of $30° C$. An accelerated life test is conducted by using voltage in conjunction with temperature as the accelerating mechanism. Ten capacitors are tested at each voltage-temperature combination, and each test is terminated after six failures are observed.

The following table gives an estimate of the mean life for each voltage-temperature combination. Would it be possible to predict, on the basis of the results of this life test, the mean life at use conditions temperature? If so, what would it be?

Temperature (°C)	Applied Voltage (V)			
	100	200	300	400
30	1500 hr	1400 hr	1100 hr	1000 hr
40	1400	1100	1000	900
50	1100	1000	900	800

REFERENCES

Aitchison, J. and J. A. C. Brown (1957), *The Lognormal Distribution*, Cambridge University Press, Cambridge, England.

Allen, W. R. (1959), Inferences from Tests with Continuously Increasing Stress, *Journal of the Operations Research Society of America*, pp. 303-312.

Anderson, T. W. (1958), *An Introduction to Multivariate Statistical Analysis*, John Wiley and Sons, New York.

Barlow, R. E. and E. M. Scheuer (1971), Estimation from Accelerated Life Tests, *Technometrics*, Vol. 13, No. 1, pp. 145-149.

Birnbaum, Z. E., J. D. Esary and A. W. Marshall (1966), A Stochastic Characterization of Wear-Out for Components and Systems, *Annals of Mathematical Statistics*, Vol. 37, pp. 816-825.

Epstein, B. and M. Sobel (1954), Some Theorems Relevant to Life Testing from an Exponential Distribution, *Annals of Mathematical Statistics*, Vol. 25, pp. 373-381.

Goldberg, (1964), Comprehensive Failure Mechanisms Theory, in *Physics of Failure in Electronics*, 2nd Ed., pp. 25-60.

Goodman, A. (1971), Extended Iterative Weighted Least Squares: Estimation of a Linear Model in the Presence of Complications, *Naval Research Logistics Quarterly*, Vol. 18, No. 2, pp. 243-276.

Graybill, F. (1961), *An Introduction to Linear Statistical Models*, Vol. 1, McGraw-Hill Book Company, New York.

Hudson, D. J. (1968), Interval Estimation from the Likelihood Function, *Technical Memorandum*, Bell Telephone Laboratories, Inc., Holmdel, New Jersey.

Kaplan, W. (1959), *Advanced Calculus*, Addison-Wesley Publishing Company, Reading, Massachusetts.

Kempthorne, O. (1952), *The Design and Analysis of Experiments*, John Wiley and Sons, New York.

Levenbach, G. J. (1957), Accelerated Life Testing of Capacitors, *IRE Transactions*, PGRQC No. 10, pp. 9-20.

Mann, N. R. (1972), Design of Over-Stress Life-Test Experiments When Failure Times Have the Two-Parameter Weibull Distribution, *Technometrics*, Vol. 14, No. 2, pp. 437-451.

McGraw-Hill Encyclopedia of Science and Technology, Vol. 7, p. 357, McGraw-Hill Book Company, New York.

Mood, A. M. and F. Graybill (1963), *Introduction to the Theory of Statistics*, McGraw-Hill Book Company, New York.

Parzen, E. (1960), *Modern Probability Theory and Its Applications*, John Wiley and Sons, New York.

Pershing, A. V. and G. E. Hollingsworth (1964), Derivation of Delbruck's Model for Random Failure of Semiconductor Material, in *Physics of Failure in Electronics, Vol. 2, edited by M. F. Goldberg and Joseph Vaccaro*, pp. 61-67.

Pieruschka, E. (1961), Relation Between Lifetime Distribution and the Stress Level Causing Failures, LMSD-800440, Lockheed Missiles and Space Division, Sunnyvale, California.

Singpurwalla, N. D. (1968), Distribution of the Time-to-Failure Random Variable and Inferences from It, When Its Parameters Are Re-parameterized as a Function of the Environment, Ph.D. Dissertation, New York University, New York.

Singpurwalla, N. D. (1971a), Inference from Accelerated Life Tests When Observations Are Obtained from Censored Samples, *Technometrics*, Vol. 13, No. 1, pp. 161-170.

Singpurwalla, N. D. (1971b), A Problem in Accelerated Life Testing, *Journal of the American Statistical Association*, Vol. 66. No. 336, pp. 841-845.

Singpurwalla, N. D. (1973), Inference from Accelerated Life Tests Using Arrhenius Type Re-parameterizations, *Technometrics* Vol. 15, No. 2, pp. 289-299.

Singpurwalla, N. D. (1974), Estimation of a Join Point in Heteroscedastic Regression Models Arising in Accelerated Life Testing, *Communications in Statistics* (to appear).

Singpurwalla, N. D. and D. Y. Goldschen (1974), Inference from Accelerated Life Tests Using Eyring Type Re-parameterization, Naval Research Logistics Quarterly (to appear).

Sprott, D. A. and J. Kalbfleisch (1968), Examples of Likelihoods and Comparison with Point Estimates and Large Sample Approximations, *Journal of the American Statistical Association*, Vol. 64, No. 326, pp. 468-484.

Thomas, R. E. (1964), When Is a Life Test Truly Accelerated?, *Electronic Design* pp. 64-70.

Thomas, R. E. and H. C. Gorton (1963), Research towards a Physics of Aging in Electronics, *Proceedings of the Second Annual Symposium on the Physics of Failure in Electronics*.

Zelen, M. (1959), Factorial Experiments in Life Testing, *Technometrics*, Vol. 1, No. 3, pp. 269-288.

Zelen, M. (1960), Analysis of Two Factor Classifications with Respect to Life Tests, in *Contributions to Probability and Statistics, edited by I. Olkin*, Stanford University Press, Stanford, California.

CHAPTER 10

System Reliability

For the purposes of this book, a system will be defined as a given equipment configuration. The objective of this chapter is to indicate how analytical models of the reliability of a system can be derived, using probabilistic and/or statistical techniques. Although the emphasis of this book is on statistical methods, in the interest of a balanced perspective the earlier parts of this chapter discuss a few techniques that may be considered by some to be purely probabilistic.

In describing the reliability of a given system, it is necessary to specify the equipment failure process, describe how the equipments are connected, provide their rules of operation, and identify the states in which the system is classified as failed. If the system is such that maintainence is allowable, the repair mechanism must also be accounted for. An attempt will be made here to discuss how mathematical models of a system's reliability can be derived under simple assumptions regarding the above processes. It is hoped that the reader will be able to extend these concepts to other situations. For a more detailed development of the various aspects of system reliability, the reader is referred to Barlow and Proschan (1965).

From a mathematical viewpoint, the simplest hypothesis is to assume that at any instant of time the system can be in one of a possible finite number of states, and that the equipments fail according to the exponential failure model. This formulation makes it possible to employ a *Markovian approach*, which assumes that the future course of the system depends on its state at the present time and not on its past history. Carhart (1953) presented reports of a statistical nature substantiating the Markovian hypothesis, but there is at least one good additional reason for suggesting a Markov model.

Cox and Smith (1954) have established that, if each component in the system has approximately an exponential failure law, and if these components are replaced as they fail, the system failure process is Markovian. The system configuration, on the other hand, defines the manner in which the system reliability function will behave. For example, if the equipments are connected *serially*, the failure of any one of them will cause the system to fail. If, on the other hand, the equipments are connected either in *parallel redundancy* (all operating simultaneously) or in *standby redundancy* (on-line with the others off-line, waiting to be sequentially switched on when the preceding unit fails), the system will fail only if all equipments or a specified number of them fail.

10.1 SOME GENERAL STOCHASTIC PROCESSES

To be able to develop the results that appear in the following sections, it is first necessary to present some basic results in the theory of stochastic processes. Attention is confined here to Markov chains (finite state space) having either a discrete or a continuous time parameter. Most reliability problems fall into this category of stochastic processes.

A *stochastic process*, $\{X(t); t \in T\}$, is a family of random variables such that, for each t contained in the index set T, $X(t)$ is a random variable. In reliability studies the variable t is usually interpreted as time, and $X(t)$ represents the *state* of the system at time t. A discrete parameter stochastic process, $\{X(t); t = 0, 1, 2, \dots\}$, or a continuous parameter stochastic process, $\{X(t); t \geqslant 0\}$, is said to be a *Markov process* if, for any set of n time points, $t_1 < t_2 < \dots < t_n$ in the index set of the process, and any real numbers $x_1, x_2, \dots, x_n$,

$$\Pr\{X(t_n \leqslant x_n | X(t_1)) = x_1, X(t_2) = x_2, \dots, X(t_{n-1}) = x_{n-1}\}$$

$$= \Pr\{X(t_n) \leqslant x_n | X(t_{n-1}) = x_{n-1}\}.$$

A *Markov chain* is described by a sequence of discrete-valued random variables, $\{X(t_n)\}$, where t_n is discrete valued or is continuous. In other words, a Markov chain is a Markov process with a discrete state space. If t_n is discrete, then, without loss of generality, the times of state transitions can be labeled as the states of the process and will be denoted by the nonnegative integers $i = 0, 1, 2, \dots, m$. Let

$$p_{ij}{}^{n, n+1} = P\{X(n+1) = j | X(n) = i\}.$$

Then $p_{ij}{}^{n, n+1}$ is called a *transition probability*, and a Markov chain is completely determined by its transition probabilities.

A Markov chain is said to be *stationary* in time when the transition probability depends only on the time difference, that is,

$$p_{ij}{}^{n, n+1} = p_{ij}{}^{0, 1} = p_{ij}.$$

The Markov *transition probability matrix* of a Markov chain is a matrix of the p_{ij}'s, that is, $\mathbf{P}=(p_{ij})$, where

$$\left.\begin{array}{c} p_{ij} \geqslant 0 \\ \sum_{j=0}^{m} p_{ij} = 1 \end{array}\right\}, \quad i,j = 0,1,2,\ldots,m.$$

Two states, i and j, are said to *communicate*, denoted as $i \sim j$, if and only if there exist integers $n_{ij} > 0$ and $n_{ji} > 0$ such that $p_{ij}{}^{n_{ij}} > 0$ and $p_{ji}{}^{n_{ji}} > 0$, where $p_{ij}{}^{n}$ is the probability that, starting in state i, the process will be in state j after n steps.

An *ergodic set* of states is one in which all states communicate and which cannot be left once it is entered. An *ergodic state* is an element of an ergodic set. A state is called *transient* if it is not ergodic. A state i is said to be an *absorbing* state if and only if $p_{ii} = 1$.

The *period* of a state i, $d(i)$, is the greatest common divisor of all integers, $n \geqslant 1$, for which $p_{ii}{}^{n} > 0$. If $d(i) = 1$, the state i is said to be *nonperiodic*. A complete treatment of such processes, including applications, can be found in Karlin (1966). Consider the following example.

A particle can move along a line in discrete steps, starting from state 1, in such a way that the following transition matrix holds:

$$\mathbf{P} = \begin{array}{c} \\ 1 \\ 2 \\ 3 \end{array} \begin{array}{c} \text{State} \quad 1 \quad\ \ 2 \quad\ \ 3 \\ \begin{pmatrix} \frac{1}{2} & \frac{1}{2} & 0 \\ \frac{1}{4} & \frac{1}{2} & \frac{1}{4} \\ 0 & \frac{1}{2} & \frac{1}{2} \end{pmatrix} \end{array}.$$

It is easy to verify that the probability of being in state 1 after two steps is

$$p_{11} \cdot p_{11} + p_{12} \cdot p_{21} = \tfrac{1}{4} + \tfrac{1}{8} = \tfrac{3}{8}.$$

The same result could be obtained by squaring the matrix $\mathbf{P}$. In general, it is easy to verify that, if $\mathbf{P}(0)$ is the initial vector, the probability of being in the different states after n steps is

$$\mathbf{P}^{(n)} = \mathbf{P}(0)\mathbf{P}^{n}.$$

where $\mathbf{P}^{(n)}$ is the nth power of the matrix $\mathbf{P}$.

In the example under discussion, since $\mathbf{P}(0)$ is $(1,0,0)$, and since

$$\mathbf{P}^{2} = \begin{pmatrix} \frac{3}{8} & \frac{4}{8} & \frac{1}{8} \\ \frac{2}{8} & \frac{4}{8} & \frac{2}{8} \\ \frac{1}{8} & \frac{4}{8} & \frac{3}{8} \end{pmatrix}, \quad \mathbf{P}^{(2)} \text{ is } (\tfrac{3}{8}, \tfrac{4}{8}, \tfrac{1}{8}).$$

If the number of transitions becomes large, the columnar entries rapidly approach each other, suggesting that

$$\lim_{n \to \infty} P_{ij}^{(n)} = q_{ij}.$$

Although this result is not valid for all transition matrices, it can be shown that, if it is possible to go from one state to any other state in a finite number of steps, the matrix P^n approaches the matrix Q, with elements q_{ij}.

In such a case, the limiting transition probabilities can be found from the equation

$$\mathbf{XP} = \mathbf{X},$$

where $\mathbf{X}$ is a unique probability vector satisfying the condition that

$$\sum_{i=0}^{m} x_i = 1,$$

x_i being the elements of $\mathbf{X}$.

For the example under discussion, it follows (see Problem 10.1) that

$$\lim_{n \to \infty} P^{(n)} = \left(\tfrac{1}{4}, \tfrac{1}{2}, \tfrac{1}{4} \right) = X.$$

Thus, in the limit, the particle will spend 25% of its time in state 1, 50% in state 2, and 25% in state 3, irrespective of the state in which it started. The initial state determines the *transient behavior* of the process, and after several steps (transitions) it settles down to a *steady state* or a constant value.

10.1.1 Absorbing States

If a process starts in state E_i, one is often interested in determining the mean number of steps to reach an absorbing state, say E_j. This concept is useful in reliability studies to determine the mean time to failure.

Kemeny and Snell (1960, p. 46) have shown that, if E_K is a transient set of states with a matrix Q obtained by truncating P, the mean number of times that a process is in state E_j, starting in state E_i, before absorption is given by $E(N_{ij}) = n_{ij}$, where n_{ij} are the elements of the matrix

$$\mathbf{N} = (\mathbf{I} - \mathbf{Q})^{-1},$$

I being the identity matrix. Furthermore, it has been shown that

$$\mathbf{E}\left(N_{ij}^2 \right) = \mathbf{N}(2 \mathbf{N} \, dg - 1),$$

where $N\,dg$ is the product of the diagonal elements of N. The expressions $E(N_{ij})$ and $E(N_{ij}^2)$ can be used quite effectively for Markov chains that do not have any absorbing states. For example, suppose that we are interested in determining the expected number of visits to state E_j, starting from state E_i, before reaching state E_K. This can be immediately accomplished by making state E_K an absorbing state.

It is also easy to compute the probability h_{ij} that a Markov chain process will ever go to a transient state E_j, starting in a transient state E_i. This probability is given by the elements of the matrix H, where

$$H = (h_{ij}) = (N-I)N\,dg^{-1}$$

(see Kemeney and Snell, 1960, p. 61).

As an example of this, suppose that state 3 in the preceding example is an absorbing state. Thus 1 and 2 are transient states with a matrix Q, given by

$$Q = \begin{array}{c} 1 \\ 2 \end{array} \begin{pmatrix} \frac{1}{2} & \frac{1}{2} \\ \frac{1}{4} & \frac{1}{2} \end{pmatrix}.$$

The matrix $N = (I-Q)^{-1}$ is given by

$$N = \begin{array}{c} 1 \\ 2 \end{array} \begin{pmatrix} 4 & 4 \\ 2 & 4 \end{pmatrix},$$

and thus, if the particle started in state 2, the mean number of times that it would be in state 1 before absorption is two.

To obtain $E(N_{ij}^2)$, it is to be noted that, since $N\,dg = 16$, $E(N_{ij}^2)$ is given by

$$E(N_{ij}^2) = \begin{array}{c} 1 \\ 2 \end{array} \begin{pmatrix} 124 & 124 \\ 62 & 124 \end{pmatrix}.$$

It is easy to verify that the matrix H is given by

$$H = \begin{array}{c} 1 \\ 2 \end{array} \begin{pmatrix} \frac{3}{16} & \frac{4}{16} \\ \frac{2}{16} & \frac{3}{16} \end{pmatrix}.$$

10.1.2   Renewal Processes

A *renewal process* is a sequence of independent, identically distributed, nonnegative random variables, not all 0 with probability 1. Renewal theory plays a significant role in reliability, and the objective here is to capture some essential results of renewal theory that will be useful in the subsequent text.

Let $X_1, X_2, \ldots, X_n$ be a renewal process having a distribution function F, and let $F^K(x)$ be the distribution function of $T_K = X_1 + X_2 + \cdots + X_K$, the time to the Kth renewal.

Interest usually centers around $N(t)$, the number of renewals in time $(0, t)$, subject to a convention that $N(t) = 0$ if $X_1 > t$. Clearly,

$$P\{N(t) = n\} = P[\text{the time to } n\text{th renewal is before } t,$$

$$\text{and the time to the } (n+1)\text{st renewal is after } t].$$

$$\therefore P\{N(t) = n\} = P(X_1 + X_2 + \cdots + X_n \leqslant t \text{ and } X_1 + X_2 + \cdots X_{n+1} > t)$$

$$= F^n(t) - F^{n+1}(t).$$

A proof of this result is left as an exercise for the reader (see the hint in Problem 10.1).

It follows from the above result that $P(N(t) \geqslant n) = F^n(t)$. The *renewal function* $M(t)$ is defined as the expected number of renewals in $(0, t)$ that is,

$$M(t) = E[N(t)] = \sum_{k=1}^{\infty} k \; P\{N(t) = k\}$$

$$= F(t) - F^2(t) + 2F^2(t) - 2F^3(t) + 3F^3(t) - \cdots$$

$$= \sum_{n=1}^{\infty} F^n(t).$$

Since $F^n(t)$ is the *n*-fold convolution of F with itself (see Section 4.3), it follows that

$$F^{n+1}(t) = \int_0^t F^n(t-\tau) \, dF(\tau).$$

Thus

$$M(t) = F(t) + \sum_{n=1}^{\infty} \int_0^t F^n(t-\tau) \, dF(\tau)$$

$$= \int_0^t [1 + M(t-x)] \, dF(\tau).$$

If F has a density f, then

$$m(t) = \frac{d}{dt} M(t) = \int_0^t [1 + m(t-x)] f(x) dx \qquad (10.1)$$

is known as the *renewal density*. Analogously to $M(t)$, it is possible to write $m(t)$ as

$$m(t) = \sum_{n=1}^\infty f^n(t).$$

It is convenient to interpret $m(t)dt$ as the probability of a renewal occurring in $(t, t+dt)$.

The solution of the integral equation (10.1) can be obtained using Laplace transforms. Since it has been assumed that F has a density f, we define

$$L\{f(t)\} = f^*(s) = \int_0^\infty e^{-st} f(t) dt$$

as the Laplace transform of $f(t)$.

Using the convolution property of Laplace transforms discussed in Section 3.1.5, one sees that, when the Laplace transform is applied to Equation (10.1),

$$m^*(s) = f^*(s) + m^*(s) f^*(s).$$

From this it follows after some simplification that

$$L\{m(t)\} = \frac{L\{f(t)\}}{1 - L\{f(t)\}}$$

or that

$$L\{f(t)\} = \frac{L\{m(t)\}}{1 + L\{m(t)\}}$$

Thus either $m(t)$ or $f(t)$ determines the other.

For example, if $f(t)$ is taken to be the exponential density, such that $f(t) = \lambda e^{-\lambda t}$, $\lambda > 0$, then

$$L\{f(t)\} = \frac{\lambda}{s+\lambda},$$

and it follows that $L\{m(t)\} = \lambda / s$.

The inverse transform gives $m(t)=\lambda$, from which it follows that $M(t) = \int_0^t m(\tau)d\tau = \lambda t$, a well-known result.

Renewal rates for failure distributions other than the exponential have been studied and are well reported in the literature (Barlow and Proschan, 1965, p.51, and the references therein). In general, if F has mean μ_1, then, as $t\to\infty$,

$$\frac{N(t)}{t} \to \frac{1}{\mu_1}.$$

In reliability theory, a renewal process naturally arises when a large number of equipments operating independently have the same distribution of failure, or when a single equipment that fails is replaced immediately with a similar equipment. In such situations, one is normally interested in determining the expected number of failures in a given time, or the probability distribution of the time at which all the equipments fail, as well as in other related questions such as spare parts provisioning.

Since the results of this section have been given in terms of Laplace transforms that have to be inverted to obtain the final solution, the following table of inverse Laplace transforms will be useful.

$L\{f(t)\}=f^*(s)$	$f(t)$
s^{-1}	1
s^{-2}	t
s^{-n}	$t^{n-1}/(n-1)!$
$(s+\lambda)^{-1}$	$e^{-\lambda t}$
$(s+\lambda)^{-2}$	$te^{-\lambda t}$
$(s+\lambda)^{-n}$	$e^{-\lambda t}t^{n-1}/(n-1)!$
$[(s+a)(s+b)]^{-1}$	$(e^{-at}-e^{-bt})/(a-b)$

10.1.3   The Poisson Failure Process

The Poisson process was discussed in some detail in Section 4.2 as a process that gives rise to an exponential failure distribution. In this section, it is shown how the reliability of a system whose components fail according to the postulates of a Poisson process can be evaluated.

Consider a system which has n components, and assume that only one component can fail at a time. The state of the system is 0 when all the n components are operating, and is n when all the components have failed. It is assumed that at time 0 all the components are operative, and that the failure of each component takes place according to the postulates of a Poisson process (see Section 4.2.1), with a parameter λ. In Section 10.1 it was assumed that the state transitions take place at discrete points in time characterized by $n = 0, 1, 2, \ldots$. It is more realistic, however, to recognize that failures occur continuously in time; in particular, if a component is operating at time t, the conditional probability that it fails in $(t, t + dt)$ is $\lambda\, dt$. From the basic character of the process, it is possible to write down the following transition probability matrix P, which defines the process in $(t, t + dt)$:

$$
\mathbf{P} =
\begin{array}{c}
\\ 0 \\ 1 \\ 2 \\ . \\ . \\ . \\ n-1 \\ n
\end{array}
\begin{pmatrix}
\begin{array}{cccccc}
0 & 1 & 2 & \ldots & n-1 & n \\
1-\lambda & \lambda & 0 & \ldots & 0 & 0 \\
0 & 1-\lambda & \lambda & \ldots & 0 & 0 \\
0 & 0 & 1-\lambda & \ldots & 0 & 0 \\
. & & & & & \\
. & & & & & \\
. & & & & & \\
0 & 0 & 0 & \ldots & 1-\lambda & \lambda \\
0 & 0 & 0 & \ldots & 0 & 1
\end{array}
\end{pmatrix}
$$

Note that, since the process is continuous in time t, the dt terms have been left out of the matrix.

The above formulation leads directly to a system of linear homogeneous differential equations whose solution can be found using the Laplace transform method discussed in Section 4.2.1. The matrix method discussed below simplifies much of the detail presented in Section 4.2.1.

We define $\mathbf{A}$ as the differential transition matrix, where $\mathbf{A}$ is obtained as $\mathbf{P} - \delta_{ij}$, δ_{ij} being the Kronecker delta function,

$$
\delta_{ij} = \begin{cases} 1, & i = j, \\ 0, & i \neq j. \end{cases}
$$

Thus

$$
\mathbf{A} =
\begin{array}{c}
\\ 0 \\ 1 \\ . \\ . \\ . \\ n
\end{array}
\begin{pmatrix}
\begin{array}{ccccc}
0 & 1 & 2 & \ldots & n \\
-\lambda & \lambda & 0 & \ldots & 0 \\
0 & -\lambda & \lambda & \ldots & 0 \\
. & & & & \\
. & & & & \\
. & & & & \\
0 & 0 & 0 & \ldots & 0
\end{array}
\end{pmatrix}
$$

If $P_i(t)$ denotes the probability that the system is in state i at time t, a system of differential equations can immediately be written down from the above matrix as

$$P_0'(t) = -\lambda P_0(t),$$

$$P_1'(t) = \lambda P_0(t) - \lambda P_1(t),$$

$$\vdots$$

$$P_n'(t) = \lambda P_{n-1}(t) - \lambda P_n(t),$$

and $P_0(0) = 1$, $P_1(0) = P_2(0) \ldots = P_n(0) = 0$ are the initial conditions.

This system of equations is identical to Equations (4.4) and (4.5), whose solution can be obtained, using Laplace transforms, as

$$P_n(t) = \frac{e^{-\lambda t}(\lambda t)^n}{n!}, \quad n = 0, 1, 2, \ldots.$$

10.2 RELIABILITY MODELS FOR NONMAINTAINED SYSTEMS

In this section, the developments of the preceding sections are applied to series and parallel systems on which no maintenance action is possible. What will be presented here by no means covers all the situations that may be encountered in practice, and for a more detailed discussion the reader is referred to Sandler (1963).

A *series system* is a configuration of components such that the system is said to be operative if and only if all the components in the configuration are operative. Such systems are also referred to as *chain models* or *weakest-link models*, since the system fails as soon as the weakest component fails.

A *standby-redundant system* is a configuration of components such that the system is said to have failed when all the components (or a specified number of them) have failed. The word standby refers to the fact that only one component operates at a time, while the other, nonfailed components are waiting to be switched on when the operating component fails.

A *parallel-redundant system* is identical to a standby-redundant system except that all the components in the configuration operate at the same time. Such systems are also known as *rope models*, since the system fails when all the components fail and its behavior is thus akin to that of a rope, which breaks when all the fibers break.

In what follows, reliability models of the three configurations defined above are developed under the assumption that failures of the components take place in accordance with the postulates of a Poisson process, discussed in Section 4.2.1.

10.2.1   Series System

Let the series system be made up of n independently operating components, each of which fails with a Poisson parameter (failure rate) λ_i. It is implied by the postulates of the process that the probability of two or more components failing in an interval of time dt is $0(dt)$:

$$P_0(t+dt) = P_0(t)\left(1 - \sum_i^n \lambda_i dt\right) + 0(dt)$$

or

$$P_0'(t) + P_0(t)\sum_i^n \lambda_i = 0.$$

The solution of the above equation, using Laplace transforms, yields

$$P_0(t) = \exp\left(-\sum_i^n \lambda_i t\right),$$

and this is precisely the reliability function to time t. If $\lambda_i = \lambda$, for all i, then

$$R(t) = \exp(-n\lambda t)$$

or

$$1 - R(t) = F(t) = 1 - \exp(-n\lambda t),$$

and this is the distribution function of the smallest-order statistic from an exponential distribution with a parameter λ.

10.2.2   Standby-Redundant System

The assumptions under which a reliability model for such a system can be obtained are identical to those discussed in Section 10.2.1. For simplicity it will be assumed that the system is made up of only two components, with λ_1 as the failure rate of the operating equipment and λ_2 as the failure rate of the standby equipment.

If it is assumed that the switchover is perfect, the transition probability matrix for this system is given by

$$P = \begin{array}{c} \\ 0 \\ 1 \\ 2 \end{array}\begin{array}{ccc} 0 & \quad 1 & \;2 \\ \left(\begin{array}{ccc} 1-(\lambda_1+\lambda_2) & \lambda_1+\lambda_2 & 0 \\ 0 & 1-\lambda_1 & 1 \\ 0 & 0 & 1 \end{array}\right). \end{array}$$

The differential transition matrix A leads to the following system of differential equations:

$$P_0'(t) = -(\lambda_1 + \lambda_2) P_0(t),$$

$$P_1'(t) = (\lambda_1 + \lambda_2) P_0(t) - \lambda_1 P_1(t),$$

$$P_2'(t) = \lambda_1 P_1(t),$$

where $P_0(0) = 1$, $P_1(0) = P_2(0) = 0$ are the initial conditions.

The solution of the above system, using Laplace transforms, yields

$$P_0(t) = \exp(-(\lambda_1 + \lambda_2)t)$$

and

$$P_1(t) = \frac{\lambda_1 + \lambda_2}{\lambda_2} \exp(-\lambda_1 t) - \frac{\lambda_1 + \lambda_2}{\lambda_2} \exp[-(\lambda_1 + \lambda_2)t], \quad \text{if } \lambda_2 > 0.$$

The reliability of the system to time t, $R(t)$, is

$$R(t) = P_0(t) + P_1(t).$$

This result can be extended to a system having n components in standby redundancy.

In the above example, if $\lambda_1 = \lambda_2 = \lambda$, then

$$R(t) = 2\exp(-\lambda t) - \exp(-2\lambda t).$$

10.2.3 Parallel-Redundant System

Under the general assumptions discussed in Section 10.2.1, if each component has the same failure rate λ, the transition probability matrix for this system is

$$
P = \begin{array}{c}
\\
0 \\
1 \\
2 \\
. \\
. \\
. \\
n
\end{array}
\begin{pmatrix}
0 & 1 & 2 & \cdots & n \\
1 - n\lambda & n\lambda & 0 & \cdots & 0 \\
0 & 1-(n-1)\lambda & (n-1)\lambda & \cdots & 0 \\
0 & 0 & 1-(n-2)\lambda & \cdots & 0 \\
 & & & & \\
 & & & & \\
 & & & & \\
0 & 0 & 0 & \cdots & 1
\end{pmatrix}.
$$

Using the initial conditions $P_0(0)=1, P_1(0)=\cdots=P_n(0)=0$, we can verify that

$$R(t)=P_0(t)+P_1(t)+\cdots+P_{n-1}(t)$$

$$=1-[1-\exp(-\lambda t)]^n,$$

or

$$1-R(t)=F(t)=[1-\exp(-\lambda t)]^n,$$

and this is precisely the distribution function of the largest order statistic from an exponential distribution with parameter λ. If each component had a different failure rate, say λ_i, then

$$R(t)=1-\prod_{i=1}^{n}[1-\exp(-\lambda_i t)].$$

It is to be noted that, if $n=2$ and $\lambda_1=\lambda_2=\lambda$, then

$$R(t)=2\exp(-\lambda t)-\exp(-2\lambda t),$$

a result that was observed for the case of a two-component standby-redundant system.

In the standby case, since the off-line equipment either cannot fail or has a failure rate less than that of the on-line equipment, the standby system reliability will always be greater than that for parallel redundancy.

The expression $R(t)=1-[1-\exp(-\lambda t)]^n$ has some interesting properties. If $\exp(-\lambda t)$ is denoted by P,

$$R(t)=1-(1-P)^n$$

can be recognized as a binomial process. In view of this fact it can easily be generalized that, when at least k out of n components are required for a system to be in an operable state,

$$R(t)=\sum_{i=k}^{n}\binom{n}{i}P^i(1-P)^{n-i}.$$

10.2.4 Moments of the Time to System Failure

The mean time to failure of a system (MTTF) and the variance of the time to failure are often quantities of interest. Given $R(t)$, the reliability of a system to time t, one can easily show (see Problem 10.8) that, if T denotes the time to failure of a system,

$$E(T) = \int_0^\infty R(t)\,dt.$$

Since $\text{Var}(T) = E(T^2) - [E(T)]^2$, it is easy to see that

$$\text{Var}(T) = \int_0^\infty T^2 \frac{d}{dt}[1 - R(t)]\,dt - \left[\int_0^\infty R(t)\,dt\right]^2$$

if $(d/dt)[1 - R(t)]$ exists.

For example, for a two-component parallel-redundant system with $\lambda_1 = \lambda_2 = \lambda$, $R(t)$ was found to be $2\exp(-\lambda t) - \exp(-2\lambda t)$. Then

$$E(T) = \int_0^\infty 2\exp(-\lambda t)\,dt - \int_0^\infty \exp(-2\lambda t)\,dt$$

$$= \frac{2}{\lambda} - \frac{1}{2\lambda} = \frac{3}{2\lambda}.$$

10.3   RELIABILITY MODELS FOR MAINTAINED SYSTEMS

Maintained systems are systems on which maintenance action is possible during a finite interval of time. It is assumed here that the maintenance action is such as to restore the failed system to initial condition. In dealing with maintained systems, an additional figure of merit is usually of interest. The *availability* of a system is the probability that the system is in an acceptable state at any instant of time t, given that the system was fully operative at $t = 0$. In this section, reliability models of maintained systems are considered, and for simplicity it is assumed that the equipment failure process and its repair process are both Markovian. This assumption implies that the probability of completing a repair in a time interval $(t, t + dt)$, given that it was not completed at t, is $\mu\,dt + \text{o}(dt)$. A similar argument holds for the failure process. With such a formulation a system can go from one state to another with the failure times and the repair times independently, identically distributed exponentially with parameters λ and μ, respectively. The resulting Markov process, referred to as a *birth/death process*, is fundamental in queuing theory.

Another figure of merit that is useful in studying maintained systems is the *mean recurrence time*, which is the length of time that a system takes to return from a failed to an acceptable state.

10.3.1 Repair of a Single Unit

Consider a system consisting of a single unit that can be in either the failed or the operating state. If 0 designates the operating state, then, under the usual assumptions, the transition probability matrix for this system is

$$
\mathbf{P} = \begin{matrix} 0 \\ 1 \end{matrix} \begin{pmatrix} 0 & 1 \\ 1-\lambda & \lambda \\ \mu & 1-\mu \end{pmatrix}.
$$

From this matrix, following Section 10.1.3, we can immediately write the following differential equations:

$$
P_0'(t) = -\lambda P_0(t) + \mu P_1(t),
$$

$$
P_1'(t) = \lambda P_0(t) - \mu P_1(t).
$$

If the initial conditions are taken as $P_0(0) = 1$ and $P_1(0) = 0$, the Laplace transforms of the above equations yield

$$
(s+\lambda) P_0(s) - \mu P_1(s) = 1
$$

and

$$
-\lambda P_0(s) + (s+\mu) P_1(s) = 0,
$$

and this in turn gives

$$
P_0(s) = \frac{s+\mu}{s(s+\lambda+\mu)}.
$$

The inverse Laplace transform of $P_0(s)$ gives $P_0(t)$, which in effect is the availability at time t, $A(t)$. Thus

$$
A(t) = P_0(t) = \frac{\mu}{\lambda+\mu} + \frac{\lambda}{\lambda+\mu} \exp[-(\lambda+\mu)t].
$$

If the system was initially failed, the initial conditions become $P_0(0) = 0, P_1(0) = 1$, and in this case

$$
A(t) = \frac{\mu}{\lambda+\mu} - \frac{\lambda}{\lambda+\mu} \exp[-(\lambda+\mu)t].
$$

It is to be noted that, as t becomes very large, the two expressions for availability given above become equivalent. This implies that the behavior of the system becomes independent of the starting state after the system has been operating for a long period of time. A similar situation was pointed out in Section 10.1, where the particle in question made transitions at discrete points in time, and the number of transitions was large.

The average availability for a fixed period of time, $(0, t)$, is the average uptime and is given by

$$A(t) = \frac{1}{t} \int_0^t A(s)\, ds$$

$$= \frac{\mu}{\lambda + \mu} + \frac{\lambda}{(\lambda + \mu)^2 t} - \frac{\lambda}{(\lambda + \mu)^2 t} e^{-(\lambda + \mu)t}.$$

If $t \to \infty$, the availability becomes $\mu/(\lambda + \mu)$, and this is referred to as the *steady-state availability*. Often, in practice, interest centers around determining the expected number of system failures in time $(0, t)$. This information is of importance in connection with maintaining an inventory of spare parts, scheduling repairmen, planning production, and making other such managerial decisions.

Let $N_{01}(t)$ be the number of times that the system visits state 1 (the failed state), given that the system was initially at state 0. If the repair processes were instantaneous, the behavior of the system could be characterized by a simple renewal process, and hence $E[N_{01}(t)] = \lambda t$ (see Section 10.1.2). If, however, the repair process takes a finite amount of time, which is characterized by some distribution function $G(t)$, the system can be described by what is known as an *alternating renewal process*. If $F(t)$ denotes the failure distribution of the component, some general expressions can be derived for $M_{ij}(t) = E[N_{ij}(t)]$, where $N_{ij}(t)$ denotes the number of visits to state j in time $(0, t)$, given that the system was in state i at time 0. In the situation being discussed, $i, j = 0, 1$.

If the unit is "on" at time 0, the expected number of visits to the "on" state, given that the first failure occurs at time x, is $M_{10}(t - x)$.

$$\therefore \qquad M_{00}(t) = \int_0^t M_{10}(t - x)\, dF(x).$$

On the other hand, if the unit is "off" at time 0, the expected number of visits to the "on" state, given that the first repair occurs at time x, is $1 + M_{00}(t - x)$. Thus

$$M_{10}(t) = \int_0^t [1 + M_{00}(t - x)]\, dG(x).$$

The above equations can be solved using the convolution theorem for

Laplace transforms. Let

$$M_{ij}^*(s) = \int_0^\infty e^{-st} \, dM_{ij}(t)$$

denote the Laplace transform of $M_{ij}(t)$. Then it follows that

$$M_{00}^*(s) = M_{10}^*(s) F^*(s)$$

and

$$M_{10}^*(s) = G^*(s) + M_{00}^{**}(s) G^*(s),$$

or

$$M_{00}^*(s) = \frac{F^*(s) G^*(s)}{1 - F^*(s) G^*(s)}$$

and

$$M_{10}^*(s) = \frac{G^*(s)}{1 - F^*(s) G^*(s)}.$$

Similarly, it can be shown that

$$M_{11}^*(s) = \frac{F^*(s) G^*(s)}{1 - F^*(s) G^*(s)}$$

and

$$M_{01}^*(s) = \frac{F^*(s)}{1 - F^*(s) G^*(s)}.$$

If the repair and the failure distributions are exponential with parameters μ and λ, respectively, $G^*(s) = \mu/(s+\mu)$ and $F^*(s) = \lambda/(s+\lambda)$. These two equations give

$$M_{01}^*(s) = \frac{\lambda(s+\mu)}{s^2 + (\lambda+\mu)s}.$$

The inverse transform of $M_{01}^*(s)$ yields

$$M_{01}(t) - E[N_{01}(t)] = \frac{\lambda^2(1 - e^{-(\lambda+\mu)t})}{(\lambda+\mu)^2} + \frac{\lambda\mu t}{\lambda+\mu},$$

which is the expected number of system failures in $(0,t)$, given that the system was operative at time 0.

82 SYSTEM RELIABILITY

10.3.2   Series Configurations

Consider a series system consisting of two components, and assume that each equipment can fail with the same failure rate, λ, and can be repaired with the same repair rate, μ. Assume also that there is only one repair facility to service the failed component. State 0 is designated as the operating state of the system, and states 1 (only one equipment operating) and 2 (both equipments failed) are the nonoperating states. It is clear that the number of repair facilities will influence the availability of the system.

Under the usual Markovian assumptions employed here, the transition probability matrix of the system can be written as

$$\mathbf{P} = \begin{array}{c} \\ 0 \\ 1 \\ 2 \end{array} \begin{pmatrix} 1-2\lambda & 2\lambda & 0 \\ \mu & 1-(\lambda+\mu) & \lambda \\ 0 & \mu & 1-\mu \end{pmatrix}.$$

Following Section 10.1.3 we can write from this matrix the following differential equations:

$$P_0'(t) = -2\lambda P_0(t) + \mu P_1(t),$$

$$P_1'(t) = 2\lambda P_0(t) - (\lambda+\mu) P_1(t) + \mu P_2(t),$$

$$P_2'(t) = \lambda P_1(t) - \mu P_2(t).$$

If the initial conditions are given, this system of equations can be solved for the transient case, using Laplace transforms, as was done in the preceding sections.

Generally, we are interested in the steady-state solution; this gives the proportion of time that the system spends in each state after it has been in operation for some time. Since the steady-state solution is independent of the starting state, such a solution does not require that the initial conditions be known. In Section 10.1 it was pointed out that, if it is possible to go from one state to another in a finite number of steps, the elements of the transition probability matrix approach a limiting value as the number of transitions becomes large. Following a similar argument for processes whose transitions are continuous in time, we can show that

$$\lim_{t\to\infty} P_i(t) = P_i.$$

This implies that

$$\lim_{t\to\infty} P_i'(t) = 0.$$

Thus, for a steady-state solution of the system under discussion, it is necessary to set the left-hand sides of the differential equations to 0, and these, with the condition that $\sum_i P_i = 1$, give

$$P_0 = \frac{\mu^2}{\mu^2 + 2\lambda\mu + 2\lambda^2},$$

$$P_2 = \frac{2\lambda^2}{\mu^2 + 2\lambda\mu + 2\lambda^2},$$

and

$$P_1 = 1 - P_0 - P_2 = \frac{2\lambda\mu}{\mu^2 + 2\lambda\mu + 2\lambda^2}.$$

The steady-state availability, $\lim A(t)$, $t \to \infty$, is of course P_0.

Generalizations of the above procedure to series systems having n components and r repairmen ($r < n$) are possible, using the same technique, although the computations become cumbersome.

For a series system, it is to be noted that maintenance action does not influence the reliability function,

$$R(t) = \exp - \sum_{i=1}^{n} \lambda_i t,$$

but it does affect the availability function. However, for redundant systems, which are discussed next, maintenance action influences the reliability function and the expected time to system failure.

10.3.3 Parallel-Redundant Configurations

Consider a two-component parallel-redundant system having the same parameters as the series system considered in Section 10.3.2. It is assumed that only one repairman is scheduled for maintenance and that the states of the system are the same as in Section 10.3.2.

The availability is the proportion of time that the system spends in states 0 and 1; the reliability function describes the probability of reaching state 2 in some interval of time $(0, t)$, given that the system was in state 0 at time 0; and the MTTF is the average time to reach state 2 for the first time, given that the system starts in state 0.

The transition probability matrix for this system is the same as that given in Section 10.3.2, and it follows that the steady-state availability is

$$\lim_{t \to \infty} A(t) = P_0 + P_1 = \frac{\mu^2 + 2\lambda\mu}{\mu^2 + 2\lambda\mu + 2\lambda^2}.$$

To find the reliability function of this system, it is necessary to amend the transition probability matrix of the system so that the state corresponding to system failure is an absorbing state; thus state 2 is made an absorbing state. The transition probability matrix is now given by

$$\mathbf{P}= \begin{array}{c} \\ 0 \\ 1 \\ 2 \end{array} \begin{pmatrix} 1-2\lambda & 2\lambda & 0 \\ \mu & 1-(\lambda+\mu) & \lambda \\ 0 & 0 & 1 \end{pmatrix}.$$

Corresponding to this matrix, a system of differential equations can be written down, and with the initial conditions as $P_0(0)=1, P_1(0)=P_2(0)=0$, the Laplace transforms of these equations are

$$P_0(s)(s+2\lambda)-\mu P_1(s)=1,$$

$$-P_0(s)2\lambda+(s+\mu+\lambda)P_1(s)=0,$$

$$\lambda P_1(s)-sP_2(s)=0.$$

Solving for $P_1(s)$, we obtain

$$P_1(s)=\frac{2\lambda}{s^2+(3\lambda+\mu)s+2\lambda^2}.$$

Let

$$s_1=\frac{-(3\lambda+\mu)+\sqrt{\lambda+\mu^2+4\mu\lambda}}{2}$$

and

$$s_2=\frac{-(3\lambda+\mu)-\sqrt{(\lambda+\mu)^2+4\mu\lambda}}{2}.$$

Then

$$P_1(s)=\frac{2\lambda}{(s-s_1)(s-s_2)},$$

and by partial fraction expansions $P_1(s)$ can be written as

$$P_1(s)=\frac{2\lambda}{s_1-s_2}\left(\frac{1}{s-s_1}-\frac{1}{s-s_2}\right).$$

The inverse transform of $P_1(s)$ gives

$$P_1(t) = \frac{2\lambda}{s_1 - s_2}(e^{s_1 t} - e^{s_2 t}).$$

Then $P_0(s)$ can be obtained from $P_1(s)$ as

$$P_0(s) = \frac{\lambda + \mu + s}{s^2 + (3\lambda + \mu)s + 2\lambda^2} = \frac{\lambda + \mu + s_1}{(s_1 - s_2)(s - s_1)} - \frac{\lambda + \mu + s_2}{(s_1 - s_2)(s - s_2)}.$$

The inverse transform of $P_0(s)$ gives

$$P_0(t) = \frac{(\lambda + \mu + s_1)e^{s_1 t} - (\lambda + \mu + s_2)e^{s_2 t}}{s_1 - s_2}.$$

The reliability function, $R(t)$, is simply $P_0(t) + P_1(t)$, and hence

$$R(t) = \frac{(3\lambda + \mu + s_1)e^{s_1 t} - (3\lambda + \mu + s_2)e^{s_2 t}}{s_1 - s_2}$$

$$= \frac{s_1 e^{s_2 t} - s_2 e^{s_1 t}}{s_1 - s_2}.$$

If $F(t)$ is the distribution function of the time to first failure, $F(t) = 1 - R(t)$; and, by differentiating $F(t)$ with respect to t, the probability density function of the time to first failure can be obtained. The MTTF can be obtained by integrating $R(t)$ from 0 to ∞; note that s_1 and s_2 are both negative for $\lambda, \mu > 0$.

Reliability models for standby redundancy can be obtained similarly.

PROBLEMS FOR SECTIONS 10.1–10.3

10.1. For the transition probability matrix given in Section 10.1 show that

$$\lim_{n \to \infty} p^{(n)} = X = (\tfrac{1}{4}, \tfrac{1}{2}, \tfrac{1}{4}).$$

Hint: Let π_1, π_2, π_3 be the elements of X. Expand $XP = X$ in terms of simultaneous equations. Solve for π_1, π_2, and π_3, using the condition that $\pi_1 + \pi_2 + \pi_3 = 1$.

10.2. The transition probability matrix for a particle that moves along a line in discrete steps is given by

$$
\mathbf{P} = \begin{array}{c} \\ 1 \\ 2 \\ 3 \\ 4 \end{array}
\begin{array}{c} \begin{array}{cccc} 1 & \;\; 2 & \;\; 3 & \;\; 4 \end{array} \\
\left(\begin{array}{cccc}
\frac{1}{3} & \frac{1}{3} & 0 & \frac{1}{3} \\
\frac{1}{4} & \frac{1}{4} & \frac{1}{4} & \frac{1}{4} \\
0 & 0 & 1 & 0 \\
\frac{1}{4} & \frac{1}{4} & \frac{1}{2} & 0
\end{array} \right)
\end{array}
$$

a. Does this matrix have a unique probability vector X such that $XP = X$?

b. If $P(0)$ is $(\frac{1}{4}, \frac{1}{4}, \frac{1}{4}, \frac{1}{4})$, what is $P^{(2)}$?

c. If the particle starts in state 4, what is the expected number of times that it will return to state 4 before adsorption?

d. If the particle starts in state 1, what is the variance of the number of times that it will be in state 4 before absorption?

10.3. If $N(t)$ denotes the number of renewals in time $(0, t)$ for a renewal process, show that

$$P\{N(t) = n\} = F^n(t) - F^{n+1}(t),$$

where $F^n(t)$ is the distribution function of T_n, and $T_n = \sum_{i=1}^n X_i$.

Hint < If A is the event $T_n \leqslant t$, B is the event $T_{n+1} \leqslant t$, and $\bar{B}$ is the complement of B, application of the theorem concerning the union of three events gives the desired result.

10.4. For the two-component standby-redundant system discussed in Section 10.2.2, show that

$$P_1(t) = \frac{\lambda_1 + \lambda_2}{\lambda_2} \exp(-\lambda_1 t) - \frac{\lambda_1 + \lambda_2}{\lambda_2} \exp[-(\lambda_1 + \lambda_2)t], \quad \text{if } \lambda_2 > 0.$$

10.5. A standby-redundant system consisting of two components with perfect switching has failure rates λ_1 for the on-line component and λ_2 for the standby equipment. In addition, the system is exposed to a hazard, characterized by a Poisson process with a parameter λ_{12}, that is detrimental to both components. Assuming independence of the hazards, develop an expression for the reliability of this system.

10.6. For an n-component parallel-redundant system, if the failure rate of each component is λ_i, show that

$$R(t) = 1 - \prod_{i=1}^n [1 - \exp(-\lambda_i t)],$$

where $R(t)$ is the reliability function.

10.7. A two-component parallel-redundant system is such that each component has a failure rate λ_1 when both components are operative, and a failure rate $2\lambda_1$ when either of the components has failed. The increase in the failure rate is presumably caused by an increase in the supply of power to the surviving component. Obtain an expression for the reliability function of this system. Can you generalize your result to the case of n components in parallel?

10.8. If a random variable $T > 0$ has a distribution function $F(t)$, show that

$$E(T) = \int_0^\infty [1 - F(t)] \, dt.$$

Under what conditions is the above result true?

10.9. Find the mean time to failure of the systems described in Problems 10.4, 10.5, and 10.7.

10.10. For the series system consisting of two components discussed in Section 10.3.2, develop an expression for the availability when two repairmen are assigned to service failed equipments. Assume that only one repairman can service a failed equipment at a time, with a repair rate μ.

10.11. For a two-component standby-redundant system with perfect switching, develop expressions for the steady-state availability, the reliability function, and the mean time to first system failure, given that both components are operative at time 0. Assume that there is only one repairman, that off-line equipment has a zero failure rate, and that on-line equipment has a failure rate of λ. Assume that the repair rate is μ.

10.4 CONFIDENCE BOUNDS FOR SYSTEM RELIABILITY

In this section, methods of obtaining lower confidence bounds on the probability of successful operation of nonmaintained systems are surveyed, and numerical comparisons are given. It is assumed that failure data have been collected from life tests performed on prototypes of the various subsystems which make up the system, but that, for some reason, the system has not been tested as a whole. The reason may be, for example, reluctance to incur the expense or simply the fact that it is virtually impossible to test the entire system without destroying it. Methods applicable to series and/or parallel and more logically complex systems are discussed.

10.4.1 Confidence Bounds for Specific System Models

There are two types of models for which one can theoretically calculate, from subsystem failure data, confidence bounds on system reliability that are known to be optimum in some sense. For a series system made up of k independent subsystems, each having exponentially distributed failure time T, there exists a lower confidence bound that is most accurate (has the highest probability of being close to the true system reliability) for all values of system reliability among exact bounds that are unbiased. (A family of lower confidence bounds $\underline{\theta}(X)$ on θ at confidence level $1-\alpha$ is unbiased if $P[\underline{\theta}(X) \leqslant \theta'] \leqslant 1-\alpha$ for all $\theta' < \theta$ and for all values of nuisance parameters.) The restriction of unbiasedness is necessary here because of the nuisance parameters $\lambda_1,\ldots,\lambda_k$, the hazard rates for the k independent subsystems. For this model, series system reliability $R(t_m)$ at time $t_m > 0$ is equal to

$$\prod_{j=1}^{k} \exp(-t_m\lambda_j) = \exp\left(-t_m \sum_{j=1}^{k} \lambda_j\right), \quad \lambda_j > 0, j = 1,\ldots,k,$$

so that one can also think of obtaining an upper confidence bound on the system hazard rate, $\phi = \sum_{j=1}^{k}\lambda_j$.

The optimum lower confidence bound on series-system reliability, derived for two subsystems with exponential failure data by Lentner and Buehler (1963) and generalized to $k \geqslant 2$ subsystems by El Mawaziny (1965), depends on the assumption that, for the jth subsystem, n_j prototypes have been tested until r_j, with $1 \leqslant r_j \leqslant n_j$, failures occur, $j = 1,\ldots,k$. One observes, for the jth subsystem, ith smallest failure times (see Section 5.1), $t_{i,j}, i = 1,\ldots,r_j$, and computes

$$z_j = \sum_{i=1}^{r_j} t_{i,j} + (n_j - r_j)t_{r_j,j}, \quad j = 1,\ldots,k.$$

Calculation of a confidence bound based on the z_j's by El Mawaziny's method must be performed iteratively on a computer; and, if the product of the number of subsystems and the total number of failures is large, problems of loss of precision result. See Mann (1970).

The other system model for which a method of obtaining optimum confidence bounds on system reliability has been derived is one in which binomially distributed *attribute* or pass/fail data are collected. No assumptions are made concerning the form of failure-time distributions; the assumptions are that the system is serial or parallel-redundant, the subsystems and the subsystem prototype tests are independent, the failure distribution for any specified subsystem is the same for each prototype of that subsystem tested, and the duration of each test is the intended operating time for the particular subsystem in the system. For the binomially distributed data obtained for this model, Buehler (1957) defines a small-sample method of obtaining confidence bounds on system reliability. These bounds, however, like binomial confidence bounds for a single component, are conservative in general rather than exact because of the discreteness of the numbers of failures (see Section 7.4.3). In order to obtain a confidence bound that is exact in general and uniformly most accurate unbiased, a random number uniformly distributed on $(0,1)$ must be generated and used in calculating the bound. (Confidence bounds that depend on a uniform random variate in addition to the failure data will be referred to as *randomized*, while the conservative bounds that do not depend on a generated random number will be called *nonrandomized*.)

Lipow and Riley (1959) used Buehler's definition to compute and tabulate nonrandomized lower confidence bounds on series-system reliability R_s for systems containing one, two, or three subsystems and attribute subsystem data. In all of their tabulations, the number n of items tested is the same for each subsystem because of the problem of ordering the failure combinations in calculating each confidence bound. The value of n ranges from 5 to several hundred. The number of failures is not required to be the same for each subsystem and ranges from 0 to about $n/2$. The tabulated lower confidence bounds on series-system reliability are known to be optimum (to correspond to shortest intervals with $+1$ as an upper bound) among nonrandomized confidence bounds based on the subsystem attribute data for two independent subsystems when the sum of the numbers of failures is less than $2\sqrt{n}$. For three subsystems, the problem of ordering the failure-combinations prevents Lipow and Riley from claiming optimality for the bounds based on their particular failure-combinations. They do claim, however, that they are very close to the optimum nonrandomized confidence bounds.

Steck (1957) gives in graphical form the same sort of conservative nonrandomized confidence bounds for parallel-system reliability R_p for systems composed of two independent subsystems and having at most one observed failure.

(a) Approximate and Nonoptimum Exact Confidence Bounds for the Exponential Model. For each of these system models, a number of methods have been derived for approximating the optimum bounds. In some cases, the approximations were derived before the optimum bounds. In other cases, they were derived because of the problem involved in calculating the optimum bounds.

For the exponential series-system model with Type II censoring, that is, a fixed number of failures, the method of Kraemer (1963) yields bounds that depend on only the smallest $z_j = r_j \hat{\theta}_j, j = 1, \ldots, k$. This method gives confidence bounds that are very inaccurate (i.e., they exhibit a high probability of being far from the true system reliability). Numerical calculation of a confidence bound by Kraemer's method and by most of the other methods discussed for this exponential model is illustrated after the presentation of the methods.

The methods of Sarkar (1971) and Lieberman and Ross (1971), developed for the exponential series-system model, depend on combinations of failure times selected from the various subsystems, rather than the sufficient statistics, $Z_1, \ldots, Z_k$, observed as $z_1, \ldots, z_k$. As Lieberman and Ross point out, in using their method different bounds can be obtained by ordering the subsystem tests differently and thus combining the subsystem test data differently. This problem can be overcome, however, by using a computer to calculate all possible combinations of failure times. The confidence intervals of Lieberman and Ross are exact and, in general, stochastically shorter than those of either Sarkar or Kraemer. The general methodology developed by Lieberman and Ross has been extended to complex systems by Saunders (1972).

Grubbs (1971) uses what will be called here the *fiducial* approach to approximate the distribution of a series-system failure rate, given the exponential subsystem failure data; that is, he assumes that, since (see Section 5.1) the distribution of $2Z_j$ is $\lambda_j^{-1} \chi^2(2r_j)$, then λ_j, given z_j, is distributed as $\chi^2(2r_j)/2z_j, j = 1, \ldots, k$. Grubbs employs an approach that uses the mean, $\Sigma_{j=1}^k r_j / z_j$, and the variance, $\Sigma_{j=1}^k r_j / z_j^2$, of the fiducial distribution of the system failure rate, $\phi = \Sigma_j \lambda_j$, to fit a weighted chi-square distribution.

Burnett and Wales (1961) and Levy and Moore (1967) suggest for this series-system model, as well as for other models that are more logically complex than a series system, Monte Carlo simulation of the fiducial distribution of system reliability, given the subsystem failure data. Such an approach yields for the series-system model essentially the same result as the method of Grubbs; see Mann (1970) and Grubbs (1971). The fiducial approach is equivalent to obtaining Bayesian bounds on system reliability at time t_m with the prior density for the jth subsystem hazard rate, $\lambda_j > 0$, equal to $\lambda_j^{-1}, j = 1, \ldots, k$, the prior density yielding the optimum confidence bound for $k = 1$. See, for example, Section 8.4.1.

Although lower confidence bounds on system reliability obtained by the fiducial method are optimum for one subsystem and approach the optimum confidence bounds as the numbers of failures increase for all subsystems, they are conservative in general. For this series-system model, the fiducial lower bound on system reliability is always less than the corresponding optimum exact classical lower confidence bound, and sometimes by a considerable amount. Examples are given in Table 10.1.

The confidence bounds on series-system reliability derived by El Mawaziny and Buehler (1967) for the exponential model depend on the asymptotic normality of system hazard rate, given the subsystem data as it is assumed to be given in the derivation of the optimum bounds of El Mawaziny. That is, we assume that z_1 and $u_2 = z_1 - z_2, \ldots, u_k = z_1 - z_k$ are given, where the subscript 1 is arbitrarily assigned. As shown in Table 10.1, a lower confidence bound obtained by this method is always greater than the corresponding optimum bound when the number of failures is small for any subsystem. One of the most serious problems inherent in this method of obtaining lower confidence bounds on system reliability is that subsystems for which a single failure occurs are ignored. The consequences of this fact are shown in Table 10.1.

The methods of Rosenblatt (1963) and Madansky (1965) yield asymptotic lower confidence bounds on series-system reliability that, like the fiducial bounds, tend to be conservative for this model unless all r_j's are large. These methods are discussed in more detail in Section 10.4.1(b), dealing with confidence bounds based on binomial subsystem data.

The approximately optimum (AO) confidence bounds of Mann and Grubbs (1972) are based on a combination of the approach of Grubbs (1971) and that of El Mawaziny and Buehler (1967) for the exponential series-system model. One conditions on z_1 and $\mathbf{u} = (u_2, \ldots, u_k)$, as in deriving the optimum confidence bounds on R_s, and determines expressions for the approximate posterior mean and variance of the system hazard rate, ϕ. The conditional variate z_1, given $\mathbf{u}$, belongs to the Koopman-Darmois exponential family with parameter ϕ and nuisance parameters $\lambda_1, \ldots, \lambda_k$, so that, according to the discussion given by Lehmann (1959, Section 4.3), a uniformly most accurate unbiased upper confidence bound for ϕ can be obtained from z_1 and $\mathbf{u}$.

Table 10.1 Lower 50%, 75%, and 90% confidence bounds, on series-system reliability: exponential subsystem failure data, with mission time $t_m = 1$

r_j = Number of Failures	Observed Values of $z_j = r_j \hat{\theta}_j$					Confidence Level	EI Mawaziny Optimum Lower Bound	Approx. χ^2 (fiducial)	AO Bounds Approx. χ^2 Based on (10.2), (10.3), (10.4)	EI Mawaziny-Buehler Normal Approx.
	z_1	z_2	z_3	z_4	z_5					
4,2,2,3,3	9.919	15.996	26.897	26.511	62.439	.75	.485	.406	.479	.525
4,2,2,3,3	28.412	13.992	9.990	98.228	44.667	.50	.670	.569	.657	.710
1,2,2,3,3	.563	4.609	21.045	25.328	11.419	.90	.010	.006	.009	.426
1,2,2,3,3	9.976	2.381	20.275	10.044	51.118	.75	.242	.173	.233	.365
1,2,2,3,3	.563	4.609	21.045	25.328	11.419	.50	.179	.100	.156	.595
2,3,4	23.059	32.504	37.188	—	—	.50	.812	.757	.804	.831
2,3,4	75.758	16.084	57.503	—	—	.90	.671	.647	.669	.733
2,3,4	75.758	16.084	57.503	—	—	.50	.795	.766	.791	.827
2,3,4	19.753	31.249	21.321	—	—	.75	.692	.629	.688	.721
2,2,2	16.966	42.258	15.318	—	—	.50	.828	.761	.824	.863
2,2,2	16.966	42.258	15.318	—	—	.75	.764	.691	.760	.811
4,3,2	42.753	45.791	31.890	—	—	.90	.772	.725	.766	.801
4,3,2	84.794	36.989	4.165	—	—	.90	.359	.344	.358	.526
4,3,2	28.274	44.690	9.480	—	—	.90	.566	.524	.559	.658
4,3,2	171.581	29.062	11.293	—	—	.90	.649	.614	.642	.737
4,3,2	78.668	54.968	57.296	—	—	.90	.849	.815	.846	.868
4,3,2	25.527	56.469	23.476	—	—	.75	.750	.700	.746	.777

Mann and Grubbs (1972) have shown that the posterior distribution of ϕ is that of a sum of weighted chi squares which, as demonstrated in Sections 5.1 and 5.2 can be well approximated by a single weighted chi-square variate.Once the approximate conditional mean and variance of the system hazard rate are calculated, one can determine rather well the approximate weighted chi-square posterior distribution of the system hazard rate by use of a classical chi-square distribution (see Patnaik, 1949) and thus obtain approximately optimum lower confidence bounds on series-system reliability. A Wilson-Hilferty (1931) approximation of chi square by a Gaussian variate (see Sections 5.1 and 5.2) can be used to facilitate the calculations when noninteger degrees of freedom are encountered. The expressions for the conditional mean m and variance v of the system hazard rate ϕ derived by Mann and Grubbs have been simplified by Mann (1974b) and are as follows:

$$ m = \frac{\displaystyle\sum_{j=1}^{k} (r_j - 1)}{z_j} + z_{(1)}^{-1} \qquad (10.2) $$

and

$$ v = \frac{\displaystyle\sum_{j=1}^{k} (r_j - 1)}{z_j^2} + z_{(1)}^{-2}, \qquad (10.3) $$

where $z_{(1)}$ is the smallest of the z's. To approximate the optimum lower bound on $R_s(t_m)$ at confidence level $1 - \alpha$, using the Wilson-Hilferty transformation, one calculates

$$ \underline{R}_s(t_m) = \exp\left[-t_m m \left(1 - \frac{v}{9m^2} + \frac{\eta_{1-\alpha} v^{1/2}}{3m} \right)^3 \right], \qquad (10.4) $$

where η_γ is the 100γth percentile of a standard normal distribution. Confidence bounds based on Equations (10.2), (10.3), and (10.4) are compared in Table 10.1 with the optimum classical exact confidence bounds of El Mawaziny and with the fiducial bounds and the asymptotic approximation of El Mawaziny and Buehler. The data used for the comparisons were generated in conjunction with the study described by Mann (1970). The approximately optimum lower confidence bounds $\underline{R}_s(t_m)$ on $R_s(t_m)$ agree to within about a unit in the second decimal place with the optimum lower confidence bounds of El Mawaziny (1965).

Calculation of lower confidence bounds on series-system reliability $R_s(t_m)$ from a data set given in Table 10.1 will be used as an example of the implementation of some of the methods discussed. We take $r_1 = 2, r_2 = 3, r_3 = 4, z_1 = 23.059, z_2 = 32.504, z_3 = 37.188$. To obtain an AO confidence bound using (10.2), (10.3) and (10.4), we form

$$m = \frac{1}{23.059} + \frac{2}{32.504} + \frac{3}{37.188} + \frac{1}{23.059} = .22894$$

and

$$v = \frac{1}{(23.059)^2} + \frac{2}{(32.504)^2} + \frac{3}{(37.188)^2} + \frac{1}{(23.059)^2} = .00782.$$

Then $2(.22894)\phi/.00782$, given the failure data, has approximately a chi-square distribution with $2(.22894^2)/.00782 \cong 13.4$ degrees of freedom. The Wilson-Hilferty transformation works well for 3 or more degrees of freedom. Hence, for $t_m = 1$,

$$R_s(1) \geqslant \exp\left\{ -.22894\left[1 - \frac{(.00782/.22894^2)}{9} + \frac{0.0\left(\sqrt{.00782}/.22894\right)}{3} \right]^3 \right\}$$

$$= .804$$

with probability approximately .50 (since 0.0 is the 50th percentile of a standard normal distribution). The optimum 50% confidence bound obtained by the method of El Mawaziny (1965) is .812.

A fiducial bound is obtained in much the same manner as the AO confidence bound, except that, for m and v, as defined for the AO confidence bounds, one substitutes $\sum_{j=1}^{k} r_j/z_j = .28658$ and $\sum_{j=1}^{k} r_j/z^2_j = .00949$, respectively. The weighted chi-square assumption is made, and use of the Wilson-Hilferty transformation yields .757 as the fiducial lower bound for $R_s(1)$.

The El Mawaziny-Buehler approximation involves using the asymptotic conditional mean and variance of the posterior distribution of ϕ, which are, respectively, $\sum_{j=1}^{k}(r_j - 1)/z_j$ and $\sum_{j=1}^{k}(r_j - 1)/z_j^2$. The posterior distribution of ϕ is assumed to be Gaussian, which is appropriate for all r_j's sufficiently large. Thus a 50% upper confidence bound on ϕ, based on this asymptotic approximation, is given by the asymptotic mean of the posterior distribution of ϕ, or $1/23.059 + 2/32.504 + 3/37.188 = .18557$. This, since $t_m = 1$, gives $\exp(-.18557) = .831$ as a 50% lower confidence bound for $R_s(t_m)$.

Maximum-likelihood asymptotic bounds are obtained similarly (see El Mawaziny and Buehler, 1967) except that in this case the fiducial mean and variance, .28658 and .00949, of the posterior distribution of ϕ are used, and the 50% lower confidence bound on $R_s(1)$ is exp $(-.28658) = .752$. Another interpretation of the maximum-likelihood approach to obtaining asymptotic confidence bounds for $R_s(t_m)$ (see Rosenblatt, 1963) is to assume that the maximum-likelihood estimate, $R_s(1) = \prod_{j=1}^{k}(1 - r_j/n_j)$, based on numbers of failures and successes only, is asymptotically normally distributed with asymptotic mean R_s and asymptotic variance σ_R^2. Easterling (1972b) has modified this approach, using $\hat{R}_s(t_m) = \exp(-t_m \Sigma r_j / z_j)$, approximating the asymptotic variance, $\sigma_{\hat{R}}^2 = t_m^2 \Sigma_j (R_s/\theta_j)^2 \sigma_{\hat{\theta}_j}^2$, of $\hat{R}_s(t_m)$ by substituting maximum-likelihood estimates $\hat{\theta}_j = z_j/r_j, j = 1,\ldots,k$, for θ_j and fitting $\hat{R}_s(t_m)$ with a beta distribution with parameters to be estimated from the data. The asymptotic variance estimate obtained is

$$\hat{\sigma}_{\hat{R}}^2 = \sum_{j=1}^{k} \left(\frac{t_m^2 r_j}{z_j^2} \right) \cdot \left[\hat{R}_s(t_m) \right]^2.$$

For the present data one uses the results above to calculate, for $t_m = 1$, $\hat{R}_s(t_m) = \exp(-.28658) = .752$ and $\hat{\sigma}_{\hat{R}} = \sqrt{.00949}\,(.752) = \sqrt{.00536} = .0732$. Easterling approximates $n - x$, the psuedo number of successes, by $[\hat{a} + 1]$, with $\hat{a} = \hat{R}_s^2(1 - \hat{R}_s)/\hat{\sigma}_{\hat{R}}^2 - \hat{R}_s$, and $x + 1$, the psuedo number of failures plus 1, by $\hat{c} = [\hat{a} + \hat{b} + 1] - [\hat{a} + 1]$, where $\hat{b} = \hat{R}_s(1 - \hat{R}_s)^2/\hat{\sigma}_{\hat{R}}^2 + \hat{R}_s - 1$, and $[s]$ is the greatest integer less than or equal to s. Thus the appropriate beta parameters are

$$[\hat{a} + 1] = [25.41 + 1] \qquad \text{and} \qquad \hat{c} = [25.41 + 8.38 + 1] - 26 = 8.$$

The 50th percentile of a beta distribution with parameters 26 and 8 is .770, slightly higher than the lower confidence bound obtained for R_s by the asymptotic normal approximation based on the maximum-likelihood estimate.

The general maximum-likelihood method, as well as the modification of Easterling and the likelihood-ratio method, has the advantage that it can be applied to logically complex systems. As shown above, however, the maximum-likelihood method, modified or not, leads to confidence bounds on system reliability that tend to be quite conservative for small r_j's. The likelihood-ratio method requires iterative procedures, even for the series-system model; and, since it also depends on asymptotic properties of maximum-likelihood estimators, it, too, tends to give confidence bounds that do not approximate the optimum confidence bounds well unless the number of failures is large for each subsystem.

The approach of Easterling described above suggests a method by which, for an exponential model with Type II censoring, nearly optimum confidence bounds on the reliability of a parallel system, and ultimately a logically complex system, can be derived. This is discussed in Section 10.4.1(c), which deals with more general models.

Before presenting a model with binomially distributed "go/no go" data, we discuss the application of the methods of Kraemer (1963), Sarkar (1971), and Lieberman and Ross (1971) to the data set considered above. Since the method of Kraemer depends only on the smallest of the observed z_j's, it is necessary to observe only $z_1 = 23.059$. Since each observable $2Z_j\lambda_j, j = 1,\ldots,k$, has an independent chi-square distribution with $2r_j$ degrees of freedom, $R_s(t_m) \geqslant \exp[-t_m\chi^2_{.50}(2r)/46.118]$ with probability *greater than or equal to* .50, where $r = \sum_{j=1}^k r_j$. For $\sum_{j=1}^k r_j = 9$, as above, and $t_m = 1$, the lower confidence bound calculated for $R_s(1)$ is $\exp(-20.601/46.118) = .639$. This is the smallest of the lower confidence bounds yielded by any of the methods illustrated and the farthest from the optimum bound of .812.

To apply the methods of Sarkar and of Lieberman and Ross it is necessary to have more information than the values of r_1, r_2, r_3, and realizations of the sufficient statistics z_1, z_2, z_3 given above. In both of these methods the actual failure time for any given subsystem is used and combined with the failure times or differences of failure times from other subsystems according to an arbitrary ordering of failure times for all subsystems. The results obtained will vary with the ordering patterns selected. One can eliminate this variability, however, by averaging all possible combinations.

We note from the results above and those displayed in Table 10.1 that the approximately optimum lower confidence bounds on series-system reliability, for this exponential model with Type II censoring, approximate the optimum (uniformly most accurate unbiased) lower confidence bounds of El Mawaziny exceedingly well. We note also that, unlike the optimum confidence bounds, which must be obtained iteratively from equations involving computational difficulties, the AO confidence bounds are fairly simple to obtain: one simply calculates m and v by substituting observed data into (10.2) and (10.3) and then substitutes a specified t_m and the calculated m and v into (10.4). Finally, we note that none of the other methods described is designed specifically to approximate the optimum confidence bounds for all possible data sets.

Other system configurations for this fixed-number-of-failures model, for example, parallel systems and logically complex systems, are discussed in Section 10.4.1(c).

(b) Approximate and Nonoptimal Exact Confidence Bounds for the Binomial Model. For series or parallel-redundant models and the case in which only pass/fail binomially distributed data are collected for each subsystem, many methods involving large-or small-sample approximations or Bayesian techniques have been derived for obtaining nonrandomized confidence bounds on the probability of successful operation of a system. Among the large-sample methods is one derived by Madansky (1965). The extension of Madansky's method to logically complex systems has been accomplished by Myhre and Saunders (1968b). In using this method, one parameterizes so as to determine iteratively parameter values that make the negative of the logarithm of the likelihood ratio equal to one half of a specified percentile of the chi-square distribution with 1 degree of freedom; see Wilks (1938). In determining the likelihood ratio, it is necessary to maximize the joint binomial density function subject to a constraint specified by the equation relating system reliability to the subsystem reliabilities.

The principal disadvantage associated with use of the likelihood-ratio procedure is that any component exhibiting zero failures in a life-test situation is ignored in the determination of a series-system reliability confidence bound and tends to bias parallel-system confidence bounds on the high side. Thus the procedure yields lower confidence bounds that are too high whenever zero failures are observed during the life test, no matter how large the sample size for all components. This method is, in fact, not appropriate for highly reliable systems.

Another method suggested for obtaining confidence bounds on system reliability is based on the asymptotic normality of the unbiased *simulation* estimator (often equivalent to the maximum-likelihood estimator) of system reliability. Such a procedure and its theoretical rationale are discussed at length by Rosenblatt (1963). Easterling (1972a) investigated a variation of this procedure in which psuedo numbers of system tests and successes are determined from the estimate of the asymptotic variance of the maximum-likelihood estimator (MLE) of system reliability (in somewhat the same fashion as described earlier for his exponential confidence bounds). These numbers are then substituted into the incomplete beta function, and confidence bounds obtained as in determining usual optimum nonrandomized confidence bounds for a single component and binomial sampling. Easterling found his approximation to be better for several series, parallel, and series-parallel systems that he investigated than that based on the asymptotic normality of MLE's and comparable to that based on the asymptotic chi-square distribution of the logarithm of the likelihood-ratio function (which is more difficult to implement).

Woods and Borsting (1968) have approximated the distribution of a modification of $-2\ln \hat{R}_s$, with $\hat{R}_s$ the MLE of series-system reliability, by a continuity-corrected gamma distribution. The approximation appears to give exact bounds when sample sizes are large, but does not seem to perform well for sample sizes as small as 10.

Murchland and Weber (1972) describe a method similar to that detailed by Rosenblatt for obtaining confidence bounds on the reliability of a complex system. These authors, however, suggest the use of Chebyshev inequalities rather than adopting any particular assumptions concerning the distribution (asymptotic or otherwise) of the unbiased simulation estimator of system reliability, and their bounds involve an expression for the exact rather than the asymptotic variance of the estimator of R. Their procedure presents no particular computational difficulties as long as the system in question remains logically simple. Calculation of the Murchland-Weber variance estimate for the reliability estimator becomes very complicated, however, as the system increases in complexity.

A more serious difficulty inherent in this method and all methods dependent on the maximum-likelihood or simulation estimator of system reliability is similar to that mentioned for the likelihood-ratio method. It is the fact that any component exhibiting no failures during the life tests is ignored in the determination of bounds for strictly serial systems and yields a lower confidence bound on parallel-system reliability that is always 1. The maximum-likelihood and simulation estimators are also inappropriate for providing the basis for confidence bounds on the reliability of highly reliable systems.

Table 10.2 gives for series systems the optimum nonrandomized lower confidence bounds of Lipow and Riley (1959), likelihood-ratio (LR) and maximum-likelihood (ML) approximations calculated by Myhre and Saunders (1968a), and Easterling's modified maximum-likelihood (MMLI) approximations, all for data sets in which at least one failure occurs for each component. Results based on the Woods and Borsting approximation (labeled Gamma) are shown in Table 10.3, with examples of confidence bounds obtained from subsystems exhibiting no failures.

Here we illustrate the use of the methods based on the MLE (equivalent in this case to the simulation estimator) for one of the data sets shown in Table 10.3 for which one subsystem exhibits no failures: $k = 2, n_1 = 5, n_2 = 10$, with numbers of failures $x_1 = 0, x_2 = 1$. The MLE, $\hat{R}_s$, of system reliability, R_s, is $(5/5)(9/10) = .9$, and the asymptotic variance of $\hat{R}_s$ is, for $P_j = 1 - R_j, j = 1, \ldots, k,$

$$\sigma_{\hat{R}}^2 = \sum_{j=1}^{k} \left(\frac{\partial R_s}{\partial R_j} \right)^2 \operatorname{Var}\left(\hat{R}_j \right) = R_s^2 \sum_{j=1}^{k} \frac{P_j/(1 - P_j)}{n_j}.$$

Table 10.2 Nonrandomized lower confidence bounds on reliability for a series system of two or three components

	Number of Failures		Confidence Level									
			90					95				
$k=2$:	x_1	x_2	ML	LR	Opti-mum	AO	MMLI	ML	LR	Opti-mum	AO	MMLI
$n=10$	1	1	.655	.629	.607	.606	.585	.611	.571	.548	.552	.530
	1	2	.545	.529	.497	.493	.489	.495	.473	.433	.435	.436
	2	2	.456	.451	.445	.430	.441	.405	.397	.392	.382	.391
	1	4	.347	.350	.344	.335	.318	.292	.301	.298	.293	.271
	2	3	.373	.375	.354	.353	.362	.320	.326	.304	.307	.315
$n=20$	1,	2	.756	.739	.716	.728	.709	.728	.700	.677	.693	.671
	2,	2	.701	.687	.683	.678	.669	.670	.647	.643	.643	.631
	1,	3	.697	.683	.660	.675	.655	.665	.643	.620	.639	.616
	2,	3	.647	.636	.622	.628	.619	.614	.597	.582	.593	.580
	3,	3	.599	.591	.585	.582	.570	.565	.551	.544	.548	.532
$k=3$:	x_1 x_2	x_2										
$n=20$	1, 1,	1	.760	.743	.747	.741	.721	.732	.705	.709	.708	.684
	1, 1,	2	.704	.690	.693	.689	.669	.673	.651	.644	.657	.631
	1, 2,	2	.654	.643	.639	.643	.619	.621	.604	.598	.609	.580
	1, 2,	3	.605	.596	.595	.597	.587	.571	.557	.544	.554	.549
$n=30$	1, 2,	3	.723	.714	.705	.716	.669	.698	.683	.674	.691	.638
	1, 1,	1	.835	.822	.825	.820	.803	.816	.794	.796	.798	.775
	2, 2,	2	.725	.715	.712	.717	.703	.700	.685	.681	.692	.672
$n=50$	1 2	4	.805	.798	.789	.800	.788	.788	.776	.767	.781	.766
	1 1	2	.874	.865	.861	.870	.852	.860	.845	.841	.854	.833
$n=100$	1, 1,	2	.936	.931	.929	.932	.923	.929	.920	.918	.923	.913
	2, 3,	5	.866	.861	.858	.865	.856	.855	.848	.844	.854	.842

If we substitute for R_s and P_j, $j=1,\ldots,k$, their maximum-likelihood estimates, the estimate to be used for the asymptotic variance becomes

$$\hat{\sigma}_{R_s}^2 = \prod_{i=1}^{k} \left[(n_i - x_i)/n_i \right]^2 \sum_{j=1}^{k} \frac{x_j}{(n_j - x_j)/n_j}.$$

Table 10.3   Nonrandomized 90% lower confidence bounds on series-system reliability for unequal sample sizes

k = 1:	Number of Tests			Number of Failures			Opti-mum	AO	Fidu-cial	Gamma	Uniform Prior on R_s
	n_1			x_1							
	7			1			.547	.548	.548	.257	.594
	7			0			.720	.722	.722	1.00	.750
	10			4			.354	.354	.354	.383	.401
	10			1			.663	.664	.664	.386	.690
k = 2	n_1	n_2		x_1	x_2						
	15	10		0	0		.794	.795	.678	1.00	.837
	5	10		0	1		.603	.571	.475	.386	.640
k = 3	n_1	n_2	n_3	x_1	x_2	x_3					
	10	5	4	1	1	0	.403	.385	.225	.096	.454
	10	10	9	0	1	1	.588	.577	.417	.450	.589
	10	9	9	2	1	0	.504	.485	.348	.180	.505

For the present data $\hat{\sigma}_{\hat{R}_s}$ is equal to $(.9)^2(1/90) = .009$. The use of asymptotic normal theory gives, for $\alpha = .10, R_s \geqslant (.9 - 1.282\sqrt{.009}) = .887$. This is considerably larger than the optimum nonrandomized 90% lower confidence bound of .603, which was calculated by means of Buehler's method by Mann and Fertig (1972), using the ordering of failure combinations $(0,0), (0,1)$, and independently by M. Lipow.

To use Easterling's modification we first set $\hat{\sigma}_{\hat{R}}^2$ equal to $\hat{R}_s(1 - \hat{R}_s)/\hat{n}$ and solve for $\hat{n}$. This gives $\hat{n} = .9(.1)/.009 = 10$ and thus a psuedo sample size of $[\hat{n}] + 1 = 10$, with $[s]$ the greatest integer less than s. Then $\hat{n} - \hat{x} = \hat{R}(10) = 9$, and the pseudo number of survivors is $[\hat{n} - \hat{x}] + 1 = 9$. The lower confidence bound obtained for R_s by this method is then the 10th percentile of a beta distribution with parameters $10 - 1 = 9$ and $1 + 1 = 2$, or .663, a value that is much closer to the optimum nonrandomized confidence bound of .603 than the one (.887) given by the unmodified maximum-likelihood method.

The gamma approximation of Woods and Borsting involves a modification of $\hat{R}_s$, the maximum-likelihood and simulation estimator of R_s. To calculate their lower confidence bound, we first evaluate

$$\sum_{j=1}^{k} \hat{T}_j \quad \text{and} \quad \sum_{j=1}^{k} \left(\frac{\hat{T}_j}{n_j} \right), \quad \text{with} \quad \hat{T}_j = \frac{a_j x_j}{n_j} + \frac{b_j (x_j/n_j)^2}{2},$$

$$a_j = \frac{(2n_j - 3)/(n_j - 1)}{2} \quad \text{and} \quad b_j = \frac{n_j}{n_j - 1}, \quad j = 1, \dots, k.$$

This gives

$$\sum_{j=1}^{k} \hat{T}_j = (.944)(\tfrac{1}{10}) + \frac{1.111(\tfrac{1}{10})^2}{2} = .10 \quad \text{and} \quad \frac{\sum_{j=1}^{k} \hat{T}_j}{n_j} = .01.$$

Then $\hat{S} = \sum_{j=1}^{k} \hat{T}_j$ is an approximately unbiased estimator of $-\ln R_s$ with variance estimate $\sum_{j=1}^{k} \hat{T}_j / n_j$.

By an approach similar to that of Easterling an estimate $\hat{r}$ is made of the parameter for the approximate gamma variate, $\hat{r}\hat{S}/(-\ln R)$ (that is, $2r\hat{S}/(-\ln R)$ is approximately chi square with $2r$ degrees of freedom). The estimate is $\hat{r} = \hat{S}^2 / \sum_{j=1}^{k} \hat{T}_j / n_j$, which, in this case, is equal to 1. (If $\hat{r}$ is moderately large and not an integer, $2\hat{r}$ is rounded.) This gives a 90% lower confidence bound of $.2/.211 = .947$ for $\ln R_s$ and a lower confidence bound of .386 for R_s. If a degree of freedom is added, the 90% lower confidence bound on R_s is .513, a value somewhat closer to the optimum non-randomized bound of .603. The confidence bound calculated, however, is also the one that would be calculated for a single subsystem with $n = 10, x = 1$. In this case the confidence bounds calculated on the basis of either 2 or 3 degrees of freedom are both considerably smaller than the optimum nonrandomized lower 90% confidence bound of .663. This approach apparently does not perform well unless sample sizes are rather large.

Although no illustrations will be given here, the problem that no account is taken of zero failures, exhibited in all of the above methods, also applies to the confidence bounds of Murchland and Weber and to those based on the likelihood ratio.

Small-sample approximate confidence bounds on R_s for an independent series system and binomial data have been derived by Garner and Vail (1961), Connor and Wells (1962), Abraham (1962), and Lindstrom and Madden (see Lloyd and Lipow, 1962). The first two of these approaches use various methods of combining confidence bounds on subsystem reliability to obtain the desired bounds on system reliability. The other two use binomial or Poisson approximations for certain statistics. Some of these methods are sensitive to inequality of sample sizes for subsystems. Lower confidence bounds obtained by most of these approximate methods have been compared, on the basis of three sets of data, by Schick and Prior (1966) with optimum nonrandomized bounds determined by using results of Lipow and Riley (1959). The data apply to series systems composed of two subsystems, and in all of the three cases the sample sizes are equal. Only the Lindstrom and Madden values compare at all favorably with the Lipow and Riley (1959) bounds. The Lindstrom and Madden method, however, yields bounds that tend to be too low except in special cases.

Bayesian bounds on system reliability can also be obtained by choosing, for a system composed of k components, a prior density function for the jth component reliability, $R_j, j = 1, \ldots, k$, and then simulating (or obtaining through a Mellin-transform technique) the posterior distribution of system reliability, given the component failure data. Zimmer, Prairie, and Breipohl (1965) and Springer and Thompson (1968) have suggested the use of prior densities on component reliabilities that are uniform on $(0, 1)$. Mastran (1968) and Parker (1972) prefer to assign prior densities in such a way that the prior density for *system* reliability is uniform.

Results of Raiffa and Schlaifer (1961) suggest that any prior density approximating the one yielding an optimum confidence bound should be *improper* (i.e., be such that its integral cannot be made equal to 1 by affixing a constant factor), as is the prior density function $p(R) = R^{-1}$ [or, equivalently, $p(-\ln R)$ uniform on the positive half real line], associated with the optimum nonrandomized binomial confidence bound for a single component. Results of Fertig (1972) and Mann (1970) relating to series systems with exponential subsystem data obtained from Type II censoring corroborate this conjecture. Their results show further that for the exponential model the prior densities on component reliabilities (which we shall refer to henceforth as *posterior* prior densities) associated with the optimum classical lower confidence bound on series-system reliability contain component failure data. The approximately optimum lower confidence bounds on R_s, based on the conditional mean m and variance v of system hazard rate and given by formulas (10.2) and (10.3), implicitly depend on an assumption of a posterior prior density involving $z_{(1)}$ for λ_j, $j = 1, \ldots, k$, where $z_{(1)}$ is the smallest of the z's. Finally, results of Mann (1974a) indicate that both randomized and nonrandomized optimum confidence bounds for the reliability of series and parallel systems with binomial subsystem data are associated with prior densities containing present failure data.

Use of component-reliability posterior prior density functions that are not approximately optimum can yield Monte Carlo lower confidence bounds having a high probability of being inordinately far from the true system reliability. For example, if for $k>1$ a prior density for R_j of $p(R_j)=R_j^{-1}, j=1,\ldots,k$, the optimum (nonrandomized) prior density for $k=1$, is used, the lower confidence bounds obtained (referred to henceforth as *fiducial* bounds) for system reliability R_s are uniformly lower than the optimum nonrandomized lower confidence bounds. For a parallel system the difference between the two types of confidence bounds is not large unless sample sizes $n_j, j=1,\ldots,k$, are very small. However, for series systems large discrepancies can result. For example, for a series system with $n_1=10$, $n_2=9$, $n_3=9$, $x_1=2$, $x_2=1$, and $x_3=0$, where the x's are numbers of failures, an optimum nonrandomized lower confidence bound of .504 and a fiducial bound of .348 are obtained. These results are shown by Mann (1974a), and other comparisons with optimum nonrandomized confidence bounds are presented in Tables 10.3 and 10.4.

Using a bound based on uniform prior densities for the component reliabilities is equivalent to using the fiducial bound with $n_j, j=1,\ldots,k$, each increased by 1. Thus, for the data set above, the bound based on uniform prior densities for component reliabilities is .387. Bounds based on a uniform prior density for *system* reliability have been calculated by Mastran for the data sets in Table 10.3. They appear to improve as k increases with all sample sizes nearly equal.

Harris (1971) has derived a procedure for obtaining nonrandomized and optimum randomized confidence bounds on products and quotients of Poisson parameters. Since for large sample sizes and small probabilities the Poisson distribution can serve to approximate binomial distributions, Har-

Table 10.4 Nonrandomized 90% lower confidence bounds on parallel-system reliability

$k=2$:	n_1	n_2		x_1	x_2		Optimum	AO	Fiducial
	10	10		0	1		.968	.968	.956
	2	2		0	1		.54	.54	.492
	10	10		0	0		.988	.987	.976
	50	50		0	1		.9985	.9983	.9978
	20	20		0	1		.991	.990	.987
$k=3$:	n_1	n_2	n_3	x_1	x_2	x_3			
	5	5	5	0	0	0	.997	.996	.987
	10	10	10	0	0	0	.999	.998	.998

ris has suggested using his randomized results to approximate *parallel-system* reliability confidence bounds under appropriate conditions. The appropriate conditions are, of course, large sample sizes and high reliabilities for all components.

Harris compares numerically, for a selection of Poisson failure data applying to parallel systems composed of two subsystems, his Poisson lower bounds and the Poisson approximation of Buehler (1957) to the optimum nonrandomized lower confidence bounds on R_p. Harris' randomized lower confidence bounds on R_p are optimum (uniformly most accurate unbiased) for Poisson data, but his nonrandomized lower confidence bounds tend to be extremely conservative.

The method of Harris is rather difficult to implement, since it involves calculation of the percentiles of the modified Bessel function. Also, unfortunately, there is no procedure by which Harris' bounds can be made to apply to at least moderately reliable series systems with binomial subsystem data.

Mann (1974a) has used the approach of Mann and Grubbs (1972) for exponential subsystem failure data to approximate both optimum randomized and optimum nonrandomized lower confidence bounds on series-and parallel-system reliability for systems with binomial subsystem data. For sample sizes large and numbers of failures small, Mann's approximately optimum randomized lower confidence bounds on parallel-system reliability R_p agree very well with the optimum randomized Poisson bounds of Harris (see Table 10.5). To obtain the AO randomized lower confidence bounds on R_p, one first calculates

$$m_p(\delta) = \sum_{j=1}^{k} \sum_{i=x_j+1}^{n_j} \left(\frac{1}{i}\right) + \frac{.5}{x_{(1)}+\frac{1}{2}} - .5 \sum_{j=1}^{k} \frac{\delta}{x_j+\frac{1}{2}} \qquad (10.5)$$

and

$$v_p(\delta) = \sum_{j=1}^{k} \sum_{i=x_j+1}^{n_j} \left(\frac{1}{i}\right)^2 + \frac{.5}{\left(x_{(1)}+\frac{1}{2}\right)^2} - .5 \sum_{j=1}^{k} \frac{\delta}{\left(x_j+\frac{1}{2}\right)^2} \qquad (10.6)$$

with $x_{(1)}$ equal to the smallest of the x's and δ equal to both $+1$ and -1. Then $m_p' = um_p(1) + (1-u)m_p(-1)$ and $v_p' = uv_p(1) + (1-u)v_p(-1)$, where u is a realization of a random variate U uniform on $(0,1)$. Use of m_p' and v_p' with the Wilson-Hilferty (1931) transformation for the approximate noncentral chi-square distribution of $-\ln(1-R_p)$ gives (see Equation 10.4)

$$\text{Prob}\left(1 - R_p \leqslant \exp\left\{-m_p'\left[1 - \frac{v_p'}{(3m_p')^2} + \frac{\eta_\alpha \sqrt{v_p'}}{3m_p'}\right]^3\right\}\right) \cong 1-\alpha, \qquad (10.7)$$

with η_γ the 100γth percentile of a standard normal distribution.

A similar approximation to the optimum exact lower confidence bounds on R_s is obtained by calculating $m_s(\delta)$ and $v_s(\delta)$ [$m_p(\delta)$ and $v_p(\delta)$, respectively, with x_j replaced by $n_j - x_j$ in the limits of the sums and $.5/(x_{(1)} + \frac{1}{2})^s$ and $.5\delta/(x_j + \frac{1}{2})^s$ replaced by $.5/(n - x)_{(1)}^s$, and $5\delta/(n_j - x_j)^s$ where $s = 1, 2$ and $(n - x)_{(1)}$ is the smallest value of $n_j - x_j, j = 1, \ldots, k$]. The next step is to substitute into (10.7) R_s for $1 - R_p$, $\geqslant$ for $\leqslant$, $\eta_{1-\alpha}$ for η_α, $m_s' = um_s(-1) + (1 - u)m_s(1)$ for m_p' and $v_s' = uv_s(-1) + (1 - u)v_s(1)$ for v_p' to obtain the AO randomized (approximately exact) lower confidence bound on R_s. These are approximate generalizations of the exact expressions for the randomized moments of $-\ln(1 - R)$ when $k = 1$, that is, the moments corresponding to the prior density function on R yielding optimum randomized confidence bounds.

Expressions (10.5) and (10.6) are approximations for differences of polygamma functions $\Psi(\cdot)$ and $\Psi'(\cdot)$, the first and second derivatives respectively, of the logarithm of the gamma function. An example of calculation of an AO lower confidence bound on R_p by use of Equations (10.5) and (10.6) is given after the discussion and examples of the method for obtaining AO nonrandomized confidence bounds.

For independent parallel systems, the expressions for the posterior mean m_p and variance v_p of $-\ln(1 - R_p)$ associated with a prior density for $R_j, j = 1, \ldots, k$, yielding approximately optimum (A0) *nonrandomized* lower confidence bounds are given as follows when $x_{(1)}$, the smallest $x_j, j = 1, \ldots, k$, is greater than 0:

$$m_p = -\ln\left[\frac{\hat{P}(1 + .5/x^*)}{1 + .5\hat{P}/x^*}\right] \qquad (10.8)$$

and

$$v_p = \frac{1 - \exp(-m_p)}{x^* + .5}, \qquad (10.9)$$

where $\hat{P} = \prod_{j=1}^{k}(1 - \hat{R}_j)$ is the maximum-likelihood estimate of the probability of parallel-system failure, and $x^* = 1/\sum_{j=1}^{k}(1/x_j)$. Any subsystem for which $x_j = n_j$ is ignored in calculating x^*.

If $\hat{P} = x_{(1)} = 0$,

$$v_p = \psi'(x_{(1)} + 1) - \frac{1}{n'} = \psi'(1) - \frac{1}{n'} = 1.6449 - \frac{1}{n'},$$

Table 10.5 90% lower confidence bounds on parallel-system reliability, large sample sizes

Number of Failures			Buehler's Poisson Approximation	Harris' Nonrandomized Confidence Bound (Poisson Approx.)	Harris' "Randomized" Confidence Bound	AO Nonrandomized Confidence Bound ($m_1 = m_2 = m_3 = 100$)	AO "Randomized" Confidence Bound ($m_1 = m_2 = m_3 = 100$)
x_1	x_2						
3	5		$1 - 41.2/n_1 n_2$	$1 - 48.6/n_1 n_2$	$1 - 41.6/n_1 n_2$	$1 - 42.0/n^2$	$1 - 41.7/n^2$
1	4		$1 - 18.8/n_1 n_2$	$1 - 23.5/n_1 n_2$	$1 - 18.4/n_1 n_2$	$1 - 20.0/n^2$	$1 - 19.4/n^2$
2	2		$1 - 16.8/n_1 n_2$	$1 - 21.1/n_1 n_2$	$1 - 17.0/n_1 n_2$	$1 - 16.9/n^2$	$1 - 16.4/n^2$
0	0		$1 - 1.33/n_1 n_2$	$1 - 3.72/n_1 n_2$	$1 - 2.29/n_1 n_2$	$1 - 1.32/n^2$	$1 - 2.32/n^2$
x_1	x_2	x_3					
1	2	1	$1 - 19.0/n_1 n_2 n_3$ [a]	$1 - 40.0/n_1 n_2 n_3$	$1 - 27.0/n_1 n_2 n_3$	$1 - 20.7/n^3$	$1 - 27.2/n^3$
2	3	5	$1 - 133/n_1 n_2 n_3$ [a]	$1 - 186/n_1 n_2 n_3$	$1 - 145/n_1 n_2 n_3$	$1 - 129/n^3$	$1 - 146/n^3$

[a] Likelihood-ratio confidence bound substitubed, since confidence bounds based on Buehler's Poisson approximation are unavailable for $k > 2$.

where

$$n' = l \left(\frac{\prod\limits_{j=1}^{k} n_j}{\prod\limits_{x_j \neq 0} x_j} \right)^{1/l}$$

with l the number of x_j's equal to 0, and $\psi'(z)$ is the trigamma function with $\psi'(z) \cong 1/z + .5/z^2 + .1\overline{6}/z^3 - .03\overline{3}/z^5 + .0238/z^7$. In this case,

$$m_p = \ln \left[2l \left(\frac{\prod\limits_{j=1}^{k} n_j}{\prod\limits_{x_j \neq 0} x_j} \right)^{1/l} \right],$$

but the upper confidence bound obtained for $1 - R_p$ is raised to the lth power.

An upper confidence bound on $1 - R_p$ is obtained for $\hat{P} > 0$ by making use of the fact that $-2m_p \ln(1 - R_p)/v_p$ has an approximate chi-square posterior distribution with $2m_p^2/v_p$ degrees of freedom, that is, $-2\ln(1 - R_p)$ is approximately a weighted chi-square. The Wilson-Hilferty transformation of chi-square to normality can then be used by substituting values for m_p and v_p in place of those for m_p' and v_p', respectively, into (10.7). When $\hat{P}$ is equal to 0, the AO upper confidence bound for $1 - R_p$ obtained by substitution of calculated values of m_p and v_p into (10.7) must be raised to the lth power before subtracting from 1 to obtain the corresponding lower confidence bound on R_p. Numerical examples of the calculation of AO lower confidence bounds on R_p are given after the discussion of the method for obtaining AO confidence bounds on series-system reliability, and the expressions for m_p and v_p are derived in Section 10.4.2.

For independent series systems the approximate expressions for m_s and v_s, the posterior mean and variance of $-\ln R_s$, yielding the AO confidence bounds for at least moderately reliable subsystems are

$$m_s = \frac{.5(1 + 1/a)}{n^0} + \frac{\hat{P}}{1 - .5\hat{P}} \qquad (10.10)$$

and

$$v_s = \frac{.5(1 + 1/a)m_s}{n^0}, \qquad (10.11)$$

where $n^0 = n_{(1)}(1 - .5\hat{P}^2)(1 - .5\hat{P})$, with $n_{(1)}$ the smallest of $n_1,\ldots,n_k$, $\hat{P} = 1 - \Pi\hat{R}_j$, the maximum-likelihood estimate of the probability of series-system failure, and $a = n_{(1)}\Sigma_j(1/n_j)$. For use with less reliable systems, more generally applicable expressions for m_s and v_s are given by Equations (10.13), (10.14), and (10.15).

As in calculating a confidence bound by the method of Buehler (1957), if a lower confidence bound on R_s obtained when a subsystem is included is larger than the value found when the system is excluded, the latter bound is the correct one to use. For this reason all zero-failure subsystems are ignored in calculating the value of $a = n_{(1)}\Sigma_j(1/n_j)$ except for any *single* zero-failure subsystem with its sample size equal to $n_{(1)}$. It, too, is ignored, however, if at least one other subsystem that exhibits failure has sample size $n_{(1)}$ or if the lower confidence bound obtained by not ignoring it is larger than that found when it is ignored. The exclusion of some subsystems in calculating series-system confidence bounds also is necessitated when all sample sizes are large and most of the failures pertain to a single subsystem. In this case the bound obtained on the basis of all subsystem failures can be larger than that based on the single subsystem exhibiting nearly all the failures.

The expressions for m_s and v_s are in agreement with results (pointed out by Winterbottom, 1973) of Bol'shev and Loginov (1966), who show that optimum nonrandomized confidence bounds for given sets $\{n_j\}$ and $\{n_j - x_j\}$ are functions of $n_{(1)}$, $\hat{P} = 1 - \Pi_{j=1}^{k}(n_j - x_j)/\Pi_{j=1}^{k}(n_j)$, and the various n_j's. The expressions for m_s and v_s are derived in Section 10.4.2.

To obtain the AO nonrandomized lower confidence bound on R_s by fitting the posterior distribution of $-\ln R_s$ with a noncentral chi-square distribution and then using the Wilson-Hilferty transformation of chi-square to normality, we substitute appropriate values of m_s and v_s in

$$\text{Prob}\left(R_s \geqslant \exp\left\{-m_s\left[1 - \frac{v_s}{(3m_s)^2} + \frac{\eta_{1-\alpha}\sqrt{v_s}}{3m_s}\right]^3\right\}\right) = 1 - \alpha, \quad (10.12)$$

where η_γ is the 100γth percentile of a standard normal variate.

The AO randomized and nonrandomized lower confidence bounds on R_s and R_p are all based on the result, demonstrated by Mann (1974a), that $-\ln R_s$ and $-\ln(1 - R_p)$ have posterior distributions corresponding to those of sums of weighted chi-square variates. Thus, $-\ln R_s$ and $-\ln(1-R_p)$, like the system hazard rate ϕ of Section 10.4.1(a), have posterior distributions corresponding to weighted chi-square distributions. Since this result also holds for $k = 1$, it follows that $-\ln R$ and $-\ln(1 - R)$, where R

and $1 - R$ are beta variates, are both approximate weighted chi squares. This result is used in Chapter 5 to approximate distribution percentiles of beta variates with noninteger degrees of freedom.

Table 10.2 gives, for a series system with failures occurring in all subsystems, comparisons of the AO confidence bounds with the optimum nonrandomized confidence bounds calculated by Lipow and Riley (1959), those based on the likelihood-ratio (LR) and maximum-likelihood (ML) approximations, calculated by Myhre and Saunders (1968a), and those based on the modified maximum-likelihood (MMLI) approximation of Easterling (1971). Table 10.3 compares the optimum nonrandomized lower confidence bounds, AO nonrandomized confidence bounds, fiducial bounds, bounds calculated by Mastran on the basis of a uniform prior density for R_s, and bounds based on the gamma approximation of Woods and Borsting (1968). Table 10.4 applies to parallel systems and compares optimum nonrandomized and AO nonrandomized lower confidence bounds with fiducial bounds on R_p.

Table 10.5 applies to parallel systems, large sample sizes, and small numbers of failures, and gives the optimum nonrandomized Poisson lower confidence bounds on R_p of Buehler (1957); the optimum "randomized" Poisson bounds calculated by Harris (1971) from his formulas by using the expected value, .5, of a random variate uniform on $(0, 1)$; and the non-randomized version of Harris' lower confidence bounds on R_p, the bottom of the range of possible lower confidence bounds to be obtained by randomizing with his method. Also shown in Table 10.5 are the AO lower confidence bounds of Mann (1974a) corresponding to the optimum non-randomized lower confidence bounds on R_p and the optimum "randomized" lower confidence bound (calculated by setting the uniform random variate U equal to 0.5 in evaluating m_p' and v_p'). Where Buehler's Poisson approximation is inapplicable $(k > 2)$, confidence bounds based on the likelihood ratio and calculated by Harris (1971) are presented.

We give now examples of calculations of AO lower confidence bounds and comparisons with lower confidence bounds on R_s and R_p that are known to be optimum.

First, consider a data set given in Table 10.2, $k = 3, n_1 = n_2 = n_3 = 20, x_1 = x_2 = x_3 = 1$. To obtain a lower confidence bound $\underline{R}_s$ on R_s, one calculates $\hat{P} = 1 - (\frac{19}{20})^3 \cong .1426$, $a = 20(\frac{3}{20}) = 3$, and $n^0 = 20(1 - .0713)(1 - .01) \cong 18.388$, so that, from (10.10) and (10.11),

$$m_s \cong \frac{.666}{18.388} + \frac{.1426}{.9287} \cong .1898, \qquad v_s \cong .00688, \qquad \text{and} \qquad \frac{m_s^2}{v_s} \cong 5.234.$$

Then, at confidence level .95, use of the Wilson-Hilferty transformation yields $\underline{R}_s = \exp\{-.1898[1 - .0212 + 1.645(.146)]^3\} \cong .708$. The optimum

95% nonrandomized confidence bound given in Table 10.2 (from calculations of Lipow and Riley, 1959) is .709.

Next, consider two series systems with unequal sample sizes for subsystems. The first data set is one considered earlier for illustration of various asymptotic methods based on $\hat{P}$: $n_1 = 10, n_2 = 5, x_1 = 1, x_2 = 0$. Here $\hat{P} = .1$, $a = 5(\frac{1}{5} + 1/10) = 1.5$, and $n^0 = 5(.95)(.995) \cong 4.726$, so that

$$m_s \cong \frac{.8333}{4.726} + \frac{.10}{.95} = .2812 \qquad \text{and} \qquad m_s^2/v_s \cong 1.594.$$

Then the 90% AO lower confidence bound on R_s is, from (10.12), exp $\{-.2812[1 - .07 + 1.282(.264)]^3\}$, or .571. The optimum value is .603, calculated by Mann and Fertig (1972) and by M. Lipow, using the exact method of Buehler (1957), based on the ordering $(0,0)$, $(1,0)$ of failure combinations when summing over these combinations to iteratively compute the value of the bound. Note that .571 is less than .663, the 90% lower confidence bound on R associated with $k = 1, n = 10, x = 1$. Note, too, that the AO confidence bound is somewhat conservative for n_2 as small as 5.

Finally, we calculate a 90% lower confidence bound for $k = 3, n_1 = n_2 = 10, n_3 = 9, x_1 = 0, x_2 = x_3 = 1$. Since $x_1 = 0$ and $n_1 = n_2$, the first subsystem can be ignored. (It can easily be demonstrated that, if the first subsystem is included, both m_s and v_s will be smaller, with $\hat{P}$ unchanged, and that the lower confidence bound will be larger than if the subsystem is ignored.) Then $\hat{P} = .2$, $n^0 = 9(.9)(.98) = 7.938$, and $a = 9(\frac{1}{9} + \frac{1}{10}) = 1.9$. Therefore

$$m_s \cong \frac{.7631}{7.938} + \frac{.2}{.9} = .3184 \qquad \text{and} \qquad m_s^2/v_s \cong 3.311,$$

so that the AO lower confidence bound on R_s is, from (10.12), exp $\{-.3184[1 - .0336 + 1.284(.183)]^3\} = .577$. The optimum confidence bound calculated by Mann and Fertig (1972), using Buehler's exact method and the ordering of failure combinations (for the second and third subsystems) $(0,0)$, $(1,0)$, $(0,1)$, $(1,1)$, which gives the largest bound for the combination brought in after $(0,1)$, is .588. The next failure combination brought in, $(2,0)$ or $x_2 = 2, x_3 = 0$, yields lower 90% confidence bounds by both the exact method and the AO method that are larger than the optimum bound for $k = 1, n = 10, x = 2$, namely, .550. The latter 90% confidence bound (.550) is therefore the correct one to use for $n_1 = n_2 = 10, n_3 = 9, x_1 = x_3 = 0, x_2 = 2$.

To obtain an AO upper confidence bound on $1 - R_p$ for $k = 2, n_1 = 50, n_2 = 200, x_1 = x_2 = 3$, we calculate, using (10.8) and 10.9), $x^* = 1/\frac{2}{3} = 1.5$, $\hat{P} = .0009, m_p \cong -\ln[.0009(1 + .\frac{1}{3})] = 6.72$. Then, since $x^* \cong 1.5$ and $v_p \cong 1.0/2.0 = .50$, with probability approximately .95, $1 - R_p > \exp\{-6.72[1 - .0012 - 1.645(.033)]^3\} \cong .0035$. This agrees extremely well with .00366, the Poisson approximation of Buehler (1957) obtained by optimal ordering of the failure combinations. Buehler's approximation, unfortunately, is available only for $k=2$.

Next are two examples of confidence bounds for parallel-system reliability, which can be compared with optimum lower confidence bounds calculated by Steck (1957). First consider the data set $n_1 = n_2 = 10, x_1 = x_2 = 0$. Here $\hat{P} = 0$ and $l = 2$, so that $m_p = \ln(40) = 3.688$, $v_p = 1.644 - .05 = 1.594$, and the 90% upper AO confidence bound on $1 - R_p$ is $(\exp\{-3.688[1 - .0130 - 1.282(.114)]^3\})^2 = .0126$. The corresponding lower 90% confidence bound on R_p is .987, while the optimum confidence bound calculated by Steck is .988.

For $n_1 = n_2 = 2, x_1 = 0, x_2 = 1$, the optimum lower confidence bound determined by Steck is .54. The posterior mean of $-\ln(1 - R_p)$ is $\ln(8) = 2.0796$, and the posterior variance is $1.644 - .25$. Thus the calculated 90% upper confidence bound for $1 - R_p$ is $(.46)^1 = .46$, and the lower confidence bound on R_p agrees with Steck's value.

As a final example consider a parallel system composed of three independent subsystems with $n_1 = 100, n_2 = 100, n_3 = 100, x_1 = 2, x_2 = 3, x_3 = 5$. Given in Table 10.4 is the 90% lower confidence bound on $R_p, (1 - 133/100^3)$, computed iteratively by Harris (1971), using the likelihood-ratio approach (which gives very good asymptotic results when there are no zero-failure subsystems). To determine the AO lower confidence bound one calculates from (10.5), (10.6), and (10.7), $\hat{P} = .000030$, $x^* \cong 1/1.03 \cong 1.0$, $m_p \cong -\ln[(.000030)(1 + .5/1.0)] \cong 10.00$, and $v_p \cong .644$. Then the 90% upper confidence bound for $1 - R_p$ is $\exp\{-10.00[1 - .000724 - 1.282(.027)]^3\} = 129/100^3$, a result very close to the likelihood-ratio value.

Before discussing some more general models, we give an example of the calculation of an AO "randomized" confidence bound for $1 - R_p$. Let us consider the data set $n_1 = n_2 = 100, x_1 = x_2 = 0$. From (10.5), (10.6), and (10.7), with $\delta = 0$, we obtain

$$m_p(0) = \sum_{j=1}^{2} \sum_{i=1}^{100} \frac{1}{i} + \frac{.5}{.5} \cong 2[\ln(100.5) - (-.577)] + 1 \cong 11.37$$

and

$$v_p(0) = \sum_{j=1}^{2} \sum_{i=1}^{100} \frac{1}{i^2} + \frac{.5}{.5^2} \cong 2\left(\frac{-1}{100.5} + 1.6449\right) + 2 \cong 5.28.$$

More accurate evaluations of $m_p(0)$ and $v_p(0)$ are given by

$$m_p(0) = 2\psi(101) - (\tfrac{1}{2})2\psi(1 - \tfrac{1}{2}) \cong 4.60 + 1.93 \cong 11.18$$

and

$$v_p(0) = -2\psi'(101) + (\tfrac{1}{2})2\psi'(\tfrac{1}{2}) \cong -.02 + 4.93 \cong 4.91,$$

with $2 = l$ and $\tfrac{1}{2} = 1/k$ occurring in both places because $x_1 = x_2 = 0$. We have used here tabulated values of Davis (1964) for $\psi(\cdot)$ and $\psi'(\cdot)$, the approximations $\psi(z) \cong \ln(z - .5)$ and $\psi'(z) \cong 1/(z - .5)$, and the recursive relationships $\psi(z) = \psi(z + 1) - 1/z$ and $\psi'(z) = \psi'(z + 1) + 1/z$.

The number of degrees of freedom, $2m_p(0)^2 / v_p$, is sufficiently large so that the chi-square of $-2m_p(0)\ln(1 - R_p)/v_p(0)$ has converged to a normal variate. Thus the upper 90% AO "randomized" confidence bound for $1 - R_p$ is approximately $\exp(-11.18 + 1.282\sqrt{4.91})$, or $\exp(-8.37) = 2.32/100^2$. The optimum "randomized" value obtained by Harris (1971), using his iterative method, is $2.28/100^2$. To obtain the AO approximation to Harris' (conservative) nonrandomized version of the optimum randomized confidence bound for this model, one may simply use

$$m_p'(1) = m_p(0) - 2\left(\frac{.5}{.5}\right) = 9.16 \qquad \text{and} \qquad v_p'(1) = v_p(0) - 2\left(\frac{.5}{.5^2}\right) = .91.$$

A more accurate evaluation yields

$$m_p(1) \cong 2\psi(101) - (\tfrac{1}{2})2\psi(1\tfrac{1}{2}) \cong 9.12$$

and

$$v_p(1) \cong -2\psi'(101) + (\tfrac{1}{2})2\psi'(1\tfrac{1}{2}) \cong .914.$$

These values give, as the upper confidence bound on R_p, $\exp(-9.12 + 1.282\sqrt{.914}) = 3.73/100^2$. The "nonrandomized" confidence bound calculated by Harris is $3.72/100^2$.

(c) More General Exponential and Binomial Models. In Section 10.4.1(a), methods of obtaining confidence bounds on the reliability of series systems were considered for the situation in which Type II censored exponential failure data are collected for each subsystem. In the following, a procedure due to Mann and Grubbs (1974) for obtaining confidence bounds on parallel-system reliability from any specified set of such subsystem failure data is discussed. Subsequently, exponential and binomial models applying to fixed test times that may not be equal to a specified mission time are presented.

CONFIDENCE BOUNDS FOR PARALLEL-SYSTEM RELIABILITY: TYPE II CENSORING. For the model to be considered first, for the jth subsystem, $j = 1, \ldots, k$, the number of failures r_j is fixed, and the weighted sum,

$z_j = \sum_{j=1}^{r_j} t_{ij} + (n-r)t_{r_j}$, of ordered failure times is observed. It is known that the reliability $R_j(t_m)$ of the jth subsystem at time t_m is such that the posterior distribution of $-\ln R_j(t_m) = t_m\lambda_j$ is a weighted chi-square with mean $m_j(t_m) = t_m r_j/z_j$ and variance $v_j(t_m) = t_m^2 r_j/z_j^2, j = 1,\ldots,k$ (see Grubbs, 1971).

Suppose that we fit the posterior distribution of $R_j(t_m)$ corresponding to the weighted chi-square posterior distribution for $-\ln R_j(t_m)$, given r_j and z_j, with a beta distribution with parameters $n_j^* - x_j^* + 1$ and $x_j^* + 1, j = 1,\ldots,$ k, where neither n_j^* nor x_j^* is necessarily an integer. Then, as in the fiducial model with $k = 1$ and n_j and $n_j - x_j$ increased by 1,

$$m = m_j(t_m) = \psi(n_j^* + 2) - \psi(n_j^* - x_j^* + 1) \cong \ln(n_j^* + 1) - \ln(n_j^* - x_j^*)$$

and

$$v = v_j(t_m) = \psi'(n_j^* + 2) + \psi'(n_j^* - x_j^* + 1) \cong -\frac{1}{n_j^* + 1} + \frac{1}{n_j^* - x_j^*}$$

(This approximation for the differences of digamma and trigamma functions works well for x_j^* small relative to $n_j^* + 1 \geqslant 3$.)

Therefore

$$n_j^* + 1 \cong \frac{e^m - 1}{v} \cong \frac{m}{v} = \frac{z_j}{t_m}$$

and

$$n_j^* - x_j^* \cong \frac{(e^m - 1)e^{-m}}{v} \cong \frac{m}{v} - \frac{m^2}{v} = \frac{z_j}{t_m} - r_j,$$

so that $x_j^* \cong r_j - 1$ when $m = t_m r_j/z_j$ is small, $j = 1,\ldots,k$. It can be demonstrated that this approach is applicable as long as each $t_m x_j^*/z_j$ is less than about .2 and each $z_j/t_m \geqslant 3$. Now note that, from (10.2) and (10.3), the mean m_s^z and variance v_s^z of $-\ln R_s = t_m\phi = t_m\sum_{j=1}^k \lambda_j$, given r, z_1, and $\mathbf{u}$, are approximately

$$t_m\left[\sum_{j=1}^k \left(\frac{r_j}{z_j} - \frac{1}{z_j}\right) + \frac{1}{z_{(1)}}\right] \quad \text{and} \quad t_m^2\left[\sum_{j=1}^k \left(\frac{r_j}{z_j^2} - \frac{1}{z_j^2}\right) + \frac{1}{z_{(1)}^2}\right],$$

respectively, where $z_{(1)}$ is the smallest of the z_j's. Substituting the results above, one obtains

$$m_s^z \cong \sum_{j=1}^k \left[\psi(n_j^* + 2) - \psi(n_j^* - x^* + 1) - \frac{1}{n_j^* + 1}\right] + \frac{1}{n^*_{(1)} + 1}$$

and

$$v_s^z \cong \sum_{j=1}^{k} \left[-\psi'(n_j^*+2) + \psi'(n_j^*-x^*+1) - \frac{1}{(n_j^*+1)^2} \right] + \frac{1}{(n^*_{(1)}+1)^2}.$$

Thus

$$m_s^z \cong \sum_{j=1}^{k} \left[\psi(n_j^*+1) - \psi(n_j^*-x_j^*+1) \right] + \frac{1}{n^*_{(1)}+1}$$

and

$$v_s^z \cong \sum_{j=1}^{k} \left[-\psi'(n_j^*+1) + \psi'(n_j^*-x_j^*+1) \right] + \frac{1}{(n^*_{(1)}+1)^2},$$

where $n^*_{(1)}$ is the smallest of the n_j^*s, and we have used the recursive properties $\psi(y+1) = \psi(y) + 1/y$ and $\psi'(y+1) = \psi'(y) - 1/y^2$, given by Davis (1964).

As an example of this result, consider the failure data from Table 10.1 used to illustrate the various confidence-bound procedures described in Section 10.4.1(a). Here $r_1 = 2$, $r_2 = 3$, $r_3 = 4$, $z_1 = 23.059$, $z_2 = 32.504$, and $z_3 = 37.188$. We let t_m be 1. Then $n_j^* + 1 = z_j$, $j = 1, \ldots, 3$, and $x_j^* = r - 1$, $j = 1, \ldots, k$, so that

$$m_s^z = \psi(23.059) - \psi(22.059) + \psi(32.504)$$

$$- \psi(30.504) + \psi(37.188) - \psi(34.188) + \frac{1}{23.059}.$$

This is approximately equal to

$$\ln[(22.059)(31.504)(36.188)] - \ln[(21.059)(29.504)(33.188)] + \frac{1}{23.059}$$

$$= 10.126 - 9.933 + .043 = .235.$$

In like manner,

$$v_s^z \cong -\frac{1}{22.059} - \frac{1}{31.504} - \frac{1}{36.188} + \frac{1}{21.059} + \frac{1}{29.504} \frac{1}{33.504} + \left(\frac{1}{23.059} \right)^2$$

$$= .00868.$$

These values differ only slightly from $m = .229$ and $v = .00782$, calculated in Section 10.4.1(a) directly from the exponential data.

Suppose that we simply consider the x^*'s and n^*'s as ordinary binomial failure data, forgetting that the n^*'s are not integers. Then

$$\hat{P}_s = 1 - \prod_{j=1}^{k} \frac{(n_j^* - x_j^*)}{n_j^*} \cong .180 \quad \text{and} \quad n_{(1)} = 22.059,$$

so that from, (10.10) and (10.11),

$$m_s^z \cong \frac{.180}{.91} + \frac{.5(1 + 1/2.34)}{22.059(.95)(.984)} \cong .235$$

and

$$v_s^z \cong \frac{.235}{28.78} = .00816.$$

This direct binomial approximation yields results for the mean and variance of $-\ln R_s$ even closer than those above to the values based on the exponential approximation. Note that the value used for $\hat{P}_s$ is the binomial rather than the exponential maximum-likelihood estimate of series-system unreliability, $1 - R_s$.

For the given data, it seems appropriate to approximate lower confidence bounds on R_p by means of Equations (10.8) and (10.9), using $1 - \prod_{j=1}^{k} x_j^* / n_j^*$ for $\hat{R}_p$, the probability of parallel-system survival. Before this binomial approximation is employed for parallel-system reliability with exponential data, however, a check should be made to ensure that the exponential and binomial approximations give approximately the same values for m_s and v_s.

In Section 10.4.2 these results are applied to the problem of obtaining confidence bounds for logically complex systems. Now consideration is given to systems with unequal fixed test times.

SYSTEM RELIABILITY CONFIDENCE BOUNDS FOR UNEQUAL FIXED TEST TIMES. An approach similar to that used directly above can be employed to obtain confidence bounds for series and parallel systems with fixed-test-time subsystem data when the test times differ from each other and from the mission time t_m. We assume that for an exponential model $-\ln R(t_0)$ $= t_0 \lambda$ for each subsystem has mean $m(t_0)$ and variance $v(t_0)$, where t_0 is the subsystem test time. Note that the mean and variance of $-\ln R(t_m)$ are, respectively, $m(t_m) = (t_m/t_0)m(t_0)$ and $v(t_m) = (t_m/t_0)^2 v(t_0)$. Here we shall observe simply the number of failures r_j in time t_{0j} and the sample size n_j for the jth subsystem, rather than $\hat{\theta}_j = \sum_{i=1}^{r} t_{ij} + (n_j - r_j)t_{0j}$, noting from Section 5.1 that very little usable information is lost by assuming that the number of failures is binomial if $R(t_0) \geq .5$. As above, we wish to obtain the effective parameters, $n'_j - x'_j$ and $x'_j + 1$, of the posterior beta distribution of $R_j(t_m)$ when the number of failures is r_j and the sample size is n_j, that is, when both original parameters, $n_j - r_j$ and $r_j + 1$, are integers. Suppressing the subscript j, we obtain:

$$\psi(n'+1) + \psi'(n'-x') \cong \ln n' - \ln(n'-x'-1) \cong m(t_m),$$

$$-\psi'(n'+1) + \psi'(n'-x') \cong -\frac{1}{n'} + \frac{1}{n'-x'-1} \cong v(t_m).$$

Then

$$n' \cong \frac{e^{m(t_m)} - 1}{v(t_m)} \cong \frac{m(t_m)}{v(t_m)} = \frac{(t_0/t_m)m(t_0)}{v(t_0)}$$

and

$$n' - x' \cong \frac{(e^{m(t_m)} - 1)e^{-m(t_m)}}{v(t_m)} \cong \frac{(t_0/t_m)m(t_0)}{v(t_0)} - \frac{m(t_0)^2}{v(t_0)}.$$

Thus $n'_j \cong (t_{0j}/t_m)n_j$ and $x'_j \cong r_j$, $j = 1,\ldots,k$. (Note that here we have used $n' - x'$ rather than $n' - x' + 1$ as the number of successes because of the discreteness of the data.)

As in the Type II censoring model, the approximations for the differences of the digamma and trigamma functions and the approximations for the effective sample size and effective number of failures are most appropriate when the effective number of failures is small relative to the effective sample size—in this case, when $(t_m/t_0)m(t_0) = (t_m/t_0)[\psi(n+1) - \psi(n-r)]$ is small.

An alternative approach that leads to the same result is the following. If it is assumed that all of the r observed failures for any given subsystem take place in time t_0, the maximum-likelihood estimate for θ for that subsystem is simply nt_0/r and the maximum-likelihood estimate of t_m/θ is $t_m r/nt_0$, or $r/[n(t_0/t_m)]$. This approach also leads to the result that the effective number of failures x' is r and the effective sample size n' is $n(t_0/t_m)$.

Note also that (10.10), (10.11), and (10.12) together imply that an optimum nonrandomized confidence bound ρ on series-system reliability, based on n_j and x_j, $j = 1,\ldots,k$, will be equal to ρ^{t_m/t_0} if m_s is multiplied by t_m/t_0 with m_s^2/v_s held fixed. It will be shown later that effective system sample size is proportional to $n_{(1)}$, and therefore, for number of system failures fixed, multiplying $n_{(1)}$ by t_0/t_m is equivalent to multiplying $\hat{P}_s$ by t_m/t_0 and thus, for $\hat{P}_s/2$ small, to multiplying m_s by t_m/t_0 and v_s by $(t_m/t_0)^2$ if $a = n_{(1)}\sum_{j=1}^k 1/n_j$ is also held fixed. The bound obtained will therefore be essentially equal to that found by multiplying each n_j by t_0/t_m when $\hat{P}_s/2$ is small. This result applies, of course, only when the hazard rate is constant.

Let t_0/t_m be equal to 2, and consider, from Table 10.2, $n_1 = n_2 = n_3 = 50$, $x_1 = x_2 = 1$, $x_3 = 2$. The optimum nonrandomized 90% and 95% lower confidence bounds on R_s are .861 and .841, respectively with square roots .928 and .917. The corresponding confidence bounds for $n_1 = n_2 = n_3 = 100$ and $x_1 = x_2 = 1$, $x_3 = 2$ are .929 and .918, respectively. For smaller sample sizes the agreement is quite good, although not always to nearly three significant figures. For example, for $n_1 = n_2 = 10$, $x_1 = 1$, $x_1 = 2$ the 90% optimum nonrandomized lower confidence bound shown in Table 10.2 is .497 with square root .705, while the corresponding confidence bound for $n_1 = n_2 = 20$ is .716. For $n_1 = n_2 = n_3 = 20$, $x_1 = x_2 = x_3 = 1$, the 95% optimum nonrandomized lower confidence bound is .709, whereas for $n_1 = n_2 = n_3 = 30$ the corresponding confidence bound is .796. The value of $(.709)^{2/3}$ is .795. Finally, the 90% optimum nonrandomized confidence bound for $n_1 = n_2 = n_3 = 30$, $x_1 = x_2 = x_3 = 3$ is .705, and the two-thirds power of the corresponding confidence bound for $n_1 = n_2 = n_3 = 20$ is .707.

An example is now given of the calculation of an approximately optimum confidence bound for constant hazard function subsystem failure data for which the test times are different from each other. Consider the data set $k = 2$, $n_1 = 5$, $n_2 = 10$, $x_1 = 0$, $x_2 = 1$. We assume that t_{01}, corresponding to the first sample, is 385 hours, and t_{02}, corresponding to $n_2 = 10$, $x_2 = 1$ is equal to the mission time t_m of 300 hours. Then $n_2' = n_2 = 10$ and $n_1' = 385(5)/300 = 6.416$, so that our effective sample data set for $t_m = 300$ hours is $k = 2$, $n_1' = 6.416$, $n_2' = 10$, $x_1' = 0$, $x_2' = 1$. Then, for the system with these new test data, $n^0 = 6.416(.95)(.995) = 6.06$, $a = 1.6416$, and $\hat{R}_s = \frac{9}{10}$. Therefore, using (10.10) and (10.11), we compute

$$m_s(300) = \frac{.804}{6.06} + \frac{.1}{.95} = .237$$

and

$$v_s(300) = \frac{.237}{7.53} = .0314.$$

This gives a 90% AO nonrandomized lower confidence bound of .616 at $t_m = 300$ hours. For $t_m = 150$ hours the corresponding lower confidence bound on R_s is approximately $\sqrt{.616} = .796$. In these cases we must check that the confidence bounds obtained are not larger than those for $n = 10$, $x = 1$ or $n = 20$, $x = 1$. These two 90% lower confidence bounds on R_s are .663 and .819, respectively, so no modification is necessary.

10.4.2   Confidence Bounds for Logically Complex Coherent Systems

In this section, confidence bounds for the reliability of logically complex and coherent systems are treated. Coherent systems are defined here as defined by Birnbaum, Esary, and Saunders (1961): systems which, if they perform when a given set of components perform, also perform when a set of components containing the given set perform and which fail when all their components fail and perform when all their components perform.

This section, then, applies to systems whose reliability equations may be of the form, for example, $R_3 - R_3(1 - R_1)(1 - R_2)R_6 - R_6(1 - R_4)(1 - R_5)$ or very much more complicated, as long as all subsystems tested as a single unit are independent and at least the coherency requirement is satisfied.

Among the methods discussed earlier that can be applied to logically complex systems are those based on the asymptotic normality of maximum-likelihood estimators of system reliability, as well as Easterling's variation of the maximum-likelihood method, and the generalization of Myhre and Saunders (1968b, 1971) of Madansky's procedure based on the asymptotic distribution of the likelihood ratio. As stated earlier, all of these asymptotic methods tend to give inappropriate results for fixed-test-time life testing whenever there are subsystems that have been tested but have exhibited no failures. Also, asymptotic maximum-likelihood and likelihood-ratio methods, when applied to systems with exponential Type II censored failure data, are prone to yield confidence bounds that do not approximate optimum confidence bounds well for series systems. In addition, the likelihood-ratio method, requiring iterative procedures, is difficult to implement, even for simple series and parallel systems.

Saunders (1972) has generalized the method of Lieberman and Ross (1971) to logically complex systems. As in the application of this method to series systems, it is necessary to consider all possible combinations of failure times in order to avoid arbitrarily obtaining different results from the same data. Such a procedure can, of course, become rather complicated and will necessitate the use of a computer when the number of subsystems and the total number of failure times are at all large.

In the following we give an outline of the derivation, due to Mann and Fertig (1972), of a rather simple method for obtaining confidence bounds on the reliability of a logically complex system. The method can be applied to binomial and Type I censored exponential data and also to Type II censored exponential data when values of z/rt_m are sufficiently large. The problem of zero failures for some subsystems presents no difficulties, and the results approximate very well the optimum confidence bounds obtained by exact methods. The method, because it approximates optimum confidence bounds, is implicitly based on use of approximately optimum improper posterior prior densities for subsystem reliabilities, which contain subsystem test data. There is a requirement that the system be expressible as a series or parallel system composed of additional series or parallel systems.

The approach used will be to approximate the effective sample size n^* and failure number x^* and to make note of certain other values for the most basic subsystems composed of components in series or in parallel within the logically complex system. Once this is done, one can approximate effective sample sizes and failure numbers for the next most basic subsystems made up of the most basic subsystems arranged in either series or parallel, and so on. Although this procedure sounds rather complicated, determination of n^*, x^*, and other necessary values at each stage turns out to be quite simple. The approach consists, therefore, primarily of going through the logic of the system and making a few simple calculations at each stage.

(a) Determination of n^* and x^*. We consider the basic subsystems composed of components in series and give an outline of the derivation of Equations (10.10) and (10.11) for the posterior moments, m_s and v_s, of $-\ln R_s$. This derivation involves the determination of n^* for the series system as a function of $n_j, j=1,\ldots,k$, the component sample sizes. Here, as in Section 10.4.1(c), for Type II censored exponential data, x_j, is considered to be $r_j - 1$ and n_j takes the value $(z_j/t_m) - 1, j = 1,\ldots,k$, with z_j/t_m large relative to $r_j - 1$. We consider the series system to be a single component, so that $-\ln R_s$, with R_s the probability of series-system success, is assumed to have a beta distribution with parameters $n^* - x^*$ and $x^* + 1$. For a series system, this is in agreement with the method of Easterling (1972a) and assumes an improper system prior density R_s^{-1} for R_s. It can be shown (see Mann, 1974a) that m_s and v_s, the mean and variance, respectively, of $-\ln R_s$, are then given by

$$m_s = \psi(n^* + 1) - \psi(n^* - x^*) \tag{10.13}$$

and

$$v_s = -\psi'(n^* + 1) + \psi'(n^* - x^*) \tag{10.14}$$

where

$$\psi(z) \cong \ln(z - .5) \quad \text{and} \quad \psi'(z) \cong \frac{1}{z - .5}$$

are the first and second derivatives of $\ln \Gamma(z)$. Therefore

$$m_s \cong \ln(n^* + .5) - \ln(n^* - x^* - .5)$$

and

$$v_s \cong -\frac{1}{n^* + .5} + \frac{1}{n^* - x^* - .5},$$

where here we are being somewhat more precise than in Section 10.4.1(c) in our evaluation of the individual terms of the approximation of the differences between digamma and trigamma functions. (The approximations are good to two or more significant figures for $n^* - x^* \geqslant 2.0$.) From these equations we obtain, for $m_s \leqslant .5$,

and

$$n^* + .5 = \frac{\exp(m) - 1}{v} \cong \frac{m}{v} + \frac{\frac{1}{2}m^2}{v}$$

$$n^* - x^* - .5 \cong \frac{1 - \exp(-m)}{v} \cong \frac{m}{v} - \frac{\frac{1}{2}m^2}{v}$$

so that

$$x^* + 1 \cong \frac{m^2}{v}.$$

Next we make use of the result of Bol'shev and Loginov (1966) that optimum nonrandomized confidence bounds for series systems depend on

$$\hat{P}_s = 1 - \prod_{j=1}^{k} \hat{R}_j = 1 - \prod_{j=1}^{k} \frac{(n_j - x_j)}{n_j},$$

the maximum-likelihood estimator of $1 - R_s$. Thus we approximate $\hat{P}_s$ as

$$\hat{P}_s \cong \frac{m^2/v - 1}{m/v + .5m^2/v - .5}$$

Dividing numerator and denominator of the right side of this approximation by m/v and collecting terms, we obtain

$$m_s \cong \frac{\hat{P}_s}{1 - \hat{P}_s/2} + \frac{1}{n_0},$$

where n_0 is m_s/v_s. Therefore m_s^2/v_s is $n_0 m_s$. For a single component it can be shown, by setting m_s^2/v_s equal to $x + 1$, that this approach leads to

$n_0 = n(1 - \hat{P}_s/2)$. Solution of an equation of the form

$$\frac{\hat{P}_s}{1 - \hat{P}_s/2} + \frac{1}{n_0} \cong \psi(n+1) - \psi(n-x) = \sum_{n-x}^{n} \frac{1}{i}$$

(see Davis, 1964) shows that, in fact, n_0 is better approximated by $n(1 - \hat{P}_s/2)(1 - \hat{P}_s^2/2)$, and well approximated even though m_s is larger than .5. It can easily be ascertained by comparisons with tabulated values that the percentiles of a beta distribution with parameters $n - x$ and $x + 1$ can be approximated by (10.10) and (10.11) with $n_{(1)} = n$ and $a = 1$, in conjunction with the Wilson-Hilferty transformation, with results correct to about two significant figures if x is not too much more than $n/2$ and/or $n/2$ is not too small. The Wilson-Hilferty transformation is used to transform to normality the approximate weighted chi-square posterior distribution of $-\ln R_s$.

Now the question remaining when R_s represents the reliability of a series system rather than that of a single component is how to generalize the expression for $n_0 = n(1 - \hat{P}/2)(1 - \hat{P}^2/2)$, that is, how to evaluate n^*. Results of Bol'shev and Loginov (1966) and those of Section 10.4.1(c) relating to the approximately optimum exponential confidence bounds of Mann and Grubbs (1971) suggest that n_0 is a function of $n_{(1)}$, the smallest of the n_j's. Fertig (1972) shows, in fact, that for the exponential model with Type II censoring *optimum* confidence bounds depend *only* on $z_{(1)}$, that is, $(n_{(1)} + 1)t_m$ here, when $r_1 = \cdots = r_k = 1$. Weaver (1972) demonstrates similar results for binomial data when no failures occur for any subsystem. For large n^* and small $\hat{P}_s$, the results based on (10.10), (10.11), and the Wilson-Hilferty transformation will agree with the Poisson approximation of Bol'shev and Loginov when n_0 is equal to $n_{(1)}$. For $n_0 = n$ and $k = 1, n_0 = n(1 - \hat{P}_s/2)(1 - \hat{P}_s^2/2) = n$, which of course can be true only if $\hat{P}_s = 0$.

As has been shown, for $k = 1$ and m_s not too large,

$$m_s = \sum_{i=n-x}^{n} \frac{1}{i} \cong \frac{\hat{P}}{1 - \hat{P}/2} + \frac{1}{n(1 - \hat{P}/2)(1 - P^2/2)}.$$

It can also be shown that for , $k \geqslant 1$,

$$\sum_{j=1}^{k} \sum_{i=n_j-x_j}^{n_j} \frac{1}{i} \cong \frac{\hat{P}}{1 - \hat{P}/2} + \frac{\sum_{j=1}^{k} 1/n_j}{(1 - \hat{P}/2)(1 - \hat{P}^2/2)}.$$

If we add $(1/n_{(1)})/[(1 - \hat{P}/2)(1 - \hat{P}^2/2)]$ to this expression and then weight the two final terms by $.5/[n_1\Sigma_{j=1}^{k}(1/n_j)]$, we obtain (10.10). Now it can be demonstrated that, for $k \geqslant 2, .5[\Sigma_{j=1}^{k}(1/n_j)+1/n_{(1)}]/[(1-\hat{P}/2)(1-\hat{P}^2/2)]$ is, under the conditions specified here for m_s, approximately equal to $.5[\Sigma_{j=1}^{k}(1/(n_j-x_j))+1/(n-x)_{(1)}]$, where $(n-x)_{(1)}$ is the smallest of $(n_1-x_1),\ldots,(n_k-x_k)$. The latter expression can be shown, using the expression defined in the discussion following Equations (10.5) and (10.6), to be the increment added to $\Sigma_{j=1}^{k}\Sigma_{i=n_j-x_j+1}^{n_j}(1/i)$ to obtain m_s with the uniform random variate u set equal to 1. Thus the weight $1/[n_{(i)}\Sigma_{j=1}^{k}(1/n_j)]$ converts the approximation of the nonrandomized version of the optimum randomized (exact) lower confidence bound for R_s to the approximation of the optimum nonrandomized confidence bound obtainable by the exact method of Buehler (1957). The appropriate value for x^*, the effective number of failures for the series system, is $n^*\hat{P}_s$, with

$$\frac{1}{n^*} = \frac{.5\left[\sum_{j=1}^{k}(1/n_j)+1/n_{(1)}\right]}{n_{(1)}\sum_{j=1}^{k}(1/n_j)}. \tag{10.15}$$

We can use the approach just described to derive Equations (10.8) and (10.9) for m_p and v_p. The value of m_p is, in general, large, so that the approximation for m_p must be expressed in terms of a logarithm. From $m_p \cong \ln(n^*+.5)-\ln(x^*+.5)$ and for $\hat{P}_p \neq 0$,

$$m_p+\ln\left(1+\frac{.5}{x}\right) = \sum_{i=x+1}^{n}\frac{1}{i}+\ln\left(1+\frac{.5}{x}\right) \cong -\ln\left(\frac{\hat{P}_p}{1+.5\hat{P}_p/x}\right)$$

we obtain

$$m_p = -\ln\left[\frac{\hat{P}_p(1+.5/x^*)}{1+.5\hat{P}_p/x^*}\right].$$

Then, from $v_p \cong -1/(n^*+.5)+1/(x^*+.5)$ and for $\hat{P}_p \neq 0$,

$$v_p \cong 1-\frac{\exp(-m_p)}{x^*+.5}.$$

Also, it can be shown that, for $k \geqslant 1$ and $\hat{P}_p \neq 0$,

$$-\ln\left(\frac{\hat{P}_p}{1+.5\hat{P}_p/x^*}\right) \cong \sum_{j=1}^{k}\sum_{i=x_j+1}^{n_j}\frac{1}{i}+\frac{1.5}{x_{(1)}},$$

where l is the number of values of $x_{(1)}$, and that the value to be subtracted from both sides of this equation in order to make the two sides equal to m_p, the expression yielding an approximately optimum bound, is $\ln[1 + .5 \sum_{j=1}^{k}(1/x_j)]$. In other words, $x^* \cong 1/\sum_{j=1}^{k}(1/x_j)$ and $n^* = x^*/\hat{P}_p$. The value of $1.5/x_{(1)} - \ln(1 + .5/x^*)$ represents the increment added to m_p when the parameters of the beta distribution are nonintegers (with probability 1), that is, when k is greater than 1. The expression for x^* has been obtained empirically by evaluating by computer a more complicated expression involving the second and third moments of $-\ln(1 - R_p)$ for a wide variety of data sets.

Note that, when all n_j's are large and all $\hat{P}_j$'s are small, the estimator of the effective sample size proposed by Easterling (1972a), $[(1 - \hat{P}_p)/\hat{P}_p]/[\sum_{j=1}^{k}(n_j - x_j)/(x_j/n_j)]$, is equal to $[1/\sum_{j=1}^{k}(1/x_j)]/\hat{P}_p$. Note also that, for $l = 1$ and all x_j's large, the AO nonrandomized confidence bound on $1 - R_p$ is approximately equal to the nonrandomized version of the AO randomized confidence bound if all n's are large. This is in keeping with empirical results concerning the two confidence bounds.

When $\hat{P}_p$ is equal to 0, x^* must be equal to $x_{(1)} = 0$. If there is a single $x_{(1)}$ equal to 0, Equation (10.8) is applicable, since x^* and $x_{(1)}$ in $\hat{P}_p$ cancel each other. In this case, $n^* = \prod_{j=1}^{k} n_j / \prod_{x_j \neq 0} x_j$. If there are l values of $x_{(1)}$, then $x_{(1)}$ and $(x_{(1)}')^{1/l}$ in $\hat{P}_p^{1/l}$ also cancel each other. Therefore (10.8), with $x^* = lx_{(1)}$, is still applicable if $\hat{P}_p$ is raised to the power $1/l$. Compensation for raising $\hat{P}_p$ to the power $1/l$ is made by raising $\exp(m_p)$ to the power l *after* calculating m_p^2/v_p. An appropriate value for $(n^*)^l$, where n^* is the sample size for the total system, is then $(lx_{(1)})^l/\hat{P}_p = l^l \prod_{j=1}^{k} n_j / \prod_{x_j \neq 0} x_j$.

(b) Combining Subsystems. If a subsystem of $k \geqslant 2$ components connected in series is to be incorporated as one component of a parallel system, an effective number of failures, $x^* = n_s^* \hat{P}_s$, with $n_s^* = n_{(1)}/(.5 + .5/a)$ and $\hat{P}_s = 1 - \prod_{j=1}^{k} \hat{R}_j$, are available for this component. No other failure-number information is known about this series subsystem as a subsystem, nor is any other failure-number information meaningful. Hence, to evaluate the effective number of failures for a parallel system of r elements incorporating single components and both parallel and series subsystems, one merely calculates $1/\sum_{i=1}^{r} 1/x_i^*$, where x_i^* is equal to $1/\sum_j (1/x_{ij})$ or $n_i^* \hat{P}_{si}$ accordingly as the ith element is a parallel or a series subsystem. If the ith element is a single component, both expressions apply. If x_i^* for any component is 0, the effective number of failures for the parallel system is 0. The effective sample size for the parallel system is determined as in the discussion following Equations (10.8) and (10.9), that is, by dividing the effective number of failures by $\hat{P}_p$. The value of $\hat{P}_p$ for the parallel system made up of the r subsystems and components is simply $\prod_{i=1}^{r} \hat{P}_i$.

For any parallel subsystem with $k \geqslant 2$ components, $n^* = x_p^* / \hat{P}_p$ with $x_p^* = 1/\Sigma_{j=1}^{k}(1/x_j)$ and $\hat{P}_p = \Pi_{j=1}^{k}(1 - \hat{R}_j)$, or $(n_p^*)^l = l^l \Pi_{j=1}^{k} n_j / \Pi_{x_j \neq 0} x_j$ and $\hat{P}_p = 0$. To evaluate the effective number of failures for a series system incorporating r single components and parallel and series subsystems, one must compare the sample size or effective sample size of each single component and parallel subsystem with the smallest sample size within each series subsystem. Call the smallest of these sample sizes $n_{(1)}$. Also, for the ith series subsystem, call n_i^* the value $1/\Sigma_j(1/n_{ij})$; and, for the ith parallel subsystem, $n_i^* = x_i^* / \hat{P}_{pi}$ or $n_i^* = l_i (\Pi_{j=1}^{k} n_{ij} / \Pi_{x_{ij} \neq 0} x_{ij})^{1/l}$. Then the effective sample size for the total series system is $n_{(1)} \Sigma_{i=1}^{r}(1/n_i^*)$ $/[.5/n_{(1)} + .5\Sigma_{i=1}^{r}(1/n_i^*)]$. The value of $\hat{P}_s$ for this series system is $1 - \Pi_{i=1}^{r}(1 - \hat{P}_i)$. To determine the effective number of failures for the system, one simply multiplies the effective sample size by $\hat{P}_s$.

If the calculated value of m_s based on (10.10) is not less than about .5, use for m_s and v_s, instead of the values given by (10.10) and (10.11), respectively:

$$m_s = \psi(n^* + 1) - \psi(n^* - x^*) \qquad \text{and} \qquad v_s = -\psi'(n^* + 1) + \psi'(n^* - x^*).$$

To illustrate the procedure, we use data for two series elements, $n_1 = n_2 = 10$, $r_1 = r_2 = 1$, which are in parallel with an element having $n_3 = 10$, $r_3 = 2$. First, for the two series elements, we have

$$\hat{P}_s = 1 - \prod_{i=1}^{2} \hat{R}_i = 1 - (.9)(.9) = .19 \quad \text{and} \quad a = n_{(1)} \Sigma_j \left(\frac{1}{n_j} \right) = (10)(.1 + .1) = 2.$$

Then

$$n_s^* = \frac{n_{(1)}}{.5 + .5/a} = \frac{10}{.75} = 13.34 \qquad \text{and} \qquad r_s^* = n_s^* \hat{P}_s = (13.34)(.19) = 2.54.$$

Now we combine the two resulting components in parallel, that is, $n_s^* = 13.34$, $r_s^* = 2.54$ with $n_3 = 10$, $r_3 = 2$, and obtain

$$\hat{P}_p = \prod_{j=1}^{2} \left(1 - \hat{R}_j \right) = \left(\frac{1.27}{6.67} \right) \left(\frac{2}{10} \right) = .038,$$

$$r_p^* - \frac{1}{\Sigma_j(1/r_j^*)} - \frac{1}{(1/2.54 + \frac{1}{2})} - 1.11,$$

$$n_p^* = \frac{r_p^*}{\hat{P}_p} = \frac{1.11}{.038} = 29.2.$$

The equivalent final system has, therefore, $r^* = 1.11$ and $n^* = 29.2$, and we may note that $(n^* - r^*)/n^* = .962$.

The mean m and variance v of $-\ln R$ for a system with only a single component exhibiting r^* failures out of n^* tests are

$$m = \psi(n^* + 1) - \psi(n^* - r^*) \quad \text{and} \quad v = -\psi'(n^* + 1) + \psi'(n^* - r^*).$$

By using the approximations

$$\psi(z) \cong \ln(z - .5) \quad \text{and} \quad \psi'(z) \cong \frac{1}{z - .5},$$

we find in a straightforward manner that

$$m \cong \ln\left(\frac{n^* + .5}{n^* - r^* - .5}\right) \quad \text{and} \quad v \cong \frac{1}{n^* - r^* - .5} - \frac{1}{n^* + .5}.$$

With $r^* = 1.11$ and $n^* = 29.2$, we find that $m \cong .0737$ and $v \cong .00258$, and from (10.12) we obtain an approximate lower 95% confidence bound of .842 for the system reliability.

PROBLEMS FOR SECTION 10.4

10.12. Consider a system of five electronic components arranged in series. Suppose that the failure times for prototypes of all the components are exponentially distributed, and that the numbers of failures and observed values of the z's are given by $r_1 = 4$, $r_2 = r_3 = 2$, $r_4 = r_5 = 3$, $z_1 = 9.919$, $z_2 = 15.996$, $z_3 = 26.897$, $z_4 = 26.511$, $z_5 = 62.439$. For $t_m = 1$ calculate a lower 75% confidence bound on series-system reliability R_s by the method of Easterling and also by the method given by (10.2), (10.3), and (10.4). Compare your results with the tabulated values in Table 10.1 of the AO lower confidence bound and the optimum (uniformly most accurate unbiased) lower confidence bound. The 75th percentile of a standard normal distribution is .674.

10.13. For the data set given above, assume that $t_m = \frac{1}{4}$ and determine 90% lower confidence bounds on R_s by the same two methods.

10.14. For the same exponential data set with $t_m = \frac{1}{4}$, use the AO beta approximation of the posterior distribution of $1 - R_p$ to find an approximate lower 95% confidence bound on system reliability when the components are in parallel. Be sure that the value used for $\hat{P}_p$ is equal to $\Pi_{j=1}^{k}(r_j - 1)/(z_j/t_m - 1)$. Note that the approximation

$$n^*_{(1)} + 1 = \frac{z_{(1)}}{t_m} = 4(9.919) = 39.68,$$

is very nearly equal to the more precise approximation,

$$n^*_{(1)} + 1 = \frac{z_{(1)}}{t_m} + \frac{r_1 - 1}{2} - .5 = 39.68 + 1.5 - .5 = 40.68,$$

a value that is not proportional to $1/t_m$ and therefore is inappropriate. Note that the same near equality of the two approximations is true also of each of the other combinations of values of z_j/t_m and r_j.

10.15. Compare the values obtained in Problems 10.1 and 10.2 with those obtained by the AO beta approximation for lower confidence bounds on R_s. Be sure that the estimate $\hat{P}_s$ of $1-R_s$ used is $1-\Pi_{j=1}^{k}[1-(r_j-1)/(z_j/t_m-1)]$.

10.16. Assume that prototypes of three kinds of engine hardware have each been randomly subjected to life testing for a period of 24 hours, and that the observed sample sizes and binomially distributed numbers of failures are given by $n_1=10$, $n_2=8$, $n_3=12$, and $r_1=2$, $r_3=r_2=1$. Assume a constant hazard function.

a. Calculate an approximate 90% lower confidence bound on system reliability, when the three types of hardware are arranged independently in series, for mission time t_m equal to 12 hours.

b. Calculate an approximate 50% lower confidence bound on system reliability, when the three types of hardware are independently in parallel, for $t_m=12$ hours.

10.17. For $k=2$, $n_1=n_2=100$, $x_1=1$, $x_2=4$, calculate approximations to the optimum nonrandomized confidence bounds, the "randomized" confidence bounds, and the lower limit of the optimum randomized confidence bound on R_p at confidence level .90. Use Equation (10.7) with (10.8) and (10.9) and with (10.5) and (10.6) for $\delta=0$ and $u=1$. Compare with values in Table 10.5.

10.18. Show for binomial data, $k=1$, $n=10$, $x=2$, that $m_s=\Sigma_{i=8}^{10}(1/i)$, the posterior mean of $-\ln(1-p)$ corresponding to the optimum nonrandomized lower-confidence bound on $1-p$, is approximated by

$$\frac{\hat{P}}{1-\hat{P}/2}+\frac{1}{n(1-\hat{P}/2)(1-\hat{P}^2/2)},$$

where

$$P(X=x)=\binom{n}{x}p^x(1-p)^{n-x}\quad\text{and}\quad\hat{P}=\frac{x}{n}.$$

10.19. For two sets of binomial data, as given in Problem 10.18, that is, $n_1=n_2=10$, $x_1=x_2=2$, show that $\Sigma_{j=1}^{2}1/(n_j-x_j)+1/(n-x)_{(1)}$ is approximately equal to

$$\frac{\sum_{j=1}^{2}(1/n_j)+1/n_{(1)}}{(1-\hat{P}_s/2)(1-\hat{P}_s^2/2)},$$

where $(n-x)_{(1)}$ and $n_{(1)}$ are the smallest of n_1-x_1 and n_2-x_2 and of n_1 and n_2, respectively, and $\hat{P}_s=1-(n_1-x_1)(n_2-x_2)/n_1n_2$.

10.20. a. Show that, for the data described in Problem 10.18, $m_p=\Sigma_{i=3}^{10}(1/i)$ is approximately equal to $-\ln[\hat{P}_p(1+.5/x)/(1+.5\hat{P}_p/x)]$.

b. Show that the same approximation applies when $x=1$, $n=5$.

c. Show that, for the two data sets $n_1=10$, $n_2=5$, $x_1=2$, $x_2=1$, $\Sigma_{i=3}^{10}(1/i)+\Sigma_{i=2}^{5}(1/i)$ $+1.5/x_{(1)}$ is approximately equal to $-\ln[\hat{P}_p/(1+.5\hat{P}_p/x^*)]$, where $\hat{P}_p=x_1x_2/n_1n_2$, $x_{(1)}$ is the smallest of x_1 and x_2, and $l=1$, and where $x^*=1/(x_1^{-1}+x_2^{-1})$.

REFERENCES FOR SECTIONS 10.1–10.3

Barlow, R. E. and F. Proschan (1964), *Mathematical Theory of Reliability*, John Wiley and Sons, New York.

Carhart, R. R. (1953), A Survey of the Current Status of the Reliability Problem, *Research Memorandum* RM-1131, Rand Corporation.

Cox, D. R. and W. L. Smith (1954), On the Superposition of Renewal Processes, *Biometrika*, Vol. 41, pp. 91-99.

Karlin, S. (1966), *A First Course in Stochastic Processes*, Academic Press, New York.

Kemeney, J. and J. Snell (1960), *Finite Markov Chains*, D. Van Nostrand Company, Princeton, New Jersey.

Sandler, G. H. (1963), *System Reliability Engineering*, Prentice-Hall, Englewood Cliffs, New Jersey.

REFERENCES FOR SECTION 10.4

Abraham, John K. (1962), A Confidence Interval for the Reliability of Multicomponent Systems, *Proceedings of the Seventh Conference on the Design of Experiments in Army Research, Development and Testing*, ARODR 62-2, pp. 483-518.

Birnbaum, Z. W., J. D. Esary, and S. C. Saunders (1961), Multi-component Systems and Structures and Their Reliability, *Technometrics*, Vol. 3, pp. 55-77.

Bol'shev, L. N. and E. A. Loginov (1966), Interval Estimates in the Presence of Nuisance Parameters, in *Theory of Probability and Its Applications*, Vol. XI, No. 1. pp. 82-94.

Buehler, R. J. (1957), Confidence Intervals for the Product of Two Binomial Parameters, *Journal of American Statistical Association*, Vol. 52, pp. 482-493.

Burnett, T. L. and B. A. Wales (1961), System Reliability Confidence Limits, *Proceedings of the Seventh National Symposium on Reliability and Quality Control*, pp. 118-128.

Connor, W. S. and W. T. Wells (1962), Simulating Tests of a System from Tests of Its Components, *Proceedings of the Eighth National Symposium on Reliability and Quality Control*, pp. 14-16.

Davis, Philip J. (1964), Gamma Function and Related Functions, in *Handbook of Mathematical Functions*, edited by M. Abromowitz and I. A. Stegun, U.S. Government Printing Office, Washington, D.C.

Easterling, Robert G. (1972a), Approximate Confidence Limits for System Reliability, *Journal of the American Statistical Association*, Vol. 67, pp. 220-222.

Easterling, Robert G. (1972b), *Approximate Confidence Limits for System Reliability: Exponentially Distributed Component Lifetimes* (To be submitted for publication).

El Mawaziny, A. H. (1965), Chi-Square Distribution Theory with Applications to Reliability Problems, Ph.D. Thesis, Iowa State University, Ames, Iowa.

El Mawaziny, A. H. and R. J. Buehler (1967), Confidence Limits for the Reliability of a Series System, *Journal of the American Statistical Association*, Vol. 62, pp. 1452-1459.

Fertig, K. W. (1972), Bayesian Prior Distributions for Systems with Exponential Failure-Time Data, *Annals of Mathematical Statistics*, Vol. 43, pp. 1441-1448.

Garner, Norman R. and Richard W. J. Vail (1961), Confidence Limits for System Reliability, *Military Systems Design*, Vol. 7, No. 5.

Grubbs, Frank (1971), Approximate Fiducial Bounds for the Reliability of a Series System for Which Each Component Has an Exponential Time-to-Fail Distribution, *Technometrics*, Vol. 13, pp. 865-871.

Harris, Bernard (1971), Hypothesis Testing and Confidence Intervals for Products and Quotients of Poisson Parameters with Applications to Reliability, *Journal of the American Statistical Association*, Vol. 66, pp. 609-613.

Kraemer, H. C. (1963), One-Sided Confidence Intervals for the Quality Indices of a Complex Item, *Technometrics*, Vol. 5, pp. 400-403.

Lehmann, E. L. (1959), *Testing Statistical Hypotheses*, John Wiley and Sons, New York.

Lentner, M. M. and R. J. Buehler (1963), Some Inferences about Gamma Parameters with an Application to a Reliability Problem. *Journal of the American Statistical Association*, Vol. 58, pp. 670-677.

Levy, L. and A. H. Moore (1967), A Monte Carlo Technique for Obtaining System Reliability Confidence Limits from Component Test Data, *IEEE Transactions on Reliability*, Vol. R-16, pp. 69-72.

Lieberman, Gerald J. and Sheldon M. Ross (1971), Confidence Intervals for Independent Exponential Series Systems, *Journal of the American Statistical Association*, Vol. 66, pp. 837-840.

Lipow, M. and J. Riley (1959), *Tables of Upper Confidence Bounds on Failure Probability of 1, 2, and 3 Component Serial Systems*, Vols. I and II, Space Technology Laboratories AD-609-100 and AD-636-718.

Lloyd, D. K. and M. Lipow (1962), *Reliability: Management, Methods, and Mathematics*, Prentice-Hall, Englewood Cliffs, New Jersey.

Madansky, Albert (1965), Approximate Confidence Limits for the Reliability of Series and Parallel Systems, *Technometrics*, Vol. 7, pp. 495-503.

Mann, Nancy R. (1970), Computer-Aided Selection of Prior Distributions for Generating Monte Carlo Confidence Bounds on System Reliability, *Naval Research Logistics Quarterly*, Vol. 17, pp. 41-54.

Mann, Nancy R. (1974a), Approximately Optimum (Randomized and Nonrandomized) Confidence Bounds on Series-and Parallel-System Reliability for Systems with Binomial Subsystem Data, *I.E.E.E. Transactions on Reliability* (to appear).

Mann, Nancy R. (1974b), Simplified Expressions for Obtaining Approximately Optimum System-Reliability Confidence Bounds from Exponential Subsystem Data, *Journal of the American Statistical Association* (to appear).

Mann, Nancy R. and Kenneth W. Fertig (1972), Approximately Optimum (Randomized and Nonrandomized) Confidence Bounds on the Reliability of a Logically Complex Coherent System, *Rocketdyne Research Report* RR 72-02, Rocketdyne, Canoga Park, California (to be submitted for publication).

Mann, Nancy R. and Frank E. Grubbs (1972), Approximately Optimum Confidence Bounds on Series-System Reliability for Exponential Time to Fail Data, *Biometrika*, Vol. 59, pp. 191-204.

Mann, Nancy R. and Frank E. Grubbs (1974), Approximately Optimum Confidence Bounds for System Reliability Based on Component Test Data, *Technometrics* (to appear).

Mastran, David V. (1968), A Bayesian Approach for Assessing the Reliability of Air Force Re-entry Systems, *Proceedings of the ASME Reliability and Maintainability Symposium*, pp. 380-383.

Murchland, J. D. and Gunter G. Weber (1972), A Moment Method for the Calculation of a Confidence Interval for the Failure of a Probability of a System, *Proceedings of the 1972 Annual Reliability and Maintainability Symposium*, pp. 565-575.

Myhre, Janet and Sam C. Saunders (1968a), Comparison of Two Methods of Obtaining Approximate Confidence Intervals for System Reliability, *Technometrics*, Vol. 10, pp. 37-49.

Myhre, J. M. and S. C. Saunders (1968b), On Confidence Limits for the Reliability of Systems, *Annals of Mathematical Statistics*, Vol. 39, pp. 1463-1472.

Myhre, J. M. and S. C. Saunders (1971), Approximate Confidence Limits for Complex Systems with Exponential Component Lives, *Annals of Mathematical Statistics*, Vol. 42, pp. 342-348.

Parker, J. B. (1972), *Bayesian Prior Distributions for Multi-component Systems, Naval Research Logistics Quarterly*, Vol. 19, pp. 509-515.

Patnaik, P. B. (1949), The Non-central χ^2 and F Distributions and Their Applications, *Biometrika*, Vol. 36, pp. 202-232.

Raiffa, H. and R. Schlaifer (1961), *Applied Statistical Decision Theory*, Graduate School of Business Administration, Harvard University.

Rosenblatt, Joan Raup (1963), Confidence Limits for the reliability of Complex Systems, in *Statistical Theory of Reliability*, edited by Marvin Zelen, University of Wisconsin Press, Madison, Wisconsin, pp. 115-137.

Sarkar, Tapas K. (1971), An Exact Lower Confidence Bound for the Reliability of a Series System Where Each Component Has an Exponential Time to Failure Distribution, *Technometrics*, pp. 535-546.

Saunders, Sam C. (1972), On Confidence Bounds for System Reliability When the Components Have Exponentially Distributed Lives, *Boeing Document* D1180-14820-1, The Boeing Company, Seattle, Washington.

Schick, G. J. and R. J. Prior (1966), Reliability and Confidence of Serially Connected Systems, *Proceedings of the Third Space Congress*, Cocoa Beach, Florida, pp. 352-360.

Springer, M. D. and W. E. Thompson (1968), Bayesian Confidence Limits for Reliability of Redundant Systems When Tests are Terminated at First Failure, *Technometrics*, Vol. 10, pp. 29-36.

Steck, G. P. (1957), *Upper Confidence Limits for the Failure Probability of Complex Networks*, SC-433 (TR), Sandia Corporation.

Weaver, A. K. (1972), The Determination of the Lower Confidence Limit on the Reliability of a Serial System, in *Operations Research and Reliability*, edited by Daniel Grouchko, Gordon and Breach, New York.

Wilks, S. S. (1938), The Large-Sample Distribution of the Likelihood Ratio for Testing Composite Hypotheses, *Annals of Mathematical Statistics*, Vol. 9, pp. 60-62.

Wilson, E. B. and M. M. Hilferty (1931), The Distribution of Chi-Square, *Proceedings of National Academy Sciences (U.S.)*, Vol. 17, pp. 684-688.

Winterbottom, A. (1973), Lower Confidence Limits for Series System Reliability in the Attribute Case, *Journal of the American Statistical Association* (to appear).

Woods, M. and J. Borsting (1968), A Method for Computing Lower Confidence Limits on System Reliability Using Component Failure Data with Unequal Sample Sizes, *U.S. Naval Postgraduate Report* NPS55 wo/Bg 8061A.

Zimmer, W. J., R. R. Prairie, and A. M. Breipohl (1965), A Consideration of the Bayesian Approach in Reliability Evaluation, *IEEE Transactions on Reliability*, Vol. R-14, pp. 107-113.

APPENDIX

Table I The 100γth percentile or γth quantile of the χ^2 distribution with ν degrees of freedom is a number $\chi^2_{\nu,\gamma}$ [sometimes denoted by $\chi^2_\gamma(\nu)$] such that $P(\chi^2_\nu \leqslant \chi^2_{\nu,\gamma}) = \gamma$. The values $\chi^2_{\nu,\gamma}$ given in Table I *are labeled by* $\alpha = 1 - \gamma$. Thus, $\chi^2_{12,.05} = 5.22063$ is found in the column labeled $\alpha = .95$. On the other hand, $\chi^2_{12,.95} = 21.0261$ is in the column labeled $\alpha = .05$.

Table II The 100γth percentile or γth quantile of the F distribution with ν_1 and ν_2 degrees of freedom is a number $F_\gamma(\nu_1, \nu_2)$ such that $P[F(\nu_1, \nu_2) \leqslant F_\gamma(\nu_1, \nu_2)] = \gamma$. The values $F_\gamma(\nu_1, \nu_2)$ given in Table II *are labeled by* $\alpha = 1 - \gamma$. To find, say, $F_{.90}(4, 6)$, one looks on the page labeled $\alpha = 1 - \gamma = .10$ and finds $F_{.90}(4, 6) = 3.1808$. Moreover, since $F_\gamma(\nu_1, \nu_2) = [F_{1-\gamma}(\nu_2, \nu_1)]^{-1}$, only values of $\gamma > .50$ are tabulated. Thus, for finding $F_{.10}(7, 12)$, the appropriate value of $\alpha = 1 - \gamma = .90$ is not given. However, $F_{.10}(7, 12)$ is equal to $[F_{.90}(12, 7)]^{-1}$, which can be determined from Table II as $(2.6681)^{-1}$.

Table I Percentage points of the χ^2 distribution*

ν \ α	0.995	0.990	0.975	0.950	0.900	0.750
1	$392{,}704 \times 10^{-10}$	$157{,}088 \times 10^{-9}$	$982{,}069 \times 10^{-9}$	$393{,}214 \times 10^{-8}$	0.0157908	0.1015308
2	0.0100251	0.0201007	0.0506356	0.102587	0.210720	0.575364
3	0.0717212	0.114832	0.215795	0.351846	0.584375	1.212534
4	0.206990	0.297110	0.484419	0.710721	1.063623	1.92255
5	0.411740	0.554300	0.831211	1.145476	1.61031	2.67460
6	0.675727	0.872085	1.237347	1.63539	2.20413	3.45460
7	0.989265	1.239043	1.68987	2.16735	2.83311	4.25485
8	1.344419	1.646482	2.17973	2.73264	3.48954	5.07064
9	1.734926	2.087912	2.70039	3.32511	4.16816	5.89883
10	2.15585	2.55821	3.24697	3.94030	4.86518	6.73720
11	2.60321	3.05347	3.81575	4.57481	5.57779	7.58412
12	3.07382	3.57056	4.40379	5.22603	6.30380	8.43842
13	3.56503	4.10691	5.00874	5.89186	7.04150	9.29906
14	4.07468	4.66043	5.62872	6.57063	7.78953	10.1653
15	4.60094	5.22935	6.26214	7.26094	8.54675	11.0365
16	5.14224	5.81221	6.90766	7.96164	9.31223	11.9122
17	5.69724	6.40776	7.56418	8.67176	10.0852	12.7919
18	6.26481	7.01491	8.23075	9.39046	10.8649	13.6753
19	6.84398	7.63273	8.90655	10.1170	11.6509	14.5620
20	7.43386	8.26040	9.59083	10.8508	12.4426	15.4518
21	8.03366	8.89720	10.28293	11.5913	13.2396	16.3444
22	8.64272	9.54249	10.9823	12.3380	14.0415	17.2396
23	9.26042	10.19567	11.6885	13.0905	14.8479	18.1373
24	9.88623	10.8564	12.4001	13.8484	15.6587	19.0372
25	10.5197	11.5240	13.1197	14.6114	16.4734	19.9393
26	11.1603	12.1981	13.8439	15.3791	17.2919	20.8434
27	11.8076	12.8786	14.5733	16.1513	18.1138	21.7494
28	12.4613	13.5648	15.3079	16.9279	18.9392	22.6572
29	13.1211	14.2565	16.0471	17.7083	19.7677	23.5666
30	13.7867	14.9535	16.7908	18.4926	20.5992	24.4776
40	20.7065	22.1643	24.4331	26.5093	29.0505	33.6603
50	27.9907	29.7067	32.3574	34.7642	37.6886	42.9421
60	35.5346	37.4848	40.4817	43.1879	46.4589	52.2938
70	43.2752	45.4418	48.7576	51.7393	55.3290	61.6983
80	51.1720	53.5400	57.1532	60.3915	64.2778	71.1445
90	59.1963	61.7541	65.6466	69.1260	73.2912	80.6247
100	67.3276	70.0648	74.2219	77.9295	82.3581	90.1332
t_α	-2.5758	-2.3263	-1.9600	-1.6449	-1.2816	-0.6745

* Reproduced by permission of Professor E. S. Pearson from "Tables of the Percentage Points of the χ^2 Distribution," *Biometrika*, **32** (1941), pp. 188–189, by Catherine M. Thompson.

Table I (continued)

ν \ α	0.500	0.250	0.100	0.050	0.025	0.010	0.005
1	0.454937	1.32330	2.70554	3.84146	5.02389	6.63490	7.87944
2	1.38629	2.77259	4.60517	5.99147	7.37776	9.21034	10.5966
3	2.36597	4.10835	6.25139	7.81473	9.34840	11.3449	12.8381
4	3.35670	5.38527	7.77944	9.48773	11.1433	13.2767	14.8602
5	4.35146	6.62568	9.23635	11.0705	12.8325	15.0863	16.7496
6	5.34812	7.84080	10.6446	12.5916	14.4494	16.8119	18.5476
7	6.34581	9.03715	12.0170	14.0671	16.0128	18.4753	20.2777
8	7.34412	10.2188	13.3616	15.5073	17.5346	20.0902	21.9550
9	8.34283	11.3887	14.6837	16.9190	19.0228	21.6660	23.5893
10	9.34182	12.5489	15.9871	18.3070	20.4831	23.2093	25.1882
11	10.3410	13.7007	17.2750	19.6751	21.9200	24.7250	26.7569
12	11.3403	14.8454	18.5494	21.0261	23.3367	26.2170	28.2995
13	12.3398	15.9839	19.8119	22.3621	24.7356	27.6883	29.8194
14	13.3393	17.1170	21.0642	23.6848	26.1190	29.1413	31.3193
15	14.3389	18.2451	22.3072	24.9958	27.4884	30.5779	32.8013
16	15.3385	19.3688	23.5418	26.2962	28.8454	31.9999	34.2672
17	16.3381	20.4887	24.7690	27.5871	30.1910	33.4087	35.7185
18	17.3379	21.6049	25.9894	28.8693	31.5264	34.8053	37.1564
19	18.3376	22.7178	27.2036	30.1435	32.8523	36.1908	38.5822
20	19.3374	23.8277	28.4120	31.4104	34.1696	37.5662	39.9968
21	20.3372	24.9348	29.6151	32.6705	35.4789	38.9321	41.4010
22	21.3370	26.0393	30.8133	33.9244	36.7807	40.2894	42.7956
23	22.3369	27.1413	32.0069	35.1725	38.0757	41.6384	44.1813
24	23.3367	28.2412	33.1963	36.4151	39.3641	42.9798	45.5585
25	24.3366	29.3389	34.3816	37.6525	40.6465	44.3141	46.9278
26	25.3364	30.4345	35.5631	38.8852	41.9232	45.6417	48.2899
27	26.3363	31.5284	36.7412	40.1133	43.1944	46.9630	49.6449
28	27.3363	32.6205	37.9159	41.3372	44.4607	48.2782	50.9933
29	28.3362	33.7109	39.0875	42.5569	45.7222	49.5879	52.3356
30	29.3360	34.7998	40.2560	43.7729	46.9792	50.8922	53.6720
40	39.3354	45.6160	51.8050	55.7585	59.3417	63.6907	66.7659
50	49.3349	56.3336	63.1671	67.5048	71.4202	76.1539	79.4900
60	59.3347	66.9814	74.3970	79.0819	83.2976	88.3794	91.9517
70	69.3344	77.5766	85.5271	90.5312	95.0231	100.425	104.215
80	79.3343	88.1303	96.5782	101.879	106.629	112.329	116.321
90	89.3342	98.6499	107.565	113.145	118.136	124.116	128.299
100	99.3341	109.141	118.498	124.342	129.561	135.807	140.169
t_α	0.0000	$+0.6745$	$+1.2816$	$+1.6449$	$+1.9600$	$+2.3263$	$+2.5758$

For $30 < \nu < 100$, linear interpolation where necessary will give 4 significant figures. For $\nu > 100$ take $\chi^2\nu, \alpha = \frac{1}{2}(t_\alpha + \sqrt{2\nu - 1})^2$. A description of this table is given on page 529.

531

Table II Percentage points of the F distribution* ($\alpha = 0.50$)

ν_1 / ν_2	1	2	3	4	5	6	7	8	9
1	1.0000	1.5000	1.7092	1.8227	1.8937	1.9422	1.9774	2.0041	2.0250
2	0.66667	1.0000	1.1349	1.2071	1.2519	1.2824	1.3045	1.3213	1.3344
3	0.58506	0.88110	1.0000	1.0632	1.1024	1.1289	1.1482	1.1627	1.1741
4	0.54863	0.82843	0.94054	1.0000	1.0367	1.0617	1.0797	1.0933	1.1040
5	0.52807	0.79877	0.90715	0.96456	1.0000	1.0240	1.0414	1.0545	1.0648
6	0.51489	0.77976	0.88578	0.94191	0.97654	1.0000	1.0169	1.0298	1.0398
7	0.50572	0.76655	0.87095	0.92619	0.96026	0.98334	1.0000	1.0126	1.0224
8	0.49898	0.75683	0.86004	0.91464	0.94831	0.97111	0.98757	1.0000	1.0097
9	0.49382	0.74938	0.85168	0.90580	0.93916	0.96175	0.97805	0.99037	1.0000
10	0.48973	0.74349	0.84508	0.89882	0.93193	0.95436	0.97054	0.98276	0.99232
11	0.48644	0.73872	0.83973	0.89316	0.92608	0.94837	0.96445	0.97661	0.98610
12	0.48369	0.73477	0.83530	0.88848	0.92124	0.94342	0.95943	0.97152	0.98097
13	0.48141	0.73145	0.83159	0.88454	0.91718	0.93926	0.95520	0.96724	0.97665
14	0.47944	0.72862	0.82842	0.88119	0.91371	0.93573	0.95161	0.96360	0.97298
15	0.47775	0.72619	0.82569	0.87830	0.91073	0.93267	0.94850	0.96046	0.96981
16	0.47628	0.72406	0.82330	0.87578	0.90812	0.93001	0.94580	0.95773	0.96705
17	0.47499	0.72219	0.82121	0.87357	0.90584	0.92767	0.94342	0.95532	0.96462
18	0.47385	0.72053	0.81936	0.87161	0.90381	0.92560	0.94132	0.95319	0.96247
19	0.47284	0.71906	0.81771	0.86987	0.90200	0.92375	0.93944	0.95129	0.96056
20	0.47192	0.71773	0.81621	0.86830	0.90038	0.92210	0.93776	0.94959	0.95884
21	0.47108	0.71653	0.81487	0.86688	0.89891	0.92060	0.93624	0.94805	0.95728
22	0.47033	0.71545	0.81365	0.86559	0.89759	0.91924	0.93486	0.94665	0.95588
23	0.46965	0.71446	0.81255	0.86442	0.89638	0.91800	0.93360	0.94538	0.95459
24	0.46902	0.71356	0.81153	0.86335	0.89527	0.91687	0.93245	0.94422	0.95342
25	0.46844	0.71272	0.81061	0.86236	0.89425	0.91583	0.93140	0.94315	0.95234
26	0.46793	0.71195	0.80975	0.86145	0.89331	0.91487	0.93042	0.94217	0.95135
27	0.46744	0.71124	0.80894	0.86061	0.89244	0.91399	0.92952	0.94126	0.95044
28	0.46697	0.71059	0.80820	0.85983	0.89164	0.91317	0.92869	0.94041	0.94958
29	0.46654	0.70999	0.80753	0.85911	0.89089	0.91241	0.92791	0.93963	0.94879
30	0.46616	0.70941	0.80689	0.85844	0.89019	0.91169	0.92719	0.93889	0.94805
40	0.46330	0.70531	0.80228	0.85357	0.88516	0.90654	0.92197	0.93361	0.94272
60	0.46053	0.70122	0.79770	0.84873	0.88017	0.90144	0.91679	0.92838	0.93743
120	0.45774	0.69717	0.79314	0.84392	0.87521	0.89637	0.91164	0.92318	0.93218
∞	0.45494	0.69315	0.78866	0.83918	0.87029	0.89135	0.90654	0.91802	0.92698

*Reproduced by permission of Professor E. S. Pearson from "Tables of Percentage Points of the Inverted Beta (F) Distribution," *Biometrika*, **33** (1943), pp. 73–88, by Maxine Merrington and Catherine M. Thompson.

Table II (continued)

r_2 \ r_1	10	12	15	20	24	30	40	60	120	∞
1	2.0419	2.0674	2.0931	2.1190	2.1321	2.1452	2.1584	2.1716	2.1848	2.1981
2	1.3450	1.3610	1.3771	1.3933	1.4014	1.4096	1.4178	1.4261	1.4344	1.4427
3	1.1833	1.1972	1.2111	1.2252	1.2322	1.2393	1.2464	1.2536	1.2608	1.2680
4	1.1126	1.1255	1.1386	1.1517	1.1583	1.1649	1.1716	1.1782	1.1849	1.1916
5	1.0730	1.0855	1.0980	1.1106	1.1170	1.1234	1.1297	1.1361	1.1426	1.1490
6	1.0478	1.0600	1.0722	1.0845	1.0907	1.0969	1.1031	1.1093	1.1156	1.1219
7	1.0304	1.0423	1.0543	1.0664	1.0724	1.0785	1.0846	1.0908	1.0969	1.1031
8	1.0175	1.0293	1.0412	1.0531	1.0591	1.0651	1.0711	1.0771	1.0832	1.0893
9	1.0077	1.0194	1.0311	1.0429	1.0489	1.0548	1.0608	1.0667	1.0727	1.0788
10	1.0000	1.0116	1.0232	1.0349	1.0408	1.0467	1.0526	1.0585	1.0645	1.0705
11	0.99373	1.0052	1.0168	1.0284	1.0343	1.0401	1.0460	1.0519	1.0578	1.0637
12	0.98856	1.0000	1.0115	1.0231	1.0289	1.0347	1.0405	1.0464	1.0523	1.0582
13	0.98421	0.99560	1.0071	1.0186	1.0243	1.0301	1.0360	1.0418	1.0476	1.0535
14	0.98051	0.99186	1.0033	1.0147	1.0205	1.0263	1.0321	1.0379	1.0437	1.0495
15	0.97732	0.98863	1.0000	1.0114	1.0172	1.0229	1.0287	1.0345	1.0403	1.0461
16	0.97454	0.98582	0.99716	1.0086	1.0143	1.0200	1.0258	1.0315	1.0373	1.0431
17	0.97209	0.98334	0.99466	1.0060	1.0117	1.0174	1.0232	1.0289	1.0347	1.0405
18	0.96993	0.98116	0.99245	1.0038	1.0095	1.0152	1.0209	1.0267	1.0324	1.0382
19	0.96800	0.97920	0.99047	1.0018	1.0075	1.0132	1.0189	1.0246	1.0304	1.0361
20	0.96626	0.97746	0.98870	1.0000	1.0057	1.0114	1.0171	1.0228	1.0285	1.0343
21	0.96470	0.97587	0.98710	0.99838	1.0040	1.0097	1.0154	1.0211	1.0268	1.0326
22	0.96328	0.97444	0.98565	0.99692	1.0026	1.0082	1.0139	1.0196	1.0253	1.0311
23	0.96199	0.97313	0.98433	0.99558	1.0012	1.0069	1.0126	1.0183	1.0240	1.0297
24	0.96081	0.97194	0.98312	0.99436	1.0000	1.0057	1.0113	1.0170	1.0227	1.0284
25	0.95972	0.97084	0.98201	0.99324	0.99887	1.0045	1.0102	1.0159	1.0215	1.0273
26	0.95872	0.96983	0.98099	0.99220	0.99783	1.0035	1.0091	1.0148	1.0205	1.0262
27	0.95779	0.96889	0.98004	0.99125	0.99687	1.0025	1.0082	1.0138	1.0195	1.0252
28	0.95694	0.96802	0.97917	0.99036	0.99598	1.0016	1.0073	1.0129	1.0186	1.0243
29	0.95614	0.96722	0.97835	0.98954	0.99515	1.0008	1.0064	1.0121	1.0177	1.0234
30	0.95540	0.96647	0.97759	0.98877	0.99438	1.0000	1.0056	1.0113	1.0170	1.0226
40	0.95003	0.96104	0.97211	0.98323	0.98880	0.99440	1.0000	1.0056	1.0113	1.0169
60	0.94471	0.95566	0.96667	0.97773	0.98328	0.98884	0.99441	1.0000	1.0056	1.0112
120	0.93943	0.95032	0.96128	0.97228	0.97780	0.98333	0.98887	0.99443	1.0000	1.0056
∞	0.93418	0.94503	0.95593	0.96687	0.97236	0.97787	0.98339	0.98891	0.99445	1.0000

A description of these tables is given on page 529. Where necessary, interpolation should be carried out using the reciprocals of the degrees of freedom. The function $120/\nu$ is convenient for this purpose.

Table II (continued)* ($\alpha = 0.25$)

v_1 / v_2	1	2	3	4	5	6	7	8	9
1	5.8285	7.5000	8.1999	8.5810	8.8198	8.9833	9.1021	9.1922	9.2631
2	2.5714	3.0000	3.1534	3.2320	3.2799	3.3121	3.3352	3.3526	3.3661
3	2.0239	2.2798	2.3555	2.3901	2.4095	2.4218	2.4302	2.4364	2.4410
4	1.8074	2.0000	2.0467	2.0642	2.0723	2.0766	2.0790	2.0805	2.0814
5	1.6925	1.8528	1.8843	1.8927	1.8947	1.8945	1.8935	1.8923	1.8911
6	1.6214	1.7622	1.7844	1.7872	1.7852	1.7821	1.7789	1.7760	1.7733
7	1.5732	1.7010	1.7169	1.7157	1.7111	1.7059	1.7011	1.6969	1.6931
8	1.5384	1.6569	1.6683	1.6642	1.6575	1.6508	1.6448	1.6396	1.6350
9	1.5121	1.6236	1.6315	1.6253	1.6170	1.6091	1.6022	1.5961	1.5909
10	1.4915	1.5975	1.6028	1.5949	1.5853	1.5765	1.5688	1.5621	1.5563
11	1.4749	1.5767	1.5798	1.5704	1.5598	1.5502	1.5418	1.5346	1.5284
12	1.4613	1.5595	1.5609	1.5503	1.5389	1.5286	1.5197	1.5120	1.5054
13	1.4500	1.5452	1.5451	1.5336	1.5214	1.5105	1.5011	1.4931	1.4861
14	1.4403	1.5331	1.5317	1.5194	1.5066	1.4952	1.4854	1.4770	1.4697
15	1.4321	1.5227	1.5202	1.5071	1.4938	1.4820	1.4718	1.4631	1.4556
16	1.4249	1.5137	1.5103	1.4965	1.4827	1.4705	1.4601	1.4511	1.4433
17	1.4186	1.5057	1.5015	1.4873	1.4730	1.4605	1.4497	1.4405	1.4325
18	1.4130	1.4988	1.4938	1.4790	1.4644	1.4516	1.4406	1.4312	1.4230
19	1.4081	1.4925	1.4870	1.4717	1.4568	1.4437	1.4325	1.4228	1.4145
20	1.4037	1.4870	1.4808	1.4652	1.4500	1.4366	1.4252	1.4153	1.4069
21	1.3997	1.4820	1.4753	1.4593	1.4438	1.4302	1.4186	1.4086	1.4000
22	1.3961	1.4774	1.4703	1.4540	1.4382	1.4244	1.4126	1.4025	1.3937
23	1.3928	1.4733	1.4657	1.4491	1.4331	1.4191	1.4072	1.3969	1.3880
24	1.3898	1.4695	1.4615	1.4447	1.4285	1.4143	1.4022	1.3918	1.3828
25	1.3870	1.4661	1.4577	1.4406	1.4242	1.4099	1.3976	1.3871	1.3780
26	1.3845	1.4629	1.4542	1.4368	1.4203	1.4058	1.3935	1.3828	1.3737
27	1.3822	1.4600	1.4510	1.4334	1.4166	1.4021	1.3896	1.3788	1.3696
28	1.3800	1.4572	1.4480	1.4302	1.4133	1.3986	1.3860	1.3752	1.3658
29	1.3780	1.4547	1.4452	1.4272	1.4102	1.3953	1.3826	1.3717	1.3623
30	1.3761	1.4524	1.4426	1.4244	1.4073	1.3923	1.3795	1.3685	1.3590
40	1.3626	1.4355	1.4239	1.4045	1.3863	1.3706	1.3571	1.3455	1.3354
60	1.3493	1.4188	1.4055	1.3848	1.3657	1.3491	1.3349	1.3226	1.3119
120	1.3362	1.4024	1.3873	1.3654	1.3453	1.3278	1.3128	1.2999	1.2886
∞	1.3233	1.3863	1.3694	1.3463	1.3251	1.3068	1.2910	1.2774	1.2654

*Reproduced by permission of Professor E. S. Pearson from "Tables of Percentage Points of the Inverted Beta (F) Distribution," *Biometrika*, **33** (1943), pp. 73–88, by Maxine Merrington and Catherine M. Thompson.

Table II (continued)

ν_2 \ ν_1	10	12	15	20	24	30	40	60	120	∞
1	9.3202	9.4064	9.4934	9.5813	9.6255	9.6698	9.7144	9.7591	9.8041	9.8492
2	3.3770	3.3934	3.4098	3.4263	3.4345	3.4428	3.4511	3.4594	3.4677	3.4761
3	2.4447	2.4500	2.4552	2.4602	2.4626	2.4650	2.4674	2.4697	2.4720	2.4742
4	2.0820	2.0826	2.0829	2.0828	2.0827	2.0825	2.0821	2.0817	2.0812	2.0806
5	1.8899	1.8877	1.8851	1.8820	1.8802	1.8784	1.8763	1.8742	1.8719	1.8694
6	1.7708	1.7668	1.7621	1.7569	1.7540	1.7510	1.7477	1.7443	1.7407	1.7368
7	1.6898	1.6843	1.6781	1.6712	1.6675	1.6635	1.6593	1.6548	1.6502	1.6452
8	1.6310	1.6244	1.6170	1.6088	1.6043	1.5996	1.5945	1.5892	1.5836	1.5777
9	1.5863	1.5788	1.5705	1.5611	1.5560	1.5506	1.5450	1.5389	1.5325	1.5257
10	1.5513	1.5430	1.5338	1.5235	1.5179	1.5119	1.5056	1.4990	1.4919	1.4843
11	1.5230	1.5140	1.5041	1.4930	1.4869	1.4805	1.4737	1.4664	1.4587	1.4504
12	1.4996	1.4902	1.4796	1.4678	1.4613	1.4544	1.4471	1.4393	1.4310	1.4221
13	1.4801	1.4701	1.4590	1.4465	1.4397	1.4324	1.4247	1.4164	1.4075	1.3980
14	1.4634	1.4530	1.4414	1.4284	1.4212	1.4136	1.4055	1.3967	1.3874	1.3772
15	1.4491	1.4383	1.4263	1.4127	1.4052	1.3973	1.3888	1.3796	1.3698	1.3591
16	1.4366	1.4255	1.4130	1.3990	1.3913	1.3830	1.3742	1.3646	1.3543	1.3432
17	1.4256	1.4142	1.4014	1.3869	1.3790	1.3704	1.3613	1.3514	1.3406	1.3290
18	1.4159	1.4042	1.3911	1.3762	1.3680	1.3592	1.3497	1.3395	1.3284	1.3162
19	1.4073	1.3953	1.3819	1.3666	1.3582	1.3492	1:3394	1.3289	1.3174	1.3048
20	1.3995	1.3873	1.3736	1.3580	1.3494	1.3401	1.3301	1.3193	1.3074	1.2943
21	1.3925	1.3801	1.3661	1.3502	1.3414	1.3319	1.3217	1.3105	1.2983	1.2848
22	1.3861	1.3735	1.3593	1.3431	1.3341	1.3245	1.3140	1.3025	1.2900	1.2761
23	1.3803	1.3675	1.3531	1.3366	1.3275	1.3176	1.3069	1.2952	1.2824	1.2681
24	1.3750	1.3621	1.3474	1.3307	1.3214	1.3113	1.3004	1.2885	1.2754	1.2607
25	1.3701	1.3570	1.3422	1.3252	1.3158	1.3056	1.2945	1.2823	1.2689	1.2538
26	1.3656	1.3524	1.3374	1.3202	1.3106	1.3002	1.2889	1.2765	1.2628	1.2474
27	1.3615	1.3481	1.3329	1.3155	1.3058	1.2953	1.2838	1.2712	1.2572	1.2414
28	1.3576	1.3441	1.3288	1.3112	1.3013	1.2906	1.2790	1.2662	1.2519	1.2358
29	1.3541	1.3404	1.3249	1.3071	1.2971	1.2863	1.2745	1.2615	1.2470	1.2306
30	1.3507	1.3369	1.3213	1.3033	1.2933	1.2823	1.2703	1.2571	1.2424	1.2256
40	1.3266	1.3119	1.2952	1.2758	1.2649	1.2529	1.2397	1.2249	1.2080	1.1883
60	1.3026	1.2870	1.2691	1.2481	1.2361	1.2229	1.2081	1.1912	1.1715	1.1474
120	1.2787	1.2621	1.2428	1.2200	1.2068	1.1921	1.1752	1.1555	1.1314	1.0987
∞	1.2549	1.2371	1.2163	1.1914	1.1767	1.1600	1.1404	1.1164	1.0838	1.0000

A description of these tables is given on page 529. Where necessary, interpolation should be carried out using the reciprocals of the degrees of freedom. The function $120/\nu$ is convenient for this purpose.

Table II (continued)* ($\alpha = 0.10$)

v_2 \ v_1	1	2	3	4	5	6	7	8	9
1	39.864	49.500	53.593	55.833	57.241	58.204	58.906	59.439	59.858
2	8.5263	9.0000	9.1618	9.2434	9.2926	9.3255	9.3491	9.3668	9.3805
3	5.5383	5.4624	5.3908	5.3427	5.3092	5.2847	5.2662	5.2517	5.2400
4	4.5448	4.3246	4.1908	4.1073	4.0506	4.0098	3.9790	3.9549	3.9357
5	4.0604	3.7797	3.6195	3.5202	3.4530	3.4045	3.3679	3.3393	3.3163
6	3.7760	3.4633	3.2888	3.1808	3.1075	3.0546	3.0145	2.9830	2.9577
7	3.5894	3.2574	3.0741	2.9605	2.8833	2.8274	2.7849	2.7516	2.7247
8	3.4579	3.1131	2.9238	2.8064	2.7265	2.6683	2.6241	2.5893	2.5612
9	3.3603	3.0065	2.8129	2.6927	2.6106	2.5509	2.5053	2.4694	2.4403
10	3.2850	2.9245	2.7277	2.6053	2.5216	2.4606	2.4140	2.3772	2.3473
11	3.2252	2.8595	2.6602	2.5362	2.4512	2.3891	2.3416	2.3040	2.2735
12	3.1765	2.8068	2.6055	2.4801	2.3940	2.3310	2.2828	2.2446	2.2135
13	3.1362	2.7632	2.5603	2.4337	2.3467	2.2830	2.2341	2.1953	2.1638
14	3.1022	2.7265	2.5222	2.3947	2.3069	2.2426	2.1931	2.1539	2.1220
15	3.0732	2.6952	2.4898	2.3614	2.2730	2.2081	2.1582	2.1185	2.0862
16	3.0481	2.6682	2.4618	2.3327	2.2438	2.1783	2.1280	2.0880	2.0553
17	3.0262	2.6446	2.4374	2.3077	2.2183	2.1524	2.1017	2.0613	2.0284
18	3.0070	2.6239	2.4160	2.2858	2.1958	2.1296	2.0785	2.0379	2.0047
19	2.9899	2.6056	2.3970	2.2663	2.1760	2.1094	2.0580	2.0171	1.9836
20	2.9747	2.5893	2.3801	2.2489	2.1582	2.0913	2.0397	1.9985	1.9649
21	2.9609	2.5746	2.3649	2.2333	2.1423	2.0751	2.0232	1.9819	1.9480
22	2.9486	2.5613	2.3512	2.2193	2.1279	2.0605	2.0084	1.9668	1.9327
23	2.9374	2.5493	2.3387	2.2065	2.1149	2.0472	1.9949	1.9531	1.9189
24	2.9271	2.5383	2.3274	2.1949	2.1030	2.0351	1.9826	1.9407	1.9063
25	2.9177	2.5283	2.3170	2.1843	2.0922	2.0241	1.9714	1.9292	1.8947
26	2.9091	2.5191	2.3075	2.1745	2.0822	2.0139	1.9610	1.9188	1.8841
27	2.9012	2.5106	2.2987	2.1655	2.0730	2.0045	1.9515	1.9091	1.8743
28	2.8939	2.5028	2.2906	2.1571	2.0645	1.9959	1.9427	1.9001	1.8652
29	2.8871	2.4955	2.2831	2.1494	2.0566	1.9878	1.9345	1.8918	1.8568
30	2.8807	2.4887	2.2761	2.1422	2.0492	1.9803	1.9269	1.8841	1.8490
40	2.8354	2.4404	2.2261	2.0909	1.9968	1.9269	1.8725	1.8289	1.7929
60	2.7914	2.3932	2.1774	2.0410	1.9457	1.8747	1.8194	1.7748	1.7380
120	2.7478	2.3473	2.1300	1.9923	1.8959	1.8238	1.7675	1.7220	1.6843
∞	2.7055	2.3026	2.0838	1.9449	1.8473	1.7741	1.7167	1.6702	1.6315

*Reproduced by permission of Professor E. S. Pearson from "Tables of Percentage Points of the Inverted Beta (F) Distribution," *Biometrika*, **33** (1943), pp. 73–88, by Maxine Merrington and Catherine M. Thompson.

Table II (continued)

ν_1 / ν_2	10	12	15	20	24	30	40	60	120	∞
1	60.195	60.705	61.220	61.740	62.002	62.265	62.529	62.794	63.061	63.328
2	9.3916	9.4081	9.4247	9.4413	9.4496	9.4579	9.4663	9.4746	9.4829	9.4913
3	5.2304	5.2156	5.2003	5.1845	5.1764	5.1681	5.1597	5.1512	5.1425	5.1337
4	3.9199	3.8955	3.8689	3.8443	3.8310	3.8174	3.8036	3.7896	3.7753	3.7607
5	3.2974	3.2682	3.2380	3.2067	3.1905	3.1741	3.1573	3.1402	3.1228	3.1050
6	2.9369	2.9047	2.8712	2.8363	2.8183	2.8000	2.7812	2.7620	2.7423	2.7222
7	2.7025	2.6681	2.6322	2.5947	2.5753	2.5555	2.5351	2.5142	2.4928	2.4708
8	2.5380	2.5020	2.4642	2.4246	2.4041	2.3830	2.3614	2.3391	2.3162	2.2926
9	2.4163	2.3789	2.3396	2.2983	2.2768	2.2547	2.2320	2.2085	2.1843	2.1592
10	2.3226	2.2841	2.2435	2.2007	2.1784	2.1554	2.1317	2.1072	2.0818	2.0554
11	2.2482	2.2087	2.1671	2.1230	2.1000	2.0762	2.0516	2.0261	1.9997	1.9721
12	2.1878	2.1474	2.1049	2.0597	2.0360	2.0115	1.9861	1.9597	1.9323	1.9036
13	2.1376	2.0966	2.0532	2.0070	1.9827	1.9576	1.9315	1.9043	1.8759	1.8462
14	2.0954	2.0537	2.0095	1.9625	1.9377	1.9119	1.8852	1.8572	1.8280	1.7973
15	2.0593	2.0171	1.9722	1.9243	1.8990	1.8728	1.8454	1.8168	1.7867	1.7551
16	2.0281	1.9854	1.9399	1.8913	1.8656	1.8388	1.8108	1.7816	1.7507	1.7182
17	2.0009	1.9577	1.9117	1.8624	1.8362	1.8090	1.7805	1.7506	1.7191	1.6856
18	1.9770	1.9333	1.8868	1.8368	1.8103	1.7827	1.7537	1.7232	1.6910	1.6567
19	1.9557	1.9117	1.8647	1.8142	1.7873	1.7592	1.7298	1.6988	1.6659	1.6308
20	1.9367	1.8924	1.8449	1.7938	1.7667	1.7382	1.7083	1.6768	1.6433	1.6074
21	1.9197	1.8750	1.8272	1.7756	1.7481	1.7193	1.6890	1.6569	1.6228	1.5862
22	1.9043	1.8593	1.8111	1.7590	1.7312	1.7021	1.6714	1.6389	1.6042	1.5668
23	1.8903	1.8450	1.7964	1.7439	1.7159	1.6864	1.6554	1.6224	1.5871	1.5490
24	1.8775	1.8319	1.7831	1.7302	1.7019	1.6721	1.6407	1.6073	1.5715	1.5327
25	1.8658	1.8200	1.7708	1.7175	1.6890	1.6589	1.6272	1.5934	1.5570	1.5176
26	1.8550	1.8090	1.7596	1.7059	1.6771	1.6468	1.6147	1.5805	1.5437	1.5036
27	1.8451	1.7989	1.7492	1.6951	1.6662	1.6356	1.6032	1.5686	1.5313	1.4906
28	1.8359	1.7895	1.7395	1.6852	1.6560	1.6252	1.5925	1.5575	1.5198	1.4784
29	1.8274	1.7808	1.7306	1.6759	1.6465	1.6155	1.5825	1.5472	1.5090	1.4670
30	1.8195	1.7727	1.7223	1.6673	1.6377	1.6065	1.5732	1.5376	1.4989	1.4564
40	1.7627	1.7146	1.6624	1.6052	1.5741	1.5411	1.5056	1.4672	1.4248	1.3769
60	1.7070	1.6574	1.6034	1.5435	1.5107	1.4755	1.4373	1.3952	1.3476	1.2915
120	1.6524	1.6012	1.5450	1.4821	1.4472	1.4094	1.3676	1.3203	1.2646	1.1926
∞	1.5987	1.5458	1.4871	1.4206	1.3832	1.3419	1.2951	1.2400	1.1686	1.0000

A description of these tables is given on page 529. Where necessary, interpolation should be carried out using the reciprocals of the degrees of freedom. The function $120/\nu$ is convenient for this purpose.

Table II (continued)* ($\alpha = 0.05$)

v_2 \ v_1	1	2	3	4	5	6	7	8	9
1	161.45	199.50	215.71	224.58	230.16	233.99	236.77	238.88	240.54
2	18.513	19.000	19.164	19.247	19.296	19.330	19.353	19.371	19.385
3	10.128	9.5521	9.2766	9.1172	9.0135	8.9406	8.8868	8.8452	8.8123
4	7.7086	6.9443	6.5914	6.3883	6.2560	6.1631	6.0942	6.0410	5.9988
5	6.6079	5.7861	5.4095	5.1922	5.0503	4.9503	4.8759	4.8183	4.7725
6	5.9874	5.1433	4.7571	4.5337	4.3874	4.2839	4.2066	4.1468	4.0990
7	5.5914	4.7374	4.3468	4.1203	3.9715	3.8660	3.7870	3.7257	3.6767
8	5.3177	4.4590	4.0662	3.8378	3.6875	3.5806	3.5005	3.4381	3.3881
9	5.1174	4.2565	3.8626	3.6331	3.4817	3.3738	3.2927	3.2296	3.1789
10	4.9646	4.1028	3.7083	3.4780	3.3258	3.2172	3.1355	3.0717	3.0204
11	4.8443	3.9823	3.5874	3.3567	3.2039	3.0946	3.0123	2.9480	2.8962
12	4.7472	3.8853	3.4903	3.2592	3.1059	2.9961	2.9134	2.8486	2.7964
13	4.6672	3.8056	3.4105	3.1791	3.0254	2.9153	2.8321	2.7669	2.7144
14	4.6001	3.7389	3.3439	3.1122	2.9582	2.8477	2.7642	2.6987	2.6458
15	4.5431	3.6823	3.2874	3.0556	2.9013	2.7905	2.7066	2.6408	2.5876
16	4.4940	3.6337	3.2389	3.0069	2.8524	2.7413	2.6572	2.5911	2.5377
17	4.4513	3.5915	3.1968	2.9647	2.8100	2.6987	2.6143	2.5480	2.4943
18	4.4139	3.5546	3.1599	2.9277	2.7729	2.6613	2.5767	2.5102	2.4563
19	4.3808	3.5219	3.1274	2.8951	2.7401	2.6283	2.5435	2.4768	2.4227
20	4.3513	3.4928	3.0984	2.8661	2.7109	2.5990	2.5140	2.4471	2.3928
21	4.3248	3.4668	3.0725	2.8401	2.6848	2.5727	2.4876	2.4205	2.3661
22	4.3009	3.4434	3.0491	2.8167	2.6613	2.5491	2.4638	2.3965	2.3419
23	4.2793	3.4221	3.0280	2.7955	2.6400	2.5277	2.4422	2.3748	2.3201
24	4.2597	3.4028	3.0088	2.7763	2.6207	2.5082	2.4226	2.3551	2.3002
25	4.2417	3.3852	2.9912	2.7587	2.6030	2.4904	2.4047	2.3371	2.2821
26	4.2252	3.3690	2.9751	2.7426	2.5868	2.4741	2.3883	2.3205	2.2655
27	4.2100	3.3541	2.9604	2.7278	2.5719	2.4591	2.3732	2.3053	2.2501
28	4.1960	3.3404	2.9467	2.7141	2.5581	2.4453	2.3593	2.2913	2.2360
29	4.1830	3.3277	2.9340	2.7014	2.5454	2.4324	2.3463	2.2782	2.2229
30	4.1709	3.3158	2.9223	2.6896	2.5336	2.4205	2.3343	2.2662	2.2107
40	4.0848	3.2317	2.8387	2.6060	2.4495	2.3359	2.2490	2.1802	2.1240
60	4.0012	3.1504	2.7581	2.5252	2.3683	2.2540	2.1665	2.0970	2.0401
120	3.9201	3.0718	2.6802	2.4472	2.2900	2.1750	2.0867	2.0164	1.9588
∞	3.8415	2.9957	2.6049	2.3719	2.2141	2.0986	2.0096	1.9384	1.8799

*Reproduced by permission of Professor E. S. Pearson from "Tables of Percentage Points of the Inverted Beta (F) Distribution," *Biometrika*, **33** (1943), pp. 73–88, by Maxine Merrington and Catherine M. Thompson.

Table II (continued)

v_1 / v_2	10	12	15	20	24	30	40	60	120	∞
1	241.88	243.91	245.95	248.01	249.05	250.09	251.14	252.20	253.25	254.32
2	19.396	19.413	19.429	19.446	19.454	19.462	19.471	19.479	19.487	19.496
3	8.7855	8.7446	8.7029	8.6602	8.6385	8.6166	8.5944	8.5720	8.5494	8.5265
4	5.9644	5.9117	5.8578	5.8025	5.7744	5.7459	5.7170	5.6878	5.6581	5.6281
5	4.7351	4.6777	4.6188	4.5581	4.5272	4.4957	4.4638	4.4314	4.3984	4.3650
6	4.0600	3.9999	3.9381	3.8742	3.8415	3.8082	3.7743	3.7398	3.7047	3.6688
7	3.6365	3.5747	3.5108	3.4445	3.4105	3.3758	3.3404	3.3043	3.2674	3.2298
8	3.3472	3.2840	3.2184	3.1503	3.1152	3.0794	3.0428	3.0053	2.9669	2.9276
9	3.1373	3.0729	3.0061	2.9365	2.9005	2.8637	2.8259	2.7872	2.7475	2.7067
10	2.9782	2.9130	2.8450	2.7740	2.7372	2.6996	2.6609	2.6211	2.5801	2.5379
11	2.8536	2.7876	2.7186	2.6464	2.6090	2.5705	2.5309	2.4901	2.4480	2.4045
12	2.7534	2.6866	2.6169	2.5436	2.5055	2.4663	2.4259	2.3842	2.3410	2.2962
13	2.6710	2.6037	2.5331	2.4589	2.4202	2.3803	2.3392	2.2966	2.2524	2.2064
14	2.6021	2.5342	2.4630	2.3879	2.3487	2.3082	2.2664	2.2230	2.1778	2.1307
15	2.5437	2.4753	2.4035	2.3275	2.2878	2.2468	2.2043	2.1601	2.1141	2.0658
16	2.4935	2.4247	2.3522	2.2756	2.2354	2.1938	2.1507	2.1058	2.0589	2.0096
17	2.4499	2.3807	2.3077	2.2304	2.1898	2.1477	2.1040	2.0584	2.0107	1.9604
18	2.4117	2.3421	2.2686	2.1906	2.1497	2.1071	2.0629	2.0166	1.9681	1.9168
19	2.3779	2.3080	2.2341	2.1555	2.1141	2.0712	2.0264	1.9796	1.9302	1.8780
20	2.3479	2.2776	2.2033	2.1242	2.0825	2.0391	1.9938	1.9464	1.8963	1.8432
21	2.3210	2.2504	2.1757	2.0960	2.0540	2.0102	1.9645	1.9165	1.8657	1.8117
22	2.2967	2.2258	2.1508	2.0707	2.0283	1.9842	1.9380	1.8895	1.8380	1.7831
23	2.2747	2.2036	2.1282	2.0476	2.0050	1.9605	1.9139	1.8649	1.8128	1.7570
24	2.2547	2.1834	2.1077	2.0267	1.9838	1.9390	1.8920	1.8424	1.7897	1.7331
25	2.2365	2.1649	2.0889	2.0075	1.9643	1.9192	1.8718	1.8217	1.7684	1.7110
26	2.2197	2.1479	2.0716	1.9898	1.9464	1.9010	1.8533	1.8027	1.7488	1.6906
27	2.2043	2.1323	2.0558	1.9736	1.9299	1.8842	1.8361	1.7851	1.7307	1.6717
28	2.1900	2.1179	2.0411	1.9586	1.9147	1.8687	1.8203	1.7689	1.7138	1.6541
29	2.1768	2.1045	2.0275	1.9446	1.9005	1.8543	1.8055	1.7537	1.6981	1.6377
30	2.1646	2.0921	2.0148	1.9317	1.8874	1.8409	1.7918	1.7396	1.6835	1.6223
40	2.0772	2.0035	1.9245	1.8389	1.7929	1.7444	1.6928	1.6373	1.5766	1.5089
60	1.9926	1.9174	1.8364	1.7480	1.7001	1.6491	1.5943	1.5343	1.4673	1.3893
120	1.9105	1.8337	1.7505	1.6587	1.6084	1.5543	1.4952	1.4290	1.3519	1.2539
∞	1.8307	1.7522	1.6664	1.5705	1.5173	1.4591	1.3940	1.3180	1.2214	1.0000

A description of these tables is given on page 529. Where necessary, interpolation should be carried out using the reciprocals of the degrees of freedom. The function $120/v$ is convenient for this purpose.

Table II (continued)* ($\alpha = 0.025$)

ν_2 \ ν_1	1	2	3	4	5	6	7	8	9
1	647.79	799.50	864.16	899.58	921.85	937.11	948.22	956.66	963.28
2	38.506	39.000	39.165	39.248	39.298	39.331	39.355	39.373	39.387
3	17.443	16.044	15.439	15.101	14.885	14.735	14.624	14.540	14.473
4	12.218	10.649	9.9792	9.6045	9.3645	9.1973	9.0741	8.9796	8.9047
5	10.007	8.4336	7.7636	7.3879	7.1464	6.9777	6.8531	6.7572	6.6810
6	8.8131	7.2598	6.5988	6.2272	5.9876	5.8197	5.6955	5.5996	5.5234
7	8.0727	6.5415	5.8898	5.5226	5.2852	5.1186	4.9949	4.8994	4.8232
8	7.5709	6.0595	5.4160	5.0526	4.8173	4.6517	4.5286	4.4332	4.3572
9	7.2093	5.7147	5.0781	4.7181	4.4844	4.3197	4.1971	4.1020	4.0260
10	6.9367	5.4564	4.8256	4.4683	4.2361	4.0721	3.9498	3.8549	3.7790
11	6.7241	5.2559	4.6300	4.2751	4.0440	3.8807	3.7586	3.6638	3.5879
12	6.5538	5.0959	4.4742	4.1212	3.8911	3.7283	3.6065	3.5118	3.4358
13	6.4143	4.9653	4.3472	3.9959	3.7667	3.6043	3.4827	3.3880	3.3120
14	6.2979	4.8567	4.2417	3.8919	3.6634	3.5014	3.3799	3.2853	3.2093
15	6.1995	4.7650	4.1528	3.8043	3.5764	3.4147	3.2934	3.1987	3.1227
16	6.1151	4.6867	4.0768	3.7294	3.5021	3.3406	3.2194	3.1248	3.0488
17	6.0420	4.6189	4.0112	3.6648	3.4379	3.2767	3.1556	3.0610	2.9849
18	5.9781	4.5597	3.9539	3.6083	3.3820	3.2209	3.0999	3.0053	2.9291
19	5.9216	4.5075	3.9034	3.5587	3.3327	3.1718	3.0509	2.9563	2.8800
20	5.8715	4.4613	3.8587	3.5147	3.2891	3.1283	3.0074	2.9128	2.8365
21	5.8266	4.4199	3.8188	3.4754	3.2501	3.0895	2.9686	2.8740	2.7977
22	5.7863	4.3828	3.7829	3.4401	3.2151	3.0546	2.9338	2.8392	2.7628
23	5.7498	4.3492	3.7505	3.4083	3.1835	3.0232	2.9024	2.8077	2.7313
24	5.7167	4.3187	3.7211	3.3794	3.1548	2.9946	2.8738	2.7791	2.7027
25	5.6864	4.2909	3.6943	3.3530	3.1287	2.9685	2.8478	2.7531	2.6766
26	5.6586	4.2655	3.6697	3.3289	3.1048	2.9447	2.8240	2.7293	2.6528
27	5.6331	4.2421	3.6472	3.3067	3.0828	2.9228	2.8021	2.7074	2.6309
28	5.6096	4.2205	3.6264	3.2863	3.0625	2.9027	2.7820	2.6872	2.6106
29	5.5878	4.2006	3.6072	3.2674	3.0438	2.8840	2.7633	2.6686	2.5919
30	5.5675	4.1821	3.5894	3.2499	3.0265	2.8667	2.7460	2.6513	2.5746
40	5.4239	4.0510	3.4633	3.1261	2.9037	2.7444	2.6238	2.5289	2.4519
60	5.2857	3.9253	3.3425	3.0077	2.7863	2.6274	2.5068	2.4117	2.3344
120	5.1524	3.8046	3.2270	2.8943	2.6740	2.5154	2.3948	2.2994	2.2217
∞	5.0239	3.6889	3.1161	2.7858	2.5665	2.4082	2.2875	2.1918	2.1136

*Reproduced by permission of Professor E. S. Pearson from "Tables of Percentage Points of the Inverted Beta (F) Distribution," *Biometrika*, **33** (1943), pp. 73–88, by Maxine Merrington and Catherine M. Thompson.

Table II (continued)

ν_2 \ ν_1	10	12	15	20	24	30	40	60	120	∞
1	968.63	976.71	984.87	993.10	997.25	1001.4	1005.6	1009.8	1014.0	1018.3
2	39.398	39.415	39.431	39.448	39.456	39.465	39.473	39.481	39.490	39.498
3	14.419	14.337	14.253	14.167	14.124	14.081	14.037	13.992	13.947	13.902
4	8.8439	8.7512	8.6565	8.5599	8.5109	8.4613	8.4111	8.3604	8.3092	8.2573
5	6.6192	6.5246	6.4277	6.3285	6.2780	6.2269	6.1751	6.1225	6.0693	6.0153
6	5.4613	5.3662	5.2687	5.1684	5.1172	5.0652	5.0125	4.9589	4.9045	4.8491
7	4.7611	4.6658	4.5678	4.4667	4.4150	4.3624	4.3089	4.2544	4.1989	4.1423
8	4.2951	4.1997	4.1012	3.9995	3.9472	3.8940	3.8398	3.7844	3.7279	3.6702
9	3.9639	3.8682	3.7694	3.6669	3.6142	3.5604	3.5055	3.4493	3.3918	3.3329
10	3.7168	3.6209	3.5217	3.4186	3.3654	3.3110	3.2554	3.1984	3.1399	3.0798
11	3.5257	3.4296	3.3299	3.2261	3.1725	3.1176	3.0613	3.0035	2.9441	2.8828
12	3.3736	3.2773	3.1772	3.0728	3.0187	2.9633	2.9063	2.8478	2.7874	2.7249
13	3.2497	3.1532	3.0527	2.9477	2.8932	2.8373	2.7797	2.7204	2.6590	2.5955
14	3.1469	3.0501	2.9493	2.8437	2.7888	2.7324	2.6742	2.6142	2.5519	2.4872
15	3.0602	2.9633	2.8621	2.7559	2.7006	2.6437	2.5850	2.5242	2.4611	2.3953
16	2.9862	2.8890	2.7875	2.6808	2.6252	2.5678	2.5085	2.4471	2.3831	2.3163
17	2.9222	2.8249	2.7230	2.6158	2.5598	2.5021	2.4422	2.3801	2.3153	2.2474
18	2.8664	2.7689	2.6667	2.5590	2.5027	2.4445	2.3842	2.3214	2.2558	2.1869
19	2.8173	2.7196	2.6171	2.5089	2.4523	2.3937	2.3329	2.2695	2.2032	2.1333
20	2.7737	2.6758	2.5731	2.4645	2.4076	2.3486	2.2873	2.2234	2.1562	2.0853
21	2.7348	2.6368	2.5338	2.4247	2.3675	2.3082	2.2465	2.1819	2.1141	2.0422
22	2.6998	2.6017	2.4984	2.3890	2.3315	2.2718	2.2097	2.1446	2.0760	2.0032
23	2.6682	2.5699	2.4665	2.3567	2.2989	2.2389	2.1763	2.1107	2.0415	1.9677
24	2.6396	2.5412	2.4374	2.3273	2.2693	2.2090	2.1460	2.0799	2.0099	1.9353
25	2.6135	2.5149	2.4110	2.3005	2.2422	2.1816	2.1183	2.0517	1.9811	1.9055
26	2.5895	2.4909	2.3867	2.2759	2.2174	2.1565	2.0928	2.0257	1.9545	1.8781
27	2.5676	2.4688	2.3644	2.2533	2.1946	2.1334	2.0693	2.0018	1.9299	1.8527
28	2.5473	2.4484	2.3438	2.2324	2.1735	2.1121	2.0477	1.9796	1.9072	1.8291
29	2.5286	2.4295	2.3248	2.2131	2.1540	2.0923	2.0276	1.9591	1.8861	1.8072
30	2.5112	2.4120	2.3072	2.1952	2.1359	2.0739	2.0089	1.9400	1.8664	1.7867
40	2.3882	2.2882	2.1819	2.0677	2.0069	1.9429	1.8752	1.8028	1.7242	1.6371
60	2.2702	2.1692	2.0613	1.9445	1.8817	1.8152	1.7440	1.6668	1.5810	1.4822
120	2.1570	2.0548	1.9450	1.8249	1.7597	1.6899	1.6141	1.5299	1.4327	1.3104
∞	2.0483	1.9447	1.8326	1.7085	1.6402	1.5660	1.4835	1.3883	1.2684	1.0000

A description of these tables is given on page 529. Where necessary, interpolation should be carried out using the reciprocals of the degrees of freedom. The function $120/\nu$ is convenient for this purpose.

Table II (continued)* ($\alpha = 0.01$)

v_2 \ v_1	1	2	3	4	5	6	7	8	9
1	4052.2	4999.5	5403.3	5624.6	5763.7	5859.0	5928.3	5981.6	6022.5
2	98.503	99.000	99.166	99.249	99.299	99.332	99.356	99.374	99.388
3	34.116	30.817	29.457	28.710	28.237	27.911	27.672	27.489	27.345
4	21.198	18.000	16.694	15.977	15.522	15.207	14.976	14.799	14.659
5	16.258	13.274	12.060	11.392	10.967	10.672	10.456	10.289	10.158
6	13.745	10.925	9.7795	9.1483	8.7459	8.4661	8.2600	8.1016	7.9761
7	12.246	9.5466	8.4513	7.8467	7.4604	7.1914	6.9928	6.8401	6.7188
8	11.259	8.6491	7.5910	7.0060	6.6318	6.3707	6.1776	6.0289	5.9106
9	10.561	8.0215	6.9919	6.4221	6.0569	5.8018	5.6129	5.4671	5.3511
10	10.044	7.5594	6.5523	5.9943	5.6363	5.3858	5.2001	5.0567	4.9424
11	9.6460	7.2057	6.2167	5.6683	5.3160	5.0692	4.8861	4.7445	4.6315
12	9.3302	6.9266	5.9526	5.4119	5.0643	4.8206	4.6395	4.4994	4.3875
13	9.0738	6.7010	5.7394	5.2053	4.8616	4.6204	4.4410	4.3021	4.1911
14	8.8616	6.5149	5.5639	5.0354	4.6950	4.4558	4.2779	4.1399	4.0297
15	8.6831	6.3589	5.4170	4.8932	4.5556	4.3183	4.1415	4.0045	3.8948
16	8.5310	6.2262	5.2922	4.7726	4.4374	4.2016	4.0259	3.8896	3.7804
17	8.3997	6.1121	5.1850	4.6690	4.3359	4.1015	3.9267	3.7910	3.6822
18	8.2854	6.0129	5.0919	4.5790	4.2479	4.0146	3.8406	3.7054	3.5971
19	8.1850	5.9259	5.0103	4.5003	4.1708	3.9386	3.7653	3.6305	3.5225
20	8.0960	5.8489	4.9382	4.4307	4.1027	3.8714	3.6987	3.5644	3.4567
21	8.0166	5.7804	4.8740	4.3688	4.0421	3.8117	3.6396	3.5056	3.3981
22	7.9454	5.7190	4.8166	4.3134	3.9880	3.7583	3.5867	3.4530	3.3458
23	7.8811	5.6637	4.7649	4.2635	3.9392	3.7102	3.5390	3.4057	3.2986
24	7.8229	5.6136	4.7181	4.2184	3.8951	3.6667	3.4959	3.3629	3.2560
25	7.7698	5.5680	4.6755	4.1774	3.8550	3.6272	3.4568	3.3239	3.2172
26	7.7213	5.5263	4.6366	4.1400	3.8183	3.5911	3.4210	3.2884	3.1818
27	7.6767	5.4881	4.6009	4.1056	3.7848	3.5580	3.3882	3.2558	3.1494
28	7.6356	5.4529	4.5681	4.0740	3.7539	3.5276	3.3581	3.2259	3.1195
29	7.5976	5.4205	4.5378	4.0449	3.7254	3.4995	3.3302	3.1982	3.0920
30	7.5625	5.3904	4.5097	4.0179	3.6990	3.4735	3.3045	3.1726	3.0665
40	7.3141	5.1785	4.3126	3.8283	3.5138	3.2910	3.1238	2.9930	2.8876
60	7.0771	4.9774	4.1259	3.6491	3.3389	3.1187	2.9530	2.8233	2.7185
120	6.8510	4.7865	3.9493	3.4796	3.1735	2.9559	2.7918	2.6629	2.5586
∞	6.6349	4.6052	3.7816	3.3192	3.0173	2.8020	2.6393	2.5113	2.4073

*Reproduced by permission of Professor E. S. Pearson from "Tables of Percentage Points of the Inverted Beta (F) Distribution," *Biometrika*, **33** (1943), pp. 73–88, by Maxine Merrington and Catherine M. Thompson.

Table II (continued)

v_2 \ v_1	10	12	15	20	24	30	40	60	120	∞
1	6055.8	6106.3	6157.3	6208.7	6234.6	6260.7	6286.8	6313.0	6339.4	6366.0
2	99.399	99.416	99.432	99.449	99.458	99.466	99.474	99.483	99.491	99.501
3	27.229	27.052	26.872	26.690	26.598	26.505	26.411	26.316	26.221	26.125
4	14.546	14.374	14.198	14.020	13.929	13.838	13.745	13.652	13.558	13.463
5	10.051	9.8883	9.7222	9.5527	9.4665	9.3793	9.2912	9.2020	9.1118	9.0204
6	7.8741	7.7183	7.5590	7.3958	7.3127	7.2285	7.1432	7.0568	6.9690	6.8801
7	6.6201	6.4691	6.3143	6.1554	6.0743	5.9921	5.9084	5.8236	5.7372	5.6495
8	5.8143	5.6668	5.5151	5.3591	5.2793	5.1981	5.1156	5.0316	4.9460	4.8588
9	5.2565	5.1114	4.9621	4.8080	4.7290	4.6486	4.5667	4.4831	4.3978	4.3105
10	4.8492	4.7059	4.5582	4.4054	4.3269	4.2469	4.1653	4.0819	3.9965	3.9090
11	4.5393	4.3974	4.2509	4.0990	4.0209	3.9411	3.8596	3.7761	3.6904	3.6025
12	4.2961	4.1553	4.0096	3.8584	3.7805	3.7008	3.6192	3.5355	3.4494	3.3608
13	4.1003	3.9603	3.8154	3.6646	3.5868	3.5070	3.4253	3.3413	3.2548	3.1654
14	3.9394	3.8001	3.6557	3.5052	3.4274	3.3476	3.2656	3.1813	3.0942	3.0040
15	3.8049	3.6662	3.5222	3.3719	3.2940	3.2141	3.1319	3.0471	2.9595	2.8684
16	3.6909	3.5527	3.4089	3.2588	3.1808	3.1007	3.0182	2.9330	2.8447	2.7528
17	3.5931	3.4552	3.3117	3.1615	3.0835	3.0032	2.9205	2.8348	2.7459	2.6530
18	3.5082	3.3706	3.2273	3.0771	2.9990	2.9185	2.8354	2.7493	2.6597	2.5660
19	3.4338	3.2965	3.1533	3.0031	2.9249	2.8442	2.7608	2.6742	2.5839	2.4893
20	3.3682	3.2311	3.0880	2.9377	2.8594	2.7785	2.6947	2.6077	2.5168	2.4212
21	3.3098	3.1729	3.0299	2.8796	2.8011	2.7200	2.6359	2.5484	2.4568	2.3603
22	3.2576	3.1209	2.9780	2.8274	2.7488	2.6675	2.5831	2.4951	2.4029	2.3055
23	3.2106	3.0740	2.9311	2.7805	2.7017	2.6202	2.5355	2.4471	2.3542	2.2559
24	3.1681	3.0316	2.8887	2.7380	2.6591	2.5773	2.4923	2.4035	2.3099	2.2107
25	3.1294	2.9931	2.8502	2.6993	2.6203	2.5383	2.4530	2.3637	2.2695	2.1694
26	3.0941	2.9579	2.8150	2.6640	2.5848	2.5026	2.4170	2.3273	2.2325	2.1315
27	3.0618	2.9256	2.7827	2.6316	2.5522	2.4699	2.3840	2.2938	2.1984	2.0965
28	3.0320	2.8959	2.7530	2.6017	2.5223	2.4397	2.3535	2.2629	2.1670	2.0642
29	3.0045	2.8685	2.7256	2.5742	2.4946	2.4118	2.3253	2.2344	2.1378	2.0342
30	2.9791	2.8431	2.7002	2.5487	2.4689	2.3860	2.2992	2.2079	2.1107	2.0062
40	2.8005	2.6648	2.5216	2.3689	2.2880	2.2034	2.1142	2.0194	1.9172	1.8047
60	2.6318	2.4961	2.3523	2.1978	2.1154	2.0285	1.9360	1.8363	1.7263	1.6006
120	2.4721	2.3363	2.1915	2.0346	1.9500	1.8600	1.7628	1.6557	1.5330	1.3805
∞	2.3209	2.1848	2.0385	1.8783	1.7908	1.6964	1.5923	1.4730	1.3246	1.0000

A description of these tables is given on page 529. Where necessary, interpolation should be carried out using the reciprocals of the degrees of freedom. The function $120/v$ is convenient for this purpose.

Table II (continued)* ($\alpha = 0.005$)

ν_2 \ ν_1	1	2	3	4	5	6	7	8	9
1	16211	20000	21615	22500	23056	23437	23715	23925	24091
2	198.50	199.00	199.17	199.25	199.30	199.33	199.36	199.37	199.39
3	55.552	49.799	47.467	46.195	45.392	44.838	44.434	44.126	43.882
4	31.333	26.284	24.259	23.155	22.456	21.975	21.622	21.352	21.139
5	22.785	18.314	16.530	15.556	14.940	14.513	14.200	13.961	13.772
6	18.635	14.544	12.917	12.028	11.464	11.073	10.786	10.566	10.391
7	16.236	12.404	10.882	10.050	9.5221	9.1554	8.8854	8.6781	8.5138
8	14.688	11.042	9.5965	8.8051	8.3018	7.9520	7.6942	7.4960	7.3386
9	13.614	10.107	8.7171	7.9559	7.4711	7.1338	6.8849	6.6933	6.5411
10	12.826	9.4270	8.0807	7.3428	6.8723	6.5446	6.3025	6.1159	5.9676
11	12.226	8.9122	7.6004	6.8809	6.4217	6.1015	5.8648	5.6821	5.5368
12	11.754	8.5096	7.2258	6.5211	6.0711	5.7570	5.5245	5.3451	5.2021
13	11.374	8.1865	6.9257	6.2335	5.7910	5.4819	5.2529	5.0761	4.9351
14	11.060	7.9217	6.6803	5.9984	5.5623	5.2574	5.0313	4.8566	4.7173
15	10.798	7.7008	6.4760	5.8029	5.3721	5.0708	4.8473	4.6743	4.5364
16	10.575	7.5138	6.3034	5.6378	5.2117	4.9134	4.6920	4.5207	4.3838
17	10.384	7.3536	6.1556	5.4967	5.0746	4.7789	4.5594	4.3893	4.2535
18	10.218	7.2148	6.0277	5.3746	4.9560	4.6627	4.4448	4.2759	4.1410
19	10.073	7.0935	5.9161	5.2681	4.8526	4.5614	4.3448	4.1770	4.0428
20	9.9439	6.9865	5.8177	5.1743	4.7616	4.4721	4.2569	4.0900	3.9564
21	9.8295	6.8914	5.7304	5.0911	4.6808	4.3931	4.1789	4.0128	3.8799
22	9.7271	6.8064	5.6524	5.0168	4.6088	4.3225	4.1094	3.9440	3.8116
23	9.6348	6.7300	5.5823	4.9500	4.5441	4.2591	4.0469	3.8822	3.7502
24	9.5513	6.6610	5.5190	4.8898	4.4857	4.2019	3.9905	3.8264	3.6949
25	9.4753	6.5982	5.4615	4.8351	4.4327	4.1500	3.9394	3.7758	3.6447
26	9.4059	6.5409	5.4091	4.7852	4.3844	4.1027	3.8928	3.7297	3.5989
27	9.3423	6.4885	5.3611	4.7396	4.3402	4.0594	3.8501	3.6875	3.5571
28	9.2838	6.4403	5.3170	4.6977	4.2996	4.0197	3.8110	3.6487	3.5186
29	9.2297	6.3958	5.2764	4.6591	4.2622	3.9830	3.7749	3.6130	3.4832
30	9.1797	6.3547	5.2388	4.6233	4.2276	3.9492	3.7416	3.5801	3.4505
40	8.8278	6.0664	4.9759	4.3738	3.9860	3.7129	3.5088	3.3498	3.2220
60	8.4946	5.7950	4.7290	4.1399	3.7600	3.4918	3.2911	3.1344	3.0083
120	8.1790	5.5393	4.4973	3.9207	3.5482	3.2849	3.0874	2.9330	2.8083
∞	7.8794	5.2983	4.2794	3.7151	3.3499	3.0913	2.8968	2.7444	2.6210

*Reproduced by permission of Professor E. S. Pearson from "Tables of Percentage Points of the Inverted Beta (F) Distribution," *Biometrika*, **33** (1943), pp. 73–88, by Maxine Merrington and Catherine M Thompson.

Table II (continued)

v_1 / v_2	10	12	15	20	24	30	40	60	120	∞
1	24224	24426	24630	24836	24940	25044	25148	25253	25359	25465
2	199.40	199.42	199.43	199.45	199.46	199.47	199.47	199.48	199.49	199.51
3	43.686	43.387	43.085	42.778	42.622	42.466	42.308	42.149	41.989	41.829
4	20.967	20.705	20.438	20.167	20.030	19.892	19.752	19.611	19.468	19.325
5	13.618	13.384	13.146	12.903	12.780	12.656	12.530	12.402	12.274	12.144
6	10.250	10.034	9.8140	9.5888	9.4741	9.3583	9.2408	9.1219	9.0015	8.8793
7	8.3803	8.1764	7.9678	7.7540	7.6450	7.5345	7.4225	7.3088	7.1933	7.0760
8	7.2107	7.0149	6.8143	6.6082	6.5029	6.3961	6.2875	6.1772	6.0649	5.9505
9	6.4171	6.2274	6.0325	5.8318	5.7292	5.6248	5.5186	5.4104	5.3001	5.1875
10	5.8467	5.6613	5.4707	5.2740	5.1732	5.0705	4.9659	4.8592	4.7501	4.6385
11	5.4182	5.2363	5.0489	4.8552	4.7557	4.6543	4.5508	4.4450	4.3367	4.2256
12	5.0855	4.9063	4.7214	4.5299	4.4315	4.3309	4.2282	4.1229	4.0149	3.9039
13	4.8199	4.6429	4.4600	4.2703	4.1726	4.0727	3.9704	3.8655	3.7577	3.6465
14	4.6034	4.4281	4.2468	4.0585	3.9614	3.8619	3.7600	3.6553	3.5473	3.4359
15	4.4236	4.2498	4.0698	3.8826	3.7859	3.6867	3.5850	3.4803	3.3722	3.2602
16	4.2719	4.0994	3.9205	3.7342	3.6378	3.5388	3.4372	3.3324	3.2240	3.1115
17	4.1423	3.9709	3.7929	3.6073	3.5112	3.4124	3.3107	3.2058	3.0971	2.9839
18	4.0305	3.8599	3.6827	3.4977	3.4017	3.3030	3.2014	3.0962	2.9871	2.8732
19	3.9329	3.7631	3.5866	3.4020	3.3062	3.2075	3.1058	3.0004	2.8908	2.7762
20	3.8470	3.6779	3.5020	3.3178	3.2220	3.1234	3.0215	2.9159	2.8058	2.6904
21	3.7709	3.6024	3.4270	3.2431	3.1474	3.0488	2.9467	2.8408	2.7302	2.6140
22	3.7030	3.5350	3.3600	3.1764	3.0807	2.9821	2.8799	2.7736	2.6625	2.5455
23	3.6420	3.4745	3.2999	3.1165	3.0208	2.9221	2.8198	2.7132	2.6016	2.4837
24	3.5870	3.4199	3.2456	3.0624	2.9667	2.8679	2.7654	2.6585	2.5463	2.4276
25	3.5370	3.3704	3.1963	3.0133	2.9176	2.8187	2.7160	2.6088	2.4960	2.3765
26	3.4916	3.3252	3.1515	2.9685	2.8728	2.7738	2.6709	2.5633	2.4501	2.3297
27	3.4499	3.2839	3.1104	2.9275	2.8318	2.7327	2.6296	2.5217	2.4078	2.2867
28	3.4117	3.2460	3.0727	2.8899	2.7941	2.6949	2.5916	2.4834	2.3689	2.2469
29	3.3765	3.2111	3.0379	2.8551	2.7594	2.6601	2.5565	2.4479	2.3330	2.2102
30	3.3440	3.1787	3.0057	2.8230	2.7272	2.6278	2.5241	2.4151	2.2997	2.1760
40	3.1167	2.9531	2.7811	2.5984	2.5020	2.4015	2.2958	2.1838	2.0635	1.9318
60	2.9042	2.7419	2.5705	2.3872	2.2898	2.1874	2.0789	1.9622	1.8341	1.6885
120	2.7052	2.5439	2.3727	2.1881	2.0890	1.9839	1.8709	1.7469	1.6055	1.4311
∞	2.5188	2.3583	2.1868	1.9998	1.8983	1.7891	1.6691	1.5325	1.3637	1.0000

A description of these tables is given on page 529. Where necessary, interpolation should be carried out using the reciprocals of the degrees of freedom. The function $120/v$ is convenient for this purpose.

Index